Lecture Notes in Computer Science

Lecture Notes in Artificial Intelligence 14830

Founding Editor

Jörg Siekmann

Series Editors

Randy Goebel, *University of Alberta, Edmonton, Canada*
Wolfgang Wahlster, *DFKI, Berlin, Germany*
Zhi-Hua Zhou, *Nanjing University, Nanjing, China*

The series Lecture Notes in Artificial Intelligence (LNAI) was established in 1988 as a topical subseries of LNCS devoted to artificial intelligence.

The series publishes state-of-the-art research results at a high level. As with the LNCS mother series, the mission of the series is to serve the international R & D community by providing an invaluable service, mainly focused on the publication of conference and workshop proceedings and postproceedings.

Andrew M. Olney · Irene-Angelica Chounta ·
Zitao Liu · Olga C. Santos · Ig Ibert Bittencourt
Editors

Artificial Intelligence in Education

25th International Conference, AIED 2024
Recife, Brazil, July 8–12, 2024
Proceedings, Part II

 Springer

Editors
Andrew M. Olney (iD)
University of Memphis
Memphis, TN, USA

Irene-Angelica Chounta (iD)
University of Duisburg-Essen
Duisburg, Germany

Zitao Liu (iD)
Jinan University
Guangzhou, China

Olga C. Santos (iD)
UNED
Madrid, Spain

Ig Ibert Bittencourt (iD)
Universidade Federal de Alagoas
Maceio, Brazil

ISSN 0302-9743   ISSN 1611-3349 (electronic)
Lecture Notes in Artificial Intelligence
ISBN 978-3-031-64298-2   ISBN 978-3-031-64299-9 (eBook)
https://doi.org/10.1007/978-3-031-64299-9

LNCS Sublibrary: SL7 – Artificial Intelligence

This Springer imprint is published by the registered company Springer Nature Switzerland AG
The registered company address is: Gewerbestrasse 11, 6330 Cham, Switzerland

If disposing of this product, please recycle the paper.

Preface

The 25th International Conference on Artificial Intelligence in Education (AIED 2024) was hosted by Centro de Estudos e Sistemas Avançados do Recife (CESAR), Brazil from July 8 to July 12, 2024. It was set up in a face-to-face format but included an option for an online audience. AIED 2024 was the next in a longstanding series of annual international conferences for the presentation of high-quality research on intelligent systems and the cognitive sciences for the improvement and advancement of education. Note that AIED is ranked A in CORE (top 16% of all 783 ranked venues), the well-known ranking of computer science conferences. The AIED conferences are organized by the prestigious International Artificial Intelligence in Education Society, a global association of researchers and academics, which has already celebrated its 30th anniversary, and aims to advance the science and engineering of intelligent human-technology ecosystems that support learning by promoting rigorous research and development of interactive and adaptive learning environments for learners of all ages across all domains.

The theme for the AIED 2024 conference was "AIED for a World in Transition". The conference aimed to explore how AI can be used to enhance the learning experiences of students and teachers alike when disruptive technologies are turning education upside down. Rapid advances in Artificial Intelligence (AI) have created opportunities not only for personalized and immersive experiences but also for ad hoc learning by engaging with cutting-edge technology continually, extending classroom borders, from engaging in real-time conversations with large language models (LLMs) to creating expressive artifacts such as digital images with generative AI or physically interacting with the environment for a more embodied learning. As a result, we now need new approaches and measurements to harness this potential and ensure that we can safely and responsibly cope with a world in transition. The conference seeks to stimulate discussion of how AI can shape education for all sectors, how to advance the science and engineering of AI-assisted learning systems, and how to promote broad adoption.

AIED 2024 attracted broad participation. We received 334 submissions for the main program, of which 280 were submitted as full papers, and 54 were submitted as short papers. Of the full paper submissions, 49 were accepted as full papers, and another 27 were accepted as short papers. The acceptance rate for full papers and short papers together was 23%. These accepted contributions are published in the Springer proceedings volumes LNAI 14829 and 14830.

The submissions underwent a rigorous double-masked peer-review process aimed to reduce evaluation bias as much as possible. The first step of the review process was done by the program chairs, who verified that all papers were appropriate for the conference and properly anonymized. Program committee members were asked to declare conflicts of interest. After the initial revision, the program committee members were invited to bid on the anonymized papers that were not in conflict according to their declared conflicts of interest. With this information, the program chairs made the review assignment, which consisted of three regular members to review each paper plus a senior member to provide a meta-review. The management of the review process (i.e., bidding, assignment, discussion, and meta-review) was done with the EasyChair platform, which was configured so that reviewers of the same paper were anonymous to each other. A subset of the program committee members were not included in the initial assignment but were asked to be ready to do reviews that were not submitted on time (i.e., the emergency review period). To avoid a situation where program committee members would be involved in too many submissions, we balanced review assignments and then rebalanced them during the emergency review period.

As a result, each submission was reviewed anonymously by at least three Program Committee (PC) members and then a discussion was led by a Senior Program Committee (SPC) member. PC and SPC members were selected based on their authorship in previous AIED conferences, their experience as reviewers in previous AIED editions, their h-index as calculated by Google Scholar, and their previous positions in organizing and reviewing related conferences. Therefore all members were active researchers in the field, and SPC members were particularly accomplished on these metrics. SPC members served as meta-reviewers whose role was to seek consensus to reach the final decision about acceptance and to provide the corresponding meta-review. They were also asked to check and highlight any possible biases or inappropriate reviews. Decisions to accept/reject were taken by the program chairs. For borderline cases, the contents of the paper were read in detail before reaching the final decision. In summary, we are confident that the review process assured a fair and equal evaluation of the submissions received without any bias, as far as we are aware.

Beyond paper presentations, the conference included a Doctoral Consortium Track, Late-Breaking Results, a Workshops and Tutorials Track, and an Industry, Innovation and Practitioner Track. There was a WideAIED track, which was established in 2023, where opportunities and challenges of AI in education were discussed with a global perspective and with contributions coming also from areas of the world that are currently under-represented in AIED. Additionally, a BlueSky special track was included with contributions that reflect upon the progress of AIED so far and envision what is to come in the future. The submissions for all these tracks underwent a rigorous peer review process. Each submission was reviewed by at least three members of the AIED community, assigned by the corresponding track organizers who then took the final decision about acceptance.

The participants of the conference had the opportunity to attend three keynote talks: "Navigating Strategic Challenges in Education in the Post-Pandemic AI Era" by Blaženka Divjak, "Navigating the Evolution: The Rising Tide of Large Language Models for AI and Education" by Peter Clark, and "Artificial Intelligence in Education and Public Policy: A Case from Brazil" by Seiji Isotani. These contributions are published in the Springer proceedings volumes CCIS 2150 and 2151.

The conference also included a Panel with experts in the field and the opportunity for the participants to present a demonstration of their AIED system in a specific session of Interactive Events. A selection of the systems presented is included as showcases on the web page of the IAIED Society[1]. Finally, there was a session with presentations

---

[1] https://iaied.org/showcase.

of papers published in the International Journal of Artificial Intelligence in Education[2], the journal of the IAIED Society indexed in the main databases, and a session with the best papers from conferences of the International Alliance to Advance Learning in the Digital Era (IAALDE)[3], an alliance of research societies that focus on advances in computer-supported learning, to which the IAIED Society belongs.

For making AIED 2024 possible, we thank the AIED 2024 Organizing Committee, the hundreds of Program Committee members, the Senior Program Committee members, and the AIED proceedings chairs Paraskevi Topali and Rafael D. Araújo. In addition, we would like to thank the Executive Committee of the IAIED Society for their advice during the conference preparation, and specifically two of the working groups, the Conference Steering Committee, and the Diversity and Inclusion working group. They all gave their time and expertise generously and helped with shaping a stimulating AIED 2024 conference. We are extremely grateful to everyone!

July 2024

Andrew M. Olney
Irene-Angelica Chounta
Zitao Liu
Olga C. Santos
Ig Ibert Bittencourt

---

[2] https://link.springer.com/journal/40593.
[3] https://alliancelss.com/.

Organization

Conference General Co-chairs

Olga C. Santos   UNED, Spain
Ig Ibert Bittencourt   Universidade Federal de Alagoas, Brazil

Program Co-chairs

Andrew M. Olney   University of Memphis, USA
Irene-Angelica Chounta   University of Duisburg-Essen, Germany
Zitao Liu   Jinan University, China

Doctoral Consortium Co-chairs

Yu Lu   Beijing Normal University, China
Elaine Harada T. Oliveira   Universidade Federal do Amazonas, Brazil
Vanda Luengo   Sorbonne Université, France

Workshop and Tutorials Co-chairs

Cristian Cechinel   Federal University of Santa Catarina, Brazil
Carrie Demmans Epp   University of Alberta, Canada

Interactive Events Co-chairs

Leonardo B. Marques   Federal University of Alagoas, Brazil
Ben Nye   University of Southern California, USA
Rwitajit Majumdar   Kumamoto University, Japan

Industry, Innovation and Practitioner Co-chairs

Diego Dermeval   Federal University of Alagoas, Brazil
Richard Tong   IEEE Artificial Intelligence Standards
                                            Committee, USA
Sreecharan Sankaranarayanan   Amazon, USA
Insa Reichow   German Research Center for Artificial
                                            Intelligence, Germany

Posters and Late Breaking Results Co-chairs

Marie-Luce Bourguet   Queen Mary University of London, UK
Qianru Liang   Jinan University, China
Jingyun Wang   Durham University, UK

Panel Co-chairs

Julita Vassileva   University of Saskatchewan, Canada
Alexandra Cristea   Durham University, UK

Blue Sky Co-chairs

Ryan S. Baker   University of Pennsylvania, USA
Benedict du Boulay   University of Sussex, UK
Mirko Marras   University of Cagliari, Italy

WideAIED Co-chairs

Isabel Hilliger   Pontificia Universidad Católica, Chile
Marco Temperini   Sapienza University of Rome, Italy
Ifeoma Adaji   University of British Columbia, Canada
Maomi Ueno   University of Electro-Communications, Japan

Local Organising Co-chairs

Rafael Ferreira Mello   Universidade Federal Rural de Pernambuco,
                               Brazil
Taciana Pontual   Universidade Federal Rural de Pernambuco,
                               Brazil

AIED Mentoring Fellowship Co-chairs

Amruth N. Kumar   Ramapo College of New Jersey, USA
Vania Dimitrova   University of Leeds, UK

Diversity and Inclusion Co-chairs

Rod Roscoe   Arizona State University, USA
Kaska Porayska-Pomsta   University College London, UK

Virtual Experiences Co-chairs

Guanliang Chen   Monash University, Australia
Teng Guo   Jinan University, China
Eduardo A. Oliveira   University of Melbourne, Australia

Publicity Co-chairs

Son T. H. Pham   Nha Viet Institute, USA
Pham Duc Tho   Hung Vuong University, Vietnam
Miguel Portaz   UNED, Spain

Volunteer Co-chairs

Isabela Gasparini   Santa Catarina State University, Brazil
Lele Sha   Monash University, Australia

Proceedings Co-chairs

Paraskevi Topali   Radboud University, Netherlands
Rafael D. Araújo   Federal University of Uberlândia, Brazil

Awards Co-chairs

Ning Wang   University of Southern California, USA
Beverly Woolf   University of Massachusetts, USA

Sponsorship Chair

Tanci Simões Gomes   CESAR, Brazil

Scholarship Chair

Patrícia Tedesco   Universidade Federal de Pernambuco, Brazil

Steering Committee

Noboru Matsuda   North Carolina State University, USA
Eva Millan   Universidad de Málaga, Spain
Sergey Sosnovsky   Utrecht University, Netherlands
Ido Roll   Israel Institute of Technology, Israel
Maria Mercedes T. Rodrigo   Ateneo de Manila University, Philippines

Senior Program Committee Members

Laura Allen   University of Minnesota, USA
Ryan Baker   University of Pennsylvania, USA
Gautam Biswas   Vanderbilt University, USA
Nigel Bosch   University of Illinois at Urbana-Champaign, USA
Jesus G. Boticario   UNED, Spain
Bert Bredeweg   University of Amsterdam, Netherlands
Christopher Brooks   University of Michigan, USA
Guanliang Chen   Monash University, Australia

Craig Thompson	University of British Columbia, Canada
Stefan Trausan-Matu	University Politehnica of Bucharest, Romania
Maomi Ueno	University of Electro-Communications, Japan
Rosa Vicari	Universidade Federal do Rio Grande do Sul, Brazil
Vincent Wade	Trinity College Dublin, Ireland
Alistair Willis	Open University, UK
Diego Zapata-Rivera	Educational Testing Service, USA
Gustavo Zurita	Universidad de Chile, Chile

Program Committee Members

Mark Abdelshiheed	University of Colorado Boulder, USA
Kamil Akhuseyinoglu	University of Pittsburgh, USA
Carlos Alario-Hoyos	Universidad Carlos III de Madrid, Spain
Laia Albó	Universitat Pompeu Fabra, Spain
Giora Alexandron	Weizmann Institute of Science, Israel
Isaac Alpizar Chacon	Utrecht University, Netherlands
Ioannis Anastasopoulos	University of California, Berkeley, USA
Antonio R. Anaya	Universidad Nacional de Educación a Distancia, Spain
Tracy Arner	Arizona State University, USA
Burcu Arslan	Educational Testing Service, USA
Juan I. Asensio-Pérez	Universidad de Valladolid, Spain
Michelle Banawan	Asian Institute of Management, Philippines
Abhinava Barthakur	University of South Australia, Australia
Beata Beigman Klebanov	Educational Testing Service, USA
Brian Belland	Pennsylvania State University, USA
Francisco Bellas	Universidade da Coruna, Spain
Emmanuel Blanchard	Le Mans Université, France
Nathaniel Blanchard	Colorado State University, USA
Maria Bolsinova	Tilburg University, Netherlands
Miguel L. Bote-Lorenzo	Universidad de Valladolid, Spain
Anthony F. Botelho	University of Florida, USA
François Bouchet	Sorbonne Université - LIP6, France
Marie-Luce Bourguet	Queen Mary University of London, UK
Rex Bringula	University of the East, Philippines
Julien Broisin	Université Toulouse III - Paul Sabatier, France
Armelle Brun	LORIA - Université de Lorraine, France
Minghao Cai	University of Alberta, Canada
Dan Carpenter	North Carolina State University, USA

R. McKell Carter	University of Colorado Boulder, USA
Paulo Carvalho	Carnegie Mellon University, USA
Alberto Casas	UNED, Spain
Teresa Cerratto-Pargman	Stockholm University, Sweden
Cs Chai	Chinese University of Hong Kong, China
Jiahao Chen	TAL, China
Penghe Chen	Beijing Normal University, China
Thomas K. F. Chiu	Chinese University of Hong Kong, China
Heeryung Choi	Massachusetts Institute of Technology, USA
Chih-Yueh Chou	元智大工程系, Taiwan
Jody Clarke-Midura	Utah State University, USA
Keith Cochran	DePaul University, USA
Maria de Los Angeles Constantino González	Tecnológico de Monterrey Campus Laguna, Mexico
Evandro Costa	Federal University of Alagoas, Brazil
Alexandra Cristea	Durham University, UK
Jeffrey Cross	Tokyo Institute of Technology, Japan
Mihai Dascalu	University Politehnica of Bucharest, Romania
Jeanine DeFalco	University of New Haven, USA
Carrie Demmans Epp	University of Alberta, Canada
Vanessa Dennen	Florida State University, USA
Michel Desmarais	Polytechnique Montréal, Canada
M. Ali Akber Dewan	Athabasca University, Canada
Yannis Dimitriadis	University of Valladolid, Spain
Konomu Dobashi	Aichi University, Japan
Fabiano Dorça	Universidade Federal de Uberlandia, Brazil
Mohsen Dorodchi	University of North Carolina Charlotte, USA
Benedict du Boulay	University of Sussex, UK
Cristina Dumdumaya	University of Southeastern Philippines, Philippines
Nicholas Duran	Arizona State University, USA
Kareem Edouard	Drexel University, USA
Fahmid Morshed Fahid	North Carolina State University, USA
Xiuyi Fan	Nanyang Technological University, Singapore
Alexandra Farazouli	Stockholm University, Sweden
Effat Farhana	Vanderbilt University, USA
Mingyu Feng	WestEd, USA
Márcia Fernandes	Federal University of Uberlândia, Brazil
Brendan Flanagan	Kyoto University, Japan
Carol Forsyth	Educational Testing Service, USA
Reva Freedman	Northern Illinois University, USA

Cristiano Galafassi	Universidade Federal do Rio Grande do Sul, Brazil
Wenbin Gan	National Institute of Information and Communications Technology, Japan
Zhen Gao	McMaster University, Canada
Yiannis Georgiou	Cyprus University of Technology, Cyprus
Alireza Gharahighehi	KU Leuven, Belgium
Lucia Giraffa	Pontifícia Universidade Católica do Rio Grande do Sul, Brazil
Michael Glass	Valparaiso University, USA
Benjamin Goldberg	United States Army DEVCOM Soldier Center, USA
Aldo Gordillo	Universidad Politécnica de Madrid, Spain
Guher Gorgun	University of Alberta, Canada
Cyril Goutte	National Research Council Canada, Canada
Sabine Graf	Athabasca University, Canada
Monique Grandbastien	LORIA, Université de Lorraine, France
Beate Grawemeyer	Coventry University, UK
Quanlong Guan	Jinan University, China
Nathalie Guin	LIRIS - Université de Lyon, France
Ella Haig	University of Portsmouth, UK
Jiangang Hao	Educational Testing Service, USA
Yugo Hayashi	Ritsumeikan University, Japan
Yusuke Hayashi	Hiroshima University, Japan
Arto Hellas	Aalto University, Finland
Davinia Hernandez-Leo	Universitat Pompeu Fabra, Barcelona, Spain
Anett Hoppe	Leibniz Universität Hannover, Germany
Iris Howley	Williams College, USA
Sharon Hsiao	Santa Clara University, USA
Xiangen Hu	University of Memphis, USA
Lingyun Huang	Education University of Hong Kong, China
Shuyan Huang	TAL Education Group, China
Stephen Hutt	University of Denver, USA
Tsunenori Ishioka	National Center for University Entrance Examinations, Japan
Patricia Jaques	UFPEL, UFPR, Brazil
Stéphanie Jean-Daubias	Université de Lyon, LIRIS, France
Yang Jiang	Columbia University, USA
Yueqiao Jin	Monash University, Australia
David Joyner	Georgia Institute of Technology, USA
Hamid Karimi	Utah State University, USA
Tanja Käser	EPFL, Switzerland

Akihiro Kashihara	University of Electro-Communications, Japan
Enkelejda Kasneci	Technical University of Munich, Germany
Judy Kay	University of Sydney, Australia
Mizue Kayama	Shinshu University, Japan
Mohammad Khalil	University of Bergen, Norway
Hassan Khosravi	University of Queensland, Australia
Chanmin Kim	Pennsylvania State University, USA
Yeo Jin Kim	North Carolina State University, USA
Yoon Jeon Kim	University of Wisconsin-Madison, USA
Rene Kizilcec	Cornell University, USA
Elizabeth Koh	Nanyang Technological University, Singapore
Kazuaki Kojima	Teikyo University, Japan
Sotiris Kotsiantis	University of Patras, Greece
Eleni Kyza	Cyprus University of Technology, Cyprus
Sébastien Lallé	Sorbonne University, France
Andrew Lan	University of Massachusetts Amherst, USA
Mikel Larrañaga	University of the Basque Country, Spain
Elise Lavoué	Université Jean Moulin Lyon 3, LIRIS, France
Nguyen-Thinh Le	Humboldt Universität zu Berlin, Germany
Tai Le Quy	IU International University of Applied Sciences, Germany
Seiyon Lee	University of Florida, USA
Victor Lee	Stanford University, USA
Marie Lefevre	LIRIS - Université Lyon 1, France
Blair Lehman	Educational Testing Service, USA
Juho Leinonen	Aalto University, Finland
Liang-Yi Li	National Taiwan Normal University, Taiwan
Zhi Li	University of California, Berkeley, USA
Bibeg Limbu	University of Duisburg-Essen, Germany
Carla Limongelli	Università Roma Tre, Italy
Jionghao Lin	Carnegie Mellon University, USA
Allison Littlejohn	University College London, UK
Qi Liu	University of Science and Technology of China, China
Ziyuan Liu	Singapore Management University, USA
Sonsoles López-Pernas	University of Eastern Finland, Finland
Yu Lu	Beijing Normal University, China
Vanda Luengo	Sorbonne Université - LIP6, France
Collin Lynch	North Carolina State University, USA
John Magee	Clark University, USA
George Magoulas	Birkbeck, University of London, UK
Linda Mannila	Åbo Akademi University, Finland

Mirko Marras	University of Cagliari, Italy
Alejandra Martínez-Monés	Universidad de Valladolid, Spain
Goran Martinovic	J.J. Strossmayer University of Osijek, Croatia
Jeffrey Matayoshi	McGraw Hill ALEKS, USA
Diego Matos	Federal University of Alagoas, Brazil
Noboru Matsuda	North Carolina State University, USA
Guilherme Medeiros Machado	ECE Paris, France
Christoph Meinel	Hasso Plattner Institute, Germany
Roberto Angel Melendez Armenta	Universidad Veracruzana, Mexico
Agathe Merceron	Berlin State University of Applied Sciences, Germany
Eva Millan	Universidad de Málaga, Spain
Marcelo Milrad	Linnaeus University, Sweden
Wookhee Min	North Carolina State University, USA
Tsunenori Mine	Kyushu University, Japan
Sein Minn	Inria, France
Péricles Miranda	UFPE, Brazil
Tsegaye Misikir Tashu	University of Groningen, Netherlands
Tanja Mitrovic	University of Canterbury, New Zealand
Phaedra Mohammed	University of the West Indies, Trinidad and Tobago
Mukesh Mohania	IIIT Delhi, India
Bradford Mott	North Carolina State University, USA
Ana Mouta	USAL, Portugal
Chrystalla Mouza	University of Illinois at Urbana-Champaign, USA
Pedro J. Muñoz-Merino	Universidad Carlos III de Madrid, Spain
Takashi Nagai	Institute of Technologists, Japan
Tricia Ngoon	Carnegie Mellon University, USA
Huy Nguyen	Carnegie Mellon University, USA
Narges Norouzi	UC Berkeley, USA
Dr. Nasheen Nur	Florida Institute of Technology, USA
Benjamin Nye	University of Southern California, USA
Xavier Ochoa	New York University, USA
Jaclyn Ocumpaugh	University of Pennsylvania, USA
Elaine H. T. Oliveira	Universidade Federal do Amazonas, Brazil
Jennifer Olsen	University of San Diego, USA
Ranilson Paiva	Universidade Federal de Alagoas, Brazil
Viktoria Pammer-Schindler	Graz University of Technology, Austria
Luc Paquette	University of Illinois at Urbana-Champaign, USA
Abelardo Pardo	University of Adelaide, Australia
Zach Pardos	University of California, Berkeley, USA
Rebecca Passonneau	Pennsylvania State University, USA

Masaki Uto	University of Electro-Communications, Japan
Lisa van der Heyden	University of Duisburg-Essen, Germany
Kurt Vanlehn	Arizona State University, USA
Julita Vassileva	University of Saskatchewan, Canada
Esteban Vázquez-Cano	UNED, Spain
Olga Viberg	KTH Royal Institute of Technology, Sweden
Francisco Vicente-Castro	New York University, USA
Esteban Villalobos	Université Paul Sabatier, France
Alessandro Vivas	UFVJM, Brazil
Erin Walker	Arizona State University, USA
Ning Wang	University of Southern California, USA
Christabel Wayllace	New Mexico State University, USA
Stephan Weibelzahl	Private University of Applied Sciences Göttingen, Germany
Gary Weiss	Fordham University, USA
Chris Wong	University of Technology Sydney, Australia
Beverly Park Woolf	University of Massachusetts, USA
Qian Xiao	Trinity College Dublin, Ireland
Wanli Xing	University of Florida, USA
Masanori Yamada	Kyushu University, Japan
Sho Yamamoto	Kindai University, Japan
Zhang Yupei	Northwestern Polytechnical University, China
Andrew Zamecnik	University of South Australia, Australia
Jianwei Zhang	University at Albany, USA
Shu-Gang Zhang	Ocean University of China, China
Yong Zheng	Illinois Institute of Technology, USA
Gaoxia Zhu	National Institute of Education, Nanyang Technological University, Singapore
Stefano Zingaro	Università di Bologna, Italy
刘强	中国四川成都树德中学

International Artificial Intelligence in Education Society

Management Board

President

Olga C. Santos	UNED, Spain

Secretary/Treasurer

Vania Dimitrova   University of Leeds, UK

Journal Editors

Vincent Aleven   Carnegie Mellon University, USA
Judy Kay   University of Sydney, Australia

Finance Chair

Benedict du Boulay   University of Sussex, UK

Membership Chair

Benjamin D. Nye   University of Southern California, USA

IAIED Officers

Yancy Vance Paredes   North Carolina State University, USA
Son T. H. Pham   Nha Viet Institute, USA
Miguel Portaz   UNED, Spain

Executive Committee

Akihiro Kashihara   University of Electro-Communications, Japan
Amruth Kumar   Ramapo College of New Jersey, USA
Christothea Herodotou   Open University, UK
Jeanine A. Defalco   CCDC-STTC, USA
Judith Masthoff   Utrecht University, Netherlands
Maria Mercedes T. Rodrigo   Ateneo de Manila University, Philippines
Ning Wang   University of Southern California, USA
Olga Santos   UNED, Spain
Rawad Hammad   University of East London, UK
Zitao Liu   Jinan University, China
Bruce M. McLaren   Carnegie Mellon University, USA
Cristina Conati   University of British Columbia, Canada
Diego Zapata-Rivera   Educational Testing Service, Princeton, USA
Erin Walker   University of Pittsburgh, USA
Seiji Isotani   University of São Paulo, Brazil
Tanja Mitrovic   University of Canterbury, New Zealand

Contents – Part II

Contents – Part I

Full Papers

A First Step in Using Machine Learning Methods to Enhance Interaction Analysis for Embodied Learning Environments

Joyce Fonteles[1]([⊠]) [iD], Eduardo Davalos[1] [iD], T. S. Ashwin[1] [iD], Yike Zhang[1] [iD],
Mengxi Zhou[2] [iD], Efrat Ayalon[1] [iD], Alicia Lane[1] [iD], Selena Steinberg[2] [iD],
Gabriella Anton[1] [iD], Joshua Danish[2] [iD], Noel Enyedy[1] [iD],
and Gautam Biswas[1] [iD]

[1] Vanderbilt University, Nashville, TN 37240, USA
{joyce.h.fonteles,gautam.biswas}@Vanderbilt.edu
[2] Indiana University, Bloomington, IN 47405, USA

Abstract. Investigating children's embodied learning in mixed-reality environments, where they collaboratively simulate scientific processes, requires analyzing complex multimodal data to interpret their learning and coordination behaviors. Learning scientists have developed Interaction Analysis (IA) methodologies for analyzing such data, but this requires researchers to watch hours of videos to extract and interpret students' learning patterns. Our study aims to simplify researchers' tasks, using Machine Learning and Multimodal Learning Analytics to support the IA processes. Our study combines machine learning algorithms and multimodal analyses to support and streamline researcher efforts in developing a comprehensive understanding of students' scientific engagement through their movements, gaze, and affective responses in a simulated scenario. To facilitate an effective researcher-AI partnership, we present an initial case study to determine the feasibility of visually representing students' states, actions, gaze, affect, and movement on a timeline. Our case study focuses on a specific science scenario where students learn about photosynthesis. The timeline allows us to investigate the alignment of critical learning moments identified by multimodal and interaction analysis, and uncover insights into students' temporal learning progressions.

Keywords: Multimodal learning analytics · Embodied learning ·
Machine learning · Interaction analysis

1 Introduction

Embodied learning aligns with the natural ways in which humans perceive, interact, and learn from the world around them. By engaging the body in the learning process, we create richer, more immersive educational experiences where our actions, movements, and interactions contribute significantly to how we understand and internalize concepts [6]. It allows students to actively explore and

A. M. Olney et al. (Eds.): AIED 2024, LNAI 14830, pp. 3–16, 2024.
https://doi.org/10.1007/978-3-031-64299-9_1

embody knowledge through perception, awareness, and exploration of their environment. Embodiments not only enhance retention and a deeper understanding of abstract or complex concepts; it leverages the power of immersive experiences to make education more engaging and impactful [10].

Embodied learning data analysis presents a great challenge due to the complexity of monitoring student groups spatially and temporally. Conventional educational settings focus mostly on verbal communication and digital system interactions. Meanwhile, embodied learning necessitates the capture of non-verbal cues and body movements in 3D space, along with conversations and simulation logs [6]. Interaction Analysis (IA) is one of the main approaches employed by learning scientists because it can unravel deep insights and nuanced interactions captured in video data [13]. IA yields valuable insights, but its manual processes are time-consuming and demand substantial human resources. Therefore, recent advances in Machine Learning (ML) and Multimodal Learning Analytics (MMLA) make it easier to leverage algorithms to support human analysis, with the idea that the combination will allow researchers and educators to gain a nuanced understanding of how learners engage with content, facilitating feedback, assessment, and an enriched comprehension of the learning process [19].

Observing and tracking embodied learning scenarios generates substantial volumes of multimodal data, which is a reflection of the diverse sensory inputs needed to capture movement, gestures, gaze, interactions, and coordination in the context of learning and problem-solving tasks [2]. Motion tracking data, detailing the positions and orientations of body parts, enables granular analysis of physical interactions. Gaze tracking data reveals where learners direct their attention, shedding light on points of interest or challenges. Affect detection data adds a nuanced layer by gauging learners' emotional states. Complementing these, system logs record interactions with the learning platform, simulations, or virtual environments, offering timestamps and details of actions taken. Managing all of the heterogeneous multimodal data efficiently is a complex task, demanding sophisticated computational analysis. The challenge for educational methods using AI and ML lies in addressing the complexities of multimodal data collection, alignment, and analysis to derive meaningful insights into students' individual and collaborative behaviors in a timely manner.

Human researchers, familiar with the varied contexts of embodied learning data, are crucial for its interpretation. Unlike technology-centric approaches, we advocate for "AI-in-the-loop" methods, emphasizing the pivotal role of humans in the analysis and interpretation process. This study makes two primary contributions. First, it applies IA to determine the most effective modalities, analyses, and visualizations for employing AI to aid human interpretation of student behavior. Second, it introduces an interactive visual timeline that displays MMLA results, tailored to augment IA. This timeline represents students' movements, necessitating data processing from multiple cameras for accurate student re-identification and face tracking. Our findings reveal that these environments provoke emotional responses distinct from those observed in traditional computer-based learning settings. Additionally, we propose an innovative approach for discretizing gaze toward moving objects in 3D spaces, significantly contributing to the field of Artificial Intelligence in Education.

2 Background

2.1 Embodied Learning and Interaction Analysis

Embodied learning activities leverage the movement of students' bodies in the teaching and learning of conceptual and disciplinary ideas. These activities are grounded in the assumption that parallels can be drawn between bodily experiences and conceptual learning [16]. In science education, many embodied activity designs rely on educational technologies like virtual reality (VR) and mixed reality (MR) to immerse students in a particular scientific system, enabling feelings of presence. The design of embodied activities can be approached from a cognitive lens, which focuses on how individual students' movements or gestures can map onto underlying conceptual ideas or from a sociocultural lens, which considers the ways that youth interact with each other as socially situated. In this work, we take a sociocultural approach to designing these kinds of embodied learning activities [6]. This means that we are most interested in embodiment that happens between multiple learners in a social and cultural context; as youth engage with each other, they develop meaning together through their embodiment.

As youth participate in collaborative MR embodied activities, they must attend to and coordinate many modes, including gaze, movement, and speech [24]. IA empirically investigates human interactions with each other and objects in contextual settings. Learning scientists often use this analytic method to make sense of collaborative, embodied learning environments, where multiple students move together in a classroom or educational setting. IA's development is theoretically grounded in several methodologies, notably conversation analysis and ethnography, and has become popular with the proliferation of audiovisual recording technologies. The capability to capture learning activities from multifaceted views/positions and iterative playback of recordings is crucial to interaction analysis, as it allows close interrogation, which is the essence of IA. Thus, the goal of interaction analysis is to look for empirical evidence of the learning and learning process by discerning patterns/ regularities in how participants utilize resources within their natural environments and interact with each other.

2.2 Affect and Learning

In academics, the detection of learning-centric emotions – confusion, boredom, frustration, engagement, and delight – is crucial for comprehending learner behaviors and performance [20]. While state-of-the-art computer vision algorithms can successfully identify basic and learning-centered emotions, their application in embodied learning environments presents unique challenges [26]. In such settings, multiple students are often captured in a single video frame. Further, students are moving frequently, and this necessitates advanced techniques like re-identification for accurate emotion tracking [25].

A significant shortcoming in current emotion recognition datasets is their focus on undergraduate students, with little data representing children's facial expressions, which makes it hard to detect their emotions accurately [3]. Models like High-Speed Emotion Recognition, which are trained on diverse age groups, including children, offer an alternative by quantifying emotions on a continuous scale of valence and arousal [22]. These models utilize frameworks such as Russell's circumplex model and D'Mello's dynamics of emotion to translate continuous emotional states into discrete categories [9,21]. In dynamic environments, where students may walk, jump, and exhibit rapid movements, top-tier models like multitask cascaded convolutional networks (MTCNN) are preferred for face identification [28]. However, they typically lack training for rapid movements and partial occlusions, characteristic of embodied learning. Consequently, retraining existing models or deriving new ones tailored for these complex settings is imperative for the accurate identification and analysis of student emotions.

2.3 Gaze Detection and Interactions

Gaze analysis has been a cornerstone in understanding how learners engage and process information, revealing insights into cognitive functions and social interactions that are critical for collaborative learning and problem-solving [27]. Traditionally, eye-tracking required specific hardware and controlled environments, limiting its application in actual classroom settings and affecting the authenticity of observed learning behaviors. To overcome these constraints, advances in eye-tracking technology have introduced more versatile tools, such as lightweight eye-tracking glasses like the Tobii Glasses 3. These innovations allow for observation in more natural settings, although they face challenges like limited scalability and adaptability for children. Computer-vision methods, such as L2CS-Net [1] and Gaze360 [15], offer solutions that are more suitable for the dynamic nature of classrooms, even though they may compromise some on the precision of gaze data. Despite this, the trade-off is considered acceptable for educational research, where the focus is on broader data interpretation rather than pinpoint accuracy. However, applying these methods to children remains problematic because they have not been trained on their data.

In eye-tracking research, encoding gaze data into objects of interest (OOI) helps translate raw gaze points into meaningful insights by associating them with elements in the learning environment, such as teaching aids or interactive tools. The distinction between static and dynamic OOIs presents a significant challenge, requiring sophisticated tracking and analysis techniques [8]. 3D reconstruction emerges as a promising approach to address this, enabling detailed spatial analysis of gaze patterns [18]. However, the task is complex, especially when relying on monocular video capture that lacks depth information, posing hurdles for accurately mapping gaze in three-dimensional spaces.

2.4   Multimodal Visualizations

The visualization of multimodal data, which combines various modalities such as facial expressions, interactions, and contextual information, is an active area of research in education and learning analytics. Ez-zaouia's Emodash [12] contributes to this field by presenting a dashboard that visualizes learners' emotions inferred from facial expressions, alongside their system interactions during online learning sessions, reinforcing how one of the main challenges in the field is finding the appropriate level of detail and timescales for visualizations. This work builds upon previous research that explored visualizing learners' performances and behaviors using primarily systems logs dashboards, as well as the design of multimodal and contextual emotional dashboards for tutors [23].

Our design solution is distinct from the current literature. The timeline structure has been shaped by IA sessions conducted by researchers to identify key modalities and analytical approaches for interpreting student actions in embodied learning environments. Contrasting with prior research focused on computer-based learning environments, our study explores the unique dynamics of embodied learning within a MR context, where students' physical movements facilitate interaction. While our approach aligns with Ez-zaouia et al. in integrating system interactions and affect into an interactive timeline, it further examines the specific emotional responses elicited by the gamified aspects of embodied learning. We also incorporate gaze data to study students' shifting attention during activities. Our innovative presentation of multimodal data on a dynamic timeline, synchronized with video playback, is designed to enhance IA by providing AI-generated insights and interpretations to researchers.

3   Methods

3.1   Study Design

This study is part of a larger project entitled Generalized Embodied Modeling to support Science through Technology Enhanced Play (GEM-STEP). In this project, the motion-tracking technologies and mixed-reality environments display participants' movements on a projector screen, embodying complex scientific phenomena as researchers investigate individual and collaborative learning processes [5]. The GEM-STEP research team, including two of the authors, collaborated with a fourth-grade science teacher to co-design and co-facilitate a 20-day curriculum focused on food webs and photosynthesis. The participants in the following analysis consisted of two facilitators (the teacher and one researcher) and seven consented and assented students (four boys and three girls) from diverse racial and linguistic backgrounds. The students included multilingual learners, and their home languages included English, Kurdish, and Spanish.

Considering the diverse learners in our site, technology-mediated and embodied learning environments can expand access to science content where conventional text- or discourse-based learning may not. These environments expand students' sense-making resources, e.g., their bodies and emotions, often restricted within science learning contexts. This multimodal approach promotes students' agency and engagement while lowering linguistic barriers [17]. The curriculum and the models were designed based on these approaches while aligning with local science standards. In this work we focus on the photosynthesis model, a closed-loop system with a simulation screen which alternates from day to night. It features a mouse and a tomato plant with zoomed-in chloroplasts and roots (areas that cause molecule transformations). Students must move among these locations to model interactions between the molecules they are embodying (oxygen, water, sugar and carbon dioxide) and features on the screen. To streamline data collection we employed a distributed streaming framework called ChimeraPy [7], which supports rapid deployment in the classroom and provides time-alignment across multiple data modalities. We collected video from four cameras, multiple wireless microphones, screen recordings, and system logs.

3.2   Interaction Analysis by Researchers

Four authors from the learning sciences completed the IA of the videos. While all four analysts were knowledgeable about embodiment and participated in the design of the GEM-STEP project, two were present at data collection and two were not, and thus were less familiar with the context and the data. We split the videos among researchers such that two authors reviewed day 1 and two authors reviewed day 2, paired such that one researcher had familiarity with the site and one did not. In this way, we hoped that we might elicit multiple, diverse perspectives on the videos. For analysis, we selected three focal students because, on the first pass, they seemed to approach the activity in diverse ways. We focused on how each student seemed to be moving and when they seemed to understand the photosynthesis content.

During these co-watching sessions, researchers paused video-playing when observing notable events, such as instances of students' laughter or moments when a student facilitated the modeling of others. They then discussed the significance of these moments, including whether students demonstrated evidence of learning, what evidence was, and how the learning was mediated. Additionally, they discussed moments when IA supported understanding of the learning process were highlighted, including moments of contextual interactions. Moments when affect and gaze analysis could be important were also identified. For example, students displayed different emotions: one student (pseudonym Rose) was excited when the photosynthesis process was successful, while another student (pseudonym Taylor Swift) remained very calm at these moments. One student (pseudonym DaPaw) shifted his gaze and body away from the simulation screen while successfully modeling. Human analysts marked those moments as interesting to investigate how the AI findings might align or not with IA.

3.3   Design of Visual Timeline

By studying the IA methods applied to the embodied learning videos, it was apparent that a contextual AI-based analysis would require documenting the system's evolving state in MR, and developing algorithms for tracking and interpreting students' actions and engagement in the context of the scientific process being enacted. To visualize all of this information, we designed a visual timeline to strategically incorporate the multimodal data and analysis methods aligned with these requirements. It integrates data from system logs that tracked students' movements, including avatar shifts and interactions with objects of interest in the MR scene. Another IA-driven insight was the need to understand how system dynamics and variables, such as day-night transitions, impacted students' actions. Analyzing students' engagement and focus informed the inclusion of gaze, offering insights into the impact of shifts between the display screen, teacher, and peers on subsequent student actions. Moreover, the timeline also used facial data derived from the video analysis to capture and document students' emotions, acknowledging their significance for learning.

A powerful visualization driven by AI and ML algorithms can be a gateway to recognizing and interpreting students' individual and collective *aha* moments, signaling their insights and discoveries, which in turn can be interpreted in terms of their learning the science content. The results of our analysis had to be displayed in a way that merged the modalities into one visualization that should be clear and compelling to the human researchers, while avoiding complexities and clutter that could become tiresome. Our data visualization was initially inspired by Clara Peni'n and Jaime Serra's work from La Vanguardia called "Apoteosis 'Waka Waka'" [11], which visualizes different aspects of a concert on a timeline including lighting cues, visual effects, costumes, and lyrics; and extends Hervés multimedia player [14].

3.4   Analysis of System Logs

The analysis of system logs served as a primary component in establishing the nuances of student interactions and their evolution within the MR environment. These logs serve as a temporal record, tracing the sequence of avatar changes and movements made by students throughout the learning activity. By analyzing them, it is possible to discern patterns and trends in how students navigate the virtual space, interact with different elements, and transition between avatars. Understanding the spatial and temporal aspects of students' movements within the learning environment played a crucial role in gauging their existing knowledge and ongoing comprehension processes. Within this dataset, we extracted timestamped information to capture three key dimensions of data to assist in understanding students' learning processes: *(1) Students' States*: Understanding which molecules students were embodying at any given time allowed us to explore their evolving comprehension of scientific concepts much like the IA researchers did; *(2) Students' Actions*: Analyzing actions intertwined with their embodiment of molecules was fundamental in gauging their understanding of the photosynthesis process; *(3) System State*: Capturing the influencing variable of whether the simulation was in daylight or at night and tracking students' responses to this allowed us to discern not only if but also when they grasped the concept that photosynthesis requires sunlight.

3.5   Affect Detection

In this study, we analyzed students' facial expressions in the embodied learning environment for emotion recognition. Figure 1 illustrates the process, which involves face detection, followed by predicting continuous emotion scores on a valence-arousal scale. These scores are then categorized into learning-centered emotions based on Russell's circumflex of emotions [21] and D'Mello's dynamics of affective states [9]. Positive emotions are assigned to the first quadrant, intense unpleasant emotions to the second quadrant, subdued unpleasant emotions to the third quadrant, and serene pleasant emotions to the fourth quadrant.

The system was initially configured to read input from a video file, initializing tools for face detection, facial landmark extraction, and emotion recognition. Notably, we fine-tuned MTCNN [28] with thresholds of 0.8 for P-Net, R-Net, and O-Net. Additionally, we employed Dlib's facial landmark detector to precisely identify critical features on the face.

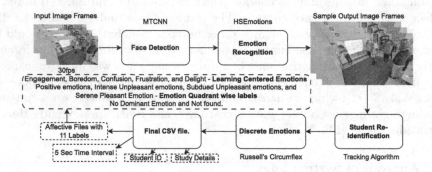

Fig. 1. Overview of Affect Detection Process

During the detailed frame-by-frame processing phase, each video frame was converted from BGR to RGB color space. MTCNN scrutinized the frame for faces, and each detected face underwent further analysis using dlib's detector for facial landmarks. The HSEmotionRecognizer processed the face region (enet_b0_8_best_afew.pt) and predicted the valance arousal values. Detailed information about each face was recorded for every frame, forming a comprehensive dataset for subsequent analysis. The system augmented the video with annotations, marking faces with bounding boxes and indicating valence and arousal scores, along with facial landmarks. This enriched video showcased the emotional analysis visually. Concurrently, it compiled this data into a structured CSV file, including the bounding boxes, valence, and arousal values. This file provided a frame-by-frame record of the emotional metrics, supporting thorough analysis and verification of the system's accuracy.

Student Re-identification. We enhanced MTCNN and HSEmotion by integrating a tracking algorithm that utilized CSV data containing frame numbers, valence, arousal, and bounding box coordinates to re-identify students across video frames. Operating at 30 fps, the algorithm maintained spatial continuity, computed bounding box centers, assigned unique IDs, and predicted positions using Euclidean distances within a set threshold. A memory component improved accuracy by compensating for minor face displacements and maintained tracking even with occlusions or movements. This approach achieved a 91% re-identification success rate, with manual adjustments addressing the remainder.

For gaze tracking and emotion categorization, the CSV data post-re-identification enabled the transformation of continuous emotion metrics into discrete states, processed in 5-s intervals to align with the frame rate. Emotions sustained over 150 frames were deemed significant, except for delight, which required 60 frames. Minimal facial expressions were labeled as "Engaged Concentration," consistent with educational emotion research [9]. Emotions were classified into Engagement, Boredom, Confusion, Frustration, and Delight, or by valence-arousal quadrants, with 11 labels in total, including 'NotFound' and 'NoDominantEmotion' for cases where faces were not visible (see Fig. 1).

3.6   Gaze Estimation

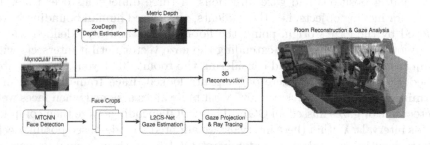

Fig. 2. Gaze Estimation Pipeline.

In GEM-STEP, we adopted a computer-vision approach for gaze estimation to observe student and teacher focus within the classroom while accommodating our logistical constraints. We encoded objects of interest (OOI) to map where participants were looking during the embodied activity, opting for a method that translated basic gaze data into more meaningful insights for our analysis. This encoding process, however, faced challenges due to the spatial nature of our learning context. To address this, we elevated our gaze analysis to a 3D perspective through room reconstruction, which provided a more accurate and physics-based approach for OOI encoding (Fig. 2).

For the 3D room reconstruction, we utilized depth estimation to transform monocular video frames into three-dimensional space – given that our camera was stationary. We employed ZoeDepth [4], a model trained on both indoor

(NYU Depth v2) and outdoor (KITTI) datasets, for its superior depth estimation capabilities. NYU Depth v2 and KITT datasets contain 1449 and 12929 RGB and depth image pairs, respectively. Through the combination of these two datasets, ZoeDepth achieves state-of-the-art (SOTA) performance in terms of metric (absolute) depth estimation – making it an excellent choice for reconstructing the room. This process allowed us to reconstruct each room frame-by-frame, aiding in the identification of both static (e.g., displays) and dynamic (e.g., students and teachers) OOIs. For our static OOIs, we labeled these using Vision6D, a 3D annotation tool.

To minimize computational demands, we initially tracked objects in 2D before mapping them into a three-dimensional context. Utilizing face bounding boxes and cropped images produced by the MTCNN and a re-identification algorithm, we then applied the L2CS-Net [1], a computer-vision model designed for gaze estimation, to calculate 3D gaze vectors for each identified face within the GEM-STEP environment. These vectors, determined by pitch and yaw measurements, were transformed into a 3D rotation matrix, denoted as R. By taking the XY centroids of the face bounding boxes and pairing them with depth information to derive a Z value, we completed the 3D translation vector t. The synthesis of R and t yielded a fully encompassing transformation matrix RT, encapsulating the origin and direction of a participant's gaze.

Armed with the RT matrix over successive frames, we tracked the students' spatial positions and gaze directions in three dimensions over time. To account for moving objects, i.e., the students, we devised human bounding boxes anchored by the gaze's origin point, the floor's plane, and a predefined width. By employing gaze ray tracing-extending the gaze vector until it intersects with an object of interest (OOI)-and considering the room's 3D layout, we were able to encode OOIs based on where participants looked. Each frame resulted in a determined OOI for every participant, with null values assigned when faces were undetectable or gazes missed all OOIs. These OOI encodings were then compiled over 5-s intervals. Within these intervals, we adopted a mode-based pooling technique to identify and select a predominant OOI for each participant's dataset throughout the given time window.

4   Results and Discussion

In our exploration of how events of IA and AI inform each other, we evaluated how our visual timeline, shown in Fig. 3, presented findings that were consistent with the interaction analysis that shaped its construction. For each video analyzed by human researchers through IA, individual student timelines were produced, allowing for new rounds of IA. The video and timeline are executed together and it is possible to navigate to specific events and zoom in/out to control the granularity of the data being shown, allowing for a more detailed or higher-level IA analysis.

The processing of system logs provided a visual representation of the molecules embodied by students in the photosynthesis process. This addressed

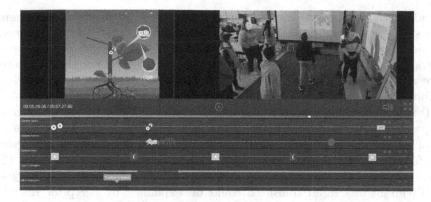

Fig. 3. Visual timeline of multimodal data for IA

one of the main concerns of researchers, the ability to identify components that informed students' understanding of the science model over time. By observing the molecules each student embodied and how long it took them to transition correctly through the model, researchers could pinpoint segments where students struggled and would benefit from scaffolds/feedback. Another aspect that interested researchers was how students moved in the real world and, subsequently, in the simulation. Changes in movement caused by attention shifts provide us with valuable information on how a student may be learning by interacting with the environment. This provides important links between the cognitive and socio-cultural aspects of learning. Since the embodied activity required them to move around and explore the virtual environment to understand how molecules were transformed, showing visual representations of the objects of interest and when students moved towards them informed if and when students figured out the correct actions that caused transformations. Furthermore, the logs also informed system transitions of day and night time, pinpointing moments to investigate if students grasped how light affects photosynthesis.

Currently, the temporal gaze information, which determined when and where students were looking, is plotted on the timeline, which provides cues for researchers to further investigate attention shifts. Gaze in this context was calculated and discretized to inform when students were looking at the display screen, the teacher, or one of their peers. Furthermore, such gaze behaviors coupled with conversational information (which we did not analyze in this paper) provide important information about student difficulties, e.g., if they were struggling with a specific transition, or if they received advice but chose to ignore it. In addition, our analysis and timeline representation allows us to study if gaze shifts were triggered by affect changes during the activity. Such patterns allow researchers to investigate more deeply relationships between students' affective states, their attention, how these might relate to previous actions, and how they inform future actions taken. The affect data revealed important insights. Delight was notably high among students during play, a rarity in traditional classrooms.

A student who grasped the concept often felt frustrated when their peers ignored their suggestions. In the collaborative part, a knowledgeable student frequently felt sad (a subdued unpleasant emotion), seemingly due to a lack of cooperation in progressing tasks, and this was accompanied by periods of boredom. However, there were cases where once actively engaged in problem-solving, many students exhibited increased positive emotions.

Temporal data presentation was important for pattern recognition. Visual comparison of timelines from three students on the first day revealed disparate initial interactions with the model. DaPaw[1] required approximately five minutes for the first transition, Rose two minutes, and Taylor Swift a mere 12 s. IA corroborated these findings, noting DaPaw's initial hesitation and suggesting Taylor Swift's rapid transition could be explained by her prior scientific understanding, which she used to direct her peers. Moreover, our ML analysis generated aggregated metrics of student learning and performance. Rose completed the photosynthesis cycle thrice, with 15 successful molecule transitions. Overall, students spent 33% of their initial time on carbon dioxide transformations and 21% of their initial time on water molecule transitions. Taylor Swift seamlessly navigated all molecule transitions, completing the cycle eight times, with each of the 44 successful transitions taking under 20 s. Conversely, DaPaw completed the cycle once, with seven successful transitions, dedicating 66% of the time to discern the correct action for the water molecule.

Following the initial evaluation, a group of 10 researchers started weekly collaborative sessions to further refine the tool using a user-centered approach. This process emphasized its capacity to highlight relevant segments for subsequent detailed examination that would have been hard to discern otherwise. The assessment revealed that representing students' state transitions as molecules they embody within the simulation, coupled with their navigational choices to get to parts of the screen that supported the transitions facilitated comprehension by mirroring the simulation's visual content. The tool was enhanced to permit selective modality display aligned with the investigators' specific research queries to mitigate cognitive overload from excessive on-screen data. Additionally, the tool was augmented to support the exploration of cooperative student dynamics by enabling simultaneous data visualization from multiple participants. In recognition of the activity's embodied nature, a functionality to alter the video's camera perspective in conjunction with the timeline also was integrated.

5   Conclusion and Future Work

Our multimodal timeline to support IA marks a significant advance in examining student interactions within mixed-reality learning settings. Leveraging machine learning, the tool captures and displays data that enriches IA, suggesting its utility in advancing IA research. It facilitates a transition from purely qualitative to mixed methods analysis by integrating quantitative data on student performance and behavioral trends over time.

---

[1] None of the names used are the students' true names.

Moving forward, the tool can be extended to accommodate different science models, broadening its applicability across diverse educational contexts, whether in embodied or computer-based learning environments and including modalities relevant to each context. An ongoing weekly meeting of IA sessions is currently assessing the timeline's usefulness and informing the inclusion of additional modalities, such as conversations that offer a more comprehensive understanding of communication dynamics. The iterative nature of IA, coupled with the versatility of the multimodal timeline, positions it as a dynamic framework that can evolve alongside emerging research questions and technological advancements, thereby fostering continued advancements in the field of IA applied to learning environments. We also hope to investigate how the timeline can be tailored towards the teachers and how the findings can be used to assist students during the learning activities.

Acknowledgement. This work was supported by the following grants from the National Science Foundation (NSF): DRL-2112635, IIS-1908632 and IIS-1908791. The authors have no known conflicts of interest to declare. We would like to thank all of the students and teachers who participated in this work.

References

1. Abdelrahman, A.A., Hempel, T., Khalifa, A., Al-Hamadi, A.: L2cs-net: fine-grained gaze estimation in unconstrained environments. In: 2023 8th International Conference on Frontiers of Signal Processing (ICFSP), pp. 98–102 (2022)
2. Andrade, A.: Understanding student learning trajectories using multimodal learning analytics within an embodied-interaction learning environment. In: Proceedings of the Seventh International Learning Analytics & Knowledge Conference (2017)
3. Ashwin, T., Guddeti, R.M.R.: Affective database for e-learning and classroom environments using Indian students' faces, hand gestures and body postures. Futur. Gener. Comput. Syst. **108**, 334–348 (2020)
4. Bhat, S.F., Birkl, R., Wofk, D., Wonka, P., Müller, M.: Zoedepth: zero-shot transfer by combining relative and metric depth (2023)
5. Danish, J., et al.: Designing for shifting learning activities. J. Appl. Instruct. Des. **11**(4), 169–185 (2022)
6. Danish, J.A., Enyedy, N., Saleh, A., Humburg, M.: Learning in embodied activity framework: a sociocultural framework for embodied cognition. Int. J. Comput.-Support. Collab. Learn. **15**, 49–87 (2020)
7. Davalos, E., Timalsina, U., Zhang, Y., Wu, J., Fonteles, J.H., Biswas, G.: Chimerapy: a scientific distributed streaming framework for real-time multimodal data retrieval and processing. In: 2023 IEEE International Conference on Big Data (BigData). IEEE (2023)
8. Davalos, E., et al.: Identifying gaze behavior evolution via temporal fully-weighted scanpath graphs. In: LAK23: 13th International Learning Analytics and Knowledge Conference, pp. 476–487. Association for Computing Machinery (2023)
9. D'Mello, S., Graesser, A.: Dynamics of affective states during complex learning. Learn. Instr. **22**(2), 145–157 (2012)
10. Enyedy, N., Danish, J.: Learning physics through play and embodied reflection in a mixed reality learning environment. In: Learning Technologies and the Body, pp. 97–111. Routledge (2014)

11. Errea, J., Gestalten (eds.): Visual journalism. Die Gestalten Verlag (2017)
12. Ez-zaouia, M., Tabard, A., Lavoué, E.: Emodash: a dashboard supporting retrospective awareness of emotions in online learning. Int. J. Hum.-Comput. Stud. **139**, 102411 (2020)
13. Hall, R., Stevens, R.: Interaction analysis approaches to knowledge in use. In: Knowledge and Interaction, pp. 88–124. Routledge (2015)
14. Hervé, N., Letessier, P., Derval, M., Nabi, H.: Amalia.js: an open-source metadata driven html5 multimedia player. In: Proceedings of the 23rd Annual ACM Conference on Multimedia Conference, pp. 709–712. ACM (2015)
15. Kellnhofer, P., Recasens, A., Stent, S., Matusik, W., Torralba, A.: Gaze360: physically unconstrained gaze estimation in the wild. In: IEEE International Conference on Computer Vision (ICCV) (2019)
16. Kersting, M., Haglund, J., Steier, R.: A growing body of knowledge: on four different senses of embodiment in science education. Sci. Educ. **30**(5), 1183–1210 (2021)
17. Lane, A., Lee, S., Enyedy, N.: Embodied resources for connective and productive disciplinary engagement [poster]. In: AERA Annual Meeting. American Educational Research Association (2024)
18. Li, T.H., Suzuki, H., Ohtake, Y.: Visualization of user's attention on objects in 3D environment using only eye tracking glasses. J. Comput. Des. Eng. **7**(2), 228–237 (2020)
19. Martinez-Maldonado, R., Echeverria, V., Santos, O.C., Santos, A.D., Yacef, K.: Physical learning analytics: a multimodal perspective. In: Proceedings of the 8th International Conference on Learning Analytics and Knowledge, pp. 375–379 (2018)
20. Pekrun, R., Stephens, E.J.: Academic emotions, p. 3–31. American Psychological Association (2012)
21. Russell, J.A.: A circumplex model of affect. J. Pers. Soc. Psychol. **39**(6), 1161 (1980)
22. Savchenko, A.V., Savchenko, L.V., Makarov, I.: Classifying emotions and engagement in online learning based on a single facial expression recognition neural network. IEEE Trans. Affect. Comput. **13**, 2132–2143 (2022)
23. Schwendimann, B.A., et al.: Perceiving learning at a glance: a systematic literature review of learning dashboard research. IEEE Trans. Learn. Technol. **10**(1), 30–41 (2017)
24. Steinberg, S., Zhou, M., Vickery, M., Mathayas, N., Danish, J.: Making sense of modes in collective embodied science activities. In: Proceedings of the 17th International Conference of the Learning Sciences-ICLS 2023, pp. 1218–1221. International Society of the Learning Sciences (2023)
25. Tang, S., Andriluka, M., Andres, B., Schiele, B.: Multiple people tracking by lifted multicut and person re-identification. In: Proceedings of the IEEE Conference on Computer Vision and Pattern Recognition, pp. 3539–3548 (2017)
26. TS, A., Guddeti, R.M.R.: Automatic detection of students' affective states in classroom environment using hybrid convolutional neural networks. Educ. Inf. Technol. **25**(2), 1387–1415 (2020)
27. Vatral, C., Biswas, G., Cohn, C., Davalos, E., Mohammed, N.: Using the dicot framework for integrated multimodal analysis in mixed-reality training environments. Front. Artif. Intell. **5**, 941825 (2022)
28. Zhang, K., Zhang, Z., Li, Z., Qiao, Y.: Joint face detection and alignment using multitask cascaded convolutional networks. IEEE Signal Process. Lett. **23**(10), 1499–1503 (2016)

"I Am Confused! How to Differentiate Between…?" Adaptive Follow-Up Questions Facilitate Tutor Learning with Effective Time-On-Task

Tasmia Shahriar⬤ and Noboru Matsuda[✉]⬤

North Carolina State University, Raleigh, NC 27695, USA
{tshahri,noboru.matsuda}@ncsu.edu

Abstract. Within the learning-by-teaching paradigm, students, who we refer as *tutors*, often tend to dictate what they know or what to do rather than reflecting on their knowledge when assisting a teachable agent (TA). It is vital to explore more effective ways of fostering tutor reflection and enhancing the learning experience. While TAs can employ static follow-up questions, such as "Can you clarify or explain more in detail?" to encourage reflective thinking, the question arises: Can Large Language Models (LLMs) generate more adaptive and contextually-driven questions to deepen tutor engagement and facilitate their learning process? In this paper, we propose ExpectAdapt, a novel questioning framework for the TA using three stacked LLMs to promote reflective thinking in tutors, thereby, facilitating tutor learning. ExpectAdapt generates adaptive follow-up questions by directing tutors towards an expected response based on the tutor's contributions using conversation history as a contextual guide. Our empirical study with 42 middle-school students demonstrates that adaptive follow-up questions facilitated tutor learning by effectively increasing problem-solving accuracy in the learning-by-teaching environment when compared to tutors answering the static follow-up questions and no follow-up questions at all.

Keywords: Learning by teaching · conversational questions · large language model · in-context learning

1 Introduction

Students learn more by assisting a teachable agent (TA)—a synthetic peer they can iteratively teach—compared to solitary learning [1]. This phenomenon is known as *tutor learning* [2–4]. In our work, we address students who teach a TA as *tutors*. Empirical studies reported that tutors often tend to dictate what they know, instead of reflecting on their understanding and critical thinking that results in a limited benefits from learning-by-teaching [5, 6]. The TA can promote tutors' reflective thinking by persistently asking follow-up questions [7–10]. Yet, automatically generating such follow-up questions is challenging due to the expertise required to formulate such questions and varying levels of prior knowledge among tutors. Effective questions must be tailored to individual

A. M. Olney et al. (Eds.): AIED 2024, LNAI 14830, pp. 17–30, 2024.
https://doi.org/10.1007/978-3-031-64299-9_2

tutors' understanding, while simultaneously pushing their cognitive boundaries, maintaining discourse coherence, context relevance, and inextricably bound to the conceptual content of the subject matter [11, 12].

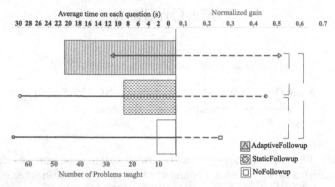

Fig. 1. ExpectAdapt wins in terms of more effective time-on-task. Tutors spent more time on average to answer adaptive follow-up questions compared to static questions (shown in barplot) that helped them achieve the same gain (shown in dotted **blue line**) by teaching significantly fewer problems (shown in solid **darkred line**) to the teachable agent (Color figure online)

Can we instruct Large Language Models (LLMs) in such a way that it can generate questions to engage tutors in critical thinking in a learning-by-teaching environment?

In this paper, we propose ExpectAdapt, a novel follow-up questioning framework for the TA. ExpectAdapt consists of two LLMs. The first LLM generates an ideal tutor's response (to TA's question) that is reflective of tutor's critical thinking. The second LLM generates a follow-up question (relevant to the conversation history) if the student's response to TA's question is not satisfactory relative to the ideal response. Figure 1 shows that tutors spent significantly more time on average to answer the expectation tailored adaptive (or ExpectAdapt for short) follow-up questions compared to static questions that only prompted tutors to explain more. Furthermore, spending more time on answering ExpectAdapt follow-up questions helped tutors *achieve the same learning gain by teaching fewer problems* to the TA compared to tutors who answered static questions. Additionally, tutors who engaged with ExpectAdapt follow-up questions achieved higher learning gains compared to those who did not answer any follow-up questions.

In this paper, we address following research questions. RQ1: Does answering ExpectAdapt follow-up questions help tutors learn? RQ2: Is ExpectAdapt follow-up questions more effective than the static follow-up questions?

Our main contributions are: (I) We propose ExpectAdapt that employs prompt engineering techniques to configure LLMs in a manner that enables them to generate contextually relevant follow-up questions. (II) We conduct an empirical evaluation study that showed the effectiveness of our proposed ExpectAdapt framework. (III) The ExpectAdapt questioning framework offers scalability, as it can be easily adapted to various problem-solving domains with minimal need for expert annotations.

2   APLUS: The Learning-By-Teaching Environment

Our study extends the traditional APLUS (Artificial Peer Learning Using SimStudent) where tutors assist SimStudent (the teachable agent) how to solve linear algebraic equations [13, 14]. Figure 2 displays the user interface of APLUS. Whenever tutor enters a linear equation to teach (Fig. 2-a), SimStudent tries to solve one step at a time by consulting its knowledge base that consists of production rules once learned like, "if [*conditions*] hold then perform [*a solution step*]." In APLUS, the *solution step* allows four basic math operations: *add*, *subtract*, *multiply*, and *divide by* a term. If SimStudent has a production that can apply, it seeks feedback from the tutor. If the tutor agrees, it proceeds to the next step. If the tutor disagrees, it asks a focal question, "Why am I wrong?". The tutor is expected to provide their textual explanation in a chat box (Fig. 2-c). If SimStudent does not have a production to apply, it requests the tutor to demonstrate the next step. After tutor demonstrates the solution step, it asks another focal question, "Why should we do it?". In the traditional APLUS, SimStudent does not ask follow-up questions after tutor's response to the focal questions.

Apart from teaching, tutor can quiz SimStudent anytime to evaluate how well SimStudent has learned thus far by observing the SimStudent's performance on the quiz. Quiz topics include one-step equations (level 1), two-step equations (level 2), equations with variables on both sides (level 3), and a final challenge that contains equations with variables on both sides (level 4) (Fig. 2-g). SimStudent works on a single quiz level at a time. Upon successfully passing a level, the subsequent level is unlocked.

Tutors may also review the resource tabs that include problem bank, unit overview, introduction video and worked out examples at any time (Fig. 2-d). The teacher agent, Mr. Williams (Fig. 2-h), provides on-demand, voluntary hints on how to teach. For example, if the tutor repeatedly teaches one-step equations, Mr. Williams might provide the hint, "SimStudent failed on the two-step equation. Teaching similar equations will help him pass that quiz item".

3   ExpectAdapt Framework

3.1   Motivation

In our past studies with APLUS, tutors often exhibited a tendency to neglect or inadequately respond to SimStudent's focal questions [6, 9]. We also found that tutors who could explain elaborately using conceptual terms learned significantly more than tutors who could not provide such responses irrespective of their prior knowledge [6]. Roscoe [8] linked tutors' inability to provide accurate, elaborated and sense-making response with their infrequent reflective behavior.

Building upon these insights, we design ExpectAdapt framework to generate questions directing tutors towards an elaborated response. ExpectAdapt generates follow-up questions tapping on the aspects of the elaborated response that tutor has not conveyed just yet throughout the current conversation history. Our work is closely aligned with AutoTutor's expectation & misconception-tailored dialogue (aka, *EMT* dialogue) [15]. AutoTutor is a computer-based tutor that attempts to simulate the dialogue moves of a human tutor to help students learn. However, our approach distinguishes itself by

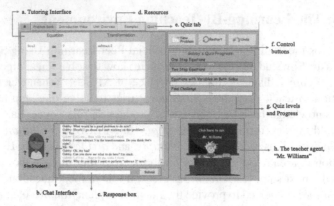

Fig. 2. APLUS interface with SimStudent in the bottom left corner

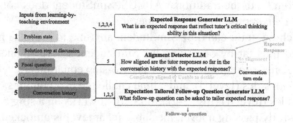

Fig. 3. ExpectAdapt framework with three stacked LLM

detecting a misalignment between tutors' response and the expected response to mimic a teachable agent's effort to bridge knowledge gaps when encountering unclear concepts. This process resembles the way a student clarifies ambiguities in a textbook, avoiding excessively corrective questioning. Additionally, AutoTutor relies heavily on scripted authoring tools, demanding significant human expertise to design various misconception cases and dialogue scenarios. In contrast, our LLM-based framework aims to reduce this substantial human effort, making it a more efficient and scalable solution.

ExpectAdapt consists of three modules: (1) Expected response generator, (2) Alignment detector, and (3) Expectation tailored follow-up question generator. All these three modules are implemented using OpenAI API key for the GPT-3.5-turbo model. Figure 3 shows the ExpectAdapt framework with three stacked LLM.

We utilized various prompt engineering techniques such as few-shot demonstrations using chain-of-thought [16–19], role prompting [20], and adding extra context in the prompt [21, 22]. We intentionally avoided the term "teachable agent" in our prompts, opting instead for the term "student". This choice is grounded in the hypothesis that LLMs may encounter difficulties in assuming the role of a teachable agent, a scenario presumably less prevalent in their pretraining datasets—LLMs excel with more common terms from pretraining [23]. For researchers aiming to replicate our results, we recommend substituting "teachable agent" with "student" shown in the prompts.

3.2   Expected Response Generator

The Expected Response Generator (ERG) LLM outputs an accurate explanation that reflects critical thinking by using domain relevant concepts to the question asked by the teachable agent. Zhang *et al.* [20] reported that providing the LLM with a specific role to play, such as a helpful assistant or a knowledgeable expert can be particularly effective in ensuring that the model's output align with the desired output. To ensure that the generated expected response is accurate we provide the ERG LLM with the role of a tutor who is expert in the domain in the task instruction part of the prompt. Figure 4 shows the prompt used for the ERG LLM.

In general, LLMs learn to perform a new task by conditioning on a few *input/output* demonstrations, a phenomenon called *in-context* learning [16]. One of the key drivers of in-context learning is the distribution of the input text *specified by* the demonstrations [24]. We hypothesized that problem state, solution step at discussion, correctness of the solution step, and the focal question asked by the TA are necessary and sufficient components to capture the entire specification of *input*. We further hypothesize that this input specification facilitates accurate generation of expected response. We based our formulation of *output*, which is the expected responses, on the definition of reflective responses defined in [6] as "A reflective response is either descriptive or reparative in its intonation and elaborates in favor or disfavor of a solution step using relevant conceptual terms." We included eight demonstrations as few-shot examples, adhering to findings that LLM performance declines with more than eight examples [24, 25].

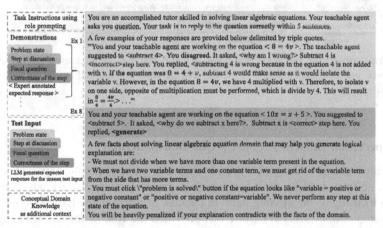

Fig. 4. Overview of the prompt components and the resultant prompt used for the expected response generation.

To enhance the LLM's ability to generate expected responses across problem states and solution steps that were not covered in the demonstrations, we incrementally integrated conceptual knowledge of algebraic domain addressing errors made by the LLM that we call *assertions* [22]. For instance, LLM generated an output, "dividing by 4 is wrong when the equation is $-4v = 6+3v$ since the coefficient is -4 not 4". Such error was prevented including an assertion like, "You cannot divide when an equation has two variable terms". Adding this assertion modified LLM output as, "there are two variable terms in this equation. If the equation was $-4v = 6$, dividing by -4 would make sense". We used the greedy decoding strategy by setting the temperature to 0 in ERG LLM. Since the current work is primarily focused on the question generation, we encourage interested readers to refer to this paper [22].

We conducted a survey with 12 in-service middle school teachers to assess the quality of the generated expected responses. The survey data revealed that LLM generated responses are (1) relevant to the *input* specification, (2) elaborate optimal solution step, and (3) sound in terms of using concept terms in its reasoning. The survey further confirmed that including assertions in the prompt improved the accuracy of expected responses by 15% over solely relying on demonstrations [22].

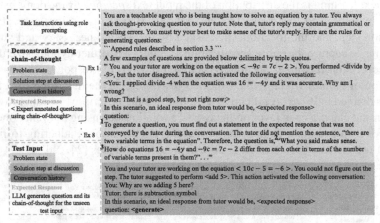

Fig. 5. Overview of the prompt components and the resultant prompt used for the expectation tailored question generation. The output components are marked using numbers in the demonstration.

3.3 Expectation Tailored Follow-Up Question Generator

To generate a question, the goal is to find any missing stem from the expected response that was not conveyed by the tutor during the conversation and ask an open-ended question focusing on the missing stem. In the task instruction, the Expectation Tailored Follow-up Question Generator LLM was provided the role of the teachable agent to encourage curiosity driven questions. The task instruction also includes a set of rules to generate the question. The rules are as follows ("you" in the rules refer to the teachable agent role-playing LLM):

- To generate a question, you must find out a missing stem from the expected response that was not covered in the conversation history and generate a question.
- If you have previously asked question about a missing stem but tutor did not provide a relevant response, find another missing stem, and generate a question.
- If you have asked questions about every possible stem from the expected response, then say, *no question*.

To further mold the LLM output, we design eight few-shot demonstrations using the chain-of-thought prompting technique [25]. These demonstrations consist of *input* comprising the problem state, the solution step discussed, and the conversation history for that step. To create more realistic conversation histories, we drew from the data collected during past studies using APLUS [26, 27] that are available publicly in Datashop [28]. This allowed us to incorporate realistic instances containing grammatical or spelling errors in tutors' responses. The *output* consists of three parts: (1) the chain-of-thought to find the missing stem from the expected response, (2) relevant acknowledgement or summarization from the conversation history to maintain conversational context, and (3) the formulated question focusing on the missing stem. Figure 5 shows the prompt used for question generation. The three parts of the output are marked using numbers in the figure.

Finally, to encourage diverse reasoning pathways and prevent the TA from asking repetitive questions, we deliberately set the temperature of the LLM decoder to 0.5 instead of using a greedy decoding approach at 0.

3.4 Alignment Detector

Our initial observation revealed that questions generated by the question generator tend to be overly corrective or out of conversational context. This happened when (1) tutor disagreed with the TA's suggested step and had a different solution step in mind than the expected response, (2) tutor's response diverged completely from the expected response, and (3) expected response was erroneous due to hallucinations [29].

An instance of (1) is shown below:

A tutor and SimStudent are working on the equation $3 - 2x = 4$. SimStudent suggested to perform add 3, but the tutor disagreed. This action activated the following conversation:

SimStudent: Why am I wrong?
Tutor: You have to add 2 because we have -2 in the equation.
SimStudent: I understand you are suggesting to add 2. Why can't I perform subtract 3?

In this case, SimStudent asking "Why can't I perform subtract 3?" is overly corrective. This question was suggested by the question generator because the expected response in this scenario was: "Add 3 is not correct because 3 is added with $- 2x$. The equation can also be written as $-2x + 3 = 4$. To undo the $+ 3$ and isolate $-2x$, we must perform the opposite of $+ 3$ which is subtract 3 on both sides. Subtract 3 will result in $-2x+3-3 = 4-3$".

Johns [30] identified that correcting tutors' contribution is limiting for learning. These observations inspired us to incorporate the alignment detector that indicates if tutor's response and the expected response are (1) completely aligned, (2) not aligned, or (3) unable to detect. We designed eight few-shot demonstrations including chain-of-thoughts covering three scenarios to design the prompt for the alignment detector LLM.

We show sample questions generated by ExpectAdapt framework in Table 1.

Table 1. Examples of questions generated by ExpectAdapt with corresponding Bloom's Taxonomy levels. Labels are author-coded for demonstration only and do not reflect an assessment of question quality or effectiveness.

Bloom's Taxonomy levels	Questions generated by ExpectAdapt
Remember	What is the coefficient in this equation?
Understand	I see that we have 7c and -9c in the equation. Can you explain what makes them like terms? How do we identify like terms in an equation?
Apply	After dividing both sides of the equation by 3," what will be the final equation? How does this step help us isolate v on its own?
Analyze	I am confused! How can we differentiate between the coefficient and the constant in an equation? Can you provide some examples to clarify this concept?
Evaluate	It appears we can cancel out the addition of 3 by subtracting 3 as well as adding -3. Can you explain why they are same?
Create	None found

4   Method

The central research questions are: (1) Does answering adaptive follow-up questions help tutors learn? (2) Are adaptive questions more effective than static follow-up questions?

To address these research questions, we conducted a semi-secondary data analysis study where two sets of empirical data were combined: (1) The data collected from a past study [6], and (2) the data that we collected from a new study that we conducted. For clarity, we call the past study **Study A** and the newly conducted study **Study B.**

Study A included two conditions of APLUS: **No**Followup condition where tutors only answered the focal question and TA never asked follow-up questions, and **Static**Followup condition where tutors answered focal questions along with static "explain more" follow-up questions. There were 16 and 17 participants in the **No**Followup and the **Static**Followup conditions respectively.

Study B, which we recently conducted, involved only one condition, **Adaptive**Followup condition, where tutors answered focal questions along with adaptive follow-up questions generated by ExpectAdapt. Nine 6th to 8th-grade middle school students from local areas were recruited through a study flyer shared within the previous participants' network (aka purposive and snowball sampling). Participants received monetary compensation. The study was conducted online where APLUS was accessed through Zoom screen-sharing.

Consequently, the current analysis compares three conditions with the total of 42 middle school students involved in three conditions. Study B followed the same format and used the same measures as Study A. That is, participants took a pre-test for 30 min on the first day of the study. Immediately after taking the pre-test, all participants watched a 10-min tutorial video on how to use APLUS. Participants were informed in the video that their goal was to help their TA pass the quiz. Participants were free to use

APLUS for three days for a total of 2 h or to complete their goal (i.e., passing the quiz), whichever came first. Upon completion, participants took a 30-min post-test.

The pre- and post-tests were isomorphic, and each consisted of 22 questions: 10 questions on solving the equation and 12 multiple-choice questions to measure the proficiency of algebra concepts. Details on the test items can be found in our previous paper [6]. We utilized a binary scoring system for each test items, i.e., answers were marked strictly as either correct or incorrect. Test scores are normalized as the ratio of participant's score to the maximum score.

One-way ANOVA with the normalized pre-test score and condition confirmed no condition difference; $M_{\text{NoFollowup}} = 0.63 \pm 0.24$ vs. $M_{\text{StaticFollowup}} = 0.60 \pm 0.18$ vs. $M_{\text{AdaptiveFollowup}} = 0.62 \pm 0.25$; $F(2,39) = 0.06, p = 0.94$. We controlled the time on task. A one-way ANOVA confirmed no condition difference on the minutes participants spent on APLUS; $M_{\text{NoFollowup}} = 215$ vs $M_{\text{StaticFollowup}} = 242$ vs. $M_{\text{AdaptiveFollowup}} = 204$; $F(2,39) = 0.66, p = .52$. To maintain consistency with StaticFollowup condition in our previous study, we purposefully limited the ExpectAdapt framework to generate a maximum of three follow-up questions for each focal question.

In the following analysis, we use the learning outcome data (normalized pre-and post-test scores) along with participant's interaction data collected by APLUS, interface actions, TA inquiries, and participants' responses.

5 Results

5.1 Tutors in Follow-Up Question Modes had Higher Test Score Improvement than NoFollowup Mode, whereas AdaptiveFollowup Tied with StaticFollowup

We conducted a repeated-measures ANOVA with test score as a dependent variable, whereas test-time (pre- vs. post-test) as the within-subject and condition (NoFollowup vs. StaticFollowup vs. AdaptiveFollowup) as the between-subject independent variables. There was an interaction between test-time and condition; $F(2, 39) = 3.05, p = 0.05$. A simple main effect of condition (paired t-test with test-time as the independent variable) revealed that tutors in follow-up conditions showed a reliable increase from pre- to post-test (StaticFollowup: paired-$t(16) = 3.86, p < 0.05$ and AdaptiveFollowup: paired-$t(8) = 2.87, p < 0.05$); but no reliable increase in the NoFollowup condition (paired-$t(15) = 1.2$, $p = 0.24$). We further ran ANCOVA analysis with the normalized post-test as dependent variable and condition as the independent variable while controlling the normalized pre-test. No condition effect was found between AdaptiveFollowup and StaticFollowup tutors; $F(1,23) = 0.03, p = .88$. However, there was a condition difference between AdaptiveFollowup vs NoFollowup; $F(1,22) = 3.90, p = .06$ and StaticFollowup vs NoFollowup tutors; $F(1,30) = 5.20, p < .05$.

The results suggests that *tutors who answered any kind of follow-up questions (static or adaptive) ended up having a higher post-test improvement than the tutors who did not answer follow-up questions.* The result also suggests that *tutors who answered adaptive follow-up questions ended up having the same post-test improvement as the tutors who answered static follow-up questions.*

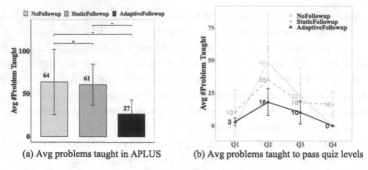

Fig. 6. Average problems taught by tutors across conditions

5.2 Tutors in AdaptiveFollowup Passed Quiz Levels by Teaching Fewer Problems Compared to StaticFollowup and NoFollowup

Tutors in AdaptiveFollowup and StaticFollowup conditions tied on post-test. This outcome prompted an investigation into whether the time spent on APLUS was comparable across these conditions. Therefore, we first analyzed the number of problems taught by tutors across all conditions. A one-way ANOVA with number of problems taught as dependent variable and condition as independent variable revealed a main-effect of condition; $F(2,39) = 5.3$ $p < .01$ $M_{NoFollowup} = 64 \pm 38$ vs. $M_{StaticFollowup} = 61 \pm 24$ vs. $M_{AdaptiveFollowup} = 27 \pm 16$. We ran pairwise T-tests with Bonferroni correction as the post-hoc analysis. The results revealed that the average problems taught by AdaptiveFollowup tutors were statistically different compared to StaticFollowup ($t(22.0) = 4.30$, $p < .05$) and NoFollowup ($t(22.0) = 3.38$, $p < .05$) tutors, whereas, the average problems taught by StaticFollowup vs. NoFollowup tutors were not different ($t(25.0) = 0.27$, $p = .27$). Therefore, the data suggests that *StaticFollowup and NoFollowup tutors taught equal number of problems on average, whereas AdaptiveFollowup tutors taught significantly fewer number of problems compared to the other two conditions.*

This finding led us to question whether this difference influenced their ability to pass quiz levels. When we ran one-way ANOVA with the maximum quiz level passed by tutors across conditions, we found no condition effect; $M_{NoFollowup} = 2$ vs. $M_{StaticFollowup} = 2$ vs. $M_{AdaptiveFollowup} = 2$ $F(2,39) = .50$ $p = .61$. We further visualized the number of problems taught before tutors could pass a quiz level across conditions, illustrated in Fig. 6. As shown in the plot, *tutors in AdaptiveFollowup condition passed the second quiz level with much fewer problems (18) taught in average compared to StaticFollowup (35) and NoFollowup (48) tutors.*

5.3 AdaptiveFollowup Tutors Spent More Time on Average to Answer TA Questions that Helped them Teach Problems Accurately

Why did AdaptiveFollowup tutors pass the quiz by teaching fewer problems? Our naïve hypothesis conjectures that tutors in the AdaptiveFollowup condition spent more time on answering the questions that facilitated accurate problem-solving, thereby enabling them to pass the quiz with fewer problems taught.

To test this hypothesis, we began by calculating the average time spent by tutors on various activities while teaching each problem within APLUS. Activities include question answering (QA), teaching (T), reviewing resource tabs like quiz (Qu), example (Ex), unit overview (Uo), problem bank (Pb), and introduction video (Iv). We conducted separate mixed model analysis with each of the activity duration per problem as a dependent variable while condition as fixed factor and tutors as random factor (shown in Table 2). The data suggest that *AdaptiveFollowup tutors spent reliably more time on QA (21.4s on average) compared to Static (8.9s) and NoFollowup (2.8s) tutors per problem.*

Our next aim is to understand how time spent on these activities relates to problem-solving accuracy, measured as the percentage of correctness (*%correctness*). *%correctness* was calculated per problem based on the ratio of correctly demonstrated steps and feedback to the total number of steps and feedback provided for that problem. We employed a linear regression model with *%correctness* as the dependent variable and prior groupings based on pre-test scores as the first term, followed by significant activities found in our previous mixed model analysis (i.e. QA, Ex, and Qu), condition, and their interactions. The result revealed a significant interaction between time spent on QA and condition; $F_{Condition:QA}$ $(2, 2279) = 9.5$, $p < .05$ and Example tab and condition; $F_{Condition:Example}$ $(2, 2283) = 5.5$, $p < .05$. Other interaction terms were not main effects. The regression model suggests that *spending 1 min more on adaptive follow-up questions results in 7.8% increase in problem-solving accuracy in APLUS, which is a notable correlation given that AdaptiveFollowup tutors spent 10 min on QA in average for all the problems taught in APLUS as shown in* Fig. 6a.

6 Discussion

Our first research question was, RQ1: Does answering expectation tailored adaptive follow-up questions help tutors learn? Our data revealed that the post-test improvement is higher for tutors who answered any follow-up questions (both static and adaptive) compared to tutors who only answered the focal questions followed by no follow-up. Tutors who answered adaptive follow-up questions ended up having the same post-test improvement as the tutors who answered static follow-up questions.

Our next research question was, RQ2: Are adaptive follow-up questions more effective than static follow-up questions? Our data revealed that tutors spent more time answering the adaptive follow-up questions than static follow-up questions. We also found a strong correlation between time spent on answering adaptive follow-up questions and accuracy in solving problems. This observation suggests that *spending more time on adaptive questions helped tutors solve problems more accurately, which resulted in teaching fewer problems in APLUS for passing the quiz levels.*

This efficiency in learning was particularly evident in the early quiz levels as shown in Fig. 6b. The relatively low number of problems taught before passing the first quiz level can be attributed to the APLUS design. APLUS allows tutors to pass the first level automatically if they quiz SimStudent after launching the app for the first time, as advised in the intro video. Similarly, teaching problems more accurately in the second and third levels increases the likelihood of SimStudent passing the fourth level without requiring additional problems taught. This is because fourth level involves equations with variables on both sides that have similar difficulties to the third level. The fewer number of problems taught at the second and third levels suggests that *AdaptiveFollowup tutors enhanced their problem-solving accuracy effectively by engaging with adaptive questions.* In contrast, StaticFollowup tutors, despite engaging with static questions, did not achieve the same level of problem-solving accuracy, leading to their need for teaching more problems to pass the quiz.

In this paper, we proposed ExpectAdapt, an expectation tailored follow-up question framework for teachable agents using large language models and showed that tutors learned efficiently by answering adaptive questions. In this work, we narrow our focus to learning outcomes to assess the efficacy of our framework. One of our future works includes delving into the cognitive level of the adaptive questions.

Table 2. Average time spent (s) across different activities per problem.

	Condition						F	p
	NoFollowup		StaticFollowup		AdaptiveFollowup			
	M	SD	M	SD	M	SD		
QA	2.8[a]	1.3	8.9[b]	5.1	21.4[c]	8.0	$[\frac{2}{32}]$ 26.82	<.05
Qu	13.1[a]	6.3	13.5[a]	9.1	49.2[c]	24.4	$[\frac{2}{29}]$ 55.35	<.05
Ex	6.9[a]	6.2	3.1[a]	2.8	13.9[c]	15.2	$[\frac{2}{35}]$ 6.52	<.05
Uo	0.8	0.9	1.7	1.7	3.3	6.1	$[\frac{2}{31}]$ 2.98	0.07
Pb	1.7	1.6	0.9	0.8	3.1	3.2	$[\frac{2}{43}]$1. 83	0.17
Iv	1.2	2.0	0.6	0.7	2.3	3.2	$[\frac{2}{55}]$1.23	0.29
T	61.1	33.3	46.0	42.7	67.8	15.2	$[\frac{2}{31}]$1.94	0.15

Means that do not share superscripts differ significantly at $p < 0.05$ in the post-hoc analysis.

The observation that tutor learning outcomes were comparable between static and adaptive questions is intriguing. One possible explanation is that teaching many problems with "shallow" question answering could be as effective for tutor learning as teaching fewer problems with elaborated question answering. In other words, the quantity of problems tackled could be as crucial as their quality. Further research could explore the optimal balance between problem quantity and quality, as well as the differential impacts these factors have on learning gains in educational settings.

7 Conclusion

We proposed ExpectAdapt, a novel follow-up questioning framework for the teachable agent using large language models that generates follow-up questions adapting based on tutors' contributions to the conversation history. We found that adaptive follow-up questions facilitated tutor learning by ensuring productive use of instructional time. Our current data demonstrated that tutors interacting with ExpectAdapt's questions exhibited greater improvement from pre- to post-test than interacting with focal questions only. We also found that while tutors achieved equivalent learning outcomes when responding to adaptive as opposed to static questions, the former demanded a higher level of engagement, as evidenced by extended question answering durations. This extended duration with adaptive question answering correlated with improved problem-solving abilities within the APLUS learning environment.

Our research provides strong evidence supporting the use of large language models to generate adaptive follow-up questions. Furthermore, in-context learning capabilities of these models provide an opportunity to incorporate expert knowledge. This results in more polished and insightful question generation with minimal efforts and easily scalable across educational domains.

Acknowledgment. This research was supported by the Institute of Education Sciences, U.S. Department of Education, through grant No. R305A180517 and National Science Foundation grant Nos. 2016966 and 2112635 to North Carolina State University. The opinions expressed are those of the authors and do not represent views of the Institute or the U.S. Department of Education and NSF.

References

1. Okita, S.Y., J. Bailenson, Schwartz, D.L.: Mere belief of social action improves complex learning. In: Barab, K.H.S., Hickey, D. (eds.) Proceedings of the International Conference for the Learning Sciences, in press, Lawrence Erlbaum, New Jersey (2008)
2. Roscoe, R.D., Chi, M.T.H.: Understanding tutor learning: knowledge-building and knowledge-telling in peer tutors' explanations and questions. Rev. Educ. Res. **77**(4), 534–574 (2007)
3. Chi, M.T.H., et al.: Learning from human tutoring. Cogn. Sci. **25**, 471–533 (2001)
4. Graesser, A.C., Person, N.K., Magliano, J.P.: Collaborative dialogue patterns in naturalistic one-to-one tutoring. Appl. Cogn. Psychol. **9**(6), 495–522 (1995)
5. Roscoe, R.D.: Opportunities and barriers for tutor learning: Knowledge-building, metacognition, and motivation. University of Pittsburgh (2008)
6. Shahriar, T., Matsuda, N.: What and How You Explain Matters: Inquisitive Teachable Agent Scaffolds Knowledge-Building for Tutor Learning. Cham: Springer Nature Switzerland (2023)
7. Roscoe, R.D., Chi, M.: Tutor learning: the role of explaining and responding to questions. Instr. Sci. **36**(4), 321–350 (2008)
8. Roscoe, R.D.: Self-monitoring and knowledge-building in learning by teaching. Instr. Sci. **42**(3), 327–351 (2014)
9. Shahriar, T., Matsuda, N.: "Can you clarify what you said?": studying the impact of tutee agents' follow-up questions on tutors' learning. in International Conference on Artificial Intelligence in Education. Springer (2021)
10. Peterson, D.S., Taylor, B.M.: Using higher order questioning to accelerate students' growth in reading. Read. Teach. **65**(5), 295–304 (2012)
11. Otero, J., Graesser, A.C.: PREG: elements of a model of question asking. Cogn. Instr. **19**(2), 143–175 (2001)
12. Azevedo, R., Cromley, J.G.: Does training on self-regulated learning facilitate students' learning with hypermedia? J. Educ. Psychol. **96**(3), 523–535 (2004)

13. Matsuda, N., et al.: Learning by Teaching SimStudent – An Initial Classroom Baseline Study comparing with Cognitive Tutor, In: Biswas, G., Bull, S. (eds.) Proceedings of the International Conference on Artificial Intelligence in Education, pp. 213–221. Springer, Berlin, Heidelberg (2011)
14. Li, N., et al.: Integrating representation learning and skill learning in a human-like intelligent agent. Artif. Intell. **219**, 67–91 (2015)
15. Graesser, A.C.: Conversations with AutoTutor help students learn. Int. J. Artif. Intell. Educ. **26**(1), 124–132 (2016)
16. Brown, T., et al.: Language models arc few-shot learners. Adv. Neural. Inf. Process. Syst. **33**, 1877–1901 (2020)
17. Radford, A., et al.: Language models are unsupervised multitask learners. OpenAI blog **1**(8), 9 (2019)
18. Devlin, J., et al.: Bert: Pre-training of deep bidirectional transformers for language under-standing. arXiv preprint arXiv:1810.04805 (2018)
19. Liu, J., et al.: What Makes Good In-Context Examples for GPT-\$3 \$? arXiv preprint arXiv: 2101.06804 (2021)
20. Zhang, Z., et al.: VISAR: A Human-AI Argumentative Writing Assistant with Visual Programming and Rapid Draft Prototyping. arXiv preprint arXiv:2304.07810 (2023)
21. Liu, J., et al.: Generated knowledge prompting for commonsense reasoning. arXiv preprint arXiv:2110.08387 (2021)
22. Shahriar, T., Matsuda, N., Ramos, K.: Assertion Enhanced Few-Shot Learning: Instructive Technique for Large Language Models to Generate Educational Explanations. arXiv preprint arXiv:2312.03122 (2023)
23. Razeghi, Y., et al.: Impact of pretraining term frequencies on few-shot numerical reasoning. in Findings of the Association for Computational Linguistics: EMNLP 2022. (2022)
24. Min, S., et al.: Rethinking the role of demonstrations: What makes in-context learning work? arXiv preprint arXiv:2202.12837 (2022)
25. Wei, J., et al.: Chain-of-thought prompting elicits reasoning in large language models. Adv. Neural. Inf. Process. Syst. **35**, 24824–24837 (2022)
26. Matsuda, N., Lv, D., Zheng, G.: Teaching how to teach promotes learning by teaching. Int. J. Artificial Intelligence Educ. 1–32 (2022)
27. Matsuda, N., et al.: Studying the effect of tutor learning using a teachable agent that asks the student tutor for explanations, In: Sugimoto, M., et al. (eds.) Proceedings of the International Conference on Digital Game and Intelligent Toy Enhanced Learning (DIGITEL 2012), pp. 25–32. IEEE Computer Society, Los Alamitos, CA (2012)
28. Koedinger, K.R., et al.: A Data Repository for the EDM community: The PSLC DataShop, In: Romero, C., et al. (eds.) Handbook of Educational Data Mining, CRC Press: Boca Raton, FL (2010)
29. Ling, C., et al.: Beyond One-Model-Fits-All: A Survey of Domain Specialization for Large Language Models. arXiv preprint arXiv:2305.18703 (2023)
30. Johns, J.P.: The relationship between teacher behaviors and the incidence of thought-provoking questions by students in secondary schools. J. Educ. Res. **62**(3), 117–122 (1968)

An Automatic Question Usability Evaluation Toolkit

Steven Moore[1]([✉]) [iD], Eamon Costello[2] [iD], Huy A. Nguyen[1] [iD], and John Stamper[1] [iD]

[1] Carnegie Mellon University, Pittsburgh, PA 15213, USA
StevenJamesMoore@gmail.com
[2] Dublin City University, Dublin 09Y0A3, Ireland

Abstract. Evaluating multiple-choice questions (MCQs) involves either labor-intensive human assessments or automated methods that prioritize readability, often overlooking deeper question design flaws. To address this issue, we introduce the Scalable Automatic Question Usability Evaluation Toolkit (SAQUET), an open-source tool that leverages the Item-Writing Flaws (IWF) rubric for a comprehensive and automated quality evaluation of MCQs. By harnessing the latest in large language models such as GPT-4, advanced word embeddings, and Transformers designed to analyze textual complexity, SAQUET effectively pinpoints and assesses a wide array of flaws in MCQs. We first demonstrate the discrepancy between commonly used automated evaluation metrics and the human assessment of MCQ quality. Then we evaluate SAQUET on a diverse dataset of MCQs across the five domains of Chemistry, Statistics, Computer Science, Humanities, and Healthcare, showing how it effectively distinguishes between flawed and flawless questions, providing a level of analysis beyond what is achievable with traditional metrics. With an accuracy rate of over 94% in detecting the presence of flaws identified by human evaluators, our findings emphasize the limitations of existing evaluation methods and showcase potential in improving the quality of educational assessments.

Keywords: Question Evaluation · Multiple-Choice Questions · Question Quality

1 Introduction

Multiple-choice questions (MCQs) are the most commonly utilized assessment format across educational settings, spanning both traditional classroom environments and digital e-learning platforms [6]. Their versatility allows for assessing a broad spectrum of learning outcomes, ranging from simple recall to complex analytical skills, in many learning domains [17]. Besides offering grading efficiency and objectivity, MCQs enable the targeting of specific misconceptions through carefully crafted alternative answer options, known as distractors. However, the development of high-quality MCQs demands a rigorous approach to ensure reliability, validity, and fairness, essential for accurately measuring learners' knowledge and competencies [25].

Recent advances in natural language processing (NLP) have sought to alleviate the burden and time-consuming nature of MCQ authoring, enabling the rapid generation of questions at scale. These technologies facilitate the generation of hundreds

A. M. Olney et al. (Eds.): AIED 2024, LNAI 14830, pp. 31–46, 2024.
https://doi.org/10.1007/978-3-031-64299-9_3

of MCQs within minutes from sources such as document files or direct text requests [15]. Despite these advances, the rise in machine-generated MCQs has not uniformly translated to an improvement in quality. Machine-generated questions produced by state-of-the-art large language models (LLMs) often mirror the inaccuracies commonly found in human generated questions [7]. Such methods raise concerns regarding trust, authenticity, and diversity, potentially leading educators to be hesitant about adopting them without comprehensive evaluation [10].

Among the various MCQ evaluation techniques proposed in the literature, human judgment remains the gold standard, but typically faces challenges with subjectivity, time efficiency, and scalability [21]. Commonly used NLP metrics such as BLEU or METEOR, on the other hand, are much more efficient and scalable, but tend to focus on superficial features like readability and fail to align with human assessments or evaluate the *pedagogical* value of MCQs [12]. The effectiveness of MCQs are only as good as their design, requiring rigorous evaluation to ensure they serve as effective tools for assessing learning.

To address this gap, our research aims to establish a standardized and rigorous automated technique for MCQ evaluation. We begin by demonstrating the limitations of current NLP-based evaluation metrics, highlighting their lack of correlation with common errors found in MCQs. Then we introduce an automated evaluation technique, Scalable Automatic Question Usability Evaluation Toolkit (SAQUET), designed for comprehensive and standardized quality assessment of MCQs across multiple domains. Leveraging the 19 criteria of the Item-Writing Flaws (IWF) rubric [26], a proven and standardized instrument, SAQUET evaluates the structural and pedagogical quality of MCQs. We evaluate SAQUET across two datasets encompassing 271 MCQs from five diverse fields: Chemistry, Statistics, Computer Science, Humanities, and Healthcare.

The primary contributions of our work include: (1) providing empirical evidence on the inadequacy of prevalent MCQ quality evaluation metrics; (2) introducing SAQUET, an open-source tool capable of domain agnostic MCQ evaluation; and (3) compiling the most extensive and varied open dataset of MCQs annotated with IWF, providing opportunity for future research in educational assessment.

2 Related Work

2.1 Generating Multiple-choice Questions

MCQs are widely recognized for their utility, but are susceptible to pattern matching and guessing [26]. However, with careful design, these issues can be mitigated, making well-crafted MCQs effective tools for evaluating a wide range of cognitive skills [15]. Crafting such high-quality questions is complex, with even the most advanced methods, including NLP and LLM based approaches, producing errors such as incorrect answers and nonsensical distractors [3]. For example, recent research showed that only 70% of machine generated MCQs for common subjects were deemed acceptable and clearly worded by human reviewers, with around 50% of the distractors being considered ineffective [22]. Furthermore, a 2023 survey revealed a hesitancy among educators to adopt these AI-powered tools, indicating a lack of trust in the generated MCQs [2].

Previous research has identified flaws in MCQs across various domains and teaching levels, including high-stakes standardized tests developed by psychometricians and domain experts [26]. These MCQs often find repeated use in test banks, practice sets, and training materials over the years. Consequently, there is a need for ongoing quality evaluation of these pre-existing questions, not just newly generated ones. This can complement analyses based on student performance data, such as those offered by Item Response Theory [1]. However, evaluating MCQs before their implementation is crucial to avoid exposing poorly designed questions to learners, which can impede their learning [23]. Crafting high-quality questions remains a significant challenge, and evaluating their quality poses an even greater one, demanding consistency, scalability, and consideration of the questions' application contexts.

2.2 Automated MCQ Quality Evaluation

Over the past decade, automated MCQ quality evaluation has relied on metrics such as BLEU, METEOR, and ROUGE [21]. These metrics primarily assess similarity to a gold standard without considering educational value or effectiveness in evaluating student knowledge [18]. While previous research states these "standardized" metrics facilitate comparison across studies, they involve numerous hyper-parameters that can vary by task and are often insufficiently reported, complicating precise comparisons and replications [16]. Moreover, prior work has demonstrated that these metrics do not sufficiently align with human evaluation [12, 27]. To align more closely with human evaluation while maintaining scalability, alternative automated approaches have explored metrics like perplexity, diversity, grammatical error, complexity, and answerability [24, 28]. These have been applied to both machine- and human generated questions, offering a broader evaluation that extends beyond mere readability to include aspects critical for educational assessments.

When evaluating MCQs, perplexity assesses a language model's ability to predict question and answer text based on its training data [4]. Lower scores suggest more coherent questions and answers with predictable language patterns, whereas higher scores indicate complexity or atypical text, suggesting the questions could be unclear or poorly structured. Diversity evaluates the range in vocabulary, structure, and content across generated texts, ensuring a variety of questions and answers and reducing repetition [13]. A higher diversity score indicates greater uniqueness among MCQs, avoiding repetitive phrases and templated patterns. Grammatical errors pinpoint grammar violations, such as incorrect verb tense or spelling, quantified for each MCQ.

Complexity is typically assessed through cognitive complexity, using Bloom's Revised Taxonomy to assign difficulty levels to MCQs based on the cognitive skills required to answer them [11]. Bloom's Revised Taxonomy categorizes cognitive skills ranging from recall (remembering) to higher-order skills (creating), with questions demanding higher-order thinking deemed more cognitively complex [7]. Answerability measures how accurately a question can be answered, using the provided context or common knowledge. Recently, LLMs such as GPT-4 have been used to automate this evaluation metric [24]. Specifically, the Prompting-based Metric on ANswerability (PMAN) strategy employs three prompts to evaluate a question's quality by how well an LLM can answer it, demonstrating that it aligns with human judgments [27].

2.3 Human MCQ Quality Evaluation

Despite the growth of automated methods for evaluating MCQ quality, human evaluation is still considered the benchmark for accuracy [11, 21]. However, this approach can be subjective, relying on vague metrics like "difficulty" or "acceptability" that are based on intuition [12]. Such evaluations are not only challenging to standardize but also difficult to scale, replicate accurately, and are time-intensive. A more objective alternative that has proven effective for over 15 years is the Item-Writing Flaws (IWF) rubric [26]. Comprising 19 criteria, the IWF rubric evaluates MCQs across any domain, focusing on pedagogical aspects beyond mere readability and surface-level features. It has been successfully applied to MCQs in diverse fields, including standardized medical exams, chemistry, and computer science MOOCs, demonstrating its utility in ensuring quality educational assessments [5, 19, 23].

3 Methods

3.1 Item-Writing Flaws (IWF) Rubric

In our study, we adopted the 19-criteria IWF rubric, a tool that has been validated and employed in prior research [5, 19, 23, 26]. The rubric is designed to be universally applicable across domains, encompassing both pedagogical considerations and factors related to human test-taking abilities. Unlike traditional metrics that primarily assess readability, the IWF rubric includes criteria that address a broader range of question quality aspects, such as unintentional hints, cues, and modality. Table 1 outlines each of the 19 criteria, providing guidance on avoiding specific flaws and ensuring adherence to the rubric's standards. Previous research indicates an MCQ with zero or one IWF can generally be considered acceptable for use, particularly in contexts such as formative assessments [26]. Conversely, an MCQ that exhibits two or more IWFs is classified as unacceptable for use. However, instructors might prioritize avoiding specific IWFs based on their use cases to align best with their learning objectives.

3.2 Technical Overview of SAQUET

Previous efforts to automate the application of the IWF rubric have explored two main strategies, using either a rule-based approach or the well-known GPT-4 model [19]. The rule-based approach demonstrated superior performance to the GPT-4-based method for most criteria across all domains used in the previous study. Building upon these findings, this current work enhances the rule-based methodology by integrating advanced methods and incorporating selective GPT-4 interventions. One of our primary objectives was not only to improve the quality of criteria classifications, but also to preserve the tool's ability to be applied across various domains, ensuring scalability and rapid processing for a large volume of MCQs. The automatic detection of the 19 IWF criteria outlined in Table 1 falls into three distinct categories: text-matching techniques, NLP-based information extraction, and enhancements provided by GPT-4.

The first category includes eight criteria: *None of the Above, All of the Above, Fill-In-The-Blank, True or False, Longest Answer Correct, Negative Worded, Lost Sequence,*

Table 1. The 19 Item-Writing Flaw rubric criteria used in this study.

Item-Writing Flaw	An Item Is Flawed If...
Longest Option Correct	The correct option is longer and includes more detailed information than the other distractors, as this clues students to this option
Ambiguous Information	The question text or any of the options are written in an unclear way that includes ambiguous language
Implausible Distractors	Any included distractors are implausible, as good items depend on having effective distractors
True or False	The options are a series of true/false statements
Absolute Terms	It contains he use of absolute terms (e.g. never, always, all) in the question text or options
Complex or K-type	It contains a range of correct responses that ask students to select from a number of possible combinations of the responses
Negatively Worded	The question text is negatively worded, as it is less likely to measure important learning outcomes and can confuse students
Convergence Cues	Convergence cues are present in the options, where there are different combinations of multiple components to the answer
Lost Sequence	The options are not arranged in chronological or numerical order
Unfocused Stem	The stem is not a clear and focused question that can be understood and answered without looking at the options
None of the Above	One of the options is "none of the above", as it only really measures students ability to detect incorrect answers
Word Repeats	The question text and correct response contain words only repeated between the two
More Than One Correct	There is not a single best-answer, as there should be only one answer
Logical Cues	It contains clues in the stem and the correct option that can help the test-wise student to identify the correct option
All of the Above	One of the options is "all of the above", as students can guess correct responses based on partial information
Fill in the Blank	The question text omits words in the middle of the stem that students must insert from the options provided
Vague Terms	It uses vague terms (e.g. frequently, occasionally) in the options, as there is seldom agreement on their actual meaning
Grammatical Cues	All options are not grammatically consistent with the stem, as they should be parallel in style and form
Gratuitous Information	It contains unnecessary information in the stem that is not required to answer the question

and *Vague terms*. Given the nature of these criteria, foundational programming techniques like string matching are primarily used for identification. However, to enhance accuracy we implemented several modifications, such as adjusting threshold parameters, incorporating checks for various question formats, expanding the list of keywords for matching, and lemmatizing the text to normalize word forms. For example, the *True or False* criteria underwent significant alterations to accommodate Yes/No questions. The *Fill-In-The-Blank* criteria required adjustments to avoid misclassification of Computer Science MCQs, which often use the underscore character. Improvements like refined pattern matching were applied to the *Lost Sequence* criteria, enabling the detection of cases not identified in the initial dataset.

The second category encompasses five criteria: *Implausible Distractors, Word Repeats, Logical Cues, Ambiguous or Unclear,* and *Grammatical Cues*. These criteria are addressed using foundational NLP techniques, including word embeddings, Named Entity Recognition (NER), and Transformer models like RoBERTa [21]. NER plays a pivotal role in analyzing *Word Repeats, Logical Cues,* and *Grammatical Cues* by allowing us to identify and compare nouns and verbs used in the MCQ. This approach enhances our ability to detect grammatical consistency, identify repeated words, and recognize synonyms. For tackling *Ambiguous Information* and *Implausible Distractors*, our attempts to incorporate GPT-4 faced challenges, as its outputs were often excessively critical, leading to a high rate of misclassifications. To address this, we instead integrated additional linguistic metrics, such as query well-formedness scores [8], and leveraged updated word embeddings to refine the evaluation.

The final category includes six criteria: *Absolute Terms, More Than One Correct, Complex or K-Type, Gratuitous Information, Unfocused Stem,* and *Convergence Cues*. This category utilizes NLP techniques similar to the previous ones, enhanced by the integration of GPT-4 API calls for additional verification. For example, simple word matching was insufficient for the *Absolute Terms* criteria, as the context in which terms like "impossible" are used needs further analysis by GPT-4 to determine their impact on answer validity. Modifications were applied to the *Convergence Cues* and *Complex or K-Type* criteria, incorporating GPT-4 for final verification check to improve accuracy. The criteria *Unfocused Stem* and *Gratuitous Information*, both of which involve lexical richness [11], benefited from GPT-4 interventions, significantly reducing false positives detected in pilot tests by better evaluating question stems for learner comprehension. Finally, the *More Than One Correct* criteria was enhanced to not only attempt at answering questions but also to discern whether a question allows for multiple correct responses or is a select-all-that-apply type. We have open-sourced the code and datasets used in this work[1].

3.3 Datasets

We utilized two datasets of MCQs previously tagged with the IWF criteria to evaluate SAQUET. The first dataset, derived from [5], encompasses MCQs in Computer Science, Humanities, and Healthcare, sourced from prominent MOOC platforms, such as Coursera and edX. The second dataset, from [19], contains student-generated MCQs from

---

[1] https://github.com/StevenJamesMoore/AIED24

Chemistry and Statistics courses. Both datasets contained MCQs with two to five answer choices each. Additionally, both datasets were evaluated by two human experts, with past studies reporting high inter-rater reliability via Kappa scores. Due to IRB permissions and formatting challenges, not all questions from these initial datasets were included in our present study. Additionally, we made minor corrections to address errors in the datasets, such as mislabeled criteria. For example, one adjustment involved reevaluating Computer Science, Humanities, and Healthcare questions to ensure True/False questions were not mistakenly flagged under the *Longest Option Correct* criteria, particularly when "False" was the correct answer.

For developing SAQUET, we initially used a subset of 25 questions, 5 from each domain, which were not included in the final evaluation dataset. Our final dataset comprised 271 MCQs across the five domains, all tagged with the 19 IWF criteria, offering a varied pool of questions for analysis. This contrasts with previous IWF research, which often focuses on a single domain [23, 26].

3.4 Evaluation

To evaluate the effectiveness of commonly employed automated techniques for assessing question quality, we applied five popular linguistic quality metrics to the 271 MCQs in our dataset: perplexity, diversity, grammatical error, complexity, and answerability. Perplexity scores were generated using a GPT-2 language model, aligning with methodologies from recent research [28]. We measured diversity through the Distinct-3 score, which quantifies the average number of unique 3-g per MCQ [13]. Grammatical errors were identified using the widely recognized Python Language Tool [20], tallying the grammatical inaccuracies in each question as done in prior research [24]. For complexity assessment, we adopted Bloom's Revised Taxonomy, assigning each MCQ a level from 0 (lowest, 'remember') to 5 (highest, 'create'), which serves as a common indicator of complexity and difficulty [11, 17]. A highly precise classifier was employed to automatically determine the Bloom's level for each question [6]. Answerability was evaluated using GPT-4, employing the strategy of the Prompting-based Metric on ANswerability (PMAN) approach [27]. This involved following the strategy of crafting specific prompts that instructed GPT-4 to choose an answer for each MCQ.

For the evaluation of SAQUET, we referenced gold standard human evaluations for our dataset. The overall match rate between our method and the human evaluations is calculated to reflect the general accuracy of our tool in classifying MCQs according to the IWF criteria. To tackle this multi-label classification challenge, we use the exact match ratio, necessitating correct identification of all labels for a match, and Hamming Loss, which calculates the average proportion of incorrect labels, offering detailed insights into our classification's accuracy on a holistic level [9]. We further assess performance using the F1 score of each criteria, which balances precision (the accuracy of positive predictions) and recall (the completeness of positive predictions) [4]. A high F1 score indicates both high precision and high recall, signifying effective identification of an IWF without excessive false positives or negatives. The micro-averaged F1 score aggregates outcomes across all criteria, offering a consolidated view of performance for the entire dataset [14]. Analysis is conducted not just on the aggregate dataset, but also segmented by domain. This allows us to identify domain-specific performance variations and areas for refinement. Where possible, we compare our results with metrics reported in prior studies using similar datasets and evaluation metrics, providing context for SAQUET's performance [5, 19].

4 Results

4.1 Limitations of Traditional Metrics in Evaluating Educational MCQs

For each of the five domains, we categorized the MCQs into two groups: one group includes MCQs with zero or one IWF and the other comprises MCQs with two or more. This classification helps differentiate between questions that are considered acceptable (zero or one IWF) and those deemed unacceptable (two or more IWF), thereby allowing for a more precise analysis given the constraints of our dataset in accordance with previous research [19, 26]. We then assessed these questions using five linguistic quality evaluation metrics, as detailed in Table 2. Our analysis revealed that, across all metrics, the performance of MCQs in each domain either matched or exceeded ones found in recent research. For comparison, [4] reported that human generated MCQs, based on Wikipedia articles and science textbooks, had average perplexity scores of 18 to 84 and diversity scores between .78 and .82. Similarly, [24] determined that the average answerability score for human generated MCQs, on the topic of middle and high school reading comprehension, was .726.

Table 2. Comparison of five common evaluation metrics for question quality across five domains, categorized by IWF Count. A circumflex (^) denotes a superior score achieved by questions with a higher IWF count in each metric.

Domain	IWF	N	Perplexity ↓	Diversity ↑	Grammatical Error ↓	Cognitive Complexity ↑	Answerability ↑
Chemistry	0–1	35	47.65	0.961	0.400	0.057	0.743
	2 +	15	57.46	$0.962^{(^)}$	$0.333^{(^)}$	$0.133^{(^)}$	0.733
Statistics	0–1	32	46.02	0.928	0.375	0.719	0.531
	2 +	18	$27.51^{(^)}$	0.888	0.444	$1.333^{(^)}$	$0.611^{(^)}$
Computer Science	0–1	62	30.73	0.927	2.129	1.145	0.806
	2 +	38	41.56	0.917	3.605	$1.500^{(^)}$	0.605
Humanities	0–1	18	47.64	0.955	0.375	1.313	0.875
	2 +	6	$28.24^{(^)}$	0.939	0.375	1.250	$1.000^{(^)}$
Healthcare	0–1	25	30.25	0.955	0.400	1.200	0.960
	2 +	22	$27.72^{(^)}$	$0.957^{(^)}$	$0.182^{(^)}$	$1.682^{(^)}$	0.909

Our analysis revealed that student-generated questions in the Chemistry and Statistics domains had relatively high perplexity scores, but in Statistics, questions with 2+

An Automatic Question Usability Evaluation Toolkit   39

IWFs exhibited a lower perplexity. The diversity metric revealed a ceiling effect, where variations are minimal across different question sets from all domains. High diversity scores are expected, as the MCQs were sourced from diverse origins and authors, such as MOOCs or digital textbooks. The impact of IWFs on a question's answerability varied, where in some cases the presence of IWFs did not reduce, and might have even enhanced, the likelihood of the LLM to correctly answer the questions.

Grammatical errors were relatively low across all domains except Computer Science, where the code syntax posed unique challenges for this criteria, contributing to higher error rates [6]. Interestingly, in both Chemistry and Healthcare, questions with more IWFs (2+) showed a lower average number of grammatical errors, suggesting a nuanced relationship between IWF count and grammatical precision. Initially we expected questions with fewer IWF would have fewer grammatical mistakes, but those may have been overlooked by the human evaluators. Cognitive complexity, measured by Bloom's Revised Taxonomy levels, was also generally higher for questions with 2+ IWFs across all domains except for Humanities, where the difference was marginal, indicating these questions with more flaws tend to engage higher-order cognitive skills.

These findings demonstrate the potential for commonly used metrics to paint an overly optimistic picture of question quality. Even questions with multiple flaws can score well on perplexity, diversity, and grammatical precision, suggesting they are well-crafted and clear. However, this can be misleading, as these metrics may not capture deeper issues such as false information, incorrect assumptions, or inaccuracies in content. For example, Fig. 1 shows a question that achieved an acceptable evaluation across all five metrics, yet it is clearly a poorly student generated question that contains three IWFs: *implausible distractors, logical cues*, and *grammatical cues*.

What is protons?
  A) positively charged particles
  B) sum the number of protons and neutrons
  C) negatively charged subatomic particles
  D) he discovered the charge of electron

Perplexity: 27.56
Diversity: 1.0
Grammatical Error: 1
Complexity: 0 (remember)
Answerability: 1

Fig. 1. A student generated MCQ from the Chemistry dataset consisting of three IWFs on the left, with the associated linguistic quality evaluation metrics on the right.

4.2 Performance of Automated IWF Classification Across Domains

The 19 IWF criteria were automatically applied to all 271 MCQs for a total of 5,149 classifications. While the overall accuracy is slightly skewed due to most of the questions containing a few flaws and thus being classified as 0 for a given criteria, the total accuracy was 94.13%, which treats each criteria classification individually. We achieved an exact match ratio of 38%, which indicates that 103 of the questions were evaluated the same across all 19 criteria between SAQUET and the different human evaluators. The Hamming Loss was 5.9%, indicating a small amount of misclassification regarding the flaws. While we only used half of the data from [19] consisting of 100 MCQs, it is our closest comparable. As such, compared to their leading rule-based method, we achieved a 3.26% overall classification accuracy improvement, a 13% higher exact match ratio, and 3.1% lower Hamming Loss.

On average, SAQUET ($M = 1.75$, $SD = 1.26$) was more likely to classify a MCQ as having more IWFs compared to the human evaluators ($M = 1.31$, $SD = 1.11$). The most IWFs assigned to a single question by both was 5. In Table 3, we present the IWF classifications from the human evaluators compared to SAQUET for all five domains.

Table 3. The number of identified flaws (N) and F1 performance scores for human evaluations (Hum) versus SAQUET (SAQ) across the five domains. A dash (-) signifies the absence of a flaw in a domain as determined by human evaluation, precluding F1 score calculation.

Item-Writing Flaws		Chemistry (50)		Statistics (50)		Computer Science (100)		Humanities (24)		Healthcare (47)	
		Hum	SAQ	Hum	SAQ	Hum	SAQ	Hum	SAQ	Hum	SAQ
Longest Option Correct	N	5	8	3	7	27	27	8	8	16	15
	F1	0.77		0.60		0.96		1.00		0.97	
Ambiguous Information	N	12	12	14	18	12	21	0	2	2	0
	F1	0.58		0.50		0.24		0.00		0.00	
Implausible Distractors	N	9	8	8	6	3	15	3	7	8	3
	F1	0.24		0.86		0.33		0.20		0.00	
True or False	N	2	2	1	0	9	10	4	4	11	11
	F1	1.00		0.00		0.95		1.00		1.00	
Absolute Terms	N	2	1	0	1	9	6	9	9	5	4
	F1	0.67		0.00		0.40		0.89		0.44	
Complex or K-type	N	2	4	4	8	15	12	0	1	4	5
	F1	0.67		0.67		0.81		0.00		0.89	
Negatively Worded	N	0	0	2	4	10	14	0	1	11	11
	F1	-		0.67		0.83		0.00		0.91	
Convergence Cues	N	2	3	9	7	7	11	0	0	1	4
	F1	0.00	0.63	0.44	-	0.00					
Lost Sequence	N	3	3	14	15	2	2	0	0	0	0
	F1	1.00	0.97	0.50	-	-					
Unfocused Stem	N	0	1	8	10	8	5	0	0	0	0
	F1	0.00	0.89	0.62	-	-					
None of the Above	N	6	5	1	1	6	6	0	0	0	0

(continued)

Table 3. (*continued*)

Item-Writing Flaws		Chemistry (50)		Statistics (50)		Computer Science (100)		Humanities (24)		Healthcare (47)	
		Hum	SAQ	Hum	SAQ	Hum	SAQ	Hum	SAQ	Hum	SAQ
	F1	0.91	1.00	1.00	-	-					
Word Repeats	N	1	1	1	1	7	11	0	0	4	11
	F1	1.00	1.00	0.56	-	0.53					
More Than One Correct	N	0	2	0	11	8	24	3	10	1	17
	F1	0.00	0.00	0.38	0.46	0.11					
Logical Cues	N	4	3	2	1	2	8	0	0	0	1
	F1	0.29	0.67	0.00	-	0.00					
All of the Above	N	1	1	1	1	2	2	0	0	2	3
	F1	1.00	1.00	1.00	-	0.80					
Fill in the Blank	N	2	2	0	0	2	2	0	0	2	2
	F1	1.00	-	1.00	-	1.00					
Vague Terms	N	0	0	0	1	3	2	0	1	3	3
	F1	-		0.00		0.80		0.00		1.00	
Grammatical Cues	N	2	1	3	1	0	1	0	1	0	1
	F1	0.67		0.00		0.00		0.00		0.00	
Gratuitous Information	N	0	2	3	5	0	3	0	2	0	0
	F1	0.00		0.50		0.00		0.00		-	
Micro-Averaged	F1	0.59		0.65		0.62		0.66		0.67	
IWF totals		53	59	74	98	132	182	27	46	70	91

The F1 scores reveal the effectiveness of SAQUET across the five domains for each criterion. Compared to the rule-based implementation in [19], our approach improved the F1 score across multiple criteria for Chemistry and Statistics questions. Performance on the *None of the Above* criteria was notably strong, as reflected by high F1 scores, indicating precise classification with minimal misclassifications. Other criteria, such as *More Than One Correct*, showed subpar performance across all domains, with frequent incorrect classifications and often overestimating its presence. The micro-averaged F1 scores provide a consolidated view of SAQUET's accuracy across all 19 criteria and allow for a domain-wise comparison of classification efficacy.

Taking the categorization of all MCQs as acceptable (zero or one IWF) or unacceptable (two or more IWF), we compared the SAQUET's classifications with those made by human evaluators. This comparison aimed to see if the overall categorization matched, despite potential misclassifications of specific IWF criteria. Figure 2 presents a confusion matrix for this acceptability classification, indicating human evaluators deemed

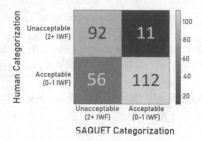

Fig. 2. A confusion matrix for the categorization of questions as acceptable or unacceptable based on their IWF by the human evaluation and SAQUET.

168 questions acceptable and 103 questions unacceptable. SAQUET matched 204 of these MCQ categorizations, with 112 classified as acceptable and 92 as unacceptable, achieving a 75.3% match rate with human evaluations.

5 Discussion

Our results demonstrate that traditional metrics used for assessing the quality of questions, especially multiple-choice, might not adequately reflect their true quality. We observed that questions with various errors, indicated by Item-Writing Flaws, which could either simplify the answering process for students or lead to confusion, often receive high scores from commonly used linguistic quality metrics. To address this gap, we introduced SAQUET, a method designed to capture these more complex aspects of question quality while remaining automated and scalable. By benchmarking against human expert evaluations, we show that SAQUET has the potential to provide a more precise and detailed assessment of question quality compared to these linguistic quality metrics. Furthermore, our contribution to the field of assessment quality evaluation research extends to making both SAQUET and our comprehensive dataset publicly available[i].

Recent efforts in NLP have aimed to shift away from traditional readability metrics like BLEU, METEOR, or ROGUE when evaluating the quality of MCQs, yet these metrics continue to be employed in recent works [2, 4, 21]. In our study, we explored alternative linguistic quality metrics (perplexity, diversity, grammar, complexity, answerability) that are also commonly used and offer a different approach to question evaluation, particularly in response to the inadequacies of previous readability metrics [12, 16, 27]. Our findings reveal that even questions with obvious flaws can be evaluated as higher quality according to these metrics. This discrepancy may still hold for machine generated questions from older models, but the improved linguistic capabilities of recent LLMs mean that more machine generated questions are likely to be deemed high quality by these standards. Recent studies have pointed out that despite the grammatical correctness of LLM outputs, the MCQs generated can suffer from issues like implausible distractors or vague wording [7, 21].

SAQUET has the advantage of operating without training data, addressing the significant challenge of sourcing IWF-tagged question datasets. Although research utilizing the IWF rubric is widespread, access to such datasets is often restricted. Importantly, SAQUET's application extends beyond assessing newly crafted questions; it is equally effective in evaluating existing question sets and machine- or human generated questions alike. This capability allows educators to pinpoint and address flaws in current questions they might be using, potentially adjusting or replacing them to suit their needs. In this study, we achieved an exact match ratio of 38% in a complex multi-label classification task with 19 binary labels, which serves as a strong baseline for future research and evaluation. When compared to human evaluations, SAQUET showed a propensity to identify IWFs more frequently. We prefer this stricter approach of identifying MCQ flaws while prioritizing false positives over false negatives, thereby ensuring only the highest quality questions are utilized for educational purposes.

For the criteria based primarily on text matching, such as *True or False, All of the Above*, or *Longest Option Correct*, one might intuitively expect perfect accuracy. However, our findings indicate that these criteria can manifest in nuanced forms, demonstrating the importance of datasets that capture a broader spectrum of these errors. For instance, True/False MCQs might also appear as Yes/No choices or contain explanation text that follows the option, complicating their identification. Similarly, interpretations of what constitutions *Longest Option Correct* can vary among human evaluators, as it did in our study. In Chemistry and Statistics this flaw was applied to questions if the second-longest option was not nearly as long (at least 80%) as the longest. In contrast, for Computer Science, Humanities, and Healthcare, a stricter interpretation was applied that flagged any question where the correct answer exceeded others in length by even a single character.

Other flaws like *More than one Correct*, which relied heavily on GPT-4, presented significant challenges, notably impacting the overall exact match ratio. This flaw saw a misclassification for 50 out of 271 questions (18.5%), making it the most problematic. The challenge arose from GPT-4's difficulty in reliably identifying the correct answer for an MCQ, frequently failing to determine if a single correct option exists. However, this limitation is not inherently negative, as it does not imply the question is flawed, just that the LLM has the inability to solve it [18, 27]. This highlights the ongoing challenge of accurately evaluating complex question criteria and the limitations of current AI in navigating such nuances, further emphasizing the need for refined and open approaches along with diverse datasets in the evaluation process.

6 Limitations and Future Work

In our study, we introduced SAQUET, an automated method for evaluating questions, employing multiple criteria that leverage LLMs like GPT-4. While outperforming traditional automated MCQ evaluation metrics, this approach comes with inherent limitations, including the black box nature of LLMs, their potential for unanticipated changes, and the risk of bias in their outputs. To mitigate these issues and enhance this work's reliability and cost-effectiveness, we utilized a specific version of GPT-4 through the `gpt-4-0125-preview`[2] API. This approach aimed to standardize the evaluation process

---

[2] https://platform.openai.com/docs/models/gpt-4-and-gpt-4-turbo

and ensure reproducibility by generating consistent outputs from predefined prompts. We further supported transparency and reproducibility by open-sourcing our code[1]. Expanding our dataset to include a greater number and diversity of questions across additional domains would likely reveal further limitations and areas for improvement in our current evaluation criteria.

For future work, we aim to enhance the evaluation techniques for the 19 IWF criteria, with a particular focus on those that currently show weaker performance. Acquiring additional datasets of MCQs annotated with IWFs will be crucial in validating and demonstrating the effectiveness of our method. We encourage educators, researchers, and practitioners to engage with our work, offering their insights and improvements to refine the criteria further, as we have done. Such collaboration would contribute to developing a more educationally robust metric enriched by collective expertise. As LLMs advance, we anticipate that our methodology will too, achieving greater accuracy for certain criteria and providing detailed feedback on how to correct identified flaws.

7 Conclusion

In this study, we highlight the limitations of current metrics for assessing question quality, particularly their oversight of deeper question attributes beyond mere surface characteristics. Through analyzing a dataset of MCQs spanning five varied domains, we illustrate that these prevalent linguistic quality metrics fall short in effectively differentiating between flawed and flawless questions. This gap demonstrates the need for a novel metric capable of comprehensive question quality evaluation. In response, we refined an alternative evaluation method that retains both automation and scalability by assessing MCQs against a detailed 19-criteria Item-Writing Flaws rubric. Upon validating this method to our dataset, we demonstrated its effectiveness across various domains and identified the criteria that were most and least effective. Our findings reveal the potential to significantly enhance question quality assessment, paving the way for more accurate and educationally valuable evaluations.

References

1. Azevedo, J.M., Oliveira, E.P., Beites, P.D.: Using learning analytics to evaluate the quality of multiple-choice questions: A perspective with classical test theory and item response theory. Int. J. Inf. Learn. Technol. **36**(4), 322–341 (2019)
2. Bhowmick, A.K., Jagmohan, A., Vempaty, A., Dey, P., Hall, L., Hartman, J., Kokku, R., Maheshwari, H.: Automating Question Generation From Educational Text. In: Artificial Intelligence XL. pp. 437–450 Springer Nature Switzerland, Cham (2023)
3. Bitew, S.K., Deleu, J., Develder, C., Demeester, T.: Distractor generation for multiple-choice questions with predictive prompting and large language models. In: RKDE2023, the 1st International Tutorial and Workshop on Responsible Knowledge Discovery in Education Side event at ECML-PKDD (2023)
4. Bulathwela, S., Muse, H., Yilmaz, E.: Scalable Educational Question Generation with Pretrained Language Models. In: Wang, N., Rebolledo-Mendez, G., Matsuda, N., Santos, O.C., and Dimitrova, V. (eds.) Artificial Intelligence in Education. pp. 327–339 Springer Nature Switzerland, Cham (2023). https://doi.org/10.1007/978-3-031-36272-9_27

5. Costello, E., Holland, J.C., Kirwan, C.: Evaluation of MCQs from MOOCs for common item writing flaws. BMC Res. (2018). https://doi.org/10.1186/s13104-018-3959-4
6. Doughty, J. et al.: A Comparative Study of AI-Generated (GPT-4) and Human-crafted MCQs in Programming Education. In: Proceedings of the 26th Australasian Computing Education Conference. pp. 114–123 ACM, Sydney NSW Australia (2024)
7. Elkins, S., Kochmar, E., Cheung, J.C.K., Serban, I.: How Teachers Can Use Large Language Models and Bloom's Taxonomy to Create Educational Quizzes. In: Proceedings of the AAAI Conference on Artificial Intelligence (2024)
8. Faruqui, M., Das, D.: Identifying Well-formed Natural Language Questions. In: Proceedings of the 2018 Conference on Empirical Methods in Natural Language Processing. pp. 798–803 (2018)
9. Ganda, D., Buch, R.: A survey on multi label classification. Recent Trends Program. Lang. 5(1), 19–23 (2018)
10. Kasneci, E., Seßler, K., Küchemann, S., Bannert, M., Dementieva, D., Fischer, F., Gasser, U., Groh, G., Günnemann, S., Hüllermeier, E.: ChatGPT for good? On opportunities and challenges of large language models for education. Learn. Individ. Differ. 102274 (2023)
11. Kurdi, G., Leo, J., Parsia, B., Sattler, U., Al-Emari, S.: A systematic review of automatic question generation for educational purposes. Int. J. Artif. Intell. Educ. 30 (2020)
12. van der Lee, C., Gatt, A., van Miltenburg, E., Krahmer, E.: Human evaluation of automatically generated text: Current trends and best practice guidelines. Comput. Speech Lang. 67, 101151 (2021). https://doi.org/10.1016/j.csl.2020.101151
13. Li, J., Galley, M., Brockett, C., Gao, J., Dolan, W.B.: A Diversity-Promoting Objective Function for Neural Conversation Models. In: Proceedings of the 2016 Conference of the North American Chapter of the Association for Computational Linguistics: Human Language Technologies. pp. 110–119 (2016)
14. Lipton, Z.C., Elkan, C., Narayanaswamy, B.: Thresholding classifiers to maximize F1 score. stat. 1050, 14 (2014)
15. Lu, X., Fan, S., Houghton, J., Wang, L., Wang, X.: ReadingQuizMaker: A Human-NLP Collaborative System that Supports Instructors to Design High-Quality Reading Quiz Questions. In: Proceedings of the 2023 CHI Conference on Human Factors in Computing Systems. pp. 1–18 ACM, Hamburg Germany (2023). doi.org/https://doi.org/10.1145/3544548.3580957
16. Mathur, N., Baldwin, T., Cohn, T.: Tangled up in BLEU: Reevaluating the Evaluation of Automatic Machine Translation Evaluation Metrics. In: Proceedings of the 58th Annual Meeting of the Association for Computational Linguistics. pp. 4984–4997 (2020)
17. Monrad, S.U., et al.: What faculty write versus what students see? Perspectives on multiple-choice questions using Bloom's taxonomy. Med. Teach. 43, 575–582 (2021)
18. Moon, H., Yang, Y., Yu, H., Lee, S., Jeong, M., Park, J., Shin, J., Kim, M., Choi, S.: Evaluating the Knowledge Dependency of Questions. In: Proceedings of the 2022 Conference on Empirical Methods in Natural Language Processing (2022)
19. Moore, S., Nguyen, H.A., Chen, T., Stamper, J.: Assessing the Quality of Multiple-Choice Questions Using GPT-4 and Rule-Based Methods. In: Responsive and Sustainable Educational Futures. pp. 229–245 Springer Nature Switzerland, Cham (2023)
20. Morris, J.: Python Language Tool, github.com/jxmorris12/language_tool_python (2022)
21. Mulla, N., Gharpure, P.: Automatic question generation: a review of methodologies, datasets, evaluation metrics, and applications. Prog. Artif. Intell. 12(1), 1–32 (2023)
22. Nasution, N.E.A.: Using artificial intelligence to create biology multiple choice questions for higher education. Agric. Environ. Educ. 2, 1 (2023)
23. Pham, H., Besanko, J., Devitt, P.: Examining the impact of specific types of item-writing flaws on student performance and psychometric properties of the multiple choice question. MedEdPublish. 7, 225 (2018)

24. Raina, V., Gales, M.: Multiple-Choice Question Generation: Towards an Automated Assessment Framework, http://arxiv.org/abs/2209.11830 (2022)
25. Scully, D.: Constructing multiple-choice items to measure higher-order thinking. Pract. Assess. Res. Eval. **22**, 1, 4 (2019)
26. Tarrant, M., Knierim, A., Hayes, S.K., Ware, J.: The frequency of item writing flaws in multiple-choice questions used in high stakes nursing assessments. Nurse Educ. Today **26**(8), 662–671 (2006)
27. Wang, Z., Funakoshi, K., Okumura, M.: Automatic Answerability Evaluation for Question Generation, http://arxiv.org/abs/2309.12546 (2023)
28. Wang, Z., Valdez, J., Basu Mallick, D., Baraniuk, R.G.: Towards Human-Like Educational Question Generation with Large Language Models. In: Rodrigo, M.M., Matsuda, N., Cristea, A.I., Dimitrova, V. (eds.) Artificial Intelligence in Education, pp. 153–166. Springer International Publishing, Cham (2022)

Evaluating Behaviors of General Purpose Language Models in a Pedagogical Context

Shamya Karumbaiah[1]([envelope])([ID]), Ananya Ganesh[2], Aayush Bharadwaj[1], and Lucas Anderson[1]

[1] University of Wisconsin-Madison, Madison, WI 53706, USA
{shamya.karumbaiah,abbharadwaj,lcanderson4}@wisc.edu
[2] University of Colorado Boulder, Boulder, CO 80309, USA
ananya.ganesh@colorado.edu

Abstract. General-purpose Language Models (LMs) bypass the need for task-specific model training by allowing textual prompts to specify a downstream task (e.g., assessment, feedback generation). One of the main benefits of using a prompt-based learning method is that it circumvents the need for supervised data and training on the downstream task. However, in high-stakes settings like education, LMs need to be evaluated rigorously on the specific downstream tasks before putting them in front of the students. Unlike traditional supervised learning models that are evaluated for a specific task, LMs are often evaluated on benchmark data and tasks that may not reflect the downstream use in education. Hence, we first present arguments for contextual evaluation of LMs. Next, we present a framework for behavior analysis - an alternative approach for model evaluation. Behavior analysis involves defining LM behaviors and designing tests (e.g., invariance to irrelevant perturbations). Using a case study of assessing science ideas in student essays, with past data from ecologically valid contexts, we illustrate how behavior analysis allowed for the identification of LM failures that are likely to go unnoticed in tests for generalization. By making the LMs more transparent for scrutiny, this study suggests a way to improve LM reliability and trustworthiness. Future studies will work with education stakeholders in translating their implicit expectations of desired model behaviors into explicitly defined tests, thereby building their agency and trust in educational AI.

Keywords: Large language models · Evaluation · Context · Behavior analysis · Stakeholders · Education · Trust · Equity

1 Introduction

Recent advances in language modeling open up new possibilities for developing task-specific models (e.g., for assessment) using general-purpose Language Models (LMs) without supervised data or training [7]. This approach may be particularly attractive for educational applications and contexts where it is hard to collect large enough data to build reliable models (e.g., for providing high-quality descriptive feedback). However, a limitation of using LMs in high-stakes settings

© The Author(s), under exclusive license to Springer Nature Switzerland AG 2024
A. M. Olney et al. (Eds.): AIED 2024, LNAI 14830, pp. 47–61, 2024.
https://doi.org/10.1007/978-3-031-64299-9_4

like education is the lack of rigorous evaluation. Unlike traditional supervised learning models that are evaluated for a specific task, LMs are often evaluated on benchmark data and tasks that may not reflect the downstream use (e.g., movie review data vs. middle school students' science essays; predicting sentiment in a movie review vs. assessing science ideas; [20]). Moreover, recent research in algorithmic bias has shown the harms of not evaluating models within a pedagogical context and student population [5]. Hence, we argue that there is a need for contextual evaluation of LMs (see discussion in Sect. 2).

In this paper, we investigate an alternative approach for model evaluation that allows us to situate LM errors in a pedagogical context (presented in Sect. 3; [14,15,18]). This approach leverages: 1) software engineering practices in analyzing system behaviors, 2) stakeholder expertise in their context (e.g., teachers as pedagogical experts), and 3) evolving research on LM limitations (e.g., factually incorrect or self-contradicting text, and struggles with negation and numerical ability). Akin to a regression test suite in software engineering, a set of test cases is generated to evaluate the model behaviors for potential failures [15]. For example, one test case could check if an LM can identify a correct response if some of its words are replaced with their synonyms. Despite being a simple capability, we will see later that this is not always the case. Likewise, another test could check if an LM can identify a correct response when students use acronyms. Stakeholders could also design test cases that are particularly relevant to their context (e.g., for bilingual students switching between languages). We illustrate the behavior analysis approach for model evaluation with a case study of LMs (Flan-T5 and RoBERTa) used to provide automated feedback on middle-school students' written science explanations.

Contribution. The current study contributes to empirical and methodological research on LM reliability by investigating the feasibility of using behavior analysis for model evaluation. The advantages of using behavior analysis are threefold. First, it allows for a rigorous evaluation of unsupervised general-purpose LMs for a specific downstream task in high-stakes settings like education when there is limited annotated data. Second, it could also be used to build stakeholder agency in evaluating harmful model failures in their pedagogical context or for a specific student population. A model behavior can serve as a contract on which stakeholders build their trust [4]. Third, stakeholder-specified model behaviors and results of behavior analysis could be used to improve models to be better suited for stakeholder needs in their context. We expect future studies on behavior analysis with stakeholders to contribute to the equitable, responsible, and trustworthy use of artificial intelligence in classrooms.

2   Need for Contextual Evaluation of Language Models

General-purpose LMs bypass the need for task-specific model training by allowing textual prompts to specify a downstream task (e.g., assessment, feedback generation) in a format familiar to LMs from their training. For example, we can assess the correctness of a student's response, "Energy can't be created or destroyed." by prompting the LM to fill in the blank in a continuing sentence "This explanation is _" with an evaluative adjective or with descriptive feedback. Instead of task-specific training with fully supervised learning or finetuning a pre-trained model, general-purpose LMs such as BLOOM, Flan-T5, GPT4, and Llama2 are trained on large datasets to learn general-purpose features of the language it is modeling.

One of the main benefits of using a prompt-based learning method is that it circumvents the need for supervised data and training on the downstream task. In a supervised learning approach, we learn the parameters θ of the model $P(y|x; \theta)$ given the input x and output y (e.g., pairs of student responses and human annotations of assessment or feedback). Prompt-based learning, in contrast, uses an LM that models $P(x; \theta)$, the probability of text x, to predict the output y. Prompting engineering involves designing a prompt function $prompt(x)$, such that $prompt(x) = y$. In the assessment example above, the prompt function $prompt(x)$ may involve two steps. First, an input x_i is transformed to $x_i' = [x_i]$. *This explanation is* [Z] by applying a template. Next, the LM $P(x_i'; \theta)$ is used to generate output y_i to fill the slot [Z]. The prompt function, $prompt(x)$, can be engineered to output y as an evaluative adjective (e.g., "correct") or descriptive feedback (e.g., "correct. Please support your claim with evidence."). Simply put, instead of providing supervised data to learn from, general-purpose LMs require reformulation of the downstream task to use text generation as the prediction.

However, in high-stakes settings like education, LMs need to be evaluated more rigorously on specific downstream tasks. Due to the unsupervised nature of LM training, evaluations are limited to benchmark testing on a wide variety of tasks [3,19]. Benchmarks are composed of diagnostic test datasets on which a model's ability to perform language understanding or generation can be tested. As an example, the GLUE and SuperGLUE benchmarks test for general-purpose linguistic knowledge [20], as well as reasoning skills through tasks like natural language inference and question answering, and text classification skills through tasks like sentiment analysis. More recently, benchmarks have also been developed for testing an LM's knowledge of specific application domains such as medicine, law, or science [3].

Even when benchmark tests match by the task type (e.g., text classification), the nature of the language (e.g., movie reviews vs. science essays by middle school students), or the construct predicted (e.g., sentiment in a movie review vs. the accuracy of science ideas) may vary significantly. If there is enough annotated data for a task in a relevant context (e.g., human assessment of middle school students' science explanations), LMs could be evaluated using standard generalization estimation (using metrics such as AUC and F1 scores). However, applications seeking prompt-based approaches for their benefits with unsupervised learning are likely to not have enough annotated data for reliable evaluations.

Moreover, generalization estimates themselves tend to be an overestimation of real-world performance [13]. In most cases, the test data for estimating generalization are likely limited in representing variations in the real world. It is not uncommon for data to come from a dominant context convenient for data collection(e.g., middle-class undergraduates; [6]). Hence, generalization estimates are also likely to ignore the biases in training as the test set drawn from the same population as training reflects similar biases. Even when tested with data from the deployment context [16], which is often laborious, the population distribution may shift over time [12].

In conclusion, we argue that in a high-stakes setting like education with language that may differ from LM training (e.g., movie review vs. middle school students' science essays) and constructs that may vary from benchmark testing (e.g., sentiment analysis vs. science idea assessment), there is a need for rigorous evaluation of general-purpose LMs to test for critical failures. Moreover, since model capabilities and in turn, the notions of model failures differ by context (e.g., assess essays for science understanding vs. language), we argue that such evaluation should also be stakeholder-driven, wherein stakeholders define desired model behaviors in their pedagogical context. In this paper, we empirically investigate the feasibility of behavior analysis as a method for model evaluation.

3   The Behavior Analysis Framework

Behaviors capture desiderata i.e., what stakeholder expects a model to do. By making the behaviors explicit, we open up the LMs for stakeholder scrutiny, improving model transparency and allowing stakeholders to build trust by understanding where the LMs succeed or fail. This approach leverages stakeholders' rich contextual knowledge (e.g., how middle school students write science essays) to translate their implicit expectations of desired model behaviors (e.g., understanding acronyms) into explicitly defined tests (e.g., comparable performance in inferring science ideas in essays after replacing scientific terms with their acronyms). Thus, the definition of behaviors and tests need to be free from technical jargon to break the high technical knowledge barrier often assumed in understanding AI systems.

The literature in Natural Language Processing (NLP) on behavior analysis (e.g., [15,18]) broadly defines three *types of behaviors*: 1) *minimum function* (MF), wherein stakeholders can specify basic functionalities expected of the system akin to unit tests in software testing (e.g., tag incorrect science ideas for feedback), 2) *directional change* (DIR), wherein stakeholders can test system behaviors expected to shift in a desired direction based on the specified condition (e.g., negated correct ideas are tagged as incorrect), and 3) *invariance* (INV), wherein stakeholders can test system behaviors expected to remain unchanged (e.g., using scientific acronyms such as PE for potential energy should not change the assessment). Similarly, unless LMs have linguistic biases, inferences of LMs need to stay invariant between monolingual and multilingual writing.

To facilitate test ideation, Ribeiro and colleagues [15] provide a list of linguistic capabilities that are generally applicable across NLP tasks such as classification, summarization, translation, and question-answering. These include *Knowledge* (of basic concepts relevant to the task; e.g., acceptable explanation

Table 1. The Behavior Analysis Framework. Example actions, supports, outputs, and decisions are in italics.

Step	Action	Framework Support
1	Specify a downstream task and provide a few examples *tag unacceptable science explanations for feedback (e.g., energy can be created)*	Instruction/prompt templates to model the task using a local language model *"classify if the following is an acceptable or unacceptable explanation of {concept}\n{text}"*
2	Define capabilities *a) model has the necessary knowledge* *b) model is robust to spelling errors* *c) model is fair to bilingual learners* *d) model doesn't hold contradictory beliefs*	Capabilities to choose from *a) knowledge* *b) robustness* *c) fairness* *d) negation*
3	Define behaviors corresponding to capabilities *a) tag acceptable explanations as acceptable* *b) spelling errors shouldn't change the result* *c) Spanish words shouldn't change the result* *d) negating should reverse the result*	Behaviors to choose from *a) minimum function* *b) invariance* *c) invariance* *d) directional change*
4	Generate test data to match model behavior expectations *a) "kinetic energy is the energy of motion"* ✓ *b) "KE is energy -> enrgy of motion"* ✓ *c) "KE is energy -> energía of motion"* ✓ *d) "KE is energy of motion -> rest"* ✗	Perturbations to choose from based on model behavior and capability *a) no perturbation (minimum function)* *b) swapping or deleting characters (invariance)* *c) words translated to Spanish (invariance)* *d) adding "not" or swapping a word with its antonym (directional)*
5	Analyze test results and generate hypotheses on harmful and beneficial model behaviors *c) model is unfair to bilingual students (linguistic bias)*	Failure rates by capability and behavior *c) more than half of bilingual test cases are incorrectly marked as unacceptable; 60% more failures when text contains any Spanish word(s)*
6	Test iteratively and refine hypotheses *c) linguistic bias persists across other non-English languages*	Modify capabilities and generate test data *c) modify capability to include other languages spoken by bilingual students*

of potential energy), *Robustness* (to spelling errors, paraphrased explanations, acronyms of scientific terms, etc.), *Taxonomy* (e.g., synonyms and antonyms of relevant words), *Negation* (not holding contradictory beliefs; e.g., negated acceptable explanation is unacceptable), and *Fairness* (comparable performance across subgroups; e.g., monolingual and bilingual essays).

Under these broad principles of behaviors and capabilities, stakeholders design tests that are meaningful for their context (see Table 1 for detailed steps). To generate test data, they can start with an example of student writing and add perturbations to reflect the test (e.g., swap letters to introduce spelling errors). We illustrate this approach for behavior analysis with a case study below (see Sects. 4 and 5).

4 Case Study: Context, Data, Models

Using the context of scaffolding students' inquiry learning of science explanations with Large Language Models (LLMs), we present a proof of concept for using behavior analysis to identify LLM errors.

4.1 Learning Context: Scaffolding Inquiry Learning of Scientific Explanations

Writing scientific explanations is an authentic science practice that is central to science learning [17]. Past research demonstrates that students struggle with learning to write "good" scientific explanations (e.g., causal, complete, backed with evidence, and connected to scientific principles; [1]). Meanwhile, teachers have limited capacity to provide timely feedback to mitigate these struggles [10]. recent research has investigated distributed scaffolding using NLP to meet students' in-the-moment needs in improving their scientific explanations [2]. Accordingly, NLP has been used to automatically score students' essays, provide feedback for revision, and report student progress to teachers to help students.

We studied scaffolding students' learning of science explanations in a 3-week inquiry-based learning unit on energy. As part of the activity, students were asked to collaborate on designing a mechanical system (e.g., a safe and fun roller coaster), testing their designs in a virtual simulation. As they experimented with different designs, they were expected to hypothesize what potential energy, kinetic energy, and law of conservation of energy are, how they relate to each other, and other physical measures such as height, mass, and friction. Writing prompts encouraged students to make their understanding more concrete by providing clear and precise explanations of their emerging science ideas. After writing and submitting a draft of their essay, students received automated feedback about the science ideas they included (or not) from an NLP system. Then, they were allowed to discuss the feedback with peers and revise their writing, which they then resubmitted for feedback.

4.2 Data Collection, Curation, and Annotation

Data for this study was collected from 6th-, 7th-, and 8th-grade students, as well as undergraduates in the midwestern United States ($N = 679$ students). The 6th-grade data was collected from 220 consenting students in three teachers' classes from a middle school located on the fringe of a small city. The 7th-grade data was collected from one teacher's class from a large city ($n = 30$ students), whereas the 8th-grade data was from one teacher's classes in a semi-rural school district ($n = 118$ students). Finally, the data from undergraduate students was collected from one professor's pre-service teachers' physics class ($n = 329$ students). While a large number of students participated in the unit, we had essay data from a total of 632 students. To create a set of essays for our analyses, we selected or combined parts of different students' essays to ensure that the science content in each essay was aligned with what students should have learned during the roller coaster curriculum and explained in their design essays. This subset of data was created to make human annotation of the essays manageable. In the end, we had 76 essays for our analyses, which were created from a total of 105 different student essays. The average word count across these essays is approximately 196 words. The range of word counts varies significantly, with the longest essay containing 441 words and the shortest at 78 words.

We chose to investigate three basic energy concepts: potential energy (PE), kinetic energy (KE), and the law of conservation of energy (LCE). We define two categories of explanations as outcome labels for the inference task: acceptable and unacceptable. We mark an essay as acceptable for a concept if it contains at least one acceptable explanation. An *acceptable* definition involves standalone phrase(s) that describe the concept accurately. This definition could be explicit, abstract, and generalizable (e.g., PE is the energy an object has due to its position) or applied in the context of the activity, in this case, the design of a roller coaster (e.g., PE is the energy stored in the car at the top of the initial drop.) *Unacceptable* definitions, on the other hand, are those that are inaccurate (e.g., PE is energy in motion) or insufficient, including measurements (e.g., PE at the top of the rollercoaster is 4.9 J), equations (e.g., PE = m*h*9.8), comparisons (e.g., there is more PE at the top of the hill than the bottom), or other observations (e.g., PE changes into KE as the car goes down the hill.)

The essays were annotated by a team of six undergraduate research assistants. These annotators were trained through a pilot annotation process, where they annotated a subset of the essays in pairs, and discussed disagreements in detail to refine the annotation guidelines. Following the training process, the dataset of 76 essays was divided into three groups of 25–26 essays, and each group was double-annotated by two annotators. Each annotator only received essays that they had not encountered in the training stage. The agreement is highest for LCE definitions, both as per Cohen's kappa (0.96) and accuracy of 98%. For PE definitions, while the proportion of essays on which both annotators agree is high at 92%, Cohen's kappa is much lower at 0.35. We see similar trends for KE as well, although the kappa value of 0.59 indicates strong agreement in this case. Given that Cohen's kappa takes into account the base rates of the labels being annotated, the lower kappa can partially be attributed to the low base rates of the acceptable label under both the PE and KE concepts (only 10% of all labels). To arrive at a single set of labels that can then be used for model training, both annotators engaged in a discussion over the essays on which their respective codes disagreed.

4.3   Large Language Models - Generalization Estimation

We investigate two Large Language Models (LLMs) in this paper - Flan-T5 and RoBERTa. Fine-tuned on a diverse set of tasks (about 1800 tasks across 473 datasets), Flan-T5 [9] excels in multiple classification benchmarks, making it a suitable LLM for our context. RoBERTa [8], on the other hand, is an LLM already being used widely for classification tasks in educational data mining.

Moreover, compared to proprietary LLMs such as GPT4, open-source alternatives like Flan-T5 and RoBERTa allow for local use of the LLM within our servers, addressing significant privacy concerns over sharing student data.

Flan-T5. The baseline model for Flan-T5 uses a zero-shot learning approach using the Flan-T5 base version from the HuggingFace Transformers library [21]. The prompt format was derived from the official guideline released by the Flan-T5 authors for classification tasks. The resulting prompt for our task was: "According to the following essay, classify the student's definition of {concept} as Acceptable or Unacceptable \n{text}". Here, "{concept}" is a placeholder for one of the three physics concepts -Potential Energy, Kinetic Energy, or the Law of Conservation of Energy-and "{text}" is replaced with the actual text of the student essay. This configuration instructed the model to evaluate the student's understanding of a specified concept based on their essay. The Flan-T5 model, leveraging its sequence-to-sequence capabilities, then generates a text response that is one of the labels to be predicted.

RoBERTa. We compare the prompt-based modeling approach (using a general-purpose LM such as Flan-T5) to RoBERTa, an LM that has been used more commonly in educational NLP in recent years. To establish a zero-shot baseline for our assessment task using the RoBERTa model, we make use of a similar NLP task for which training data is abundantly available, namely the *Natural Language Inference* (NLI) task. Given two statements, one called the premise and the other called the hypothesis, the NLI task (Bowman et al., 2015) requires models to reason about the relationship between the two sentences, specifically, predict whether the premise entails, contradicts, or is neutral to the hypothesis. We use a pre-trained RoBERTa model that has been finetuned for the NLI task as our zero-shot classification model for automated assessment. We reformat our data in the following way to make it similar to the NLI task: each essay is treated as a premise, and the hypothesis states that the essay contains an acceptable definition of the concept of interest. A prediction of entailment would mean that the model believes that given the essay, it can be deduced that the science concept is defined in an acceptable manner. The model then predicts the probabilities of entailment, neutrality, and contradiction between the premise and hypothesis. In our evaluation metric, an essay is marked as 'acceptable' if the entailment probability exceeds 0.5; otherwise, it is deemed 'unacceptable.' We develop three separate models for each of the three science concepts.

Table 2 presents the performance of the baseline Flan-T5 and RoBERTa models. We compare the model performances with the majority baseline which always predicts the dominant class, in this case, the label Unacceptable. Given the imbalance in our dataset, this tells us how a model that is only learning shallow signals such as the label distribution will perform. Without any supervised training, the zero-shot baselines of the general-purpose LMs have an F1 score ranging from 0.43 to 0.53 for the best models. The generalization estimates raise concerns about the reliability of the LLMs in assessing students' science explanations, pointing to the need for rigorous evaluation. Next, we present behavior analysis to discover LM errors and biases not obvious in generalization estimation and isolate the areas where the model may fail, especially when there is limited annotated data.

Table 2. Cross-validated results in assessing students' science explanations.

Model	PE		KE		LCE	
	Acc	F1	Acc	F1	Acc	F1
Majority	0.894	0.472	0.868	0.464	0.723	0.419
Zero-shot Baseline						
Flan-T5	0.434	**0.526**	0.447	**0.517**	0.368	0.331
RoBERTa	0.566	0.195	0.632	0.333	0.579	**0.427**

5 Case Study: Behavior Analysis Setup and Findings

Using the framework presented in Sect. 3 and Table 1, we conduct behavior analyses of Flan-T5 and RoBERTa models. In Table 3, we present an illustrative list of capabilities and behaviors designed by the annotators to evaluate LLMs used to assess science ideas for feedback (see Sect. 4 for context). A total of 567 test cases were designed for each model. The distribution across capabilities is knowledge (349), taxonomy (38), robustness (141), negation (30), and fairness (9). The distribution across behaviors is MF (89), INV (257), and DIR (221). In summary, we test the LLMs for:

Knowledge: In this capability, we begin by testing the model's minimum functions (MF) on each of the concepts and labels, by designing tests that show acceptable and unacceptable explanations of PE, KE, and LCE. Additionally, since we are interested in identifying at least one acceptable explanation, we test the model's response to stay invariant (INV) when acceptable explanations are followed by unacceptable explanations, but change directionally (DIR) when unacceptable explanations are followed by acceptable explanations.

Robustness: Here, we test for some linguistic characteristics we expect to see in middle schoolers' writing that should not change LLM outputs (INV). These include testing for perturbations such as spelling errors, paraphrased definitions, replacing scientific terms with acronyms, and adding unrelated sentences after acceptable or unacceptable explanations.

Taxonomy: In this capability, we test for LLM behaviors after replacing words in the original explanations with synonyms (INV), antonyms (DIR), and negated antonyms (INV), to make sure that the model responds appropriately to these perturbations.

Negation: Here, we test the LLMs' ability to deal with negation (DIR), such as adding "not" at the appropriate location in the definitions. LLMs are shown to fail simple negation tests like these, staying invariant [11].

Fairness: Finally, we test the LLMs to make sure that students who may not have English as their first language are graded fairly. We replace certain words with their Spanish translation to test if the model's assessment remains unchanged (INV).

Table 3 also presents the results of analyzing Flan-T5 and RoBERTa as the pass rates i.e., the proportion of test cases that passed (e.g., 0.80 refers to 80% pass rate). Flan-T5 seems to have better knowledge priors in some concepts (80% pass rate for KE and 100% for LCE) as compared to RoBERTa (0% for acceptable definitions and 100% for unacceptable definitions across all concepts). One explanation for this difference could be that Flan-T5's large corpora and a mixture of tasks during training give them more generalizable information on physics concepts. Another explanation is the mismatch in the nature of the NLP task (i.e., NLI vs. classification) with RoBERTa without supervised training. We also observe that Flan-T5 is likely to be more lenient in marking responses as acceptable when they are not (e.g., 22% pass rate for LCE unacceptable) leading to high false positives, while RoBERTa makes high false negative errors. However, when unacceptable definitions are added before or after an acceptable explanation, Flan-T5 fails to recognize the acceptable definition (26%-36% pass rate for PE and KE). A similar trend is visible when random unrelated sentences are added to test for robustness (40% pass rate for PE and 50% for KE and LCE). These findings are in line with research that shows how sensitive LLMs are to unrelated changes to the input [22], prioritizing one label over another.

In comparison, both Flan-T5 and RoBERTa perform fairly well with other forms of perturbations that are intended to mimic some variations in middle school writing in this context. This includes invariance to spelling errors (67% pass rates except for LCE), paraphrasing (from 67% to 100% pass rates), and using acronyms (64% to 80% pass rates). However, since RoBERTa fails many of the minimal function behaviors for acceptable definitions, it is unreliable, to begin with. If additional domain-related training improves RoBERTa's MF behaviors, we could begin to trust it with linguistic variations in student writing. There are also some concept-level differences in failures related to linguistic variations. For example, Flan-T5 is more sensitive to misspellings (33%) in LCE as compared to KE (67%).

Flan-T5 struggles to perform well when LCE explanations are negated (20% and 40% pass rates) or perturbed with synonyms (25% pass rate) and antonyms

Table 3. A list of capabilities (cap), behaviors, and descriptions of test cases designed for expected LLM behaviors while assessing science ideas in students' written scientific explanations. The rightmost columns contain the results (pass rates) for Flan-T5 and RoBERTa models, with red indicating higher failure.

Cap	Behavior + Description	Example: Expected Model Behavior	Flan-T5			RoBERTa		
			PE	KE	LCE	PE	KE	LCE
k n o w l e d g e	**MF:** Basic explanation of concepts that are acceptable	Potential energy is energy at rest: Acceptable	0.20	0.80	1.00	0.00	0.00	0.00
	MF: Explanation of concepts that are unacceptable	Kinetic energy is energy at rest: Unacceptable	0.80	0.70	0.22	1.00	1.00	1.00
	INV: Acceptable explanation followed by unacceptable	Potential energy is energy at rest. Potential energy is the energy lost as the car goes down the hill: Acceptable → Acceptable	0.32	0.36	0.68	0.00	0.00	0.00
	DIR: Unacceptable explanation followed by acceptable	Potential energy is the energy lost as the car goes down the hill. Potential energy is energy at rest: Unacceptable → Acceptable	0.26	0.36	0.67	0.08	0.00	0.67
r o b u s t n e s s	**INV:** Misspelling of words (swapping or deleting characters) should not change the label	Potential → Potentl energy is energy → enrgy that a body has because of its position → posiion relative to other bodies → bodeis: Acceptable → Acceptable	0.67	0.67	0.33	0.67	0.67	0.67
	INV: Paraphrased explanation should not change the label	Potential energy is energy that a body has because of its position relative to other bodies. → Energy that a body has because of its position relative to other bodies is potential energy: Acceptable → Acceptable	0.83	1.00	0.71	0.67	0.67	0.71
	INV: Using acronyms should not change the label	Kinetic energy → KE is energy at rest: Unacceptable → Unacceptable	0.67	0.80	0.64	0.67	0.67	0.64
	INV: Adding random unrelated sentences should not change the label	Potential energy is energy at rest. Roller coasters are fun: Acceptable → Acceptable	0.40	0.50	0.50	0.55	0.55	0.70
t a x o n o m y	**INV:** Replacing words with synonyms should not change the label	The law of conservation of energy says that energy can be created and destroyed → made and taken: Unacceptable → Unacceptable	0.67	0.83	0.25	0.67	0.67	0.63
	DIR: Replacing words with antonyms should flip the acceptable label	Kinetic energy is energy that the car has because it is moving → still: Acceptable → Unacceptable	0.50	0.50	0.33	0.67	0.67	0.67
n e g a t e	**DIR:** Negation of acceptable explanation is unacceptable	Potential energy is not energy at rest: Acceptable → Unacceptable	1.00	0.80	0.40	1.00	1.00	1.00
	INV: Negation of unacceptable explanation is unacceptable	Kinetic energy is the work needed to accelerate → slow down the rollercoaster car: Unacceptable → Unacceptable	0.80	0.60	0.20	1.00	1.00	1.00
f a i r	**INV:** Replacing certain words with its Spanish translation should not change the label	Potential energy is energy that a body has because of its position relative to → relativa a other bodies. Acceptable → Acceptable	0.67	1.00	1.00	0.00	0.00	0.00

(33% pass rate). These results indicate that Flan-T5 is likely to hold contradictory beliefs at the same time, i.e., "energy cannot be created or destroyed" and "energy can be created or destroyed" - a finding that is in line with recent discoveries on LLM limitations [11]. In contrast, RoBERTa seems to behave as expected the majority of the time, especially when handling negation (100% pass rate across all negation tests). Lastly, introducing a few relevant Spanish words entirely throws off the RoBERTA model (0% pass rate for all concepts), while Flan-T5 handles this condition well (e.g., 100% pass rate for KE and LCE), likely due to its exposure to relatively more multilingual data.

6 Discussion

Using a case study of assessing science ideas in student essays, we illustrate how behavior analysis helps us identify areas where LMs succeed and fail. For instance, we observe that in this context, Flan-T5 struggles to remain invariant to negation and additions of unrelated sentences to the input. In comparison, RoBERTa fails to remain invariant to multilingual writing. However, both Flan-T5 and RoBERTa do well in remaining robust to other relevant perturbations, such as spelling errors and acronyms. In this way, behavior analysis allowed us to identify LM failures that went unnoticed in our tests for generalization. Designing relevant capabilities also allowed for specifying phenomena (e.g., peculiarities in middle school students' writing such as spelling errors, age-appropriate vocabulary, and adding unrelated sentences) that are likely to be underrepresented in the LLM training data due to added protections on sharing student data (e.g., under FERPA).

Moreover, the behavior analysis framework with capabilities (e.g., robustness, fairness, knowledge) and test types (i.e., MF, INV, DIR) facilitated test ideation in the research team, an observation yet to be tested with stakeholders. Once we created an initial set of MF tests, we could scale the testing with INV and DIR tests with no additional annotated data. They allow us to test changes in the model behavior with perturbations (e.g., spell errors, acronyms) and between subgroups. This opens up new possibilities for fairness tests in contexts that have been historically underrepresented in educational AI research. For example, future research with teachers serving linguistically minoritized multilingual students can test if teachers' rich knowledge of how multilingual students write could be used to generate templates and perturbations to generate test cases automatically. Behavior analysis with templates has the potential to break down constraints of the current data-hungry approaches for evaluating LMs that limit its testing in low-data contexts by allowing stakeholders to generate test cases at scale on desired model behaviors.

There are a few limitations to the behavior analysis approach described here. First, we handcrafted relevant test cases as single sentences, while the model may need to work on larger essays with multiple concepts explained all at once. To improve the relevance of behavior analysis, one could choose to work with a current set of student essays and add the same perturbations to them instead.

Second, since the capabilities, behaviors, and test cases were designed by the research team, it is likely that not all capabilities and behaviors, in turn, errors may be meaningful or worthwhile to investigate. Hence, stakeholder-driven testing is necessary to identify high-stakes failures to increase model transparency and accountability. This approach is intended to build stakeholder expertise and trust in LM to both make decisions on whether an LM is suitable in their context and if it is, how to adapt pedagogical practices to enhance LM benefits and mitigate their harms.

Future studies with teachers will explore how they generate error hypotheses and test them out systematically to understand where an LM used for a downstream task (e.g., assessment, feedback generation) succeeds or fails, assess the benefits and harms, and make decisions on whether it is suited to their pedagogical context. Our pilot test with one teacher suggests the need for a well-designed, easy-to-use interface that reduces the cognitive load by grouping and sorting test cases, and filtering out low-value test cases (e.g., those that don't sound like middle school students). It makes us optimistic to know that similar approaches to behavior analysis have been successfully used in other domains such as content moderation [18] and commercial sentiment analysis [15] among stakeholders without prior knowledge of model evaluation. Future work could also explore how the stakeholder-specified behaviors and tests could be used to improve the LLM to make it more relevant to a local pedagogical context. For instance, can finetuning optimize a model for the pass rate of behavior tests? Finally, although we explored behavior analysis as an approach for contextual evaluation of LMs, this approach can be used more widely in not only other NLP systems but also other AI systems if the inputs could also be translated to text (e.g., using text replays of interaction logs from AI tutors).

7   Conclusion

Using the context of scaffolding students' inquiry learning of science explanations with LLMs, we present a proof of concept on using behavior analysis for contextual evaluation of general-purpose LMs. The three major affordances of this approach lie in: 1) evaluating LMs for educational applications and contexts where it is hard to collect large enough data, 2) identifying failures in model behaviors that are likely to go unnoticed in tests for generalization, and 3) localizing potential root causes of the failures (e.g., lack of domain knowledge vs. sensitivity to linguistic variations). The behavior analysis framework investigated in this paper with capabilities (e.g., robustness, fairness, knowledge) and behaviors (i.e., minimum functions, invariance, directional change) facilitated test ideation in the research team and replication across new models (e.g., Flan-T5, RoBERTa) and constructs in the context (e.g., PE, KE, LCE). Research in other domains shows that behavior analysis has the potential to lower the technical knowledge barrier to support stakeholders in translating their implicit expectations of desired model behaviors into explicitly defined tests. Future work will explore how behavior analysis could serve as a tool for education stakeholders such as teachers to evaluate LM-powered AI systems in pedagogically meaningful ways to improve transparency, accountability, and trust.

Acknowledgments. Support for this research was provided by the University of Wisconsin-Madison, Office of the Vice Chancellor for Research with funding from the Wisconsin Alumni Research Foundation. We thank Sadhana Puntambekar and Dana Gnesdilow for the data and Bowen Wang and Minghao Zhou for their support with modeling.

References

1. Berland, L.K., Hammer, D.: Framing for scientific argumentation. J. Res. Sci. Teach. **49**(1), 68–94 (2012)
2. Gerard, L., Linn, M.C., Berkeley, U.C.: Computer-based guidance to support students' revision of their science explanations. Comput. Educ. **176**, 104351 (2022)
3. Hendrycks, D., et al.: Measuring massive multitask language understanding (2020)
4. Jacovi, A., Marasović, A., Miller, T., Goldberg, Y.: Formalizing trust in artificial intelligence: prerequisites, causes and goals of human trust in AI. In: Proceedings of the 2021 ACM Conference on Fairness, Accountability, and Transparency, pp. 624–635 (2021)
5. Karumbaiah, S., Lan, A., Nagpal, S., Baker, R.S., Botelho, A., Heffernan, N.: Using past data to warm start active machine learning: does context matter? In: LAK21: 11th International Learning Analytics and Knowledge Conference, pp. 151–160 (2021)
6. Kimble, G.A.: The scientific value of undergraduate research participation (1987)
7. Liu, P., Yuan, W., Jinlan, F., Jiang, Z., Hayashi, H., Neubig, G.: Pre-train, prompt, and predict: a systematic survey of prompting methods in natural language processing. ACM Comput. Surv. **55**(9), 1–35 (2023)
8. Liu, Y., et al.: Roberta: a robustly optimized bert pretraining approach. arXiv preprint arXiv:1907.11692 (2019)
9. Longpre, S., et al. The flan collection: designing data and methods for effective instruction tuning. arXiv preprint arXiv:2301.13688 (2023)
10. McNeill, K.L., Katsh-Singer, R., González-Howard, M., Loper, S.: Factors impacting teachers' argumentation instruction in their science classrooms. Int. J. Sci. Educ. **38**(12), 2026–2046 (2016)
11. Nardo, C.: The waluigi effect (mega-post). Less Wrong (2023)
12. Quinonero-Candela, J., Sugiyama, M., Schwaighofer, A., Lawrence, N.D.: Dataset Shift in Machine Learning. MIT Press, Cambridge (2008)
13. Recht, B., Roelofs, R., Schmidt, L., Shankar, V.: Do imagenet classifiers generalize to imagenet? In: International Conference on Machine Learning, pp. 5389–5400. PMLR (2019)
14. Ribeiro, M.T., Lundberg, S.: Adaptive testing and debugging of NLP models. In: Proceedings of the 60th Annual Meeting of the Association for Computational Linguistics, vol. 1: Long Papers), pp. 3253–3267 (2022)
15. Ribeiro, M.T, Wu, T., Guestrin, C., Singh, S.: Beyond accuracy: behavioral testing of NLP models with checklist. arXiv preprint arXiv:2005.04118 (2020)
16. Sendak, M.P., Gao, M., Brajer, N., Balu, S.: Presenting machine learning model information to clinical end users with model facts labels. NPJ Dig. Med. **3**(1), 41 (2020)

17. Next Generation Science Standards Lead States. Next generation science standards: For states, by states. Appendix D: All standards, all students: Making the Next Generation Science Standards accessible to all students (2013)
18. Suresh, H., et al.: Kaleidoscope: semantically-grounded, context-specific ml model evaluation. In: Proceedings of the 2023 CHI Conference on Human Factors in Computing Systems, pp. 1–13 (2023)
19. Wang, A., Singh, A., Michael, J., Hill, F., Levy, O., Bowman, S.R.: Glue: a multi-task benchmark and analysis platform for natural language understanding (2018)
20. Warstadt, A., Singh, A., Bowman, S.R.: Neural network acceptability judgments. Trans. Assoc. Comput. Linguist. **7**, 625–641 (2019)
21. Wolf, T., et al.: Huggingface's transformers: state-of-the-art natural language processing. arXiv preprint arXiv:1910.03771 (2019)
22. Wu, T., Ribeiro, M.T., Heer, J., Weld, D.S.: Errudite: scalable, reproducible, and testable error analysis. In: Annual Meeting of the Association for Computational Linguistics, pp. 747–763 (2019)

Federated Learning Analytics: Investigating the Privacy-Performance Trade-Off in Machine Learning for Educational Analytics

Max van Haastrecht[1]([⊠])[iD], Matthieu Brinkhuis[2][iD], and Marco Spruit[1,3][iD]

[1] Leiden Institute of Advanced Computer Science (LIACS), Leiden University, Leiden, The Netherlands
{m.a.n.van.haastrecht,m.r.spruit}@liacs.leidenuniv.nl
[2] Department of Information and Computing Sciences, Utrecht University, Utrecht, The Netherlands
m.j.s.brinkhuis@uu.nl
[3] Department of Public Health and Primary Care (PHEG), Leiden University Medical Center (LUMC), Leiden, The Netherlands
m.r.spruit@lumc.nl

Abstract. Concerns surrounding privacy and data protection are a primary contributor to the hesitation of institutions to adopt new educational technologies. Addressing these concerns could open the door to accelerated impact, but current state-of-the-art approaches centred around machine learning are heavily dependent on (personal) data. Privacy-preserving machine learning, in the form of federated learning, could offer a solution. However, federated learning has not been investigated in-depth within the context of educational analytics, and it is therefore unclear what its impact on model performance is. In this paper, we compare performance across three different machine learning architectures (local learning, federated learning, and central learning) for three distinct prediction use cases (learning outcome, question correctness, and dropout). We find that federated learning consistently achieves comparable performance to central learning, but also that local learning remains competitive up to 20 local clients. We conclude by introducing FLAME, a novel metric that assists policymakers in their assessment of the privacy-performance trade-off.

Keywords: privacy · machine learning · federated learning · trust

1 Introduction

Driven by the promise of analytics to enable learning environment optimisation, education is now more datafied than ever [25]. The large-scale collection of learner data raises concerns regarding ethics, privacy, fairness, and trustworthiness [7,22]. Research tends to focus on the data protection measures educational

A. M. Olney et al. (Eds.): AIED 2024, LNAI 14830, pp. 62–74, 2024.
https://doi.org/10.1007/978-3-031-64299-9_5

institutions should implement to convince learners that they can be trusted as data fiduciaries [11]. Examples of suggested measures are limiting the boundaries of access to student data, pseudonymisation and anonymisation of learner records, and using automated bias mitigation. However, approaches that assume that personal data has already been collected fail to address a fundamental question: Did we have to collect the data in the first place?

It is not trivial to motivate which, if any, educational optimisations would warrant an intrusion of student privacy. Institutes that hold student privacy in high regard may believe that collecting personal learning data is never warranted [18]. This puts educational analytics research in an uncomfortable position, as methods and applications commonly rely heavily on personal data. Machine learning models such as deep neural networks predicting learning outcomes [23] and transformers facilitating student knowledge tracing [19] are deeply dependent on the availability of large amounts of data. On the surface, it seems that these data-hungry machine learning models are incompatible with a policy of preserving privacy. However, in recent years we have seen the development of machine learning architectures that promise the performance of machine learning without the threats to privacy posed by institute access to personal data.

Privacy-preserving machine learning architectures such as federated learning [15], where only model parameters are shared with a centrally coordinating party, offer a promising future direction for educational analytics. Along with local learning, where nothing is shared, and central learning, where everything is shared, federated learning is among the major machine learning architectures to consider from a privacy perspective. We have recently seen the first studies investigating the promise of federated learning for educational analytics [5,8]. However, to our knowledge, no study has systematically compared local learning, federated learning, and central learning across different datasets and use cases. This is a significant gap in the literature when we consider that privacy-preserving techniques could be the key to giving control back to students [4].

In this paper, we hope to take a first step in systematically investigating the promise of federated learning for learning analytics, which we term 'federated learning analytics'. We compare the performance of local learning, federated learning, and central learning across three distinct use cases: learning outcome prediction, question correctness prediction, and dropout prediction. Our methodology is geared at answering our main research question: How does the privacy-performance trade-off for machine learning algorithms manifest itself in different educational analytics use cases?

2   Background

Preserving the privacy of learners while actively collecting their data has long been recognised as a major challenge. It is evident that students should never be considered simply as sources of data, but rather as collaborators whose learning and development we are trying to serve [20]. However, although the importance of formulating and employing ethical and privacy principles was recognised early on, privacy concerns regularly played second fiddle due to the "enthusiasm for the possibilities offered by learning analytics" [16]. New legislation surrounding data protection introduced new perspectives. Besides ethical and privacy concerns, legal concerns began to drive decisions made at educational institutions. In the educational privacy framework DELICATE [3], the section on legitimacy contains the question: "Which data sources do you have already, and are they not enough?" Questions like these represented a major change of mindset. Researchers and practitioners recognised that collecting particular types of data is never warranted, and that "learning analytics is justifiable just to the extent that it does indeed promote autonomy" [18].

Basic organisational and technical controls can help to preserve student privacy, but it is questionable whether this is sufficient to gain students' trust. Prinsloo and Slade [16] convincingly argue that "the power to harvest, analyse and exploit data" lies completely with the educational institution, rather than the student. The authors outline the importance of transparency towards students and of giving students the possibility to access and update their own information. The issue with these measures is that they still require the student to entrust multiple stakeholders with their personal data, keeping alive the privacy power imbalance between the student and the data fiduciary.

Levelling out the power balance is exactly what decentralised approaches have attempted to do in recent years, by enabling the sharing of student data in a way that can enhance both privacy and security. Students thus regain some ownership over their data, helping to restore the power balance. Yet, using a decentralised architecture also introduces challenges. The most prominent of these is how to maintain performant algorithms when not all data is available in one central data store. A study of several anonymisation and differential privacy techniques found that in a GPA prediction task accuracy could drop from 76% to anywhere between 45%–63% [9]. Novel methods such as deep learning and transformers are notorious for requiring immense datasets to tune their parameters. How can we continue using these machine learning architectures when we do not have the data they so desperately need in one central location?

McMahan et al. (2017) [15] introduced the concept of federated learning, where learning occurs over a federation of users, referred to as clients. Rather than having to share data and parameters, clients train their model on local data and only share the parameter values of their model with the coordinating server. By averaging the parameters of all local clients, the resulting global model obtains better performance than if local clients operated independently. Figure 1 visualises the scenarios of local learning, federated learning, and central learning.

Decentralised machine learning could be the key towards privacy-preserving, trustworthy educational analytics [4]. Yet, few studies have investigated this promising area. Guo and Zeng [8] use federated learning in the context of educational data analysis. They consider the task of dropout prediction in the KDD Cup 2015 dataset, achieving accuracy within a couple of percentage points of the central learning scenario. However, the authors do not make their code available and do not report performance metrics other than a figure showing accuracy progression over epochs. This concern about their work was voiced by a recent

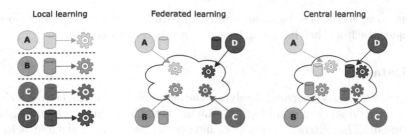

Fig. 1. Visualisation of various machine learning architectures (based on [24]). In the local learning scenario, both data and parameters remain at the client. Federated learning only shares model parameters, whereas for central learning both data and parameters are shared with a centrally coordinating server.

federated learning paper using the KDD Cup 2015 dataset [5]. Fachola et al. [5] achieve an accuracy of 81.7% for central learning and show that using federated learning an accuracy of around 80% can be achieved, even when data is spread over more than 50 clients. A downside is that the reported accuracy of 81.7% is only two percentage points higher than the proportion of dropouts in the dataset of 79.3%. Accuracy is not the right choice of metric for this dataset. If we want to draw meaningful conclusions about the potential of federated learning analytics, we need to consider multiple datasets and performance metrics.

3   Methodology

This section describes the metrics we used to compare the performance of different models, the three datasets (OULAD, EdNet, and KDD Cup 2015) employed in our experiments, and the details of our federated learning algorithm.

3.1   Metrics

Two commonly used metrics to evaluate model performance are accuracy and F_1. Accuracy represents the fraction of correctly predicted records. The F_1 score is the harmonic mean of precision (true positives divided by all predicted positives) and recall (true positives divided by all actual positives). Both metrics should be used with caution when dealing with imbalanced datasets, as they are heavily influenced by whether the majority class is labelled as positive or negative.

A metric that is less explicitly sensitive to class imbalance is the Area Under the ROC Curve (AUC). The curve in question is a plot of the true positive rate (equal to recall) on the y-axis and the false positive rate (false positives divided by all actual negatives) on the x-axis. The curve is drawn by determining the true positive rate and the false positive rate at different classification thresholds, meaning AUC requires the probability estimates of a model for its calculation. Because AUC is based on probability outputs, rather than the 0–1 classification output, it can provide more fine-grained insight into whether a model is truly learning to separate positive from negative instances. AUC does suffer from its own issues, such as that it can be biased towards certain classifiers.

3.2 Datasets

The Open University Learning Analytics Dataset (OULAD) [12], contains demographic data on students and logs of student activity within a virtual learning environment. The outcome variable of interest is the result a student achieved for a course, which can be pass, distinction, fail, or withdrawal. OULAD forms the basis for studies varying from the creation of predictive models identifying at-risk students [10] to the investigation of the role of demographics in virtual learning environments [17]. We use the work of Waheed et al. [23] as our baseline for comparison, as the authors provide a detailed description of the features they use, allowing us to conduct a replication that closely matches their process. They turn the original classification problem with four potential outcomes into four separate binary classification tasks (pass = 0 & fail = 1, pass = 0 & withdrawn = 1, fail = 0 & distinction = 1, pass = 0 & distinction = 1). Table 1 reports the accuracy and F_1 score achieved for each of these tasks.

Table 1. Descriptive statistics of the three datasets we investigate in this paper: OULAD, EdNet, and KDD Cup 2015. We additionally indicate state-of-the-art (SOTA) results for each, where the OULAD metrics are divided into PF (pass-fail), PW (pass-withdrawn), FD (fail-distinction), and PD (pass-distinction).

	OULAD [23]	EdNet [19]	KDD Cup 2015 [6]
Use case	learning outcome	question correctness	dropout
# Students	32,593	784,309	200,902
# Records	10,655,280	95,293,926	13,545,124
% Pos. class	PF: 31% fail	66% correct	79% dropout
	PW: 40% withdrawn		
	FD: 30% distinction		
	PD: 20% distinction		
SOTA	PF: Acc. = 0.845 F_1 = 0.719	Acc. = 0.725	F_1 = 0.929
	PW: Acc. = 0.947 F_1 = 0.943	AUC = 0.791	AUC = 0.909
	FD: Acc. = 0.864 F_1 = 0.770		
	PD: Acc. = 0.805 F_1 = 0.749		

EdNet is a knowledge tracing dataset containing data from users of a self-study platform [2]. Rather than having a single outcome variable per user, EdNet involves predicting for each completed multiple-choice question whether a user answered it correctly. The prediction task of EdNet is temporal in nature, explaining why papers tackling this dataset tend to employ time-series machine learning models such as transformers [1]. We use the SAINT+ transformer model [19] as our baseline for comparison, as this is the model with the current state-of-the-art performance. The authors use a version of EdNet with newer user data that is not publicly available. Yet, since the prediction task and features are identical, their results can still serve as a useful benchmark.

The final dataset we consider was used for the KDD Cup 2015 challenge. This dataset contains information on student interactions within a Massive Open Online Course (MOOC) environment. The goal is to predict student dropout, with a distinguishing characteristic being that 79% of the enrolled students dropped out. The dataset is thus highly imbalanced, explaining why KDD Cup 2015 papers tend to focus on reporting AUC and F_1 scores, rather than accuracy [6,13].

3.3 Federated Learning

Federated learning was proposed as a communication-efficient way to use all available data on individual devices to train a global model, without users having to share their personal data [15]. Another common use case for privacy-preserving machine learning is that of a group of hospitals working together to create better predictive models for the detection of illnesses [24]. The sensitivity of health data, along with the extensive legislation limiting data sharing in medical settings, provides a clear motivation for the need for a parameter-sharing infrastructure without a centrally coordinating party. A recent study in the educational field investigated a transfer learning approach and voiced concerns regarding the relevance of decentralised approaches for education [7]. Hence, we should ask to what extent decentralised machine learning contexts appear in educational environments.

Guo and Zeng [8] and Fachola et al. [5] envision a network of schools that are part of a federation sharing model parameters. These schools are part of the same governing body, but have separate physical locations, possibly in different countries. From a legal and privacy perspective, it can then be worthwhile to employ federated learning to obtain optimal insight into student behaviour without needing to share student data across schools. The use case considered in both papers is dropout prediction using the KDD Cup 2015 dataset, meaning each student has a single outcome variable per course. Federated learning on the level of the classroom or the individual is likely not realistic here, since the majority of students have fewer than five course outcomes. For the KDD Cup 2015 dataset, we will therefore investigate federated learning performance up to a maximum of 100 local clients, corresponding to roughly 2,000 students per client. OULAD is comparable to the KDD Cup 2015 dataset, with the exception that it also contains demographic information. For OULAD, we similarly analyse up to 100 local clients, corresponding to roughly 300 students per client.

For the EdNet setting, where a single student can answer thousands of questions in their self-study process, federated learning with individual students as local clients is more realistic. Nevertheless, since single users potentially have only one answered question within EdNet, it is not algorithmically practical to have local clients comprising one user. In our experiments, we will investigate the performance of local and federated learning up to a maximum of 100 local clients, corresponding to around 100 users per client when working with a randomly selected subset of 10,000 students.

4   Results

The Python code used to produce the outcomes of this section and detailed results per dataset are available on GitHub[1]. Our federated learning code adheres to the FedAvg algorithm of McMahan et al. (2017) [15]. Central learning experiments were conducted using the machine learning library scikit-learn and the gradient boosting libraries XGBoost and CatBoost. We used Pytorch as the deep learning library for our federated learning algorithm and exclusively used XGBoost with default settings as our local learning classifier.

4.1   Central Learning

Table 2 presents our central learning results using 10-fold cross-validation with an 80-20 train-test split. Our best results were achieved using CatBoost (OULAD and KDD Cup 2015) and XGBoost (EdNet). Table 2 shows that we managed to achieve comparable performance to the current state-of-the-art.

Table 2. Comparison of our central learning results to the results of Table 1, where the value between brackets represents the performance difference with earlier work.

OULAD			EdNet		KDD Cup 2015	
	Acc	F_1	Acc	AUC	F_1	AUC
PF	0.862 (+0.017)	0.751 (+0.032)	0.720 (-0.005)	0.757 (-0.035)	0.925 (-0.003)	0.881 (-0.028)
PW	0.933 (-0.014)	0.914 (-0.011)				
FD	0.893 (+0.029)	0.820 (+0.050)				
PD	0.810 (+0.005)	0.199 (-0.551)[a]				

[a]The precision and recall figures for the pass-distinction case reported in Table 3 of Waheed et al. [23] are incommensurate with their reported accuracy. We contacted the authors for clarification in August 2023, but have not received a response. Since the accuracy reported by Waheed et al. [23] is lower than ours, it is surprising that their reported precision and recall, and thus their F_1 score, are significantly higher.

Since Waheed et al. [23] extensively describe the features they engineered, we were able to reproduce these features and use them as input for OULAD classification. For the EdNet prediction task, we created lag features for previous user question correctness to turn the time series prediction task into a classification task. This enabled us to utilise the regular machine learning and gradient boosting libraries we used for OULAD and KDD Cup 2015. For the KDD Cup 2015 dataset, we designed student activity features similar to those of OULAD.

[1] https://github.com/MaxvanHaastrecht/Federated-Learning-Analytics.

4.2   Local Learning and Federated Learning

For our local and federated learning scenarios, we divided students randomly over clients. For OULAD federated learning, we used a neural network with two hidden layers of sizes 30 and 10, a learning rate η of 0.02, a cross-entropy loss function with the Adam optimiser, the number of communication rounds R set to 50, the number of local epochs per round $E = 2$, and a batch size of 64. Figure 2 shows that both federated learning and local learning perform worse than the central learning scenario. However, whereas local learning accuracy drops significantly as we progress from 10 to 100 local clients, federated learning accuracy remains roughly constant.

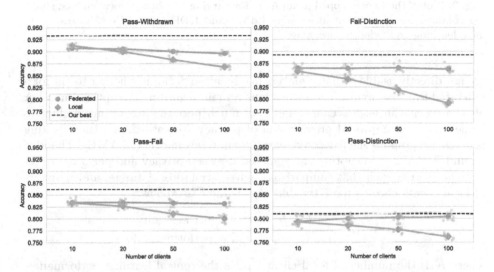

Fig. 2. Plot of the bootstrapped mean accuracy for varying numbers of clients, showing comparisons of our local learning, federated learning, and central learning results.

Figure 3 summarises the results from our EdNet and KDD Cup 2015 experiments. For KDD Cup 2015, we used the exact same federated learning settings as with OULAD. For EdNet, we changed the batch size to 128, as is used in earlier work [1], and lowered the number of communication rounds R from 50 to 20. We additionally used hidden layer sizes of 16 and 8, rather than 30 and 10, since EdNet feature engineering resulted in fewer input features for the network. Since the EdNet dataset is comparatively large, it is common practice to work with a random subset of the dataset in experimental settings such as our federated learning context [14,26]. We work with a random subset of 10,000 students and indicate the AUC of our best central learning model in Fig. 3.

4.3   Federated Learning Analytics Metric (FLAME)

Our numerical results provide an indication of the performance of federated learning compared to local learning and central learning. However, our results

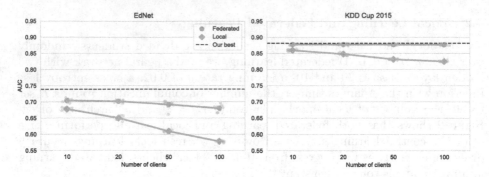

Fig. 3. Plot of the bootstrapped mean AUC for varying numbers of local clients, show-ing comparisons of our central learning EdNet and KDD Cup 2015 AUC results to local learning and federated learning.

are not directly usable by policymakers in education deciding whether to opt for a federated learning architecture. Questions remain regarding the optimal number of local clients in each scenario and how much performance we are willing to trade off for an improved preservation of privacy. To ease the decision-making process, we propose the federated learning analytics metric (FLAME). The idea behind FLAME is to capture the trade-off between privacy and performance in a single metric, such that comparisons across scenarios, datasets, and numbers of local clients become more tenable. We define FLAME as:

$$\text{FLAME} = \frac{1 - \frac{1}{K}}{1 + (p_c - p_f)} = \frac{\text{privacy gain}}{1 + \text{performance loss}},$$

where K is the number of local clients, p_c is the central learning performance, and p_f is the federated learning performance. For institutions considering to move from central learning to federated learning, p_c will be a known quantity. For institutions that do not have a centralised architecture, p_c can be estimated based on the literature or through simulations. FLAME is suited to be used for performance metrics ranging between [0,1], such as accuracy, F_1, and AUC. The numerator captures the gain in privacy achieved by employing an architecture with local clients. The denominator captures the loss in performance.

Figure 4 shows the FLAME values for EdNet and KDD Cup 2015, where AUC is the relevant performance metric. FLAME values for the local learn-ing scenario are also shown, which can be calculated by replacing the federated learning performance in the FLAME formula with local learning performance. Taking EdNet as an example, we observe that for federated learning FLAME peaks at 50 clients, whereas for local learning FLAME peaks at 20 clients. By more explicitly incorporating the privacy-performance trade-off, FLAME there-fore clarifies differences between algorithms in a way the pure AUC scores of Fig. 3 cannot.

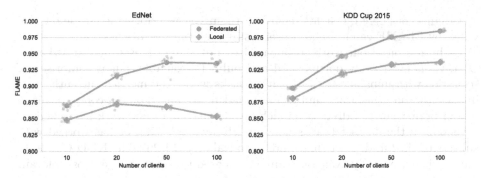

Fig. 4. FLAME values for EdNet and KDD Cup 2015, where AUC is the performance metric. In the case of 50 local clients, AUC loss must be less than 0.0315 to achieve a FLAME higher than 0.95.

5   Discussion

Our results demonstrate the potential of federated learning to preserve privacy and performance in educational contexts. For OULAD, we observed that our federated learning algorithm achieved comparable accuracy to earlier results for three out of four scenarios considered, even when the number of local clients was set to 100. For the KDD Cup 2015 dataset, federated learning matched our best results, again up to 100 local clients. Federated learning also significantly outperformed local learning for all three datasets. When dividing data over 100 local clients, the average accuracy gain for OULAD was 4.32% and the average AUC gains for EdNet and KDD Cup 2015 were 0.1017 and 0.0518, respectively.

Our FLAME values in Fig. 4 demonstrated that local learning and federated learning warrant serious consideration in settings where dividing data over 20 or more clients is realistic. However, the answer to student privacy concerns can never be purely technological. Federated learning is promising, but it carries with it additional security risks and questions whether student's perceptions of these technologies are as positive as their theoretical benefits. Yet, given the increasing tensions between the datafication of education and the privacy concerns of students, privacy-preserving machine learning architectures may offer the path of least resistance towards a bright future for educational analytics.

Federated learning is perhaps the most commonly used privacy-preserving machine learning strategy, but certainly not the only one. We did not cover other paradigms within this paper, such as split learning [21], swarm learning [24], and transfer learning [7]. In future work, it will be crucial to compare the privacy-performance trade-off for various approaches. We should be aware that in contexts where performance takes precedent, combining strategies (e.g., federated learning and split learning [21]) might be the optimal choice, whereas in contexts where privacy is paramount, a local learning approach that fosters stakeholder trust could provide the perfect fit. Regardless of the privacy-preserving paradigms considered, insights regarding the privacy-performance trade-off provided by FLAME can serve as a useful starting point for discussion.

A limitation of our work is that all benchmarking datasets had drawbacks. OULAD is extensively documented and publicly available, but is comprised of scenarios with imbalanced classification tasks where the metrics currently used in the literature (accuracy and F_1) are inadequate for thorough comparisons of model performance. EdNet is publicly available, but recent work has relied on a version of the dataset that is not publicly available [19], or has worked with subsets of the full dataset that hinder replicability [14,26]. The KDD Cup 2015 dataset is not publicly available from a dedicated website, and the most relevant publications covering this dataset in recent years only report model accuracy [5,8], when this is a highly imbalanced dataset with 79% of students dropping out. These drawbacks are not ideal, but we strongly believe these datasets offer an accurate representation of currently available benchmarks. Still, we require better benchmark datasets and accompanying research in the future.

6   Conclusion and Future Work

With education becoming more datafied than ever, researchers interested in optimising learning environments are increasingly faced with questions regarding ethics, privacy, fairness, and trustworthiness. Decisions to intrude on student privacy should be taken with the utmost caution. There are legitimate concerns whether any type of optimisation warrants the collection of sensitive learner data. Within this context, privacy-preserving machine learning that respects privacy while maintaining model performance is an intriguing recent development. However, until now, we lacked rigorous investigations of the impact of privacy-preserving architectures on educational analytics model performance.

We compared algorithm performance across three architectures (local learning, federated learning, central learning) for three different prediction use cases (learning outcome, question correctness, dropout). In doing so, we provided a comprehensive image of what can be achieved with privacy-preserving architectures. We found that even when dividing data over 100 clients, federated learning can compete with state-of-the-art results. A major finding was that although for 50 or more clients federated learning outperformed local learning, differences were often not significant when dividing data over 20 or fewer clients. This points to the importance of considering local learning as a privacy-preserving strategy for educational analytics. Future work will need to investigate how students, teachers, and other stakeholders view federated learning, since the relative complexity of privacy-preserving machine learning may diminish trust. Nevertheless, the datafication of education combined with the clear wish of students to preserve privacy signal a promising future for federated learning analytics.

References

1. Choi, Y., et al.: Towards an appropriate query, key, and value computation for knowledge tracing. In: Proceedings of the Seventh ACM Conference on Learning @ Scale, L@S 2020, pp. 341–344. ACM (2020). https://doi.org/10.1145/3386527. 3405945
2. Choi, Y., et al.: EdNet: a large-scale hierarchical dataset in education. In: Bittencourt, I.I., Cukurova, M., Muldner, K., Luckin, R., Millán, E. (eds.) AIED 2020. LNCS (LNAI), vol. 12164, pp. 69–73. Springer, Cham (2020). https://doi.org/10. 1007/978-3-030-52240-7_13
3. Drachsler, H., Greller, W.: Privacy and analytics: it's a DELICATE issue a checklist for trusted learning analytics. In: Proceedings of the 6th International Learning Analytics & Knowledge Conference, LAK 2016, pp. 89–98. ACM, Edinburgh (2016). https://doi.org/10.1145/2883851.2883893
4. Ekuban, A., Domingue, J.: Towards decentralised learning analytics (Positioning Paper). In: Companion Proceedings of the ACM Web Conference 2023, WWW 2023, pp. 1435–1438. ACM, Austin (2023). https://doi.org/10.1145/3543873. 3587644
5. Fachola, C., Tornaría, A., Bermolen, P., Capdehourat, G., Etcheverry, L., Fariello, M.I.: Federated learning for data analytics in education. Data 8(2), 43 (2023). https://doi.org/10.3390/data8020043
6. Feng, W., Tang, J., Liu, T.X.: Understanding dropouts in MOOCs. In: Proceedings of the 33rd AAAI Conference on Artificial Intelligence, AAAI 2019, vol. 33, pp. 517–524. PKP Publishing Services, Honolulu (2019). https://doi.org/10.1609/aaai. v33i01.3301517
7. Gardner, J., Yu, R., Nguyen, Q., Brooks, C., Kizilcec, R.: Cross-institutional transfer learning for educational models: implications for model performance, fairness, and equity. In: Proceedings of the 2023 ACM Conference on Fairness, Accountability, and Transparency, FAccT 2023, pp. 1664–1684. ACM, Chicago (2023). https:// doi.org/10.1145/3593013.3594107
8. Guo, S., Zeng, D.: Pedagogical data federation toward education 4.0. In: Proceedings of the 6th International Conference on Frontiers of Educational Technologies, ICFET 2020, pp. 51–55. ACM, Tokyo (2020). https://doi.org/10.1145/3404709. 3404751
9. Gursoy, M.E., Inan, A., Nergiz, M.E., Saygin, Y.: Privacy-preserving learning analytics: challenges and techniques. IEEE Trans. Learn. Technol. 10(1), 68–81 (2017). https://doi.org/10.1109/TLT.2016.2607747
10. Hlosta, M., Zdrahal, Z., Zendulka, J.: Ouroboros: early identification of at-risk students without models based on legacy data. In: Proceedings of the 7th International Learning Analytics & Knowledge Conference, LAK 2017, pp. 6–15. ACM, Vancouver (2017). https://doi.org/10.1145/3027385.3027449
11. Jones, K.M.L., Rubel, A., LeClere, E.: A matter of trust: higher education institutions as information fiduciaries in an age of educational data mining and learning analytics. J. Am. Soc. Inf. Sci. 71(10), 1227–1241 (2020). https://doi.org/10.1002/ asi.24327
12. Kuzilek, J., Hlosta, M., Zdrahal, Z.: Open university learning analytics dataset. Sci. Data 4(1), 170171 (2017). https://doi.org/10.1038/sdata.2017.171
13. Li, W., Gao, M., Li, H., Xiong, Q., Wen, J., Wu, Z.: Dropout prediction in MOOCs using behavior features and multi-view semi-supervised learning. In: 2016 International Joint Conference on Neural Networks, IJCNN, pp. 3130–3137. IEEE, Vancouver (2016). https://doi.org/10.1109/IJCNN.2016.7727598

14. Long, T., et al.: Improving knowledge tracing with collaborative information. In: Proceedings of the Fifteenth ACM International Conference on Web Search and Data Mining, WSDM 2022, pp. 599–607. ACM 2022). https://doi.org/10.1145/3488560.3498374

15. McMahan, B., Moore, E., Ramage, D., Hampson, S., Arcas, B.A.: Communication-efficient learning of deep networks from decentralized data. In: Proceedings of the 20th International Conference on Artificial Intelligence and Statistics, pp. 1273–1282. PMLR (2017)

16. Prinsloo, P., Slade, S.: Student privacy self-management: Implications for learning analytics. In: Proceedings of the 5th International Learning Analytics & Knowledge Conference, LAK 2015, pp. 83–92. ACM, Poughkeepsie (2015). https://doi.org/10.1145/2723576.2723585

17. Rizvi, S., Rienties, B., Khoja, S.A.: The role of demographics in online learning; a decision tree based approach. Comput. Educ. **137**, 32–47 (2019). https://doi.org/10.1016/j.compedu.2019.04.001

18. Rubel, A., Jones, K.M.L.: Student privacy in learning analytics: an information ethics perspective. Inf. Soc. **32**(2), 143–159 (2016). https://doi.org/10.1080/01972243.2016.1130502

19. Shin, D., Shim, Y., Yu, H., Lee, S., Kim, B., Choi, Y.: SAINT+: integrating temporal features for EdNet correctness prediction. In: Proceedings of the 11th International Learning Analytics & Knowledge Conference, LAK 2021, pp. 490–496. ACM, Irvine (2021). https://doi.org/10.1145/3448139.3448188

20. Slade, S., Prinsloo, P.: Learning analytics: ethical issues and dilemmas. Am. Behav. Sci. **57**(10), 1510–1529 (2013). https://doi.org/10.1177/0002764213479366

21. Thapa, C., Arachchige, P.C.M., Camtepe, S., Sun, L.: SplitFed: when federated learning meets split learning. In: Proceedings of the 36th AAAI Conference on Artificial Intelligence, vol. 36, pp. 8485–8493. AAAI Press, Palo Alto (2022). https://doi.org/10.1609/aaai.v36i8.20825

22. van Haastrecht, M., Brinkhuis, M., Peichl, J., Remmele, B., Spruit, M.: Embracing trustworthiness and authenticity in the validation of learning analytics systems. In: LAK23: 13th International Learning Analytics and Knowledge Conference, LAK 2023, pp. 552–558. Association for Computing Machinery, New York (2023). https://doi.org/10.1145/3576050.3576060

23. Waheed, H., Hassan, S.U., Aljohani, N.R., Hardman, J., Alelyani, S., Nawaz, R.: Predicting academic performance of students from VLE big data using deep learning models. Comput. Hum. Behav. **104**, 106189 (2020). https://doi.org/10.1016/j.chb.2019.106189

24. Warnat-Herresthal, S., et al.: Swarm Learning for decentralized and confidential clinical machine learning. Nature **594**(7862), 265–270 (2021). https://doi.org/10.1038/s41586-021-03583-3

25. Williamson, B., Bayne, S., Shay, S.: The datafication of teaching in higher education: critical issues and perspectives. Teach. High. Educ. **25**(4), 351–365 (2020). https://doi.org/10.1080/13562517.2020.1748811

26. Yang, Y., et al.: GIKT: a graph-based interaction model for knowledge tracing. In: Hutter, F., Kersting, K., Lijffijt, J., Valera, I. (eds.) ECML PKDD 2020. LNCS (LNAI), vol. 12457, pp. 299–315. Springer, Cham (2021). https://doi.org/10.1007/978-3-030-67658-2_18

Towards the Automated Generation of Readily Applicable Personalised Feedback in Education

Zhiping Liang, Lele Sha, Yi-Shan Tsai, Dragan Gašević,
and Guanliang Chen[✉]

Centre for Learning Analytics, Monash University, Melbourne, Australia
{zhiping.liang,Lele.Sha,Yi-Shan.Tsai,Dragan.Gasevic,
Guanliang.Chen}@monash.edu

Abstract. Providing personalised feedback to a large student cohort is a longstanding challenge in education. Recent work in prescriptive learning analytics (PLA) demonstrated a promising approach by augmenting predictive models with prescriptive capabilities of explainable artificial intelligence (XAI). Although theoretically sound, in practice, not all predictive features can be leveraged by XAI to prescribe useful feedback. It remains under-explored as to how to engineer such predictive features that can be used to prescribe personalised and actionable feedback. To address this, we proposed a learning activity-based approach to design features that are informative to both predictive and prescriptive performance in PLA. We conducted empirical evaluations of the quality of PLA-generated feedback compared to feedback written by experienced teachers in a large-scale university course. Four rubric criteria, including Readily Applicablility, Readability, Relational, and Specificity, were designed based on previous research. We found that: (i) By adopting learning activity-based features, PLA generates high quality feedback without sacrificing predictive performance; (ii) Most experienced teaching staff rated PLA-generated feedback as readily applicable to the course; and (iii) Compared to teacher-written feedback, the quality of PLA-generated feedback is consistently rated higher (with statistical significance) in all four rubric criteria by experienced teachers. All code is available via our GitHub repository (https://github.com/CoLAMZP/AIED-2024-AutoFeedback).

Keywords: Automated feedback · Prescriptive learning analytics · Predictive models

1 Introduction

Personalised feedback is widely acknowledged to be beneficial for student learning [9]. However, due to limited teaching resources, providing personalised feedback to a large student cohorts remains a challenge. Recently, a new body of research in prescriptive learning analytics (PLA) has emerged as a new promising direction [32]. PLA can be conceptualised as a pipeline consisting of a

A. M. Olney et al. (Eds.): AIED 2024, LNAI 14830, pp. 75–88, 2024.
https://doi.org/10.1007/978-3-031-64299-9_6

predictive step and prescriptive step. An outcome prediction (e.g., whether a student is at risk of failing a course or not) is first forecasted in the predictive step, then during prescriptive step, the features that make significant contributions to the at-risk prediction outcome are identified by explainable AI (XAI), and then altered into different simulated values which will result in an improved prediction outcome, i.e., from being at-risk to non-risk [24]. While theoretically sound, PLA principally relied on the input features of predictive models to generate meaningful feedback. Importantly, not all predictive features yield useful feedback in practice. For instance, students can not change their past academic performance despite it being a contributing factor to the at-risk prediction. We argue that, it is crucial to feed to the PLA pipeline features that are both predictive and prescriptive, so that at-risk students could first be accurately identified, then prescribe meaningful feedback. Inspired by the learner-centered feedback framework [26], which proposes a comprehensive set of effective artefact attributes for the feedback content and focuses on supporting learners with actionable future learning tasks, we proposed a learning activity-based approach to engineer features that may enable both accurate at-risk predictions and meaningful feedback generation. Concretely, we investigate the following Research Question:

> **RQ**: To what extent can learning activity-based features enable PLA to simultaneously produce accurate predictions and high-quality feedback in authentic course settings?

To answer the research question, we engineered 42 learning activity-based features that are indicative to both student performance and actionable future learning tasks. These features are input to the PLA pipeline of predictive and prescriptive modeling. To ensure a thorough assessment on predictive performance, we feed the features into 5 widely used predictive models and adopted a total of four evaluation metrics to assess predictive performance. Then, the best-performing models are coupled with the latest prescriptive model Diverse Counterfactual Explanations (DiCE) to generate personalised feedback tailored to each student. We compared the quality of machine-generated feedback with teacher-written feedback using a quality rubric consisted of four critical dimensions, including Readily Applicablility, Readability, Relational, and Specificity, in line with prior literature [12,16,22]. Our main findings are as follows: (i) By adopting learning activity-based features, PLA generates high-quality feedback without sacrificing predictive performance; (ii) Experienced teaching staff generally rated PLA-generated feedback as readily applicable to the course; and (iii) Compared to teacher-written feedback, the quality of PLA-generated feedback is consistently rated higher in all four rubric criteria by experienced teachers, and the improvement in feedback quality is statistically significant in all four criteria.

2   Related Work

Conceptualisations of Feedback. Feedback is an essential component for students to understand their learning outcomes and facilitate their study [7].

Recent research has redefined the definition of effective feedback, shifting from transmission-focused feedback that emphasises information delivery to learner-centred feedback that focuses more on the student's learning process [26]. Unlike transmission-focused feedback, which doesn't explicitly require or expect learners to improve based on the feedback comments [1], learner-centered feedback, extensively proven to be more beneficial [8], offers actionable information for improvement and fosters learners' ability to independently seek and utilize feedback skills during the learning process [26]. Building on this foundation, a recent study [26] conducted a review of the proposed feedback publications and summarized a learner-centered feedback framework. This framework provides a detailed explanation of the attributes associated with feedback content (e.g. The feedback content needs to (Attribute 1) provide actionable information for future study, (Attribute 2) highlight strengths of performance, etc.) that could be used as guidelines for feedback to support its practical application in PLA generation of feedback.

Automated Feedback in Education. The early works of developing automated feedback systems largely relied on expert-crafted rules which automatically maps to pre-defined feedback segments [19,35]. Despite being widely adopted in practice [19], rule-based feedback systems typically lack robustness to change (e.g., if there are any changes to the course structure or context) and therefore cannot scale to courses taught in various forms using Learning Management Systems (LMS). Moreover, the creation of feedback rules and templates heavily depend on domain expertise. This reliance means that significant input from knowledgeable professionals in the relevant field is necessary to establish the rules and parameters that guide the feedback process. As a result, while rule-based systems provide a structured and consistent approach to feedback, they often do not account for the diverse learning needs, learning designs and progress rates [17,33]. This limitation highlights the need for more adaptive and personalized approaches in the development of automated feedback systems. Alternatively, researchers are also investigating automated feedback generation by transformer-based large language models (LLM), most recently generative LLM (i.e., GPT models) [15,31]. However, skeptics often question about their reliability and appropriateness when adopted in an educational context given their widely detected bias for traditionally marginalised population [15], lack of transparency [15,20], uncontrollability/hallucination [20]. These growing concerns have undoubtedly hindered the practical adoption of LLM-based approaches in education.

Prescriptive Learning Analytics-Based Feedback. In a different vein, prescriptive feedback generation recently emerged as a promising candidate for personalised feedback generation. By coupling predictive models with the state-of-the-art explainable AI approaches [11], educators can gain transparent insights about the student learning progress and performance, and may subsequently prescribe actions for students to take. However, these prescriptive techniques, such as those based on SHAPley value to explain predictions [29], typically generate explanations on a high number of feature inputs of a predictive model, often resulting in verbose and overly-complex feature explanations [4,11]. As a result, additional feature-simplification mechanism (e.g., Anchors-based [25]) needs to be adopted to reduce the prediction model as a rule-based system. Recently, prescriptive approaches developed via counterfactual-based explanation have emerged as a new breakthrough [30]. In comparison, counterfactual-based explanation [18] is more robust in that it only generate explanation for a minimal set of features that would result in an alternative predictive outcome in the model, which successfully avoids verbose and overly-complex feature explanations. Despite the advancement made [24,32], the practical adoption of counterfactual-based approach in educational feedback practices is still limited due to two reasons: (i) PLA-based approaches require the input features to be simultaneously effective for producing highly accurate at-risk predictions (in predictive modeling) and generating meaningful feedback for students (in prescriptive modeling). It remains largely under-explored in terms of effective ways to engineer such features; and (ii) To the best of our knowledge, no evaluation of counterfactual-based feedback has been done in the authentic course setting.

3   Methodology

3.1   Course Context

We examined the effectiveness of feedback generation in the context of large-scale offerings (avg. 2,145 students per year) of an introductory programming course at the Monash University. We selected this course because it had one of the highest course failure rates (around 18%) in the school it was offered by. Such high failure in introductory programming has been widely reported in previous studies [34]. Therefore, prescriptive feedback was much needed in this course to guide student learning and avoid course failure. The course was carried out in a weekly manner (12 weeks per offering), and delivered through Ed Discussion, a LMS system commonly used in STEM education. Students involved in the course were expected to participate in an in-person workshop session, pre-workshop, and post-workshop learning activities on this platform. To ensure real-world applicability, we adopted the two most recent offerings of the same course at 2023 semester 1 (1,561 Students) and 2023 semester 2 (584 Students). The PLA model was trained on the first semester data, and tested separately on the second upcoming semester data. Since the study was conducted in authentic course settings, we could not know the at-risk status of semester 2 students in advance as the semester had not been concluded when the study was conducted.

3.2   Feedback Generation

Machine-Generated Feedback via PLA. To ensure both predictive and prescriptive performance, we strategically engineered an extensive set of features based on a systematic collection of all the student learning activities in pre-workshop, workshop, and post-workshop sessions, This resulted in a total of 98 features (see appendix[1]). We posit that, these detailed and diverse learning activity features could empower PLA to effectively detect at-risk students behavioral patterns, while informing the needed behavioral change to reverse

Table 1. A summary of the feature used in this study. * represents that the feature value is a binary value

Category	Feature Description
Course Contents	The fraction of specific types of course materials (Including: (F1):Pre-workshop, (F2):Workshop, (F3):Applied, (F4):Post-workshop) from week$_n$ accessed by the student
	The total amount of time the student spent viewing specific types of course materials (Including: (F5):Pre-workshop, (F6):Workshop, (F7):Applied, (F8):Post-workshop) from week$_n$
	(F9) : The total minutes a student has dedicated to viewing slides during week$_n$
	(F10): The number of slides a student has viewed during week$_n$
Forum	(F11): The total minutes a student has dedicated to viewing the thread posted on the discussion forum during week$_n$
	The ratio of specific forum-related events (Including: (F12):View thread, (F13):Post thread, (F14):Post comment, (F15):Post answer) to the total forum-related events during week$_n$
	The number of specific forum-related events (Including: (F16):View thread, (F17):Post thread, (F18):Post comment, (F19):Post answer) presented by the student during week$_n$
Assessment	Indicate whether the student attempted the assessment (Assessment types, including: (F20):Quiz, (F21):Challenge) during week$_n$*. '1' to represent attempted and '0' otherwise
	The number of times a student has attempted the assessment (Assessment types, including: (F22):Quiz, (F23):Challenge) during week$_n$
	The score of the assessment (Assessment types, including: (F24):Quiz, (F25):Challenge, (F26):Assignment 1) attempted by a student during week$_n$. (Assignment score will only be available at the end of Week 7; therefore, it will be set as 0 from Week 1 to Week 6)
	The average time interval between two attempted assessment (Assessment types, including: (F27):Quiz, (F28):Challenge) during Week$_n$ in minutes
Others	(F29):The longest consecutive active days on the LMS during week$_n$. (F30):Indicate whether the student had any interactions on LMS during week$_n$*. '1' to represent interactions and '0' otherwise. (F31):The number of times the student accesses the LMS during week$_n$. (F32):The number of days that the student has accessed the LMS during week$_n$. (F33):The number of learning sessions a student has during week$_n$. (F34):The average number of accesses made by a student during one learning session during week$_n$. (F35):The total minutes a student has dedicated to engaging with the LMS platform during week$_n$. (F36):The ratio of events on Monday to the total events for week$_n$. (F37):The average duration of the learning session during week$_n$ in minutes. (F38):The longest duration of a learning session during week$_n$ in minutes. (F39):The average time interval between two learning sessions during week$_n$ in hours. (F40):The average time interval between two consecutive active days during week$_n$ in hours. (F41):Ratio of a student's accesses to the LMS platform between 7 a.m. and 6 p.m. during week$_n$ compared to the total number of accesses. (F42):Ratio of a student's accesses to the LMS platform between 7 p.m. and 6 a.m. during week$_n$ compared to the total number of accesses

---

[1] https://bit.ly/47VrcGh.

the at-risk status [26]. Next, learning activities that occurred only once, could not be repeated, or were deemed to have minimal impact on learning outcomes and learning skill development (after consulting a senior teaching staff) were removed, e.g., 'Number of weeks of inactivity' is removed. This resulted in a final list of 42 features for each week, as detailed in Table 1. These features were then used as input into the PLA-pipeline consisting of a feature pre-processing step, predictive step, and prescriptive step. As detailed below:

1. Feature pre-processing. In line with prior studies [23], we pre-processed the 42 features by: (i) Normalising feature value via `Z-score` normalization so all features were within the value range of -100 to 100; (ii) To further boost predictive performance [21], we applied `SelectKBest`[2] to select top-k features that were most important to the at-risk prediction label, we tuned k within the range of $[5, 10, 15, 20, 25, 30, 35]$ using a Decision Tree model until the optimal predictive performance was achieved, yielding the final top-25 features as input to the predictive modelling step.

2. Predicting at-risk students. To gain a thorough understanding of predictive power of the 25 features. We assessed the predictive performance using four widely adopted accuracy metrics in prior studies [11, 14, 23]: Accuracy, F1-score, AUC, and True Negative Rate (TNR). We included five widely used predictive models including Bayes-based (Naive Bayes [10]), kernel-based (SVM [6]), tree-based (Random Forest [3]), gradient-based (XGBoost [5]), and neural network based (Multi-layer perceptron [36] with 256 hidden layer and ReLU activation). Apart from XGBoost(implemented using dmlc XGBoost library[3]), all other models were implemented using the scikit-learn library [21]. For all predictive models, we applied the hyperopt library [2] to find the best model hyper-parameters to optimize the performance. To ensure the model can effectively identify at-risk students and not biased towards predicting the non-at-risk majority label, we balanced the class between at-risk and non-at-risk using the pre-defined parameter (e.g., 'scale_pos_weight' for XGBClassifier), which was used to control the balance of positive and negative weights for unbalanced datasets.

3. Prescribing actionable feedback for at-risk students. For students who were predicted 'at-risk' by the best-performing predictive model, we adopted the Diverse Counterfactual Explanations (DiCE) [18] as our prescriptive model. DiCE is superior to other PLA-based models since it generates candidate minimal counterfactual explanations about the input feature for an alternative prediction (i.e., the minimal features to be changed to become not at-risk) as detailed in Sect. 2. We randomly selected 90 students predicted to be at-risk from the predictive results as inputs for the prescriptive model, aiming to generate corresponding counterfactual explanations for each of them, and such counterfactual explanations can be used as the basis to provide feedback for students to alternate their learning behaviour and result in improved outcomes. We followed similar DiCE settings as those used in [32], including

---

[2]   Scikit-learn package (2023), https://scikit-learn.org/stable/.
[3]   https://xgboost.readthedocs.io/en/stable/#.

the specification that the value needs to be constrained within a range (e.g., The fraction of workshop material accessed by the student should always be within 0 to 100 %.) and selecting the counterfactual explanations that entail minimal feature value change as the basis for feedback generation. Recent research emphasized on transitioning from *feedback as information* to *feedback as a learner-centred process* to promote learners' achievements [8, 37]. To enhance the usefulness of machine-generated feedback for students, we therefore adopted a learner-centred feedback framework to guide the framing of DiCE explanations as feedback [27]. We distilled a set of rules from the framework[4], and convert DiCE output into natural language feedback using the latest generative AI (GPT4.0), similar to prior PLA studies [32]. See an example provided in Table 2. Lastly, we generated feedback at mid-term (week 7), in line with the original course design, so that feedback is written based on sufficiently informative learning data, while leaving sufficient time for student to improve upon receiving the feedback. Our prompt input for GPT4 was as follows: "Please act as a teacher to provide feedback for students. The feedback must adhere to the following rules: [Learner-centered feedback rules[5]]. The feedback is based on the students' higher-than-median learning activities: [Commendable learning activities], and the under-performed learning activities with the corresponding suggestions: [Counterfactual feature value changes].".

Table 2. An example of machine-generated feedback based on DiCE counterfactual explanation. We use the format 'feature index (Week Number)' to indicate both the feature and the corresponding week.

	F35(Week7)	F38(Week7)	F32(Week5)	F33(Week5)	Label
Original	0.2	0.1	6	22	At-Risk
Counterfactual	332	156	–	–	No Risk
Feedback	Dear [Student Name], I hope this message finds you well. I am writing to deliver feedback on your recent performance and offer insights to help you leverage your potential fully. Firstly, I commend your active participation during Week 5, where you accessed the LMS on 6 d and had a total of 22 learning sessions. This level of engagement is commendable as it demonstrates a commitment to your learning journey. Well done on maintaining this consistent connection! However, I've noticed in Week 7 that your time spent on the LMS was only 0.2 min, and your longest learning session lasted just 0.1 min. Engaging deeply with course materials is essential for comprehension and success. To mirror the commitment you showed in Week 5, I suggest aiming for a total of approximately 332 min in the LMS and a continuous learning session of at least 156 min. Allocating time efficiently will give you the opportunity to participate more actively in discussions and solidify your understanding of the material. Could you share what might have caused the decrease in your engagement during Week 7? I believe that by reflecting on this, together, we can identify strategies to assist in managing your study time effectively. Remember, I am here to support you. Please feel free to ask any questions or seek further clarification. Warm regards, [Your Name]				

Human-Written Feedback. To ensure that human-written feedback was comparable to the machine-generated feedback, we presented the same 42 learning

---

[4] https://bit.ly/3u7nziI.
[5] https://bit.ly/42mVEHU.

activities from semester 2 2023 week 1 – week 7 in an excel sheet as a 'learning profile' of a student, and shared these learning activities with experienced teachers based on which they can provide feedback to the student; see the sample student learning activities in the digital appendix[6]. We invited five experienced teachers (who taught 3–6 offerings of the course) to write feedback based on students learning profile. We instructed teachers to follow the learner-centered feedback framework (similar to the instructions given to GPT) see details in appendix[7]. Each teacher was allocated 2 hours to write feedback for randomly selected (non-duplicated) at-risk student profiles from 90 students who were predicted to be at-risk. Each at-risk student will receive 1 feedback. Teachers wrote 15–24 feedback messages, resulted in 89 messages in total. Each teacher received a $100 dollar gift card after the session.

3.3 Feedback Evaluation Procedure

Participants. Four additional teaching staff who did not participate in the feedback writing session evaluated all the machine-generated and teacher-written feedback. Participating teachers received a $50 dollar gift card at the end of a feedback evaluation session.

Rubric Criteria. To assess the quality of generated feedback, we adapted a feedback rubric based on prior studies in learner-centered feedback [12,16,22]. A senior researcher specialising feedback practice was consulted during the development of the feedback rubric.

- Readily applicable [28]: In the best judgement of the teacher, whether the feedback could be readily applicable in the authentic course context to help student learning, i.e., 0 if not readily applicable, 1 if somewhat applicable, 2 if readily applicable.
- Readability [13,16]: Rate the readability of feedback concerning grammar, word choice and coherence, i.e., 0 if not readable, 1 if understandable but had issues with fluency and coherence, 2 if entirely fluent, coherent, and easy to understand.
- Relational [27]: To what extent does the feedback utilises specific languages or tones to encourage students and build relationships with students? i.e., 0 if not relational, 1 if somewhat relational, 2 if fully relational.
- Specificity [22]: To what extent is the feedback specific and pointing out areas of strengths and weakness to be improved upon? i.e., 0 if no relevant specific details for improvement, 1 if specific but not actionable, 2 if specific and actionable for improvement.

---

[6] https://bit.ly/4bnaJNL.
[7] https://bit.ly/3u7nziI.

Rubric Marking Process. Each teacher participating in the evaluation process evaluated an average of 23 machine-generated and 23 teacher-written feedback messages, from a pool of 89 machine-generated and 89 teacher-written feedback messages. These feedback messages assigned to each teacher in a random and non-duplicated manner, without informing them about the method used to write these feedback. After an 1 hour introduction session explaining the student profile, marking rubrics and marking process, the teachers proceeded to evaluate feedback based on the rubric criteria.

Rubric Results Analysis. Given that feedback within either teacher-written or machine-generated group were independent samples, we adopted the Mann-Whitney U test to calculate the statistical differences between teacher-written and machine-generated feedback with respect to each rubric criteria.

4   Results

Predictive Performance. Table 3 summarises the performance of the predictive models using data collected until week 3, 7, and 9, respectively. The full results for all weeks can be found in appendix[8]. Overall, we observed all predictive models achieved a high predictive performance after week 3 across different

Table 3. Predictive model results. The best result of each metric in each week is bold.

Week	Model	Accuracy	F1-Score	AUC	TNR
Week 3	XGBClassifier	0.8128	0.8900	**0.6430**	0.4103
	Support Vector Machine	0.8111	0.8898	0.6204	0.3590
	Naive Bayes	0.6534	0.7653	0.6536	**0.6538**
	Random Forests	0.7400	0.8377	0.6442	0.5128
	Multilayer Perceptron	**0.8388**	**0.9101**	0.5553	0.1667
Week 7	XGBClassifier	0.8804	0.9294	**0.8011**	**0.6923**
	Support Vector Machine	**0.9029**	**0.9449**	0.7438	0.5256
	Naive Bayes	0.8094	0.8852	0.7005	0.5513
	Random Forests	0.8925	0.9380	0.7648	0.5897
	Multilayer Perceptron	0.8856	0.9353	0.6959	0.4359
Week 9	XGBClassifier	0.8666	0.9199	**0.8147**	**0.7436**
	Support Vector Machine	0.8925	0.9382	0.7540	0.5641
	Naive Bayes	0.8856	0.9324	0.8149	0.7179
	Random Forests	0.8752	0.9270	0.7656	0.6154
	Multilayer Perceptron	**0.9047**	**0.9457**	0.7556	0.5513

---

[8] http://bit.ly/3Us9AhY.

metrics, generally in line with prior predictive work [11,14,23], i.e., as course progresses with more student trace data available, the predictive performance increases. This indicated that learning activity-based features may be effective for predictive tasks. Particularly, at week 7 when the feedback was provided, XGBClassifier achieved a True Negative Rate of 0.69, indicating that 69% of at-risk students could be successfully identified in advance. Subsequently, feedback generated at the prescriptive step successfully reached 69% of at-risk students. Based on this result, we selected XGBClassifier as our predictive model to generate prescriptive feedback.

Feature Importance Analysis. After applying DiCE to generate counterfactual explanations for the at-risk predictions made by XGBClassifier, we were interested to investigate what features were considered important by DiCE (i.e., the frequency of a feature which DiCE adopted to generate counterfactual explanations), and whether these "prescriptive" features (determined by DiCE) overlapped with the predictive features. So we identified the top-10 prescriptive features in all DiCE explanations, and the top-10 predictive features in SelectKBest results, and plotted their frequency of adoption by DiCE in Fig. 1. We observed that, there were three features that were important to both predictive and prescriptive modelling (i.e., the three green bars in the middle), indicating that certain learning activity features could be beneficial to both predictive (selectKBest) and prescriptive (DiCE) performance. Still, the majority of the top-10 predictive features (seven dark grey bars at the right) were rarely utilised by DiCE (3% or less), indicating that predictive features tended to be of limited relevance to prescriptive modelling.

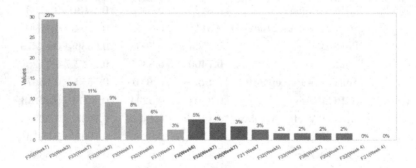

Fig. 1. The top-10 features most frequently changed by DiCE, and top-10 features in SelectKBest, with their corresponding percentage over all changes. We use the format 'feature index (Week Number)' to indicate the feature and the corresponding week. Grey at left: top-10 DiCE-changed features. Dark grey at right: Top 10 predictive features identified by SelectKBest. Green at middle: overlap of three features between DiCE Top 10 and SelectKBest Top 10. (Color figure online)

Human vs. Machine-Generated Feedback. The rubric results are summarised in Table 4. Overall, machine-generated feedback was highly rated by teaching staff, particularly in the categories of specificity, readability, and relational. This indicated that the teachers generally thought machine-generated feedback could be easily comprehended by students and inform specific learning actions while building encouraging relationship with students. Although the readily applicable criteria was the lowest among all four criteria, most teachers indicated that machine-generated feedback (around 73%) can be readily applicable (score = 2). When comparing with human-written feedback, we observed that all machine-generated feedback achieved a higher score with the relational and specificity criteria having the largest improvement. The improvement in machine-generated feedback quality compared to human-written feedback was statistically significant in all four criteria. This indicates that learning activity-powered PLA may be able to generate high-quality feedback in authentic course settings. Besides, apart from readily applicable criteria, machine-generated feedback displayed a lower standard deviation than human-written feedback in the rest of three criteria, indicating that students may be able to receive high-quality feedback more consistently from machine than human.

Table 4. Result of avg. rubric score of human-written feedback, avg. rubric score of machine-generated feedback, and statistical significance test analysis result using MWU:Mann-Whitney U test.

Rubric Criteria	Human-written (Std.)	Machine-generated (Std.)	MWU Stats.Value	MWU P-Value
Readily applicable	1.39 (0.49)	1.71 (0.50)	2679.5	<0.0001*
Readability	1.87 (0.34)	1.98 (0.15)	3515.5	0.0055*
Relational	1.31 (0.65)	2.00 (0)	1646.5	<0.0001*
Specificity	1.04 (0.45)	1.90 (0.40)	852.5	<0.0001*

5   Discussion

In this study, we investigated the use of PLA to generate personalised feedback in authentic course settings. We showed that by adopting learning activity-based features, PLA may achieve both high predictive accuracy and high-quality feedback. The quality of machine-feedback is consistently higher than human-feedback, with all four rubric displaying statistically significant improvement.

Implication. First, given the finding that PLA achieved high predictive and prescriptive performance together, we posit that it is indeed feasible to generate high-quality feedback by using learning activity-based features, without sacrificing predictive performance. Therefore, future studies in PLA should focus on engineer features that are useful for both predictive and prescriptive modelling. Second, our PLA approach can automatically provide high-quality feedback to

69% of students at risk of failure in an authentic course setting (based on predictive model results Table 3 week 7). Given the high student-teacher ratio in large-scale courses, PLA could be readily adopted in practice to reduce teachers' workload and assist in feedback practices. Third, despite that both PLA models and teachers received the same student learning activities, PLA-generated feedback outperformed teacher-written feedback in all four quality criteria (with statistical significance). This suggested the potential effectiveness of PLA in providing higher quality feedback compared to the teacher, and may be adopted in practice to support student learning. Given that our learning activity-based PLA approach is not specific to any course setting (i.e., courses other than introductory programming may also have learning activity traces), our approach could potentially be adopted at a larger scale and support personalised feedback provision across different disciplines.

Limitation and Future Work. First, although PLA generated high-quality feedback and is mostly rated to be readily applicable by experienced teachers, we have not conducted a study to validate the effectiveness of such feedback for students' learning. Given the promising finding, as a next step, we plan to deploy the PLA system to the course learning platform and allow students to receive PLA-generated feedback to support their learning. We plan to conduct studies to gain more in-depth understanding about how PLA-generated feedback support various student learning processes, and the effectiveness of automated feedback to support students' short-term and long-term learning skills development. Second, we acknowledge that there may be other features (other than learning activity-based) that could further boost the predictive and prescriptive performance of PLA. For future research, we plan to further engage with teaching staff to explore new predictive and prescriptive features to further improve the PLA performance.

References

1. Ajjawi, R., Boud, D., Henderson, M., Molloy, E.: Improving feedback research in naturalistic settings. In: Henderson, M., Ajjawi, R., Boud, D., Molloy, E. (eds.) The Impact of Feedback in Higher Education, pp. 245–265. Springer, Cham (2019). https://doi.org/10.1007/978-3-030-25112-3_14
2. Bergstra, J., Yamins, D., Cox, D.: Making a science of model search: hyperparameter optimization in hundreds of dimensions for vision architectures. In: International Conference on Machine Learning, pp. 115–123. PMLR (2013)
3. Breiman, L.: Random forests. Mach. Learn. **45**, 5–32 (2001)
4. Capuano, N., Rossi, D., Ströele, V., Caballé, S.: Explainable prediction of student performance in online courses. In: Guralnick, D., Auer, M.E., Poce, A. (eds.) TLIC 2023. LNCS, vol. 767, pp. 639–652. Springer, Heidelberg (2023). https://doi.org/10.1007/978-3-031-41637-8_52
5. Chen, T., Guestrin, C.: Xgboost: a scalable tree boosting system. In: Proceedings of the 22nd ACM SIGKDD International Conference on Knowledge Discovery and Data Mining, pp. 785–794 (2016)

6. Cristianini, N., Shawe-Taylor, J.: An Introduction to Support Vector Machines and Other Kernel-Based Learning Methods. Cambridge University Press, Cambridge (2000)
7. Dawson, P., et al.: What makes for effective feedback: staff and student perspectives. Assess. Eval. High. Educ. **44**(1), 25–36 (2019)
8. Dawson, P., et al.: Technology and feedback design. In: Spector, M., Lockee, B., Childress, M. (eds.) Learning, Design, and Technology, pp. 1–45. Springer, Cham (2018). https://doi.org/10.1007/978-3-319-17727-4_124-1
9. Deeva, G., Bogdanova, D., Serral, E., Snoeck, M., De Weerdt, J.: A review of automated feedback systems for learners: classification framework, challenges and opportunities. Comput. Educ. **162**, 104094 (2021)
10. Friedman, N., Geiger, D., Goldszmidt, M.: Bayesian network classifiers. Mach. Learn. **29**, 131–163 (1997)
11. Jang, Y., Choi, S., Jung, H., Kim, H.: Practical early prediction of students' performance using machine learning and explainable AI. Educ. Inf. Technol. **27**(9), 12855–12889 (2022)
12. Jia, Q., Cui, J., Xiao, Y., Liu, C., Rashid, P., Gehringer, E.F.: All-in-one: multi-task learning bert models for evaluating peer assessments. arXiv preprint arXiv:2110.03895 (2021)
13. Jia, Q., et al.: Insta-reviewer: a data-driven approach for generating instant feedback on students' project reports. International Educational Data Mining Society (2022)
14. Karalar, H., Kapucu, C., Gürüler, H.: Predicting students at risk of academic failure using ensemble model during pandemic in a distance learning system. Int. J. Educ. Technol. High. Educ. **18**(1), 63 (2021)
15. Kasneci, E., et al.: Chatgpt for good? on opportunities and challenges of large language models for education. Learn. Individ. Differ. **103**, 102274 (2023)
16. van der Lee, C., Gatt, A., van Miltenburg, E., Krahmer, E.: Human evaluation of automatically generated text: current trends and best practice guidelines. Comput. Speech Lang. **67**, 101151 (2021)
17. Maier, U., Klotz, C.: Personalized feedback in digital learning environments: classification framework and literature review. Comput. Educ. Artif. Intell. **3**, 100080 (2022)
18. Mothilal, R.K., Sharma, A., Tan, C.: Explaining machine learning classifiers through diverse counterfactual explanations. In: Proceedings of the 2020 Conference on Fairness, Accountability, and Transparency, pp. 607–617 (2020)
19. Pardo, A., et al.: Ontask: delivering data-informed, personalized learning support actions (2018)
20. Patel, C.R., Pandya, S.K., Sojitra, B.M.: Perspectives of chatgpt in pharmacology education, and research in health care: a narrative review. J. Pharmacol. Pharmacotherapeut. 0976500X231210427 (2023)
21. Pedregosa, F., et al.: Scikit-learn: machine learning in Python. J. Mach. Learn. Res. **12**, 2825–2830 (2011)
22. Pinger, P., Rakoczy, K., Besser, M., Klieme, E.: Implementation of formative assessment-effects of quality of programme delivery on students' mathematics achievement and interest. Assess. Educ. Principles Policy Pract. **25**(2), 160–182 (2018)
23. Ramaswami, G., Susnjak, T., Mathrani, A.: On developing generic models for predicting student outcomes in educational data mining. Big Data Cogn. Comput. **6**(1), 6 (2022)

24. Ramaswami, G., Susnjak, T., Mathrani, A.: Supporting students' academic performance using explainable machine learning with automated prescriptive analytics. Big Data Cogn. Comput. **6**(4), 105 (2022)
25. Ramaswami, G., Susnjak, T., Mathrani, A.: Effectiveness of a learning analytics dashboard for increasing student engagement levels. J. Learn. Anal. **10**(3), 115–134 (2023)
26. Ryan, T., Henderson, M., Ryan, K., Kennedy, G.: Designing learner-centred text-based feedback: a rapid review and qualitative synthesis. Assess. Eval. High. Educ. **46**(6), 894–912 (2021)
27. Ryan, T., Henderson, M., Ryan, K., Kennedy, G.: Identifying the components of effective learner-centred feedback information. Teach. High. Educ. **28**(7), 1565–1582 (2023)
28. Sarsa, S., Denny, P., Hellas, A., Leinonen, J.: Automatic generation of programming exercises and code explanations using large language models. In: Proceedings of the 2022 ACM Conference on International Computing Education Research, vol. 1, pp. 27–43 (2022)
29. Shapley, L.S., et al.: A value for n-person games (1953)
30. Smith, B.I., Chimedza, C., Bührmann, J.H.: Individualized help for at-risk students using model-agnostic and counterfactual explanations. In: Education and Information Technologies, pp. 1–20 (2022)
31. Stasaski, K., Ramanarayanan, V.: Automatic feedback generation for dialog-based language tutors using transformer models and active learning. In: 34th Conference on Neural Information Processing Systems, Vancouver (2020)
32. Susnjak, T.: Beyond predictive learning analytics modelling and onto explainable artificial intelligence with prescriptive analytics and chatgpt. Int. J. Artif. Intell. Educ. 1–31 (2023)
33. Troussas, C., Papakostas, C., Krouska, A., Mylonas, P., Sgouropoulou, C.: Personalized feedback enhanced by natural language processing in intelligent tutoring systems. In: Frasson, C., Mylonas, P., Troussas, C. (eds.) ITS 2023. LNCS, pp. 667–677. Springer, Heidelberg (2023). https://doi.org/10.1007/978-3-031-32883-1_58
34. Van Petegem, C., et al.: Pass/fail prediction in programming courses. J. Educ. Comput. Res. **61**(1), 68–95 (2023)
35. Varank, İ, et al.: Effectiveness of an online automated evaluation and feedback system in an introductory computer literacy course. Eurasia J. Math. Sci. Technol. Educ. **10**(5), 395–404 (2014)
36. Widyahastuti, F., Tjhin, V.U.: Predicting students performance in final examination using linear regression and multilayer perceptron. In: 2017 10th International Conference on Human System Interactions (HSI), pp. 188–192. IEEE (2017)
37. Winstone, N., Boud, D., Dawson, P., Heron, M.: From feedback-as-information to feedback-as-process: a linguistic analysis of the feedback literature. Assess. Eval. High. Educ. **47**(2), 213–230 (2022)

Identifying and Mitigating Algorithmic Bias in Student Emotional Analysis

T. S. Ashwin$^{(\boxtimes)}$ and Gautam Biswas$^{(\boxtimes)}$

Vanderbilt University, Nashville, TN, USA
ashwindixit9@gmail.com, gautam.biswas@vandderbilt.edu

Abstract. Algorithmic bias in educational environments has garnered increasing scrutiny, with numerous studies highlighting its significant impacts. This research contributes to the field by investigating algorithmic biases, i.e., selection, label, and data biases in the assessment of students' affective states through video analysis in two educational settings: (1) an open-ended science learning environment and (2) an embodied learning context, involving 41 and 12 students, respectively. Utilizing the advanced High-speed emotion recognition library (HSEmotion) and Multi-task Cascaded Convolutional Networks (MTCNN), and contrasting these with the commercially available iMotions platform, our study delves into biases in these systems. We incorporate real student data to better represent classroom demographics. Our findings not only corroborate the existence of algorithmic bias in detecting student emotions but also highlight successful bias mitigation strategies. The research advances the development of equitable educational technologies and supports the emotional well-being of students by demonstrating that targeted interventions can effectively diminish biases.

Keywords: Algorithmic Bias · Classification Bias · Open-ended Learning Environment · Embodied Learning · Bias Mitigation Techniques · Emotion Recognition Bias · Affective Computing · Facial Expressions

1 Introduction

In the digital era, where algorithms govern a spectrum of decisions from the mundane to the pivotal their neutrality and objectivity, especially in the machine learning and artificial intelligence domains, are being critically reevaluated [14]. Algorithmic bias, or the systematic skew that leads to unfair outcomes by favoring certain groups over others, has emerged as a significant concern. This concern is particularly pronounced in education applications, where such biases can amplify existing inequalities and compromise the integrity of educational fairness [3].

Recognizing the multifaceted nature of algorithmic bias is crucial. It manifests through various stages of the machine learning life cycle, encompassing historical, representational, measurement, aggregation, learning, evaluation, and

A. M. Olney et al. (Eds.): AIED 2024, LNAI 14830, pp. 89–103, 2024.
https://doi.org/10.1007/978-3-031-64299-9_7

deployment biases [18]. Each category, with its unique attributes, contributes to the complexities of ensuring equitable AI systems. Historical biases reflect entrenched societal inequities, while representation biases arise from data that fails to encapsulate the diversity of the target population. Measurement biases skew reality through oversimplified proxies, and aggregation biases ignore the nuances of diverse data subsets. Learning biases occur due to model choices that neglect fairness, evaluation biases emerge from unrepresentative benchmark datasets, and deployment biases arise when models are misapplied in practice. These biases present a formidable challenge in education, necessitating a nuanced understanding and a proactive stance to mitigate their multifarious impacts and steer towards a more equitable educational landscape [14].

Several research studies on algorithmic bias exist within the education domain. They have tried to understand the bias in applications such as dropout prediction and automated essay scoring, with a notable focus on racial, ethnic, and gender disparities, primarily within the United States [1]. While these studies have illuminated the differential effectiveness of algorithms across diverse student demographics, the exploration into vision data and affective state recognition, like student emotions and engagement, is relatively nascent. Noteworthy is the research indicating performance disparities in affect detection models that use the log data between rural and urban students, highlighting the need for contextually tailored approaches [15]. In the emotion recognition domain, efforts to tackle algorithmic bias are increasing, yet a comprehensive analysis of bias and the development of mitigation strategies are still in their infancy. Prevailing studies predominantly rely on preprocessed databases, characterized by predominantly frontal facial images captured in controlled environments, rather than the natural, varied settings of real life scenarios [11]. This approach tends to emphasize overtly expressive faces, usually associated with distinct actions or emotions, and often fails to capture the subtleties and complexities of spontaneous emotional expressions. Furthermore, these studies frequently utilize cloud-based algorithms or advanced techniques such as Generative Adversarial Networks (GANs) for data augmentation or bias mitigation, concentrating on specific demographic disparities or attribute biases [9]. Nevertheless, a significant gap remains: the educational domain, rich with context-specific nuances of facial expressions influenced by the learning environment, cultural background, and individual student experiences, is largely underrepresented in these datasets and methodologies. Furthermore, existing research often restricts its focus to primary emotions – happiness, sadness, anger, surprise, fear, and disgust-while neglecting academic emotions such as confusion, frustration, and boredom. This limitation represents a significant research gap in the study of bias within vision data-driven student emotion recognition, thereby hindering a comprehensive understanding of the interplay between cognition, emotion, and learning.

Addressing this gap, our study collects data from two distinct learning environments: an open-ended learning environment named Betty's Brain [4], and an embodied learning environment named GEM-STEP (Generalized Embodied Modeling - Science through Technology Enhanced Play) [6]. We broaden the scope of bias analysis to encompass both basic and educational emotions, with a primary focus on *confusion*, thereby enriching our analysis of students' affective states. This paper emphasizes confusion as a pivotal educational emotion characterized by a state of disequilibrium, arising from cognitive conflict when learners encounter challenges in assimilating new information with prior knowledge [7]. Confusion, indicative of student engagement, presents an opportune moment for educators to provide targeted support [17]. Recognizing its critical role in learning, our investigation concentrates on understanding algorithmic biases in confusion in this paper.

To examine bias, a validated emotion recognition method is essential. Affect-Net, a vast dataset of facial expressions, and the AFFDEX algorithm, integrated into the iMotions platform, are both trained on this extensive database, mitigating cultural and human labeling biases. Consequently, we selected iMotions for benchmarking because it is trained on multiple comprehensive datasets and offers detailed emotion and action unit analysis. In contrast, other methods that use the same dataset often lack such granularity, omitting action units or education-specific emotions. Furthermore, our study employs two novel techniques to address bias, contributing to the advancement of more equitable and precise emotion recognition within educational contexts.

The Key contributions of the paper are:

1. *Algorithmic Bias Analysis:* The study analyzes selection, label, and data biases in emotion recognition within educational settings.
2. *Emotions and Integration of Classroom Data:* Real student data from two different learning environments enrich the generalizability of our analyses. Additionally, it considers not only basic emotions but also learning-centered emotions, primarily confusion.
3. *Bias Mitigation Techniques:* The research introduces and applies state-of-the-art methods for identifying and mitigating biases, thereby enhancing the accuracy and fairness of emotion recognition in education.

2 Literature Review

Recent studies in higher education have leveraged log data to predict student dropout and course failure, revealing biases related to gender, ethnicity, race, and socioeconomic factors [5,12,21]. Some studies, such as [13] and [19], have focused on predicting academic achievements and evaluating speech, highlighting language-based and cross-cultural biases. A unique study by [15] explored geographical biases in detecting student affect using log data.

While existing research on algorithmic bias in education emphasizes factors like gender, ethnicity, and academic performance using log, text, or speech data, a notable gap exists regarding biases in vision data. Addressing this gap is essential for a more comprehensive understanding of algorithmic biases in educational contexts. Numerous studies have explored emotions, though not exclusively within the education domain. Many of these investigations have relied on publicly available data to address biases. In [11] study, the focus was on Cross-Database Emotion Classification bias. [20] extended the scope to include age, gender, and racial bias, and considered Cross-Database Emotion Classification. On the other hand, [9] emphasized racial bias as a primary factor.

Several studies focus on basic emotions, often excluding learning-centered emotions. [20] expanded their dataset to include expressions like smiling. These studies primarily used image-based data and computer vision. [11] analyzed four child-focused datasets. [20] used two adult datasets: RAF-DB, with diverse subjects and conditions, and CelebA, featuring celebrities with annotated facial attributes. [9] employed the CAFE dataset, with diverse children aged 2–8, and AffectNet for adult facial emotion recognition. However, none of these datasets were from the education domain. Most datasets in the education domain include undergraduate or graduate populations [2,10]. This highlights a gap in emotion recognition within the education domain, where identifying and addressing bias remains under-explored. Current methods to address algorithmic emotion recognition from facial expressions are limited, with few focusing on mitigating overgeneralization and emphasizing important features [20].

This study aims to bridge these gaps by analyzing data from open-ended and embodied learning environments to understand basic and learning-centered emotions, and to investigate bias in computer vision data within educational contexts.

3   Learning Environment and Dataset

The data for this study is taken from two distinct learning environments. The first learning environment, Betty's Brain, is an Online Educational Learning Environment (OELE) designed for middle school students [4]. It utilizes a learning-by-teaching approach, and students actively engage in building causal models of scientific processes to teach a virtual character, the Teachable Agent (TA). This interactive platform provides students with resources such as a science book with hypermedia pages, enabling them to acquire knowledge by identifying relevant concepts and establishing causal relations between these concepts to model a scientific process (e.g., climate change, thermoregulation). Students also have opportunities to quiz the TA to check their learning progress, and the results, typically motivate the students to learn more and help improve their TAs performance. A Mentor agent (MA) observes the students' interactions with the TA and intervenes when it believes the student is having difficulties. In addition, students can also query the MA.

OELE Data: The data collection included 41 students, 12 males and 29 females, in the age range of 10–12. Students worked in the Betty's Brain environment for three days, approximately 40 min a day. This produced a total of around 5000 min of screen-recording videos. The video data, captured through iMotions using the laptop's webcam, maintained a resolution of 1092*614 at a frame rate of 30 frames per second. Emotion recognition was systematically performed on each visible face within the image frames to comprehensively analyze emotional states.

In GEM-STEP, the second learning environment, students learn by working in small groups to enact a scientific process (e.g., movement of molecules in solids, liquids, and gas; or enacting the photosynthesis process). Students physically move around the classroom space, engaging in play-acting scenarios related to scientific concepts and processes, and witness their activities being projected into a computer simulation on a screen in front of the room. The simulated environment includes avatars and entities that represent key information, and direct students' attention to critical aspects of the science content. Overall, this represents a mixed-reality environment that captures students' physical activities and maps them onto the simulation of the scientific process.

Participant details: Data was collected over two days, with two groups each day. On Day 1, Group 1 had 3 girls and 2 boys, while Group 2 had 3 girls and 2 boys. On Day 2, Group 1 included 2 girls and 4 boys, and Group 2 had 2 girls and 3 boys. Each session lasted about 25 min. Videos, captured by four high-resolution cameras at the study area's corners, had a resolution of 1920*1080, processed at 30 frames per second (fps). All four cameras were utilized for emotion recognition, employing face detection and re-identification to track students' emotions as they played and moved in various directions. For both studies, the Institutional Review Board's approval was obtained, and all necessary participant consent procedures and formalities were diligently followed. Sample image screenshot of students from both the learning environment is shown in Fig. 1. The student population distribution was as follows: 60% White, 25% Black, 9% Asian, and 5% Hispanic.

4 Methodology

In our methodology, inspired by Baker and Hawn's framework [3], we navigate the complex landscape of algorithmic bias in education through three key subsections. We begin by focusing on emotion recognition, a critical dimension of bias. Following Baker's progression, we identify unknown bias by scrutinizing misclassifications, moving from a lack of awareness to a comprehensive understanding of bias in specific contexts and among particular individuals.

The second phase involves meticulously examining misclassifications, aligning with Baker's concept of transitioning from *unknown bias to known bias*. This step allows us to pinpoint instances of bias, contributing valuable insights to the discourse on algorithmic fairness in education.

The third and final section addresses bias, where we actively work towards fairness in the algorithm. Drawing on the evolving understanding within the machine learning community, we aim to rectify identified biases, contributing to the broader goal of creating a more equitable learning environment.

In essence, our streamlined methodology systematically progresses from investigating emotion recognition biases and uncovering misclassifications to actively addressing and rectifying bias in pursuit of algorithmic fairness in education, aiming to foster equal opportunities (*fairness to equity*) for all learners.

(a) OELE Environment	(b) Embodied Learning Environment

Fig. 1. Sample image snapshot from learning environments

4.1 Emotion Recognition

Emotion recognition in the embodied learning environment differs from that in the OELE environment. In the case of OELE, we utilized data collected from iMotions, which included image frames captured at 30fps. Additionally, for each student, there was a corresponding CSV file containing information on basic emotions, confusion, and 25 selected action units. The CSV file facilitated a detailed analysis of emotional states. To ensure synchronization, the videos and CSV files were time-aligned using iMotions. As part of our methodology, we also applied HSEmotion to identify and recognize the same set of emotions. This dual approach allowed us to assess and identify potential misclassifications, HSEmotion (High-Speed Face Emotion Recognition) is used to predict the valence and arousal values. The architecture utilized the pre-trained model facial emotion recognition [16].

Confusion Annotation: As HSEmotion lacks native recognition for confusion emotion, we performed annotation and model retraining specifically for this emotion category. Using the Computer Vision Annotation Tool (CVAT), we meticulously annotated images independently by two different annotators. The inter-rater reliability, measured by Cohen's Kappa, was 0.71. For model training, we considered 500 instances of confusion emotion where both annotators completely agreed ($\kappa = 1$). To enrich the dataset, we applied data augmentation, increasing the sample size by 10 times. The resulting training accuracy achieved on HSEmotion was 94.3%.

Emotion Recognition in Embodied Learning Environment. Emotion recognition in an embodied learning environment poses unique challenges due to dynamic student movements and frequent occlusions. Our methodology addresses these challenges in three steps: face recognition using MTCNN, emotion recognition, and student re-identification. We employed MTCNN with fine-tuned thresholds (0.8 for P-Net, R-Net, and O-Net) to improve face detection accuracy across diverse skin tones, expressions, and lighting conditions. This setup successfully detected faces at a rate of 30 frames per second. Following face detection, emotional states were identified using the HSEmotion method (mentioned in the previous subsection). The detected emotions and face coordinates were recorded in a CSV file for each frame.

Student Re-Identification: While MTCNN and HSEmotion are adept at facial detection and emotion recognition, respectively, they do not inherently perform re-identification-a crucial requirement in our study due to multiple students within the video frames. A robust re-identification strategy was imperative to track each student's emotions over time. Our approach involved the development of a sophisticated tracking algorithm that processed the CSV data obtained post-emotion recognition. This data comprised frame numbers, basic emotion, and face bounding box coordinates (x, y, width, height), providing comprehensive information for each detected face.

Acknowledging the principle of spatial continuity, our algorithm hypothesized that an individual's position changes incrementally between consecutive frames, considering the rapid frame rate of 30 frames per second (equivalent to 33.33 milliseconds per frame). Under this assumption, significant movement of a student within a single frame was deemed unlikely. The algorithm commenced by loading the data into a DataFrame and computing the center of each bounding box. This step was critical to establishing a reliable reference point for tracking individual movements across frames. To facilitate the tracking process, the algorithm introduced new columns in the DataFrame designated for unique Student IDs and the historical center positions ('Old_Center_X' and 'Old_Center_Y') of each identified face. Furthermore, it prepared columns for the predicted future positions ('Predicted_X' and 'Predicted_Y'), capitalizing on the inferred motion trajectory of each individual.

The core of the re-identification process relied on calculating the Euclidean distances between the centers of detected faces in consecutive frames. A distance threshold was established to determine whether a detected face in the current frame corresponded to any face from the previous frames. To account for potential changes in the pace or direction of the individuals, the algorithm incorporated a memory mechanism, considering a few past frames (denoted by memory_frames), and predicted future positions based on the velocities calculated in the X and Y coordinates. For each frame processed, if the center of a detected face was within the distance threshold of the predicted position of a previously identified individual, it was deemed the same individual. The corresponding Student ID was then assigned, and the historical center positions for that Student ID were updated. In cases where no match was found within the distance threshold, a new Student ID was assigned, indicating the detection of a new individual in the scene.

Upon the completion of the tracking process across all frames, the algorithm consolidated the valence and arousal scores for each Student ID, thereby crafting an emotional profile for each student throughout the video. The final output comprised Student IDs, bounding box coordinates, valence, arousal data, and frame numbers. This method facilitated the accurate re-identification of students with a success rate of 91%. For the remaining 9%, manual corrections were made, ensuring the integrity and continuity of the student tracking.

4.2   Identifying the Misclassification

In both studies, we employed the iMotions platform with the AFFDEX emotion recognition engine and HSEmotion, renowned for its efficacy on the AffectNet dataset. Data synchronized at 30 frames per second was processed using both models. Upon encountering misclassifications, selected frames were extracted for comparative analysis. This analysis entailed a manual review of each misclassified instance to ascertain whether it was a genuine error in emotion recognition or a result of extraneous factors such as occlusion. For the embodied learning dataset, we utilized Multi-task Cascaded Convolutional Networks (MTCNN) to crop facial images, which were then sequentially fed into iMotions for emotion classification.

4.3   Addressing the Bias

In this study, we investigate the implications of bias in facial expression recognition through a comparative analysis using three distinct methodologies, as detailed in [3]. Notably, the base architecture has been altered from $ResNet$ to $EfficientNet - B0$, and the labels are mapped to sensitivity, associating emotions considered in this study. The detailed implementation of these modifications is explained below. Initially, we establish a baseline using the $EfficientNet - B0$ model, a modification of the commonly used $ResNet$ architecture, particularly adapted for its efficiency and accuracy in various recognition tasks. This model is trained with a Cross-Entropy loss function to predict the facial expression label y_i for each input image (x_i), where the loss is calculated as $(L_{exp}(x_i) = -sum_{k=1}^{K} 1_{[y_i=k]} logp_k)$. Here, (p_k) denotes the predicted probability of the input (x_i) belonging to class (k), and $(1[cdot])$ represents the indicator function. This baseline sets the stage for evaluating the performance enhancements and bias mitigation effects introduced by the subsequent methodologies.

To address bias, we introduce two sophisticated approaches: the Attribute-aware Approach and the Disentangled Approach. The Attribute-aware Approach, inspired by concepts from previous work, incorporates sensitive attribute information directly into the classification process [20]. In this method, an attribute vector (s_i) is transformed through a fully connected layer to match the feature vector $(\phi(x_i))$ obtained from the EfficientNet-B0 backbone, and the resultant combined feature is fed into the classification layer. This strategy is designed to assess how the inclusion of explicit attribute information influences the recognition of facial expressions and its potential to reduce bias. On the other hand, the Disentangled Approach aims to extract a feature representation $(\phi(x_i))$ that is devoid of any sensitive attribute information (s_i). This is achieved by splitting the network into branches: a primary branch for expression recognition and parallel branches that employ a confusion loss (L_{conf}) and an attribute predictive Cross-Entropy loss (L_s) to ensure the sensitive attributes cannot be predicted from $(\phi(x_i))$. These branches share the network layers until the final fully connected layer, where specific task-oriented branches are formed. The overall loss is a combination of the expression recognition loss, the attribute predictive loss, and the confusion loss, weighted by a factor (α), ensuring that the final feature representation effectively encapsulates facial expression information while actively discarding sensitive attribute data. This balanced approach not only aids in mitigating bias but also enhances the robustness and fairness of the facial expression recognition task.

5 Results and Discussion

We utilized iMotions' emotion recognition engine (AFFDEX) and a retrained HSEmotion for emotion classification across two learning environments, identifying roughly 9.18 million class labels. Of these, 83.21% were consistent across both systems, while the remaining 16.79% differed, representing over 1,50,000 face image frames with misclassifications. Due to the vast number of misclassifications, manual verification was impractical. Consequently, we used iMotions results as a benchmark, investigating the misclassifications by HSEmotion and vice versa. This approach enabled us to quantify the total misclassifications for each emotion in our study. The distribution of emotions from both learning environments is as follows: Happy (23%), Neutral (20%), Sad (12%), Surprise (8%), Anger (7%), Disgust (7%), Fear (5%), Confusion (18%).

Given that the image frames include faces and the Facial Action Coding System (FACS) [8] provides a well-defined framework, our initial focus was on comprehending facial expressions using action units and assessing their influence on emotion classification. iMotions provides action units data, along with basic emotions in the CSV file. For our retrained HSEmotion model, we utilized Grad-CAM to visualize the selected face regions crucial for classification. This approach aids in understanding how specific facial features contribute to emotion predictions.

Emotions that are observed dominantly during learning are learning-centered emotions such as confusion, boredom, frustration, delight, and engagement. Instances of specific basic emotions like peak fear or peak anger are less frequent. Despite this, some images are detected as anger during student frustration or as disgust and sadness during boredom. Further, the embodied learning data had more than 40% instances of basic emotions where the students were surprised, sad, angry, happy/joy, and so on. Hence, There are ample instances classified under basic emotions.

In the misclassified basic emotions, common observations include surprise being misclassified with fear and vice versa, sadness being misclassified with confusion, anger being misclassified as fear and sadness, and fear also being misclassified as confusion, as evident in the confusion matrix heatmap 2 (b).

Other instances of misclassification occur when the face is not visible, either almost out of the frame or blurred due to the rapid movements of students in the embodied learning environment. Uncertain images, such as the transition from closed to opened eyes or the shift from one emotion to another, also contribute to misclassification in in-between frames. These nuanced scenarios highlight the challenges faced in accurately classifying emotions in dynamic learning environments.

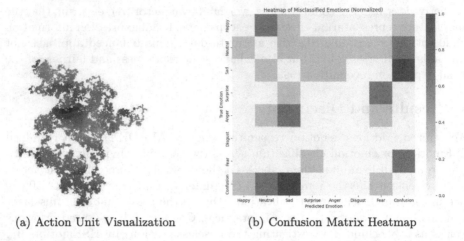

(a) Action Unit Visualization   (b) Confusion Matrix Heatmap

Fig. 2. Action unit and confusion matrices visualization

In each case, we analyzed the action unit values in the CSV file and noticed that not all action units were appropriately recognized for the intended emotion. For instance, Action Unit 4 was not identified, leading to a higher probability of sadness rather than confusion. Similarly, Action Units 20 and 7 were not recognized for fear, resulting in misclassification as surprise.

To comprehend the relationship between facial action units and their impact on emotion recognition, we referred to the iMotions documentation to understand the criteria for classification. Through examination of the AFFDEX 2.0 emotion recognition model within iMotions, we identified specific action units that play a significant role in determining certain emotions. To explore if there is any superset or subset relationship among these action units, the findings are presented in Fig. 3.

Similar to iMotions, HSEmotion also encountered misclassifications, primarily attributed to issues with properly recognizing action units. The use of Grad-CAM visualizations further highlighted these discrepancies. While misclassifications for confusion were relatively fewer due to prior training on educational data, the challenges persisted for other emotions, as depicted in Fig. 3.

In Fig. 2 (a), a slightly modified representation of Grad-CAM is presented, addressing concerns about full-face visibility to adhere to ethical considerations. It is evident from the visualization that none of the dominant action units are effectively considered (one eye slightly open and a part of the nose) for the classification process in this frontal face image frame.

Due to the misclassification at the feature level, we scrutinized the databases on which iMotions and HSEmotion systems were trained, specifically the Denver Intensity of Spontaneous Facial Action (DISFA), Affect in-the-wild (AFFwild 2), AffectNet, Audio/Visual Emotion Challenge (AVEC), Acted Facial Expressions in the Wild (AFEW), Video Level Group AFfect (VGAF), Video Game

Facial Animation (VGAF), and Engage Wild. The emergence of data, label, and selection biases is significant, largely attributable to the mismatch between the datasets' origin and their application within educational contexts. These biases manifest distinctly: Data bias arises from demographic discrepancies, as these databases, while comprehensive, do not cater specifically to the nuances of students' profiles, particularly those within the 10-14 age bracket or those with darker skin tones, resulting in a non-representative sample for the educational settings. Label bias is highlighted by the fact that expressions in these datasets, typically portrayed by adults, may not seamlessly align with the genuine emotional expressions of students in learning environments, underscoring the necessity for meticulous and context-aware annotation of action units. Selection bias is underscored by the fact that these datasets, not originally curated for educational purposes, stem from environments and contexts that starkly contrast with typical educational settings such as Betty's Brain and GEM-STEP, potentially compromising the systems' proficiency in accurately deciphering student emotions during educational activities.

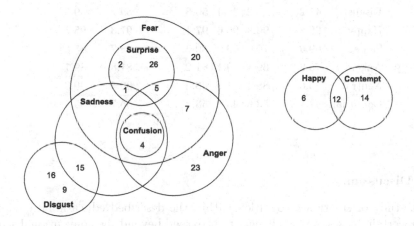

Fig. 3. The subset superset relation of facial action units with respect to emotions

Results on Addressing Bias. In this study, we explored methods to reduce bias in facial expression recognition using an EfficientNet-B0 model. The table shows the performance of different methods like Baseline, Attribute-aware (AW), Disentangled (DE), Data Augmentation Baseline (DA Baseline), Data Augmentation with Attribute-aware (DAWA), and Data Augmentation with Disentangled (DADE). The Baseline model sets a starting point with a 68.3% mAP. The AW approach improved recognition for emotions like Surprise, Fear, and Confusion, raising the mAP to 70.4%. The DE approach did well for Neutral but lowered the overall mAP to 64.6%, suggesting it might remove some useful features. Adding Data Augmentation (DA) raised the Baseline's mAP to 77.0%, showing the value of a varied dataset. DAWA, combining DA and AW, reached the highest mAP of 78.1%, balancing data diversity and attribute sensitivity.

DADE also did well, with a 78.0% mAP, showing that removing sensitive attributes while augmenting data can effectively reduce bias. Performance details reveal significant improvements in recognizing Sad and Anger with DAWA and DADE. However, DE had some issues, possibly due to the complexity of removing sensitive attributes while keeping important features. While we've addressed bias with these methods, results could be further improved by using data more representative of this specific distribution and balancing the dataset (Table 1).

Table 1. Accuracy of model after addressing bias

in %	Baseline	AW	DE	DA Baseline	DAWA	DADE
mAP	68.3	70.4	64.6	77.0	78.1	78.0
Surprise	83.4	87.7	84.7	91.1	90.8	90
Fear	44.6	52.1	44.6	59.8	61.3	59.2
Disgust	45.2	45.2	36.4	56.8	59.2	59.5
Happy	96.8	96.8	96.6	97.1	97.3	95.1
Sad	69.5	70.6	60.9	80.4	88.7	85.5
Anger	73.2	69.1	59.1	81.2	81.8	89.1
Neutral	86.6	88.4	91.9	96.4	90.4	90.3
Confusion	47.1	53.6	43.1	53.4	55.7	55.5

5.1   Discussion

In the study of emotion recognition within the described educational environments, certain biases, and challenges that extend beyond the conventional scope of data, label, and selection bias were observed. These observations highlight the complexity and nuanced nature of implementing emotion recognition technology in diverse educational settings.

Algorithmic Bias and Misclassification Concerns: The emotion recognition technology demonstrated a higher rate of misclassification for students with darker skin tones, suggesting potential algorithmic biases in the system. This could be due to the model's training on datasets not representative of this demographic, leading to its inadequate performance. Specifically, the recognition of Action Unit 5 (AU5), associated with the upper facial movements, was notably low in the embodied learning data. This indicates that the model might not be adequately trained or tuned to capture the subtleties of these facial expressions, especially in the dynamic context of an embodied learning environment.

Contextual Bias and Technical Limitations: In the embodied learning environment, the detection of AUs was particularly challenging for students. This issue was compounded by inaccuracies in face tracking within the HSEmotion system, hinting at technical limitations or contextual biases within the technology. These inaccuracies could stem from various factors, such as lighting conditions, facial features, or the model's inability to generalize across educational contexts and demographics.

Performance Inconsistency Across Demographics: The technology's performance inconsistency, particularly regarding facial expression and emotion recognition for different skin tones, underscores the importance of ensuring diversity and representativeness in the training data. This is crucial to avoid perpetuating or exacerbating pre-existing biases.

Intrusiveness and Ethical Considerations: The high misclassification rate of confusion in iMotions might be indicative of selection bias, as most of the datasets used were composed of non-student facial expressions. This not only points to a potential selection bias but also raises ethical concerns regarding the appropriateness and relevance of the data used for training models intended for educational settings.

Unexplored Biases and Limitations in the Study: It's acknowledged that the study has not delved deeply into dissecting biases related to race, age, and gender. Given the relatively small sample size of 53 students, it's challenging to draw definitive conclusions or generalize findings across these dimensions. This limitation underscores the need for larger, more diverse datasets to understand and mitigate the various biases effectively.

6   Conclusion

This study researched algorithmic biases, focusing on selection, label, and data biases in emotion recognition within educational settings. Leveraging the retrained HSEmotion, this research meticulously evaluated and contrasted inherent biases in these novel methods against the biases in the commercially available iMotions platform, across emotions, including happy, sad, disgust, fear, anger, surprise, confusion, and neutral. The study offered a realistic and nuanced analysis of algorithmic biases by integrating real student data that accurately reflects classroom diversity. The study's thorough methodologies effectively pinpointed and addressed fundamental biases, demonstrating substantial potential in diminishing these biases through specific strategies. Notably, the results validated the presence of algorithmic bias in student emotion recognition systems while also showcasing viable paths to mitigate such biases significantly. This research enriches the dialogue on creating fairer educational technologies and underscores the importance of fostering the emotional well-being of students. Future research should diversify and enrich emotion recognition datasets, enhance real-time bias monitoring and adjustment mechanisms, and closely examine the integration of these technologies within educational frameworks to ensure ethical usage and maximization of pedagogical benefits.

Acknowledgement. This research was supported by the National Science Foundation AI Institute Grant No. DRL-2112635. Any opinions, findings and conclusions or recommendations expressed in this material are those of the authors and do not necessarily reflect the views of the National Science Foundation.

References

1. Anderson, H., Boodhwani, A., Baker, R.S.: Assessing the fairness of graduation predictions. In: Proceedings of the 12th International Conference on Educational Data Mining, pp. 488–491 (2019)
2. Ashwin, T., Guddeti, R.M.R.: Affective database for e-learning and classroom environments using Indian students' faces, hand gestures and body postures. Futur. Gener. Comput. Syst. **108**, 334–348 (2020)
3. Baker, R.S., Hawn, A.: Algorithmic bias in education. Int. J. Artif. Intell. Educ. 1–41 (2021)
4. Biswas, G., Segedy, J.R., Bunchongchit, K.: From design to implementation to practice a learning by teaching system: betty's brain. Int. J. Artif. Intell. Educ. **26**, 350–364 (2016)
5. Christie, S.T., Jarratt, D.C., Olson, L.A., Taijala, T.T.: Machine-learned school dropout early warning at scale. In: International Educational Data Mining Society (2019)
6. Danish, J.A., Enyedy, N., Saleh, A., Humburg, M.: Learning in embodied activity framework: a sociocultural framework for embodied cognition. Int. J. Comput.-Support. Collab. Learn. **15**, 49–87 (2020)
7. D'Mello, S., Graesser, A.: Dynamics of affective states during complex learning. Learn. Instr. **22**(2), 145–157 (2012)
8. Ekman, P., Friesen, W.V.: Facial action coding system. Environ. Psychol. Nonverbal Behav. (1978)
9. Fan, A., Xiao, X., Washington, P.: Addressing racial bias in facial emotion recognition. arXiv preprint arXiv:2308.04674 (2023)
10. Gupta, A., D'Cunha, A., Awasthi, K., Balasubramanian, V.: Daisee: towards user engagement recognition in the wild. arXiv preprint arXiv:1609.01885 (2016)
11. Howard, A., Zhang, C., Horvitz, E.: Addressing bias in machine learning algorithms: a pilot study on emotion recognition for intelligent systems. In: 2017 IEEE Workshop on Advanced Robotics and its Social Impacts (ARSO), pp. 1–7. IEEE (2017)
12. Lee, H., Kizilcec, R.F.: Evaluation of fairness trade-offs in predicting student success. arXiv preprint arXiv:2007.00088 (2020)
13. Li, X., Song, D., Han, M., Zhang, Y., Kizilcec, R.F.: On the limits of algorithmic prediction across the globe. arXiv preprint arXiv:2103.15212 (2021)
14. Mehrabi, N., Morstatter, F., Saxena, N., Lerman, K., Galstyan, A.: A survey on bias and fairness in machine learning. ACM Comput. Surv. (CSUR) **54**(6), 1–35 (2021)
15. Ocumpaugh, J., Baker, R., Gowda, S., Heffernan, N., Heffernan, C.: Population validity for educational data mining models: a case study in affect detection. Br. J. Edu. Technol. **45**(3), 487–501 (2014)

16. Savchenko, A.: Facial expression recognition with adaptive frame rate based on multiple testing correction. In: International Conference on Machine Learning, pp. 30119–30129. PMLR (2023)
17. Sullins, J., Graesser, A.C.: The relationship between cognitive disequilibrium, emotions and individual differences on student question generation. Int. J. Learn. Technol. **9**(3), 221–247 (2014)
18. Suresh, H., Guttag, J.: A framework for understanding sources of harm throughout the machine learning life cycle. In: Equity and Access in Algorithms, Mechanisms, and Optimization, pp. 1–9 (2021)
19. Wang, Z., Zechner, K., Sun, Y.: Monitoring the performance of human and automated scores for spoken responses. Lang. Test. **35**(1), 101–120 (2018)
20. Xu, T., White, J., Kalkan, S., Gunes, H.: Investigating bias and fairness in facial expression recognition. In: Bartoli, A., Fusiello, A. (eds.) ECCV 2020. LNCS, vol. 12540, pp. 506–523. Springer, Cham (2020). https://doi.org/10.1007/978-3-030-65414-6_35
21. Yu, R., Li, Q., Fischer, C., Doroudi, S., Xu, D.: Towards accurate and fair prediction of college success: evaluating different sources of student data. In: International Educational Data Mining Society (2020)

Navigating Self-regulated Learning Dimensions: Exploring Interactions Across Modalities

Paola Mejia-Domenzain[1]([✉])(iD), Tanya Nazaretsky[1]([✉])(iD), Simon Schultze[2](iD), Jan Hochweber[2](iD), and Tanja Käser[1]([✉])(iD)

[1] EPFL, Lausanne, Switzerland
{paola.mejia,tanya.nazaretsky,tanja.kaeser}@epfl.ch
[2] St. Gallen University of Teacher Education, Lausanne, Switzerland

Abstract. Self-regulated learning (SRL) has been extensively studied using self-reported measures, such as surveys, and more recently, behavioral measures, such as trace data. While both modalities offer insights into SRL, their relationship remains ambiguous. Although previous research has compared these modalities, there has been limited work on integrating them and exploring the interplay of dimensions across modalities. To address this gap, we adopt a multimodal perspective and follow a threefold approach: horizontal, vertical, and integrated analyses. We identify behaviors per dimension from both data sources in the horizontal analysis. We then assess the alignment of dimensions across modalities in the vertical analysis. Finally, in the integrated analysis, we uncover the intricate interplay between dimensions across modalities using Canonical Correlation Analysis. For this purpose, we design and conduct a study with 79 participants interacting with an Intelligent Tutoring System. We find limited agreement in the vertical comparison between modalities. However, the integrated analysis reveals a moderate correlation, highlighting the complex relationship between behavioral actions and self-reported SRL perceptions.

Keywords: Self-Regulated Learning · Multimodal Analysis · Clustering · Canonical Correlation Analysis · Trace Data · Survey Data

1 Introduction

Self-regulated learning (SRL) is a multi-dimensional construct that encompasses learners' ability to plan, monitor, control, and regulate aspects of their learning process [5,18,19]. Gaining insight into these dimensions is not straightforward, especially when considering different sources such as contextual, behavioral, and psychological data [17]. Determining the relationship and interplay between these modalities has the potential to offer a holistic perspective on SRL, thus providing educators and researchers with nuanced information to design better instructional interventions.

A. M. Olney et al. (Eds.): AIED 2024, LNAI 14830, pp. 104–118, 2024.
https://doi.org/10.1007/978-3-031-64299-9_8

There is extensive research studying SRL using a wide range of data modalities [8,9,16,25,27]. The current landscape is comprehensively mapped in the SRL process, Multimodal data and Analysis (SMA) grid [17], organizing studies into four categories: unimodal (single dimension, single modality), horizontal (multiple dimensions, single modality), vertical (single dimension, multiple modalities), and integrated (multiple dimensions, multiple modalities). One of the identified gaps is the limited research into integrated approaches that concurrently investigate cognitive and metacognitive processes using both trace data and survey data.

Numerous studies have explored SRL using a horizontal approach that focuses primarily on singular modalities such as surveys and trace data [17]. Previous studies have often relied on self-reported questionnaires, such as the Motivation Strategies for Learning Questionnaire (MSLQ), to measure students' SRL strategies and get insight into students' perceptions and beliefs [3,4,19,27]. While surveys are context-independent measures, facilitating the comparison between different groups and studies, they can be influenced by self-report biases and provide only a static snapshot of behavior [4,8,19]. In contrast, trace data from learning platforms provides real-time granular insights into behavior. A large body of research has focused on understanding SRL using trace data by studying micro-level processes [22–24] or by identifying profiles [15,16,28]. Nevertheless, translating external actions recorded by trace data into insights about internal processes remains challenging. Trace data indicators might not completely capture the nuances of SRL behaviors, as they can be influenced by a range of factors or misinterpreted due to multiple possible hypotheses about the observed online actions [8,24].

Although some studies have explored vertical and integrated approaches to compare across modalities, the relationship between trace data and survey data remains unclear. Some studies have found that trace data could reflect self-reported SRL [10,12,27], while others report misalignment [1,8,21,29]. For example, [27] found positive correlations between trace data and self-reported goal setting, but very little correlation with task strategies and time management. This suggests that some dimensions of SRL might align better than others.

Most integrated research approaches have been focused on comparing self-reported perceptions with observed behaviors in trace data, often debating the relative merits of each modality [8,25]. [26] proposed conceptualizing SRL as both an aptitude and an event, suggesting that surveys might measure aptitude, while trace data could capture events. Therefore, an intuitive yet underexplored direction is the integration of data from both sources. Current methodologies predominantly rely on univariate analyses, like Pearson's correlations or direct one-to-one comparisons [10,12,21], which do not fully exploit the broader multivariate associations between multiple measurements across modalities. In other fields, like neuroscience [30], multivariate analysis has been widely used to provide a more comprehensive perspective, uncovering joint patterns between modalities and ensuring complete use of shared information.

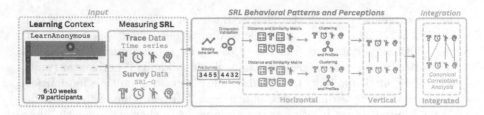

Fig. 1. Overview of our threefold multimodal approach using SRL indicators from survey and trace data. We 1) cluster each modality separately (horizontal), 2) compare their alignment (vertical), and 3) explore their interaction (integrated).

This paper bridges the aforementioned gaps by studying trace data and self-report measures from a multimodal perspective. Our multimodal approach is threefold: we conduct horizontal, vertical, and integrated analyses. Our analyses are based on a user study with 79 participants using an Intelligent Tutoring System (ITS) for 6–10 weeks. We collect self-reported SRL data through a standardized questionnaire [4] at the study's start and end, alongside recording interaction trace data on the online platform. In the horizontal analysis, we approach each modality as a separate method for collecting data on students. We employ the multi-step clustering pipeline proposed by [16] to separately cluster the survey data and trace data into multi-dimensional SRL profiles focusing on cognitive (*Elaboration*), metacognitive (*Metacognition*), and resource management (*Time Management* and *Effort*) strategies. In the vertical axis, we compare and integrate perception data from surveys and behavioral trace data, comparing each dimension separately across both modalities. For the integrated analysis, we employ Canonical Correlation Analysis (CCA) to investigate the interrelations between SRL dimensions captured through both modalities. Our main contributions lie in comparing and integrating the two data modalities to address two research questions: What is the degree of alignment between SRL behavioral dimensions within and across modalities (**RQ1**)? How do the dimensions across modalities interact and correlate with each other (**RQ2**)?

Our results present a novel perspective on the dynamics of SRL dimensions, both within and across modalities. While there is little agreement vertically across modalities, in our integrated approach, we find a moderate correlation between modalities, shedding light on the interaction between behavioral features and self-reported measures.

2   Methods

Our multimodal, threefold approach (see Fig. 1) is based on SRL indicators extracted from survey and trace data. We first cluster survey and trace data separately for each dimension and form SRL profiles (horizontal approach). We then compare the alignment between the dimensions (vertical approach) and finally explore the interaction between modalities (integrated approach).

2.1   Learning Context

We focus our analyses on data collected in a user study with `Learnnavi`, an ITS for high-school students. `Lernnavi` offers adaptive learning and testing sessions in mathematics and language learning. On the platform, users can navigate between topics and choose the subtopic they want to work on. Within each subtopic, users can choose between "learning sessions" where the content is presented adaptively or "test sessions" where their knowledge is tested. In both sessions, users have to solve tasks and at the end of the session, they can review their answers or go back to the main page.

Participants. We recruited teachers from schools using `Lernnavi` to teach mathematics. While the teachers decided on the study participation of their classroom, students could (anonymously) opt out of sharing their data. A total of 79 students completed the study. Participants were on average 16.15 years old ($\sigma = 0.73$), with a gender distribution of 72% female, 25% male, and 3% non-binary. The university's ethics committee approved the study (Nr.: 090-2020).

Procedure. Participants completed an SRL questionnaire [4] (see Sect. 2.3), both at the beginning and at the end of the study. Moreover, they took a pre- and post-test covering various mathematical topics. The tests were scored by awarding +1 for correct answers and −1 for incorrect ones, then the sum scores were normalized by the number of questions to range between −1 and 1. Lastly, they were directed to engage with `Learnnavi` throughout the study duration, which varied from 6 to 10 weeks based on teacher schedules.

2.2   Measuring Self-regulated Learning

To study students' SRL behaviors and perceptions, we employ an SRL framework tailored for online and blended learning, and consequently align features from trace data with items from survey data.

SRL Framework. The concept of SRL has been studied extensively, with multiple models emerging [18]. One popular model is the general cognitive model of learning and information processing [19], categorizing students' SRL strategies into three groups: cognitive (processing information), metacognitive (planning and monitoring comprehension), and resource management (time management and effort regulation). The MSQL [19], has been designed to measure students' SRL strategies based on this model and has been extensively used in prior research [3,4,19]. However, a recent meta-analysis [5] demonstrated that only a subset of the original scales was relevant for contemporary student behaviors in the realm of online education. Consequently, [4] created an updated SRL survey (SRL-O) tailored for online learning to more accurately reflect modern student behavior in online environments.

Survey. We hence adopt the SRL-O survey [4] for our study. We use the constructs relevant to the `Lernnavi` context and measurable by the available trace data: *Effort (Regulation)* measuring perseverance, hard work commitment to study goals; *Time Management* including students planning their weekly schedule for online study; *Elaboration* measuring to what extent learners do additional work beyond the core content; and *Metacognition* evaluating students' planning, monitoring and evaluating skills. We do not measure *Help-Seeking* as `Lernnavi` does not include any online help features (e.g., discussion forums) or other type of help on-demand and we drop *Study Environment* as this construct cannot be

Table 1. Trace data features per SRL dimensions. The features colored in gray have been excluded after the data-driven validation step (see Sect. 2.3).

Dimension	Feature Name	Description
Effort	Time Online	Duration spent studying online [6]
	Number of Events	Total count of learning events [6]
Time Management	Regularity - Week Day	Consistent activity on specific week days [2]
	Regularity - Day Hour	Consistent activity at a specific daily hour [2]
Elaboration	Test Session Ratio	Ratio of test sessions to total sessions
	Navigation Ratio	Ratio of navigation events to total events
	Topic Exploration Ratio	Ration of unique topics explored to sessions
	Time Off Ratio	Ration of time spent on other pages
Metacognition	Average Response Time	Average time taken to answer questions
	Answer Review Ratio	Ratio of answers reviewed per solved question

measured through trace data. We adapt the items to the context of our study primarily by replacing "online study" with "Lernnavi". All survey items[1] are administered on a 5-point Likert scale ranging from "strongly disagree" (0) over neutral (2) to "strongly agree" (5).

Trace Data. Our feature design meticulously aligns with the dimensions outlined in the survey (SRL-O), ensuring a careful match with the subscales, while being grounded in previous work using trace data to study SRL [8,15,16]. Consequently, we create features for the following dimensions: *Effort, Time Management, Elaboration*, and *Metacognition*. Table 1 lists all features along with the measured SRL dimension and a description.

For the *Time Management* and *Effort* dimensions, we use indicators rooted in prior work on modeling SRL strategies [15,16]. We measure *Effort* using the indicators proposed by [7], including the duration spent studying online and the cumulative count of events (e.g., answering questions, navigating the platform). To measure *Time Management*, we use the intra-week and intra-day regularity features introduced by [2], assessing whether students always work on the same

---

[1] https://github.com/epfl-ml4ed/srl-navigation/blob/main/supplementary-material/srl-o-items.pdf.

hour of the day or the same day of the week. Dimensions such as *Metacognition* and *Elaboration* present a distinct challenge. Unlike *Time Management* and *Effort* where the established features can be applied in various contexts, the nature of *Metacognition* and *Elaboration* is largely context-dependent. Thus, similar to [8], we follow the domain modeling step of the evidenced-centered design (ECD) framework to create indicators. ECD ensures valid study outcomes by aligning indicators with constructs. The domain modeling phase clarifies how measures gather evidence, support claims, and address counterclaims[2].

In the SRL-O survey, an item in the *Metacognition* category states, *"I spend time trying to interpret the questions to ensure that I understand accurately what I need to do"*. Based on this statement, we hypothesize that students with higher metacognitive skills might take longer to respond to questions (quantified by a higher average answer time), reflecting deeper deliberation and understanding [4]. We quantify this by calculating the average answer time. Additionally, another item in the survey reads: *"I think about how I might enhance my work by comparing it with the sample solution"*. Thus, we also look at students' reviewing behavior by measuring the ratio of reviewed answers to total questions seen.

Regarding the *Elaboration* dimension, in SRL-O an item is *"I try to improve my understanding by doing additional work beyond the core content (e.g., doing extra problem-solving activities or additional reading)"*. Hence, we hypothesize that students who frequently navigate through different subtopics or to other pages could be seeking supplemental tasks or reading material to elaborate on their knowledge. To quantify this behavior, we calculate the navigation ratio (number of navigation events divided by total number of events), the topic exploration ratio (diversity of topics explored), and the time off ratio (time spent on other pages). Furthermore, we hypothesize that students with high elaboration skills might be more inclined to undertake the challenging tasks in the test sessions. To assess this, we calculate the test session ratio, dividing the number of test sessions by the number of sessions, including both learning and testing sessions.

2.3   SRL Behavioral Patterns and Perceptions

To ensure continuity and comparability of our research, we rely on a previously proposed SRL clustering methodology. There exists a wealth of SRL clustering approaches in the literature [8,14–16,20,22–24], each offering unique strengths and limitations. Our selected approach [16], stands out for its ability to directly map trace data features with SRL dimensions, enabling a direct vertical comparison across different data modalities. Moreover, the multi-step pipeline can be adjusted to fit different dimensions and allows the horizontal integration of dimensions into profiles.

Survey. For the survey-based analysis, we form vectors using participants' responses from the pre- and post-survey for each construct. Specifically, each

---

[2] https://github.com/epfl-ml4ed/srl-navigation/blob/main/-material/ecd-features. pdf.

vector element corresponds to a unique questionnaire item. To illustrate, consider a student s who responded with "agree" (denoted as a 4) to all four items of the *Effort* construct in the pre-survey, and with "strongly agree" (denoted as a 5) in the post-survey. This yields the vector $v_s = [4, 4, 4, 4, 5, 5, 5, 5]$ for student s. Following [16], we then compute pairwise distances DD_d on the vectors v for each construct/dimension.

Trace Data. For our trace data analysis, we first validate the theory-based allocation of features to SRL dimensions using a data-driven approach [11]. This methodology resembles exploratory factor analysis, where, after specifying a number of dimensions, features are organized and assigned to corresponding dimensions or factors. We first calculate DT_f, the Dynamic Time Warping (DTW) pairwise distances between students for each feature f. We then use the Frobenius norm to compute the pairwise distances DF between the matrices DT_f. We convert the resulting pairwise distance matrix between features into a similarity matrix employing a Gaussian kernel. Finally, we use Spectral Clustering on the similarity matrix of DF to obtain four groups of features (representing the four dimensions). The resulting cluster solution groups two features ('Topic Exploration Ratio" and "Time Off Ratio") of the *Elaboration* dimension with the *Effort* features. To prevent overlaps between dimensions, we exclude them from the subsequent analyses.

Horizontal Approach. Following [16], per modality, we first cluster students separately for each dimension. For the trace data, we obtain the distance matrix DD_d per dimension by averaging the pairwise distance matrices, DT_f, belonging to the same dimension d. For both modalities, we transform DD_d into a similarity matrix using a Gaussian kernel and then use Spectral Clustering determining the optimal number of clusters using the Silhouette score. In a second step, we employ K-Modes on the clustering outcome of each dimension to obtained multi-dimensional profiles per modality.

Vertical Approach. We compute the agreement between the survey and trace data on the clusters obtained for each dimension using the rand index. The rand index measures the similarity between two data clusterings by considering all pairs of samples and counting agreements and disagreements. It ranges from 0 to 1, where a random cluster assignment would result in an expected agreement of 0.50.

2.4 Integration

The last step of our multimodal approach (see Fig. 1) consists of studying the interplay between the two different modalities (survey and trace data). We use Canonical Correlation Analysis (CCA), a technique for multivariate exploration [30] to identify and measure the associations among two sets of variables measured on the same entity. In particular, CCA examines linear combinations of two sets of variables, X with p variables and Y with q variables ($p \leq q$), resulting in pairs of canonical variates (S_i, T_i). Each pair, is formed by specific linear combinations (e.g., $\mathbf{S_1} = a_{11}x_1 + \cdots + a_{1p}x_p$; $\mathbf{T_1} = b_{11}y_1 + \cdots + b_{1q}y_q$), maximizing the canonical correlation r_i, which measures the correlation between S_i and T_i.

The magnitude and sign of these coefficients reveal the influence and direction of the relationship between individual variables and their respective canonical variate. In our study, we denote with S the canonical variates obtained for the survey data and with T those for the trace data.

Consequently, Y denotes the eight features ($q = 8$) engineered for the trace data (see Table 1) and X the items of the survey constructs. We aggregate the trace features over all weeks of the study to simplify the analysis. For the survey data, we average the scores of all items for each construct, resulting in $p = 8$ (an average pre- and post-survey score for each of the four constructs).

3 Results

To investigate the interplay between SRL dimensions within and across modalities, we first explored the alignment between trace data and survey data clusters (RQ1), and then integrated both modalities using CCA (RQ2).

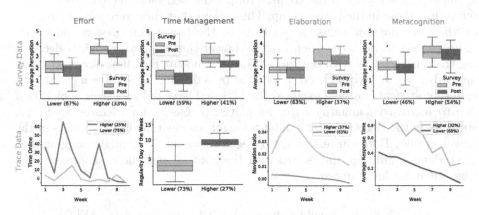

Fig. 2. Cluster per SRL dimension for the survey (top) and trace data (bottom). We obtain two groups per dimension which differ in average intensity (Higher and Lower).

3.1 SRL Dimensions Within and Across Modalities (RQ1)

In the first analysis, we followed a horizontal and vertical approach to explore and compare the SRL patterns and profiles obtained for each modality.

SRL Behavior and Perception. Figure 2 illustrates the resulting clusters for each dimension and modality. The optimal number of clusters for each dimension is two, separating participants into groups with Higher and Lower intensity. In the survey data clusters, the boxplots depict the average response within pre- and post-surveys across items associated with the respective construct. As shown in Fig. 2-top, the majority of the students were assigned to the Lower group across dimensions (e.g., 67% of participants for $Effort_{survey}$). Moreover, we observe a decrease in perception over time for the Higher group in all dimensions. This observation might be due to an initial overestimation or socially desirable response bias by students, which subsequently realigned to a more realistic assessment in the post-survey.

Figure 2-bottom shows the patterns found for the trace data exemplified by one of the features within each dimension. To visualize the time series behavior, we calculated the DTW barycenters for each cluster. The DTW barycenter is a central trajectory that characterizes the average behavior of all sequences in a group taking into account the temporal alignment of the sequences and the different sequence length. For the $Effort_{trace}$ dimension, the majority of participants were assigned to the Lower cluster, showing a more consistent lower-intensity behavior. In contrast, the minority of participants in the Higher cluster spent considerably more time on the platform and exhibited two peaks. For $Time Management_{trace}$, students in the Higher group exhibited more regular weekly patterns than those in the Lower group. This means that they were working mostly on the same days, whereas the students in the Lower cluster were more sporadic. Moreover, for $Elaboration_{trace}$, the Lower cluster shows a slightly decreasing trend contrasting with the Higher cluster students that first display a sharp increasing in the navigation ratio, followed by a decline. This pattern could indicate that students initially explored the ITS and later reduced their exploration due to increased familiarity with the platform. Lastly, for $Metacognition_{trace}$, students in the Higher cluster exhibit a longer answer time than students in the Lower cluster. Both clusters depict a declining trend. This might suggest that as students familiarize themselves with the content and question formats over time, their efficiency improves.

Table 2. Profiles found for survey (ABCDE) and trace data (ABFG)

Profile	Students		Effort	Time Management	Elaboration	Metacognition	Survey		Trace	
	Survey	Trace					Pre	Post	Pre	Post
A	38%	25%	Lower	Lower	Lower	Lower	0.31	0.30	0.33	0.41
B	13%	22%	Lower	Lower	Lower	Higher	0.22	0.42	0.30	0.22
C	15%	–	Lower	Higher	Lower	Higher	0.24	0.15	–	–
D	15%	–	Lower	Higher	Higher	Higher	0.34	0.38	–	–
E	19%	–	Higher	Higher	Higher	Higher	0.40	0.25	–	–
F	–	34%	Lower	Lower	Higher	Lower	–	–	0.32	0.19
G	–	19%	Higher	Higher	Lower	Lower	–	–	0.28	0.39

Horizontal Analysis. We integrated the separate dimensions per modality into profiles and identified four profiles for trace data and five for survey data as summarized in Table 2. Interestingly, Profile E, where students exhibit Higher behavior across all dimensions, is only present for the survey data. In contrast, Profile A, characterized by Lower behavior in every dimension, is the most common (38%) profile in the survey data, while only 25% students were assigned to that profile when using trace data. The remaining students (62% in survey data and 75% in the trace data profiles), displayed mixed behaviors belonging to heterogeneous profiles.

We also investigated the relationship between profiles and academic performance for each modality. We used a repeated-measures ANOVA[3] using the profiles as the between-subjects factor and the test time as the within-subject factor. We then proceeded with pairwise comparisons using t-tests with Benjamini-Hochberg corrections to investigate pairwise differences between profiles.

For the survey profiles, the main effects on the profiles and on time separately were not significant (profiles: $p = 0.4$, time: $p = 0.6$). However, the interaction effect between the profiles and time was significant ($p = .04$). In other words, while the profiles and time might not be significant predictors of performance on their own, their combined interaction does have a significant effect on the test scores. This can be seen in Table 2, where some profiles increase and others decrease in their scores from pre- to post- test. Pairwise comparisons revealed that only participants in profile B_{survey} improved significantly from their pre-test to their post-test ($p = 0.03$). Similarly, for the trace data profiles, the within-subject factor revealed no significant differences for either effect (profiles: $p = 0.2$, time: $p = 0.6$) and the interaction effect between profiles and time was also statistically significant ($p = .01$).

Vertical Alignment. In our alignment analysis, we investigated the potential congruence among the clusters derived from both survey and trace data for each dimension. In Fig. 2, we observe that in *Effort*, *Time Management* and *Elaboration* the majority of students were categorized into the Lower cluster in both modalities, with a similar distribution of students across clusters. Notably, for *Elaboration*, the distribution between Higher (37%) and Lower (65%) is identical for trace and survey data. Contrastingly, in the *Metacognition* dimension, more than half of the students were assigned to the Higher survey cluster (54%), whereas trace data revealed a contrasting trend, with a majority displaying Lower *Metacognition*. This contrast suggests a gap between students' self-perceived abilities and the trace data indicators. We quantified the agreement using the rand index. Indeed, the rand index revealed close-to-chance alignment for *Metacognition* ($ri = 0.52$), *Time Management* ($ri = 0.53$) and *Elaboration* ($ri = 0.54$), and a slight agreement for *Effort* ($ri = 0.67$).

---

[3] Subsequent to non significant Shapiro-Wilk and Levene's tests.

3.2   Modalities Integration (RQ2)

To further investigate the relationship between trace and survey data, we employed CCA. For the trace data, we aggregated features over the whole duration of the study and for the survey data, we used the average scores per construct. While the CCA was run with eight canonical variables, we only analyzed the four variables with the highest canonical correlations to account for the four SRL dimensions used in our survey and trace data. We found the following canonical correlations for the four variables: $r_1 = 0.67$, $r_2 = 0.51$, $r_3 = 0.48$, and $r_4 = 0.35$. A canonical correlation value of 1 would imply a perfect relationship, whereas a value of 0 would suggest no relationship. The observed canonical correlations therefore suggest that there are moderate relations between the survey and trace data. Figure 3 shows the standardized canonical coefficients per construct (survey data, left) and feature (trace data, right).

Survey. We observe that for the survey data, the pre- and post-survey answers are grouped together. Therefore, we can relate the canonical variables to the original SRL dimensions: S_1 is related to *Metacognition*, S_2 represents *Effort* and S_3 and S_4 are connected to *Time Management* and *Elaboration*, respectively. This finding suggests that the underlying constructs are relatively stable over time. In other words, the students' relative positioning within each SRL dimension remained consistent over time. For example, students with a higher perception in $Metacognition_{survey}^{pre}$ most likely also had a higher perception in $Metacognition_{survey}^{post}$ than students with an originally lower perception in $Metacognition_{survey}^{pre}$. This observation is in line with the cluster analysis for the

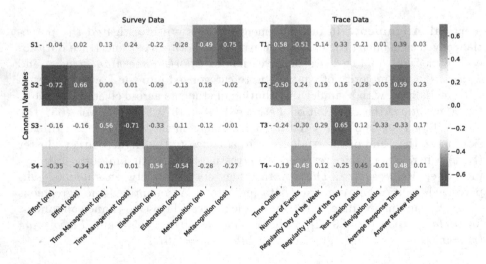

Fig. 3. Standardized coefficients per construct/feature and canonical variables for the survey (left) and trace (right) data.

survey data (see Sect. 3.1), where students were grouped by the strength of their perception (Higher versus Lower).

Interestingly, the pre- and post-survey scores for each canonical variable have inverse loading. For example, for S_3, the coefficient for $Time\ Management^{pre}_{survey}$ is positive (0.56) whereas the coefficient for $Time\ Management^{post}_{survey}$ is negative (-0.71). These inverse loadings suggest that an increase in one measure (e.g., $Time\ Management^{pre}_{survey}$) is associated with a decrease in another measure (e.g., $Time\ Management^{post}_{survey}$) when considering their relationship with the trace data. This relationship might therefore serve as an explanation for changes in perception. For example, a student who reported strong time management skills in the pre-survey and then interacted with the ITS inconsistently, would consequently report lower time management skills in the post-survey. The trace data indicator that loads the strongest on T_3 captures whether students regularly work at the same time of the day ("Regularity Hour of the Day": 0.65).

Trace Data. Moreover, regarding the feature mapping into the dimensions (see Table 1), we found that per each dimension one trace data feature loads with a high magnitude coefficient into the canonical variable associated with the corresponding dimension. "Average Response Time" loads on T_1 ($Metacognition_{survey}$), "Time Online" loads on T_2 ($Effort_{survey}$), "Regularity Hour of the Day" strongly relates to T_3 ($Time\ Management_{survey}$), and "Test Session Ratio" loads on T_4 ($Elaboration_{survey}$). This observation supports the theory and data-driven allocation of the trace features into the SRL dimensions.

Feature Interaction. The CCA sheds light on further feature interactions. For example, we observe that three features load strongly on T_1 ($Metacognition_{survey}$): "Average Response Time" (feature from $Metacognition_{trace}$), "Time Online", and "Number of Events" (features from $Effort_{trace}$). We initially hypothesized that only the $Metacognition_{trace}$ features would influence the $Metacognition_{survey}$ perception. However, the results suggest that the interaction between "Average Response Time" and absolute intensity (approximated by the $Effort_{trace}$ features) enriches the understanding of students' perception shifts regarding metacognitive skills. Similarly, the second canonical variable T_2 ($Effort_{survey}$) is not only related to the "Time Online" feature (feature from $Effort_{trace}$), but also to the different activities ("Test Session Ratio") and engagement in the activities ("Average Response Time"). Lastly, for T_4 ($Elaboration_{survey}$) we observe that apart from "Test Session Ratio" (feature from $Elaboration_{trace}$), the "Number of Events" and the "Average Response Time" feature are also related to this variable, suggesting that students' perception of elaboration might also depend on the intensity of activities done in the ITS.

4   Discussion and Implications

In this work, we used a vertical and horizontal approach to study the alignment between different SRL dimensions across modalities (**RQ1**) and an integrated approach to investigate their interplay (**RQ2**).

The horizontal approach revealed both consistency and complexity in students' profiles when combining multiple dimensions. This result is in line with previous work [16,20,28] highlighting the heterogeneous nature of students' SRL behaviors. In the vertical approach, we examined the relationship between students' perceptions obtained from survey data and actions in the learning platform. The misalignment of dimensions across modalities is coherent with prior work [1,8,21,29] and could be a result of several factors. First, students answering the survey based on aspirational behaviors or social desirability biases, rather than their actual practices. Second, differences in how students interpret and report their own behaviors versus how they manifest and are measured in digital environments [8,13]. Third, misalignment between what is measured by survey items (aptitude) and trace features (events), as discussed in [26]. Not all survey items correspond to observable behaviors within the ITS. This issue highlights an inherent difference between the data modalities: survey items can provide insight into unobservable SRL processes that cannot always be quantified using trace data. This discrepancy was a primary motivator for our integrated analysis, aimed at combining these two modalities to provide a more comprehensive understanding of SRL behavior.

Our exploration of the relationship between trace data and survey data modalities via CCA revealed a moderate relation between survey and trace data. We observed consistency in student perceptions over time, indicated by the grouping of pre- and post-survey constructs to canonical variates. The inverse loadings between pre- and post-survey dimensions mirror the decline in perceptions observed in the survey clusters, and illustrate the recalibration of students' initial views over time in response to certain experiences [19]. Using CCA we were able to relate this shift in perception to behavioral features from trace data. The CCA also revealed that there were indeed associations between the theory-based trace features and the SRL constructs measured in the survey with four features loading strongly on the expected constructs. Our findings further suggest that perceptions of certain SRL dimensions (like *Metacognition*, *Elaboration* and *Effort*) may be the result of various interacting trace features. For instance, the interplay of the "Average Response Time", "Time Online", and "Number of Events" features gives us a more holistic view of students' metacognitive skills. This underpins the multifaceted nature of students' learning experiences and the value of looking beyond isolated dimensions and integrating multiple modalities to gain a deeper understanding of student behavior.

Regarding the limitations, our study suffers from selection bias, since we only included students who have responded to both pre- and post- surveys. Moreover, it lacks external validity since it was conducted in a single context (as it is common for works analyzing students' SRL skills [1,8,10,12,21,27,29]). Nevertheless, our work is a piece of a larger puzzle of work on SRL in different modalities [8,9,16,25,27]. With this work, we aim to introduce a novel multi-modal approach that can be replicated by the AIED community in different ITS and contexts, enabling a better understanding of the nuances of the modalities. Our findings underscore the value of multimodal data and integrated approaches when investigating intricate educational phenomena like self-regulated learning.

Acknowledgments. This project was substantially co-financed by the Swiss State Secretariat for Education, Research and Innovation (SERI).

References

1. Beheshitha, S.S., Gašević, D., Hatala, M.: A process mining approach to linking the study of aptitude and event facets of self-regulated learning. In: Proceedings of LAK (2015)
2. Boroujeni, M.S., Sharma, K., Kidziński, Ł., Lucignano, L., Dillenbourg, P.: How to quantify student's regularity?. In: Proceedings of EC-TEL (2016)
3. Broadbent, J., Fuller-Tyszkiewicz, M.: Profiles in self-regulated learning and their correlates for online and blended learning students. Educ. Technol. Res. Dev. **66**(6), 1435–1455 (2018)
4. Broadbent, J., Panadero, E., Lodge, J., Fuller-Tyszkiewicz, M.: The self-regulation for learning online (srl-o) questionnaire. Metacogn. Learn. **18**(1), 135–163 (2023)
5. Broadbent, J., Poon, W.L.: Self-regulated learning strategies i& academic achievement in online higher education learning environments: a systematic review. Internet High. Educ. **27**, 1–13 (2015)
6. Chen, F., Cui, Y.: Utilizing student time series behaviour in learning management systems for early prediction of course performance. J. Learn. Anal. **7**(2), 1–17 (2020)
7. Cho, M.H., Shen, D.: Self-regulation in online learning. Dist. Educ. **34**(3), 290–301 (2013)
8. Choi, H., Winne, P.H., Brooks, C., Li, W., Shedden, K.: Logs or self-reports? misalignment between behavioral trace data and surveys when modeling learner achievement goal orientation. In: Proceedings of LAK (2023)
9. Fan, Y., et al.: Improving the measurement of self-regulated learning using multi-channel data. Metacogn. Learn. **17**(3), 1025–1055 (2022)
10. Gasevic, D., Jovanovic, J., Pardo, A., Dawson, S.: Detecting learning strategies with analytics: links with self-reported measures and academic performance. J. Learn. Anal. **4**(2), 113–128 (2017)
11. Käser, T., Busetto, A.G., Solenthaler, B., Kohn, J., von Aster, M., Gross, M.: Cluster-based prediction of mathematical learning patterns. In: Proceedings of AIED, pp. 389–399 (2013)
12. Li, Q., Baker, R., Warschauer, M.: Using clickstream data to measure, understand, and support self-regulated learning in online courses. Internet High. Educ. **45**, 100727 (2020)
13. Lim, L.A., Gasevic, D., Matcha, W., Ahmad Uzir, N., Dawson, S.: Impact of learning analytics feedback on self-regulated learning: triangulating behavioural logs with students' recall. In: Proceedings of LAK, pp. 364–374 (2021)
14. Matcha, W., et al.: Analytics of learning strategies: the association with the personality traits. In: Proceedings of LAK, pp. 151–160 (2020)

15. Mejia-Domenzain, P., Marras, M., Giang, C., Cattaneo, A.A.P., Käser, T.: Evolutionary clustering of apprentices' self- regulated learning behavior in learning journals. IEEE TLT **15**(5), 579–593 (2022)
16. Mejia-Domenzain, P., Marras, M., Giang, C., Käser, T.: Identifying and comparing multi-dimensional student profiles across flipped classrooms. In: Proceedings of AIED (2022)
17. Molenaar, I., de Mooij, S., Azevedo, R., Bannert, M., Järvelä, S., Gasević, D.: Measuring self-regulated learning and the role of AI: five years of research using multimodal multichannel data. Comput. Hum. Behav. **139**, 107540 (2023)
18. Panadero, E.: A review of self-regulated learning: six models and four directions for research. Front. Psychol. **8**, 422 (2017)
19. Pintrich, P.R., Smith, D.A., Garcia, T., McKeachie, W.J.: Reliability and predictive validity of the motivated strategies for learning questionnaire (MSLQ). Educ. Psychol. Measur. **53**(3), 801–813 (1993)
20. Poquet, O., Jovanovic, J., Pardo, A.: Student profiles of change in a university course: a complex dynamical systems perspective. In: Proceedings of LAK (2023)
21. Quick, J., Motz, B., Israel, J., Kaetzel, J.: What college students say, and what they do: aligning self-regulated learning theory with behavioral logs. In: Proceedings of LAK (2020)
22. Saint, J., Gašević, D., Matcha, W., Uzir, N.A., Pardo, A.: Combining analytic methods to unlock sequential and temporal patterns of self-regulated learning. In: Proceedings of LAK (2020)
23. Saint, J., Whitelock-Wainwright, A., Gasević, D., Pardo, A.: Trace-SRL: a framework for analysis of microlevel processes of self-regulated learning from trace data. IEEE Trans. Learn. Technol. **13**(4), 861–877 (2020)
24. Srivastava, N., et al.: Effects of internal and external conditions on strategies of self-regulated learning: a learning analytics study. In: Proceedings of LAK, pp. 392–403 (2022)
25. Van Der Graaf, J., et al.: Do instrumentation tools capture self-regulated learning? In: Proceedings of LAK (2021)
26. Winne, P.H.: Improving measurement of self-regulated learning. Educ. Psychol. **45**(4), 267–276 (2010)
27. Ye, D., Pennisi, S.: Using trace data to enhance students' self-regulation: a learning analytics perspective. Internet High. Educ. **54**, 100855 (2022)
28. Zhang, T., Taub, M., Chen, Z.: A multi-level trace clustering analysis scheme for measuring students' self-regulated learning behavior in a mastery-based online learning environment. In: Proceedings of LAK (2022)
29. Zhou, M., Winne, P.H.: Modeling academic achievement by self-reported versus traced goal orientation. Learn. Instr. **22**(6), 413–419 (2012)
30. Zhuang, X., Yang, Z., Cordes, D.: A technical review of canonical correlation analysis for neuroscience applications. Hum. Brain Mapp. **41**(13), 3807–3833 (2020)

Real-World Deployment and Evaluation of Kwame for Science, an AI Teaching Assistant for Science Education in West Africa

George Boateng[1,2(✉)], Samuel John[1(✉)], Samuel Boateng[1(✉)],
Philemon Badu[1(✉)], Patrick Agyeman-Budu[1(✉)], and Victor Kumbol[1(✉)]

[1] Kwame AI Inc., Claymont, USA
{jojo,samuel.boateng,philemon.badu,patrick.agyeman-budu,
victor}@kwame.ai, john96samuel@gmail.com
[2] ETH Zurich, Zürich, Switzerland

Abstract. Africa has a high student-to-teacher ratio which limits students' access to teachers for learning support such as educational question answering. In this work, we extended Kwame, a bilingual AI teaching assistant for coding education, adapted it for science education, and deployed it as a web app. Kwame for Science provides passages from well-curated knowledge sources and related past national exam questions as answers to questions from students based on the Integrated Science subject of the West African Senior Secondary Certificate Examination (WASSCE). Furthermore, students can view past national exam questions along with their answers and filter by year, question type, and topics that were automatically categorized by a topic detection model which we developed (91% unweighted average recall). We deployed Kwame for Science in the real world over 8 months and had 750 users across 32 countries (15 in Africa) and 1.5K questions asked. Our evaluation showed an 87.2% top 3 accuracy (n = 109 questions) implying that Kwame for Science has a high chance of giving at least one useful answer among the 3 displayed. We categorized the reasons the model incorrectly answered questions to provide insights for future improvements. We also share challenges and lessons with the development, deployment, and human-computer interaction component of such a tool to enable other researchers to deploy similar tools. With a first-of-its-kind tool within the African context, Kwame for Science has the potential to enable the delivery of scalable, cost-effective, and quality remote education to millions of people across Africa.

Keywords: Virtual Teaching Assistant · Educational Question Answering · Science Education · NLP · BERT

1 Introduction

The COVID-19 pandemic has exacerbated the already poor educational experiences of millions of students in Africa who were grappling with educational

A. M. Olney et al. (Eds.): AIED 2024, LNAI 14830, pp. 119–133, 2024.
https://doi.org/10.1007/978-3-031-64299-9_9

challenges like poor access to computers, the internet, and teachers. In 2018, the average student-teacher ratio in Sub-Saharan Africa was 35:1 which is higher compared to 14:1 in Europe [18]. In this context, students struggle to get answers to their questions. Hence, offering quick and accurate answers, outside of the classroom, could improve their overall learning experience. However, it is difficult to scale this support with human teachers.

In 2020, Boateng developed Kwame [2], a bilingual AI teaching assistant that provides answers to students' coding questions in English and French for Sua-Code, a smartphone-based online coding course [3,5]. Kwame is a deep learning-based question-answering system that finds the paragraph most semantically similar to the question via cosine similarity with a Sentence-BERT (SBERT) model [15].

We extended Kwame to work for science education in West Africa - Kwame for Science - and deployed it as a web app. Kwame for Science[1] provides passages from well-curated knowledge sources and related past national exam questions as answers to questions from students based on the Integrated Science subject of the West African Senior Secondary Certificate Examination (WASSCE). This subject is a core subject that covers various aspects of science such as biology, chemistry, physics, earth science, and agricultural science. It is mandatory for senior high school students in some West African Education Council (WAEC) member countries such as Ghana and The Gambia. Furthermore, students can view past national exam questions along with their answers and filter by year, question type (objectives, theory, and practicals), and syllabus topics (that were automatically categorized). We deployed and evaluated Kwame for Science in the real world over 8 months. This work builds upon our prior work-in-progress paper which only described the question-answering aspect of Kwame for Science and provided preliminary results for its deployment over 2.5 weeks [4].

Our key contribution is the development of a system for (1) science question answering using a state-of-the-art neural retriever - SBERT and (2) automatically categorizing syllabus topics using SBERT and a support vector machine, both for the unique context of Science education in West Africa along with its deployment and evaluation as an AI teaching assistant in an extended, real-world, large scale context (over 700 people primarily in West Africa over 8 months).

2   Related Work: Virtual Teaching Assistants

There are virtual teaching assistants (TA) such as Jill Watson [8,9], Rexy [1], a physics course TA [19], and Curio SmartChat (for K-12 science) [13]. The first work on virtual TAs was done by Professor Ashok Goel at the Georgia Institute of Technology in 2016. His team built Jill Watson, an IBM Watson-powered virtual TA to answer questions on course logistics in an online version of an Artificial Intelligence course for master's students [9]. It used question-answer pairs from past course forums. Given a question, the system finds the closest

---

[1] http://ed.kwame.ai/.

related question and returns its answer if the confidence level is above a certain threshold. Since then, various versions of Jill Watson have been developed to perform various functions (1) Jill Watson Q & A (revision of the original Jill Watson) now answers questions using the class syllabi rather than QA pairs. (2) Jill Watson SA gives personalized welcome messages to students when they introduce themselves in the course forum after joining and helps create online communities. (3) Agent Smith aids in creating course-specific Jills using the course' syllabus [8]. Jill's revision now uses a 2-step process. The first step uses commercial machine-learning classifiers such as Watson and AutoML to classify a sentence into general categories. Then the next step uses their proprietary knowledge-based classifier to extract specific details and gives an answer from Jill's knowledge base using an ontological representation of the class syllabi that they developed. Also, the responses pass through a personality module that makes Jill sound more human-like. Jill has answered thousands of questions, in one online and various blended classes with over 4,000 students over the years.

Similar work has been done by other researchers who built a platform, Rexy for creating virtual TAs for various courses also built on top of IBM Watson [1]. The authors described Rexy, an application they built that can be used to build virtual teaching assistants for different courses. They also presented a preliminary evaluation of one such virtual teaching assistant in a user study. The system has a messaging component (such as Slack), an application layer for processing the request, a database for retrieving the answers, and an NLP component built on top of IBM Watson Assistant which is trained with question-answer pairs. They use intents (the task students want to be done) and entities (the interaction context). The intents (71 of those, e.g., exam date) are defined in Rexy but the entities have to be specified by instructors as they are course-specific. For each question, a confidence score is determined and if it is below a threshold, the question is sent to a human TA to answer and the answer is forwarded to the student. The authors implemented an application using Rexy and deployed it in an in-person course (about recommender systems) with 107 students. After the course, the authors reviewed the conversation log and identified two areas of requests (1) course logistics (e.g., date and time of exams) and (2) course content (e.g., definitions and examples about lecture topics).

In the work by Zylich et al. [19], the authors developed a question-answering method based on neural networks whose goal was to answer logistical questions in an online introductory physics course. Their approach entailed the retrieval of relevant text paragraphs from course materials using TF-IDF and extracting the answer from it using an RNN to answer 172 logistical questions from a past physics online course forum. They used as data the PhysicsForum dataset from a previous physics online course consisting of past Piazza posts (2004 posts, 802 questions, 172 logistical questions eg. due dates, exam location, grading policy) and course materials (288 docs): syllabus, sections course textbook, lecture slides, emails, and forum posts. They used a two-method approach similar to Chen et al. [6]. The first is a document retriever based on DrQA and uses TF-IDF. It gets the 5 top-ranked documents. The second is an answer extractor which uses an

RNN to find the start and end sequence to use as the answer. The RNN is trained on the SQuAD dataset. They also used the BERT-based model to determine if a question is answerable based on the question-document pair. The model is fine-tuned on SQUAD 2.0 and Natural Questions which had human-generated answerability scores. They also developed an RNN-based model to predict the time that the post will get a response using information like BERT embedding of the post, the time between the last and current post, and the post type. They evaluated their approach separately for document retrieval (dr) and answer extraction (ae). They used the metrics top 1, 3, and 5 corresponding to the accuracies when the correct answer is in the top 1, 3, or 5 returned answers. They compared their system with DrQA for document retrieval and Human and IBM Watson Assistant (which was provided with the top 15 question-answer pairs). They did an ablation study for document retrieval, splitting the documents into paragraphs and using normalized cosine similarity, They assessed the use of the answerability classifier and the next post-time prediction. They also applied their system to answering 18 factual course questions using document retrieval without extracting an answer and had top 1, 3, and 5 accuracies of 44.4%, 88.9%, and 88.9% respectively.

Raamadhurai et al. developed Curio SmartChat, an automated question-answering system for K12 learning (middle school science topics) used by 20K students who asked over 100K questions [13]. Their system consists of a QA engine that retrieves answers, a content library that contains the curriculum-relevant text content and metadata (figures) and media, and a web app chat interface where students type their questions. When a student types a question, the system retrieves an answer. If it does not understand the question, it offers possible closely related questions from a bank of questions to disambiguate. If it does not find any recommended answer, it responds with small talk such as greetings. The content library contains text documents and quick definitions based on the curriculum. It also contains questions and answers tagged according to three levels of Bloom's Taxonomy: Knowledge, Understanding, and Application. They trained their custom taggers for intent and entity extraction by extending SpaCy taggers. The intent classifier decides which service will provide the response, small talk or content library, while the entity extractor will retrieve the entities the user is interested in. For example, if the query is "What is photosynthesis?", then the tagged JSON would look like "intent": "content", "entity": ["photosynthesis"]. There could be more than 1 entity but only 1 intent. They used the Universal Sentence Encoder model to encode the question and then compare it to vectors of the content via hashmap lookups. If the confidence is low, then it recommends similar questions. If the confidence is even lower, then it responds with small talk. Deployment was done with Tensorflow and Docker. Since launching the service, their system has served over 100,000 questions from around 20,000 students, mostly 13–16 years old. They mentioned that their system is mostly accurate once an answer is given because of their confidence checks but no robust evaluation was provided. For the case where similar questions were recommended or small talk responses were used, they sampled 200 random queries and 2 people individually rated them. They rated the response as valid if one of the recommendations was valid: 65% were valid, 20 questions did not have available content to answer, 107 did have and 60 of the queries were not clear. Students had several spelling mistakes posing challenges for the system.

These works are focused on answering logistical questions, except Curio SmartChat whereas ours is focused on answering content-based questions which are more challenging. Compared to Curio SmartChat which is the closest work to ours, our work uses a state-of-the-art language model (SBERT) relative to theirs (Universal Sentence Encoder). Additionally, they did not provide an evaluation of the accuracy of their system after real-world deployment using ratings from users which we do. Furthermore, our work is the first to be developed and deployed in the context of high school science education in West Africa.

3   Background: Question Answering (QA)

Within the domain of question answering, there are two main paradigms: Extractive QA and Generative QA. Extractive QA entails machine reading, which refers to selecting a span of text in a paragraph (called the context) that directly answers a question. To obtain the context, it can be directly provided such as in the task described for the SQuAD dataset [14]. Also, the most relevant passage(s) can be retrieved from various documents, a process called retrieval. This approach was used by Chen et al. [6] in their DrQA system that used TF-IDF for retrieval and LSTM for the reader which extracts the answer. BERT-based models are the current state-of-the-art [7] for implementing a reader. A similar approach was used in the physics QA system described in the previous section [19]. In Generative QA, generative models such as T5 [16] are used to generate an answer given a question as input. Various contexts can be retrieved and then passed together with the question as an input for the model to use to generate an answer. This technique is called retrieval augmented generation (RAG) [12].

In our QA system, we only use the retrieval component to obtain passages that are then displayed to the user as answers. We do not use a reader for our context of answering content-based science questions which could consist of definitions and explanations. This is because short answers such as phrases are generally not adequate but rather, whole contexts that could span multiple continuous sentences. Hence, extracting an answer span is not necessary and could even be inadequate and more error-prone. This point is also made by Zylich et al. [19] after assessing the misclassifications in their evaluation using machine reading to answer physics course questions. We as a result focus on providing whole passages (a combination of 3 sentences) as answers. Generative QA was not used given that these models were not good at generating answers at the time of this work. It is important to note that this work was done and deployed before the release of ChatGPT and GPT 3.5 API [10], a much better generative model, hence, no comparison can be made to it (see future work section about plans to integrate it).

4 Overview of Kwame for Science

Kwame for Science is a web app that consists of 2 key features: question answering and viewing past national exam questions. The question-answering (QA) feature enables students to ask science questions and receive 3 passages as answers along with a confidence score per answer which represents the similarity score (Fig. 1). In addition to the 3 passages, it displays the top 5 related past national exam questions of the Integrated Science subject and their expert answers. We picked these numbers to ensure we provided several answers (more than 1) without being overwhelming. Also, students can view the history of questions they have asked. The View Past Questions feature enables students to view past national exam questions and answers of the Integrated Science subject and filter the questions based on criteria such as the year of the examination, the specific exam, the type of questions, and topics that were automatically categorized by a topic detection model which we developed. We describe data collection and preprocessing, the QA and view past questions features and the topic detection system.

| Ask Kwame | Receive Answers | See Related Past Exam Questions | See History |

Fig. 1. Screenshots of QA feature of Kwame for Science

4.1 Data Collection and Preprocessing

Given that our goal was for Kwame to provide answers based on the Integrated Science subject of the WASSCE exam, our training data and knowledge source had to cover the topics in the WASSCE Integrated Science curriculum. We sought to use one of the approved textbooks in Ghana. Unfortunately, their copyrights did not permit such use and the publishers were unwilling to partner with us. Consequently, we searched for free and open-source books and datasets that fulfilled our needs. We came across a U.S. middle school science dataset - Textbook Questions Answering (TQA) [11] which was curated from the free and open-source textbook, CK-12. Our exploration of the dataset revealed that though it covered several of the WASSCE Integrated Science topics, it lacked others, particularly those topics that were contextual such as agricultural science. Consequently, we additionally used a dataset based on Simple Wikipedia to cover those gaps. We used Simple Wikipedia since its explanations use basic words and shorter sentences, and hence, better suited for middle school and high school students compared to regular Wikipedia [17]. We parsed the JSON files of the dataset into paragraphs as originally in the dataset. We also extracted figures that were referenced in the paragraphs so they could be returned to students along with the answers. We then split the paragraphs into groups of 3 sentences - passages - using the library sentence-splitter [2] resulting in 527K passages. We computed embeddings of the passages using SBERT and indexed them in ElasticSearch because it is open source and scalable to enable fast retrieval and runtime. These constituted the answers returned in response to questions.

Furthermore, we augmented our question-answering system by adding content specific to the curriculum. In particular, we created question-answer pairs using WASSCE questions of the Integrated Science subject that covered exams from 1993 to 2021 (except 2010 when the exam did not happen). We obtained physical copies of the exams as PDFs, scanned them, and developed software to automatically parse and extract the content using object character recognition (OCR) technology. Given the imperfection of current OCR technology, and in particular, the limitation of extracting mathematical and chemical symbols and equations present in scientific text, we hired and paid individuals to manually inspect and correct all 28 years of exam questions. We then hired and paid 4 subject-matter experts to provide answers to the exam questions because they did not originally have answers. The experts were Ghanaians and included past contestants of Ghana's National Science and Math Quiz competition and Senior High School teachers. The experts were given an annotation guidance document on how best to provide their answers, add diagrams, draw tables, and input special characters, among others to ensure that everything, particularly scientific and mathematical symbols, will be rendered properly on the web application. Furthermore, we had initial training rounds where they provided answers for some sample questions and received feedback to ensure consistency. The experts did all the answering online using Google sheets. Diagrams and images which were included in the answers were drawn by experts using third-party tools or they were obtained as open-source images found online. This process results in 3.5K question-answer pairs across 28 years of exams. After obtaining all the answers from the experts, we computed embeddings of the questions similar to the passages of the knowledge base, using SBERT, and indexed them in Elastic-Search. These constituted the related past questions (with answers) that were returned when a question was asked.

---

[2] https://pypi.org/project/sentence-splitter/.

4.2   Question Answering Feature

The question-answering (QA) feature enables students to ask science questions and receive 3 passages as answers along with a confidence score per answer which represents the similarity score (Fig. 2). In addition to the 3 passages, it displays the top 5 related past national exam questions of the Integrated Science subject and their expert answers. When a user types a question in the web app, our system computes an embedding of the question using an SBERT model that was pretrained on 215 M question-answer pairs from diverse sources. We used the SBERT model without fine-tuning it since exploratory evaluation for our science use case which entailed outputting answers for a handful of questions showed it provided adequate answers to the questions. Next, it computes cosine similarity scores with a bank of passages (which are the precomputed embeddings of passages from our knowledge source), retrieves, and returns the top 3 passages along with confidence scores to the web app and any figures or images referenced in that passage which was stored as part of passage's metadata. Additionally, it computes cosine similarity scores of the question embedding with the precomputed embeddings of the bank of past exam questions, retrieves, and returns the top 5 related questions and their expert answers, along with confidence scores. The web app then displays the answers and the related past exam questions that are above a preset similarity score threshold. If no answer is above the threshold, a message is shown saying the question could not be answered using the knowledge source of that subject. We used precomputed embeddings for fast real-time retrieval which were saved as indices in an ElasticSearch instance we hosted on the Google Cloud Platform. Users could click the answer card to see the passage highlighted in a more detailed answer which is the paragraph. Similarly, for related past question cards, they could click to open them to see the expert answer and other relevant information such as the year of the exam and the question type.

4.3   View Past Questions

The View Past Questions feature provides direct access to our data bank of past national exam questions and answers of the Integrated Science subject. Furthermore, students can filter the questions based on criteria such as the year of the examination, the specific exam, the type of questions, and by topic. All these criteria could be inferred easily from the metadata of the original exam files except the topic for which we developed a model to automatically categorize each question according to one of the syllabus topics (see Sect. 4.4). Utilizing a combination of markdown and LaTeX, we were able to properly render both questions and answers even for niche subjects that make use of equations and symbols not common to regular alphanumeric input, i.e., subjects such as Mathematics or Chemistry. Figures linked to questions or answers are also displayed alongside the question or answer. Similar to the QA feature, users could click the question cards to open them to see the expert answer and other relevant information such as the year of the exam and the question type.

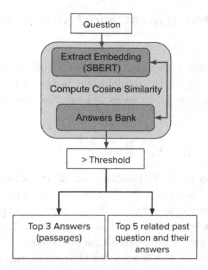

Fig. 2. Architecture of QA feature of Kwame for Science

4.4 Topic Detection Development

We trained a machine-learning topic detection model which we used to classify each of the past exam questions into one of the 48 topics in the Integrated Science subject syllabus. We parsed the official syllabus document which contained a mapping of topics to learning objectives and their corresponding content heading, content details to be covered, example teaching activities, and descriptions of evaluations. The format for these was phrases, sentences, and paragraphs which we call passages. We created a topic-passage pair for each of these mappings resulting in 928 topic-passage samples which we used as the dataset for training and testing. For each passage, we extracted TF-IDF vectors using unigrams and bigrams (baseline approach) and SBERT embeddings (main approach) as features. We then passed the features to a linear support vector machine (SVM) model. We split the data into an 80:20 train-test stratified split. We trained the model on the training set using 5-fold cross-validation and performed hyper-parameter turning to pick the best parameters. We then evaluated the trained model on the test set using unweighted average recall (UAR) given that there was a class imbalance. Our baseline approach had a UAR of 83.1% and our main approach had a UAR of 91%. The performance is good considering that it is a multi-class classification task with 48 classes showing that our topic detection system is adequate for use. We then trained the SVM model on the whole dataset. Next, we computed embeddings for all past exam questions using SBERT and then performed classification using the trained SVM model to get the topics. These were then imported into the web app as part of the view past questions feature to enable learners to see exam questions for each selected topic.

5   Real-World Deployment and Evaluation of Kwame for Science

We launched the web app in beta on 10th June 2022 which was advertised on social media. The statistics for the deployment between 10th June 2022 and 19th February 2023 (8 months) are 750 users across 32 countries (15 in Africa) and 1.5K questions asked.

5.1   Question Answering Accuracy

Users could provide feedback by upvoting or downvoting answers in response to the question "Was this helpful?". To evaluate Kwame for Science, we used the metrics top 1 and top 3 accuracies. Top 1 accuracy quantifies performance assuming only one answer was returned and voted on. Top 3 accuracy refers to the performance where for each question that received a vote, at least one answer was rated as helpful out of the 3 answers that were returned. Our evaluation results showed 72.6% top 1 accuracy (n = 197 answers), and 87.2% top 3 accuracy (n = 109 questions). The top 3 accuracy result is good, showing that Kwame for Science has a high chance of giving at least one useful answer among the 3. We also performed qualitative analysis categorizing the reasons the model incorrectly answered questions to provide insights for future improvements (Table 1). Some challenging cases occurred when there were typos in the spelling of scientific words, the questions were incomprehensible and the questions were outside the scope of the knowledge source. Also, some unhelpful answers were cases where the returned paragraph was incomplete due to issues with the dataset, and the answers were related to the question but did not directly provide an answer. These issues are similar to those encountered by Raamadhurai et al. in their real-world deployment of Curio SmartChat [13].

5.2   Usage Analytics of QA and View Past Questions

We evaluated the usage of the QA features - clicking to see detailed answers and expanding related questions - and view past questions features - filtering questions by various options and clicking to show an answer (Table 2). In total, the detailed view of answers was opened 2173 times (1.4 times per question asked). In contrast, the related past questions were expanded 1219 times (0.8 times per question asked). For the viewing past exam questions features, users most frequently used the filtering by year feature (237 times). This was followed by filtering by question type (174 times) and filtering by topic (104 times). Users could also access detailed answers to past questions by clicking to show the answer and this feature was used 931 times.

6   Challenges and Lessons

One of our biggest challenges was getting access to local textbooks to use as a knowledge source for Kwame for Science because of copyright concerns. Despite

Table 1. Reasons for unhelpful answers with examples

Reasons	Question	Unhelpful Answer
Typos in the spelling of scientific words	what is mashine learning	Learning is getting new things into your brain.
Outside the scope of the knowledge source	How is rheumatoid arthritis diagnosed	Rheumatoid arthritis is caused by autoimmunity. Autoimmunity is when the body's immune system attacks its own tissues. The most common places for this arthritis is in the wrists and the knuckles. It usually starts between the ages of 20 and 40. Children and elderly people can also get this kind of arthritis. There are medicines to slow down the process of this disease. There are also some medicines to relieve pain. It is important to treat this kind of arthritis early.
Truncated paragraph	List three excretory organs and their respective products	Examples of exocrine organs and glands:
Related to the question but does not directly provide an answer	Why is the sky blue	Blue is the color of the Earth's sky and sea. Earth looks blue when seen from outer space by astronauts.
Incomplete / Incomprehensible queries	water and glass	Light bends when it passes from air to water. The bending of the light causes the pencil to appear broken. Why do waves bend as they enter a new medium

our relentless effort to get buy-in from local publishers, their disinterest in partnering with us because of trust issues made it impossible for us to power Kwame for Science with content from local textbooks. The trust issue relates to their fear of the softcopy version of their books being redistributed publicly because of the prevalence of such behavior in Ghana. We however addressed this issue by hiring experts to provide answers to past national exam questions to augment our foreign knowledge base with contextual knowledge in the form of expert answers from local experts. This lesson about copyright issues is important for others who might be interested in creating localized educational AI products which will require huge amounts of local data which comes at a huge time and financial cost.

Another issue we had to deal with was the formatting of scientific and mathematical symbols and equations. Unfortunately, the current open-source OCR technologies we tried did not work well for extracting these symbols. Hence, manual effort was needed as an extra quality control step. This issue is impor-

Table 2. Count for Usage of QA and View Past Questions features

Events	Count
QA Feature	
Answer Detail Opened	2173
Related Question Expanded	1219
View Past Questions Feature	
Show Answer	931
Select Year	237
Select Question Type	174
Select Topic	104

tant to note especially for developing countries like Ghana and other African countries where important educational documents are mainly accessible in hard-copy format. More work is needed to develop systems that can easily convert scanned scientific documents into outputs that have the correct representation that will render properly in some standard format like markdown. Until then, significant time and monetary effort would be required to generate high-quality data for use.

Another challenge was the difficulty in getting upvotes or downvotes on responses to questions. Even though we had 1.5K questions, we received reactions to only 7.2% of those. We did not want to force users to provide a rating each time they viewed an answer as that could be annoying. Nonetheless, after the learnings from this deployment and evaluation, we decided to implement the forced rating feature to enable us to obtain a lot of data to evaluate our models in real-time and finetune them. We plan to explore approaches to nudge users rather than force them to provide ratings.

7 Limitations and Future Work

Our work did not evaluate the effect of Kwame for Science on learning outcomes. In the future, we will deploy Kwame for Science in a controlled setting (specific sets of schools) and run a randomized controlled trial to assess the impact of Kwame for Science on learning outcomes evaluated via end-of-term examinations or national exams.

We did not perform real-time classification of the topic when a question was asked. This information could be useful for generating real-time analytics of the distribution of the topics of the asked questions. Our future work will integrate the topic detection model for real-time classification. An important addition for this to work well is a pre-classification model that classifies the question as a science or non-science topic given users could ask non-scientific questions.

Furthermore, our future work will classify questions into different levels of Bloom's Taxonomy. This information will be useful to understand the types of questions students ask, e.g., definition-type questions that are in lower levels of Bloom's Taxonomy vs application-type questions in a higher level of Bloom's Taxonomy which is the ultimate goal when facilitating students' learning.

We will improve the performance of our QA system. We will use the data we collected from our real-world deployment and past exam question-answer pairs to finetune the SBERT model to improve the model's accuracy. Also, we will include a re-ranking step where we will use a BERT model and cross-attention between pairs of questions and top answers to re-rank the relevant passages returned as answers to improve the accuracy of the entire QA system. Also, we will add generative models like GPT-4 [10] to build a retrieval augmented generation (RAG) system [12] to ensure that our response directly answers users' questions.

In the future, we would explore making Kwame for Science available to African students via more accessible channels such as WhatsApp, USSD, toll-free calling and also providing answers in local languages.

8 Conclusion

In this work, we developed, deployed, and evaluated Kwame for Science which provides passages from well-curated knowledge sources and related past national exam questions as answers to questions from students. It also enables them to view past national exam questions along with their answers and filter by year, question type, and topics that were automatically categorized by a topic detection model which we developed (91% UAR). Our deployment of Kwame for Science in the real world over 8 months had 750 users across 32 countries (15 in Africa) and 1.5K questions asked. Our evaluation showed an 87.2% top 3 accuracy (n = 109 questions) implying that Kwame for Science has a high chance of giving at least one useful answer among the 3 displayed. Our usage analytics showed that the detailed view of answer cards was opened 1.4 times per question asked. Also, our categorization of the reasons for incorrectly answered questions included typos in scientific words, incomprehensible questions, incomplete passages, and limited scope of our knowledge source. Some challenges included copyright issues that affected access to local textbooks, inadequate OCR technology for parsing scanned documents, and limited user ratings. Nonetheless, with a first-of-its-kind tool within the African context, Kwame for Science has the potential to enable the delivery of scalable, cost-effective, and quality remote education to millions of people across Africa.

References

1. Benedetto, Luca, Cremonesi, Paolo: *Rexy*, a configurable application for building virtual teaching assistants. In: Lamas, David, Loizides, Fernando, Nacke, Lennart, Petrie, Helen, Winckler, Marco, Zaphiris, Panayiotis (eds.) INTERACT 2019. LNCS, vol. 11747, pp. 233–241. Springer, Cham (2019). https://doi.org/10.1007/978-3-030-29384-0_15

2. Boateng, George: Kwame: a bilingual AI teaching assistant for online SuaCode courses. In: Roll, Ido, McNamara, Danielle, Sosnovsky, Sergey, Luckin, Rose, Dimitrova, Vania (eds.) AIED 2021. LNCS (LNAI), vol. 12749, pp. 93–97. Springer, Cham (2021). https://doi.org/10.1007/978-3-030-78270-2_16

3. Boateng, G., Annor, P.S., Kumbol, V.W.A.: Suacode Africa: teaching coding online to africans using smartphones. In: Proceedings of the 10th Computer Science Education Research Conference, pp. 14–20 (2021)

4. Boateng, G., John, S., Glago, A., Boateng, S., Kumbol, V.: Kwame for science: An AI teaching assistant based on sentence-BERT for science education in west Africa. In: Proceedings of the Fourth International Workshop on Intelligent Textbooks (iTextbooks) 2022 co-located with 23d International Conference on Artificial Intelligence in Education (AIED 2022), pp. 26–30 (2022)

5. Boateng, G., Kumbol, V.W.A., Annor, P.S.: Keep calm and code on your phone: a pilot of suacode, an online smartphone-based coding course. In: Proceedings of the 8th Computer Science Education Research Conference, pp. 9–14 (2019)

6. Chen, D., Fisch, A., Weston, J., Bordes, A.: Reading Wikipedia to answer open-domain questions. arXiv preprint arXiv:1704.00051 (2017)

7. Devlin, J., Chang, M.W., Lee, K., Toutanova, K.: BERT: pre-training of deep bidirectional transformers for language understanding. arXiv preprint arXiv:1810.04805 (2018)

8. Goel, A.: AI-powered learning: making education accessible, affordable, and achievable. arXiv preprint arXiv:2006.01908 (2020)

9. Goel, A.K., Polepeddi, L.: Jill Watson: a virtual teaching assistant for online education. Tech. Rep., Georgia Institute of Technology (2016)

10. Chat completions API. https://platform.openai.com/docs/guides/text-generation/chat-completions-api

11. Kembhavi, A., Seo, M., Schwenk, D., Choi, J., Farhadi, A., Hajishirzi, H.: Are you smarter than a sixth grader? Textbook question answering for multimodal machine comprehension. In: Proceedings of the IEEE Conference on Computer Vision and Pattern recognition, pp. 4999–5007 (2017)

12. Lewis, P., et al.: Retrieval-augmented generation for knowledge-intensive NLP tasks. Adv. Neural. Inf. Process. Syst. **33**, 9459–9474 (2020)

13. Raamadhurai, S., Baker, R., Poduval, V.: Curio SmartChat: a system for natural language question answering for self-paced k-12 learning. In: Proceedings of the Fourteenth Workshop on Innovative Use of NLP for Building Educational Applications, pp. 336–342 (2019)

14. Rajpurkar, P., Zhang, J., Lopyrev, K., Liang, P.: Squad: 100,000+ questions for machine comprehension of text. arXiv preprint arXiv:1606.05250 (2016)

15. Reimers, N., Gurevych, I.: Making monolingual sentence embeddings multilingual using knowledge distillation. In: Proceedings of the 2020 Conference on Empirical Methods in Natural Language Processing. Association for Computational Linguistics (11 2020). https://arxiv.org/abs/2004.09813

16. Roberts, A., Raffel, C., Shazeer, N.: How much knowledge can you pack into the parameters of a language model? In: Proceedings of the 2020 Conference on Empirical Methods in Natural Language Processing (EMNLP), pp. 5418–5426 (2020)
17. Simple English Wikipedia. https://simple.wikipedia.org/wiki/Main_Page
18. Unesco. pupil-teacher ratio sub-saharan africa. https://data.worldbank.org/indicator/SE.PRM.ENRL.TC.ZS?locations=ZG (Feb 2020)
19. Zylich, Brian, Viola, Adam, Toggerson, Brokk, Al-Hariri, Lara, Lan, Andrew: Exploring automated question answering methods for teaching assistance. In: Bittencourt, Ig Ibert, Cukurova, Mutlu, Muldner, Kasia, Luckin, Rose, Millán, Eva (eds.) AIED 2020. LNCS (LNAI), vol. 12163, pp. 610–622. Springer, Cham (2020). https://doi.org/10.1007/978-3-030-52237-7_49

ChatGPT for Education Research: Exploring the Potential of Large Language Models for Qualitative Codebook Development

Amanda Barany[1]([✉]) [iD], Nidhi Nasiar[1] [iD], Chelsea Porter[1] [iD],
Andres Felipe Zambrano[1] [iD], Alexandra L. Andres[1] [iD], Dara Bright[2],
Mamta Shah[1,3] [iD], Xiner Liu[1], Sabrina Gao[1], Jiayi Zhang[1] [iD], Shruti Mehta[1],
Jaeyoon Choi[4], Camille Giordano[1], and Ryan S. Baker[1] [iD]

[1] The University of Pennsylvania, Philadelphia, PA 19104, USA
amanda.barany@gmail.com
[2] Consortium of DEI Health Educators, Philadelphia, PA, USA
[3] Elsevier, Philadelphia, PA 19103, USA
[4] University of Wisconsin-Madison, Madison, WI 53706, USA

Abstract. In qualitative data analysis, codebooks offer a systematic framework for establishing shared interpretations of themes and patterns. While the utility of codebooks is well-established in educational research, the manual process of developing and refining codes that emerge bottom-up from data presents a challenge in terms of time, effort, and potential for human error. This paper explores the potentially transformative role that could be played by Large Language Models (LLMs), specifically ChatGPT (GPT-4), in addressing these challenges by automating aspects of the codebook development process. We compare four approaches to codebook development – a fully manual approach, a fully automated approach, and two approaches that leverage ChatGPT within specific steps of the codebook development process. We do so in the context of studying transcripts from math tutoring lessons. The resultant four codebooks were evaluated in terms of whether the codes could reliably be applied to data by human coders, in terms of the human-rated quality of codes and codebooks, and whether different approaches yielded similar or overlapping codes. The results show that approaches that automate early stages of codebook development take less time to complete overall. Hybrid approaches (whether GPT participates early or late in the process) produce codebooks that can be applied more reliably and were rated as better quality by humans. Hybrid approaches and a fully human approach produce similar codebooks; the fully automated approach was an outlier. Findings indicate that ChatGPT can be valuable for improving qualitative codebooks for use in AIED research, but human participation is still essential.

Keywords: Large Language Models · ChatGPT · Inductive Coding · Research Methods

A. M. Olney et al. (Eds.): AIED 2024, LNAI 14830, pp. 134–149, 2024.
https://doi.org/10.1007/978-3-031-64299-9_10

1 Introduction

When examining qualitative data in education research, the process of "coding", or defining concepts and identifying where they occur in the data, is a key part of the meaning-making process [32]. Some coding projects are driven by top-down deductive approaches that apply codes from existing codebooks or frameworks [3]. Researchers have also found value in inductive, bottom-up coding techniques that ground codes in the research context (sometimes referred to as thematic analysis [5]). Inductive codes allow meaning to emerge from frequent, dominant, or significant themes in the data [4, 36]. Given these affordances, inductive codebook development has been featured in a variety of educational research contexts in the last decade, from small-scale case studies (e.g., [25]) to large-scale meta-summaries of education research (e.g., [1]).

While widely used in education research, inductive codebook development is not without its challenges. The practices for initially developing inductive codes and then applying them to the dataset are often inconsistent and there are often issues with reliability and fairness [17, 32]. To help address these issues, researchers often adopt three practices: (1) transparent procedures and documentation when creating a preliminary codebook [38], (2) pilot testing codes during codebook refinement [32], and (3) inter-coder reliability checks [30]. However, these efforts can be time-consuming (e.g., [8, 34]). Campbell et al. [7] also note a tradeoff in traditional inductive code development between time spent (efficiency), the utility of the final codebook for representing nuance, and the reliability with which coders can later code the data. As our field now works with increasingly large-scale and complex learning data, we need techniques that can maximize efficiency without limiting code utility or reliability.

A potential solution is the automation of coding processes. Though there have been decades of research attempting this (e.g., [29, 37]), existing tools have been critiqued as producing low-quality codes or codes that miss nuances that humans identify [18] and have been unable to explain the reasoning for their recommendations [26], which are significant limitations for codebook development. Large Language Models (LLMs) such as ChatGPT (GPT-4), however, draw on advanced natural language processing capabilities to process and generate human-like text based on open-ended textual input.

To support inductive coding, ChatGPT could be used in two phases of the codebook development process: (1) to identify preliminary codes from data; and (2) to test and refine codes. While scholars have already used ChatGPT for both codebook development and codebook refinement [16, 19], there has not yet been a systematic study of what benefits it can bring to each phase of this process. In this work, we tried four different ways to create a codebook for the same data: a fully manual approach, a fully automated approach, and two hybrid approaches that leverage ChatGPT, one in codebook development and one in codebook refinement. To understand when and how ChatGPT might best support the codebook development process, we compare time spent on each approach, reliability among human coders in applying the codes, ratings of codebook utility by human coders, and whether approaches yielded similar codes.

2 Automated Tools for Qualitative Data Analysis

Over time, there has been considerable interest in using natural language processing tools to support codebook development [14, 20] and automated coding [6, 15, 22]. However, qualitative coders have largely not adopted these tools, due to concerns that the codes obtained are limited in quality [18]. There has also been considerable interest in automatically coding data (given a codebook or examples) using a range of natural language processing methods (e.g., [9, 10, 14, 23, 42]), including LLMs (e.g., [11]).

Recent work has explored the use of ChatGPT for deductive coding [35] with some researchers testing the reliability of ChatGPT compared to humans or other automated coding tools [41]. Törnberg [38] proposed a step-by-step process that includes "mutual learning" between researchers and ChatGPT, where researchers iteratively refine the prompts given to ChatGPT to improve its responses and use ChatGPT's feedback to refine deductive codes and apply them reliably. Zhang et al. [44] similarly describe ChatGPT as a "tool" for data synthesis and as a "co-researcher" that can assist, challenge, or supplement coders' interpretations. Zambrano et al. [43] found that ChatGPT's explanations for coding decisions had the potential to improve construct validity, find ambiguity in codebook definitions, and help human coders achieve better interrater reliability. Other researchers have raised concerns that ChatGPT classification can be nonreproducible [30], recommending human review [28].

Two projects have explored the application of ChatGPT for inductive code development using thematic analysis. De Paoli [16] offered a 5-phase process model with LLMs that uses prompts to first identify and refine inductive codes in a dataset, and then pairs a researcher and ChatGPT to review and finalize themes. Gao et al. [19] propose that ChatGPT may have a role at three points during thematic analysis: (1) suggesting codes during open coding, (2) identifying disagreements when researchers refine and test the codes, and (3) suggesting which codes to combine during code finalization. While these studies show how ChatGPT can help develop codes, they do not test how reliably coders could apply those codes to the datasets or compare its utility at different phases of the process. The aim of our research is, therefore, to investigate whether ChatGPT is beneficial in either or both initial codebook development and refinement, in different structures of human/LLM collaboration.

3 Methods

3.1 Data Source

We conduct this research in the context of high-dosage tutoring lessons, in which trained tutors offer personalized, small-group support for mathematics learning (e.g. [13, 24]). Prior research has shown that this type of tutoring benefits students' math learning, achievement, and grades [12]. We obtained transcripts from four 60-min tutoring sessions conducted by the non-profit organization Saga Education. Sessions were conducted virtually with students in high-poverty schools in an urban region of the northeastern United States from 2022–2023. Students were 9th graders enrolled in Algebra I. Sample data from a tutoring session is shown in Fig. 1.

Id ⇓	speaker_type ⇓	text
74	tutor	In total, how many x's do we have now?
75	student	Uh, seven.
76	tutor	Exactly.
77	tutor	Now we have seven.
78	tutor	So what should the new exponent of x be?
79	student	x to the seven.
80	tutor	Exactly.

Fig. 1. A sample of transcript data from a tutoring lesson on the Saga platform.

3.2 Codebook Development

To explore the utility of ChatGPT as a tool for inductive codebook development, four approaches were applied by four different members of the research team to the same three tutoring lesson transcripts. The focus of qualitative coding in all cases was primarily to identify the instructional strategies or techniques used in the tutoring lessons, though other emergent codes were also included. For each of the approaches described below, the researchers worked independently during this stage to avoid biasing those engaging in the other approaches. The total time spent developing the four codebooks was logged by researchers to compare the efficiency of each approach.

The human-only approach applies common practices considered a gold standard for inductive coding in education research [33], while the other three approaches used ChatGPT (GPT-4) to automate some or all of the process. Within the approaches that used ChatGPT, we opted for the web-based chatbot version over the API, anticipating that future researchers might prefer the web version for more straightforward interaction with the chatbot when developing and refining a codebook. Our study design required that authors work exclusively on one codebook to avoid skewing the results due to the order of codebook development (it is expected that the next codebook developed by the same researcher using the same data would have higher quality). Therefore, although codebook development, revision, and refinement is usually a collaborative process [40], for this study each codebook developer (authors 2 to 5) worked independently to avoid biasing other members of the research team.

Human Only (H). For approach H, a researcher used the codebook development process outlined by Weston et al. [40] including stages of code conceptualization, application/review, and refinement. The researcher (author 5) first qualitatively reviewed the dataset to identify common patterns and themes that appeared across the transcripts in an inductive search for tutor/student exchanges that demonstrated specific instructional strategies or techniques. these themes were manually organized into a preliminary codebook that included tentative code names, definitions, and example quotes. The preliminary codebook was then used to reexamine the dataset to determine if revisions to the inclusion and exclusion criteria were necessary or if any new themes emerged. This process was repeated until no further additions or revisions were needed.

Human Code Development, ChatGPT Refinement (HC). For approach HC, a human coder (author 3) again engaged in qualitative review and the creation of a preliminary codebook as outlined by weston et al. [40]. The researcher then tasked chatgpt with reexamining and refining the codebook by entering the following prompt:

You are a researcher helping develop a qualitative codebook for text data of a math tutor's interactions with students. I will give you the first draft of the codebook, and then the data being coded, in batches. Please help me refine the codes and codebook, focusing on instructional strategies or techniques.

The preliminary codebook and full dataset were then entered into ChatGPT, with the three class sessions divided into 13 batches of 77–98 lines. This range was selected to maximize batch sizes (more than 75 lines), while not exceeding the processing limits of the version of ChatGPT we used (which was 4096 tokens corresponding to approximately 100 lines of our dataset). While it would be ideal for subsets to maintain a uniform line count, it was not possible to achieve this without segmenting the data in the middle of a response or explanation that required context from prior lines; breakpoints for batches were selected to minimize this type of context loss.

Every three to four batches of data, the researcher prompted ChatGPT to offer further refinement: *Please give me a refined codebook, with examples, based on all X batches of data so far.* After the final batch entry and refinement, the researcher reviewed the codebook to check for errors or inconsistencies (in response to Mesec's [27] caution that all ChatGPT output must be evaluated by a human prior to dissemination). While the refined codes and definitions were consistent with the researcher's understanding of the data, random checks of the examples showed many example quotes were hallucinations. The researcher replaced them with quotes from the original dataset.

ChatGPT Code Development, Human Refinement (CH). For approach CH, a researcher (author 2) tasked chatgpt with creating a preliminary codebook before engaging in manual re-examination and refinement, using the process in Weston et al. [40] for that stage. The researcher began with an initial review of the data to understand its structure and context but did not exhaustively analyze the dataset to identify themes. Another researcher (author 4) tested multiple prompts with ChatGPT to identify the one that could most reliably provide codes, definitions, and examples. Prompts were crafted based on best practices from existing prompt engineering frameworks (e.g., [27]), which emphasize offering ChatGPT clear, concise, and specific task descriptions to ensure the model receives the necessary information but is not confused by unnecessary details. The first 100 lines of the dataset were used for prompt development and testing.

Recognizing ChatGPT's challenges in maintaining response consistency (due to the variation of the chatbot's responses) the researcher re-evaluated each prompt across sessions with ChatGPT, using various browsers and computers. After identifying a prompt that produced consistently similar responses—where responses generated by repeated tests of the same prompt had no more than two codes that were different across runs—this test was replicated using two additional sets of 100 lines. Once the prompt's consistency was confirmed across these new sets, it was applied to the entire dataset using the same batches detailed in approach HC. The final prompt read:

Hi ChatGPT, I want to analyze the following interaction between an instructor and some students: [DATA] Please give me a codebook to analyze the instructional methodologies and the sentiment within this interaction.

The result of this process was a codebook of 8–12 themes for each batch. All themes not proposed at least three times across the 13 batches were discarded. The themes that the researcher identified as conceptually similar were grouped together into a single theme. Across the three or more examples of each code, the most straightforward definition was selected, or two were combined to create the preliminary codebook.

After the ChatGPT part of the process had been completed, the researcher (author 2) used Weston et al.'s [40] approach to codebook application/review and refinement described in approach H, repeatedly applying the codes to the dataset to refine them until no further code revisions or additions were necessary.

ChatGPT Only (C). For approach C, a researcher (author 4) used ChatGPT to develop the preliminary codebook using the procedures and prompts detailed in approach CH and used ChatGPT for the review and refinement of the codebook using the procedures and prompts detailed in approach HC. The researcher did not add any concepts or clarifications not originally provided by chatgpt. The final version of the codebook also contains three example quotes given by ChatGPT for each code. If more than three examples were provided by ChatGPT, the researcher selected the three clearest examples. No hallucinations were obtained in this condition. Beyond this review, the researcher made no additional interventions for the final version of the codebook.

3.3 Coding Procedures

Once the codebooks were finalized, four pairs of researchers (authors 6–13) who were not involved in the development process and were not familiar with the data were randomly assigned to code the tutoring lesson transcripts using one of the four codebooks. The pairs were introduced to the study design and the context of the dataset but were not told which approach produced the codebook they used. Researchers were instructed to independently mark each code as present (1) or absent (0) for each line of data based on the code's name and the inclusion/exclusion criteria provided in the definition. Upon completion of the first round of coding, coders submitted a brief survey in which they rated their codebook on a scale of 1 (lowest) to 5 (highest) for ease of use, clarity, and the mutual exclusivity and exhaustiveness of codes.

After the pairs had independently coded all lines of data, Cohen's kappa (κ) was used to assess the consistency of code applications. Research pairs then met virtually for 1-h sessions via Zoom, where they were prompted to discuss and resolve coding inconsistencies in the dataset using social moderation techniques (e.g., [21]). Codebook developers did not participate in this process or clarify any aspect of the codebook to avoid biasing the process. Coders were invited to annotate their codebooks based on what was learned from social moderation, then code a fourth lesson transcript consisting of 150 lines of new tutoring data based on their refined understanding of codes. Cohen's κ coefficients were calculated for this second round of coding.

3.4 Codebook Evaluation

To evaluate the utility of each codebook, at the end of the coding process, coders were surveyed to evaluate their perception of their codebook's ease of use, code clarity, mutual exclusivity, and exhaustiveness using a Likert Scale ranging from 1 (lowest) to 5 (highest). Criteria were chosen based on previously published principles for what constitutes a high-quality codebook [4, 5]. The level of agreement between the coders for each construct on each codebook was employed as a proxy for codebook quality. Separate researchers (authors 1 and 14) also evaluated the conceptual overlap across each of the four codebooks. One hundred pairs of codes from two codebooks for the same data point were randomly generated, with each codebook represented in at least 25 pairs and each code represented at least once. The researchers, who were not involved in codebook development, independently coded each pair as representing the same concept (1) or representing different concepts (0). While the coders reached a high percentage of agreement for the first 100 pairs, few instances of conceptual similarity occurred. Fifty-two additional code pairs with potential for conceptual similarity were purposively sampled based on review of the blinded codebooks. Both researchers then independently coded the additional pairs, obtaining a Cohen's κ coefficient of 0.86. With inter-rater reliability established, the first author reviewed all code pairs to identify every instance of code overlap across the four codebooks.

4 Results

4.1 Time Spent

Analyzing time spent can shed light on the efficiency of each of the four approaches. Table 1 gives the total time spent by the researchers on processes up to and including development of a preliminary codebook, and total time spent on codebook refinement and finalization. Time spent engaging in preliminary development varied by codebook and researcher; approach H took longest (180 min) and C the shortest (50 min).

In approaches HC and CH, which both used ChatGPT to automate code refinement, more time was spent engaging in codebook refinement than codebook development. Automated code refinement for these approaches also took more time to complete than human code refinement in approaches H or C. This may be because human intervention was still needed to make final decisions or adjustments after ChatGPT had completed the automated refinement of codes. In approach HC, the researcher noticed that ChatGPT had hallucinated example quotes to populate the final codebook and spent time replacing them with genuine quotes. In approach C, the researcher spent time merging similar codes that repeatedly emerged across the repeated prompts. This multi-step refinement may have contributed to the longer time spent. Some of the differences in time spent in each process may also relate to individual variations by the researcher.

4.2 Codebook Utility

Coders' (authors 6–13) Likert-style rankings and qualitative reflections on the codes they applied to the datasets offer preliminary insights into the quality and utility of the codebooks developed using each approach. Table 2 summarizes coders' rankings of each codebook's ease of use, code clarity, mutual exclusivity, and exhaustiveness from 1 (lowest) to 5 (highest). Approach HC and CH, which leveraged combinations of manual

Table 1. Records of time spent on codebook development.

Codebook Approach	Preliminary Codebook Development	Codebook Refinement	Total Time Spent
(1) Human	180 min	40 min	220 min
(2) Human → ChatGPT	80 min	165 min*	245 min
(3) ChatGPT → Human	107 min	60 min	167 min
(4) ChatGPT	50 min	63 min**	113 min

* 90 min spent using ChatGPT for code refinement, 75 min of human codebook revision
** 42 min spent using ChatGPT for code refinement, 21 min of human codebook revision

and automated techniques to develop and refine codes, were ranked highest for clarity and mutual exclusivity of codes, as well as for the codebooks' ease of use.

Approach H was ranked lowest for code exhaustiveness, suggesting that coders may have noticed themes in their review of data that were not represented in the codebook. Given that the researcher for approach H used only manual techniques when developing the codebook, they may have been more likely to emphasize instructional strategies that emerged in the data – the research focus – and not develop other codes (discussed further below). Approach C was ranked lowest for the mutual exclusivity of codes, suggesting that coders felt these codes had more conceptual similarities. For example, codes in the ChatGPT codebook such as *Direct Instruction* (providing direct information) compared to *Task Assignment* (directing students to a task) used similar terms to describe different concepts, and *Questioning and Check in* (asking questions to probe understanding) compared to *Metacognition* (encouraging student reflection on processes), which describe phenomena that might sometimes overlap (e.g., a check-in that induces metacognition), making it challenging to categorize them distinctly.

Table 2. Coder ratings of codebook utility.

Codebook Approach	Ease of use	Clarity of codes	Mutual Exclusivity	Exhaustiveness
(1) H	2.5	3	3	2
(2) HC	4.5	3.5	3.5	3
(3) CH	4	4	3.5	2.5
(4) C	3	3	1.5	3

4.3 Inter-rater Reliability

In the final codebooks, approach H (Human) had 9 codes, approach HC (Human → ChatGPT) had 10 codes, approach CH (ChatGPT → Human) had 8 codes, and approach C (ChatGPT) had 11 codes. Table 3 provides an overview of percent agreement and

Cohen's κ coefficients across two rounds of paired independent coding to explore whether codes from each codebook can be applied consistently and reliably. For round 1 of coding, two out of nine codes for approach H and two out of ten codes for approach HC (Human → ChatGPT) achieved a κ of 0.6 or above. Coders achieved high first-round agreement for codes related to mentorship enacted by the tutors, such as *Providing Assistance* (0.60), *Checking in or Expressing Concern* (0.71), and *Offering Greetings or Pleasantries* (0.64), and for students asking questions (*Questioning*, 0.83). One error occurred in this process: in two instances, one coder using the approach H codebook interpreted a pair of consecutive codes as a single code and applied them to the data as such; as a result, paired agreement could not be calculated for four codes against the other coder in round 1, who applied them as four independent codes.

Coders using the ChatGPT codebook saw some of the lowest agreement measures for codes in round 1 (average κ = 0.21, compared to 0.34, 0.38, and 0.26 for other codebooks), which may relate to concerns about code mutual exclusivity they identified in their final reflections. For example, when the pair met to refine their understanding of the codes before round 2 of coding, they noted that two codes (*Clarification and Reiteration* and *Corrections*) both included the term "clarification" in either their name or definition, which muddied coders' understanding of how they differed. Annotations from their discussion highlight efforts to emphasize what makes each code distinct.

For round 2 of coding, average κ coefficients for all four codebooks improved. Coders using the ChatGPT → Human codebook saw universal improvements in their agreement in round 2, with every code reaching a κ value at 0.6 or above (average κ = 0.70). Annotations from their discussion show how terms from the definitions offered concrete examples for how the code might appear in the data (e.g., *Aligning to Prior Knowledge*, "Tutor…using the word 'remember'"). However, this code pair chose to meet more times to continue refining and improving their shared understanding of the codes than the other pairs, which likely explains their improved inter-rater agreement.

Four out of nine codes for approach H and four out of ten codes for approach HC reached a κ of 0.6 or above in round 2. New codes to reach this threshold include *Comfort/consolation* and *Prompting Self-explanation* for approach H and *Clarification/rephrasing* and *Feedback* for HC. New codes in approach H did not have prior measures of agreement due to the coding inconsistency in approach H round 1.

For the ChatGPT codebook (approach C), the code *Friendly Interaction and Encouragement* reached κ of 0.6 in round 2. Some of the codes that were clarified during the pair's round 2 discussion, such as *Direct Instruction,* and *Questioning and Check-in,* saw improvement in inter-rater agreement, but did not reach the threshold for moderate agreement. Other codes saw minimal improvement or decrease in κ values.

In approach HC (4 out of 10 codes) and C (2 out of 11 codes), some κ values decreased when codes were applied to new data, and in three of the four codebooks, one or more codes did not appear in the new dataset – a common challenge when qualitatively coding relatively varied data.

Table 3. Measures of agreement across two rounds of hand coding for each item.

Approach	Code	Cohen's κ coefficient	
		Round 1	Round 2
(1) Human	1. Assistance	**0.60**	**0.62**
	2. Encouragement	0.05	0.26
	3. Checking in/concern	**0.71**	**0.80**
	4. Comfort/consolation	**	**0.61**
	5. Commendation	**	0.16
	6. Prompt self-explanation	**	**0.61**
	7. Relating/casual	0.04	0.31
	8. Scaffolding	**	0.35
	9. User interface issues	0.29	*
	Average κ	0.33	0.47
(2) Human → ChatGPT	1. Clarification/rephrasing	0.04	**0.65**
	2. Connecting to prior knowledge	0.36	*
	3. Direct instruction	0.22	0.17
	4. Engagement checks	0.45	*
	5. Feedback	0.21	**0.62**
	6. Greetings/pleasantries	**0.64**	**0.75**
	7. Guided practice	0.29	0.19
	8. Questioning	**0.83**	**0.87**
	9. Session logistics	0.32	0.15
	10. Software/tool use	0.44	0.38
	Average κ	0.38	0.47
(3) ChatGPT → Human	1. Aligning to Prior Knowledge	0.40	**0.66**
	2. Checking Understanding/Engagement	0.45	**0.60**
	3. Encouragement	0.12	**0.80**
	4. Greeting	0.43	**0.85**
	5. Guiding Feedback	0.05	**0.66**
	6. Instruction	0.21	**0.66**
	7. Technical and Logistics	0.09	**0.66**
	8. Time Management	0.31	**0.72**

(continued)

Table 3. (*continued*)

Approach	Code	Cohen's κ coefficient	
		Round 1	Round 2
	Average κ	0.26	0.70
(4) ChatGPT	1. Clarification and reiteration	0.01	*
	2. Corrections	−0.02	*
	3. Direct instruction	0.07	0.24
	4. Expressions of frustration/impatience	0.20	*
	5. Friendly interaction and encouragement	0.42	**0.67**
	6. Guided practice	0.37	0.46
	7. Metacognition	0.50	0.49
	8. Student uncertainty	0.20	−0.01
	9. Task assignment	0.30	0.30
	10. Technical problem addressal	0.21	0.28
	11. Questioning and check-in	0.08	0.55
	Average κ	0.21	0.37

* Code did not appear in sample dataset
** Coder treated two codes as a single code; agreement with second coder could not be calculated

4.4 Conceptual Overlap

The cross-codebook comparison to identify common themes resulted in 28 total pairs of codes across the four codebooks that were found to represent the same concept. Pairs aligned around nine general categories of overlap as illustrated in Table 4. Themes were identified as conceptual outliers if they appeared in only one codebook.

Four themes with universal representation across codebooks were *Checking In*, *Feedback*, *Guided Practice*, and *Technical Issues*. Three of these codes relate directly to the research focus for each approach: instructional strategies or techniques used in tutoring lessons. The fourth, *Technical Issues*, was common in the datasets, as tutors offered platform or logistical support throughout lessons. While the names and definitions had slight differences across codebooks (e.g., *Checking in/concern* versus *Questioning and check-in*), the high level of overlap suggests that all four approaches can lead to the identification of key themes if they are prevalent in the dataset.

The largest overlap between approaches was between approach HC and approach CH, the two hybrid Human-ChatGPT approaches. Seven categories were found in both codebooks, with only two categories found in HC but not CH, and only one category found in CH but not HC. Approach HC (Human → ChatGPT) had the greatest conceptual similarity to the manual approach H, finding all themes represented in the human codebook as well as three other themes. Approach CH (ChatGPT → Human) also captured most of what was seen in the human codebook (five of six themes), plus three additional themes. As such, it seems that hybrid approaches can capture most or all of what a pure human approach captures, plus additional codes. Approach C (all ChatGPT) had the lowest agreement with the other three approaches.

Overall, approaches CH and C, which used ChatGPT to automate preliminary codebook development, were more likely to generate novel themes that did not appear in any other codebooks. These themes primarily focused on student behaviors and affective states. These categories tended to be less prevalent in the dataset, and less related to the project goal. Approach C (ChatGPT only) also missed the prevalent theme *Greetings/casual*, which characterizes the casual and introductory interactions between tutor and students during sessions. All told, the fully automated approach to codebook development emerged as an outlier in terms of conceptual similarity.

Table 4. Heat map of code categories across codebooks.

Code Categories	H	HC	CH	C	Total
(1) Checking in	X	X	X	X	4
(2) Feedback	X	X	X	X	4
(3) Guided Practice	X	X	X	X	4
(4) Technical Issues	X	X	X	X	4
(5) Greetings/ casual language	X	X	X	0	3
(6) Questioning or response prompting	X	X	0	0	2
(7) Connecting to prior knowledge	0	X	X	0	2
(8) Logistics/Time Management	0	X	X	0	2
(9) Direct Instruction	0	X	0	X	2
*Student Responses	0	0	X	0	1
*Student Uncertainty	0	0	0	X	1
*Frustration or Impatience	0	0	0	X	1

* Themes that emerged as outliers in the codebooks

5   Discussion and Conclusion

In this paper, we explored when and how ChatGPT might support the process of inductive codebook development. We applied a fully manual approach, a fully automated approach, and two hybrid approaches to creating codebooks for math tutoring transcripts. The hybrid approaches involved utilizing ChatGPT for either preliminary codebook development or refinement. For each approach, we compared the time spent, ratings of codebook utility, and inter-rater reliability metrics. Lastly, we assessed whether different methods produced similar or overlapping themes.

Results indicate that automating elements of codebook development has the potential to improve the time efficiency of the process, especially when both preliminary development and final refinement are automated (approach C). However, we found that humans could not be excluded completely from the process; even when codebook refinement was automated by ChatGPT, further human refinement was still needed to address errors or inconsistencies. Thus, it may make the most sense to use automation initially and then transition to manual refinement (e.g., approach CH).

The fully automated codebook was ranked lowest for utility, had the lowest inter-rater reliability measures when applied by coders, and saw the least conceptual overlap with other codebooks. This aligns with Reiss' [29] cautions regarding the consistency and reliability of results when ChatGPT is used without human supervision. While the fully human process was able to identify many codes that were consistent across other codebooks, it also missed some themes that were represented in the two hybrid approaches (e.g., *Logistics/Time Management*). Overall, the hybrid approaches received the highest utility ratings, achieved comparable or better inter-rater reliability outcomes than other approaches, and had the highest conceptual overlap. These findings indicate that ChatGPT can be useful as both a tool and co-researcher to support "mutual learning" between humans and LLM [38, 44].

There is considerable further work to be done to understand how LLMs and humans can best work together for qualitative coding. As the first comparison of different approaches to using LLMs for inductive codebook development, our study has several limitations that can be addressed in future studies. For example, a tighter comparison could have been achieved by more precise coding instructions and by controlling the amount of time spent working on achieving inter-rater reliability.

This research offers insights into how Large Language Models can be integrated into the inductive codebook development process for qualitative data analysis in education research. Findings suggest that while automated approaches can enhance efficiency, the collaboration between human researchers and ChatGPT is most beneficial for producing high-quality, non-overlapping, and comprehensive codebooks where human coders can obtain reliable results. We hope that this study provides a foundation for further exploration and refinement of methodologies, emphasizing the potential of hybrid approaches for leveraging the strengths of automated and manual processes.

Acknowledgements. This work was supported by funding from the Learning Engineering Virtual Institute (LEVI) Engagement Hub. All opinions expressed are those of the authors.

Disclosure of Interests. The authors have no competing interests to declare that are relevant to the content of this article.

References

1. Anderson, J., Taner, G.: Building the expert teacher prototype: a metasummary of teacher expertise studies in primary and secondary education. Educ. Res. Rev. **38**, 100485 (2023). https://doi.org/10.1016/j.edurev.2022.100485
2. Bakharia, A.: On the equivalence of inductive content analysis and topic modeling. In: Eagan, B., Misfeldt, M., Siebert-Evenstone, A. (eds.) Advances in Quantitative Ethnography: First International Conference, ICQE 2019, Madison, WI, USA, October 20–22, 2019, Proceedings 1, pp. 291–298. Springer International Publishing (2019)
3. Bingham, A.J., Witkowsky, P.: Deductive and inductive approaches to qualitative data analysis. In: Vanover, C., Mihas, P., Saldana, J. (eds.) Analyzing and Interpreting Qualitative Data: After the Interview, pp. 133–146 (2021)

4. Boyatzis, R.: Transforming Qualitative Information: Thematic Analysis and Code Development. Sage, Thousand Oaks, CA (1998)
5. Braun, V., Clarke, V.: Thematic analysis. In: Cooper, H., Camic, C.M., Long, D.L., Panter, A.T., Rindskopf, D., Sher, K.J. (eds.) APA Handbook of Research Methods in Psychology, vol. 2. Research Designs: Quantitative, Qualitative, Neuropsychological, and Biological, pp. 57–71. American Psychological Association (2012)
6. Cai, Z., Siebert-Evenstone, A., Eagan, B., Shaffer, D.W., Hu, X., Graesser, A.C.: nCoder+: a semantic tool for improving recall of nCoder coding. In: Eagan, B., Misfeldt, M., Siebert-Evenstone, A. (eds.) Advances in Quantitative Ethnography. ICQE 2019. Communications in Computer and Information Science, vol. 1112. Springer (2019)
7. Campbell, J.L., Quincy, C., Osserman, J., Pedersen, O.K.: Coding in-depth semistructured interviews: problems of unitization and intercoder reliability and agreement. Sociol. Meth. Res. **42**(3), 294–320 (2013)
8. Castleberry, A., Nolen, A.: Thematic analysis of qualitative research data: is it as easy as it sounds? Curr. Pharm. Teach. Learn. **10**(6), 807–815 (2018)
9. Chen, N.C., Drouhard, M., Kocielnik, R., Suh, J., Aragon, C.R.: Using machine learning to support qualitative coding in social science: shifting the focus to ambiguity. ACM Trans. Interact. Intell. Syst. **8**(2), 1–20 (2018)
10. Cher, P.H., Lee, J.W.Y., Bello, F.: Machine learning techniques to evaluate lesson objectives. In: International Conference on Artificial Intelligence in Education, pp. 193–205. Springer International Publishing (2022)
11. Cochran, K., Cohn, C., Rouet, J.F., Hastings, P.: Improving automated evaluation of student text responses using GPT-3.5 for text data augmentation. In: International Conference on Artificial Intelligence in Education, pp. 217–228. Springer Nature Switzerland, Cham (2023). https://doi.org/10.1007/978-3-031-36272-9_18
12. Cook, P.J.: Not Too Late: Improving Academic Outcomes for Disadvantaged Youth. Northwestern University Institute for Policy Research Working Paper, 15-01 (2015)
13. Cook, P.J., Dodge, K., Farkas, G., Fryer, R.G., Guryan, J., Ludwig, J., Steinberg, L.: The (surprising) efficacy of academic and behavioral intervention with disadvantaged youth: results from a randomized experiment in Chicago, Working Paper No. 19862. National Bureau of Economic Research (2014). https://doi.org/10.3386/w19862
14. Crowston, K., Allen, E.E., Heckman, R.: Using natural language processing technology for qualitative data analysis. Int'l. J. of Soc. Res. Methodol. **15**(6), 523–543 (2012)
15. Crowston, K., Liu, X., Allen, E.E.: Machine learning and rule-based automated coding of qualitative data. In: Proc. Amer. Soc. Inf. Sci. Technol. **47**(1), 1–2 (2010). https://doi.org/10.1002/meet.14504701328
16. De Paoli, S.: Performing an inductive thematic analysis of semi-structured interviews with a large language model: an exploration and provocation on the limits of the approach. Soc. Sci. Comp. Rev. 08944393231220483 (2023)
17. Eagan, B.R., Rogers, B., Serlin, R., Ruis, A.R., Arastoopour Irgens, G., Shaffer, D.W.: Can we rely on IRR? Testing the assumptions of inter-rater reliability. In: International Conference on Computer Supported Collaborative Learning, Jan (2017)
18. Gao, J., Choo, K.T.W., Cao, J., Lee, R.K.W., Perrault, S.: CoAIcoder: examining the effectiveness of AI-assisted human-to-human collaboration in qualitative analysis. ACM Trans. Comp.-Hum. Interact. **31**(1), 1–38 (2023)
19. Gao, J., et al.: CollabCoder: A GPT-powered workflow for collaborative qualitative analysis. arXiv preprint arXiv:2304.07366 (2023). https://doi.org/10.48550/arXiv.2304.07366
20. Gauthier, R.P., Wallace, J.R.: The computational thematic analysis toolkit. In: Proceedings of the ACM on Human-Computer Interaction, 6(GROUP), pp. 1–15 (2022)
21. Herrenkohl, L.R., Cornelius, L.: Investigating elementary students' scientific and historical argumentation. J. Learn. Sci. **22**(3), 413–461 (2013)

22. Leech, N.L., Onwuegbuzie, A.J.: Beyond constant comparison qualitative data analysis: using NVivo. Sch. Psychol. Q. **26**(1), 70–84 (2011)
23. Liew, J.S.Y., McCracken, N., Zhou, S., Crowston, K.: Optimizing features in active machine learning for complex qualitative content analysis. In: Proceedings of the ACL 2014 Workshop on Language Technologies and Computational Social Science, pp. 44–48 (2014)
24. Linzarini, A., et al.: Identifying and supporting children with learning disabilities. In: Bugden, S., Borst, G. (eds.) Education and the Learning Experience in Reimagining Education: The International Science and Evidence based Education Assessment. UNESCO MGIEP, New Delhi (2022)
25. Liu, L.: Using generic inductive approach in qualitative educational research: a case study analysis. J. Educ. Learn. **5**(2), 129–135 (2016)
26. Marathe, M., Toyama, K.: Semi-automated coding for qualitative research: A user-centered inquiry and initial prototypes. In: CHI '18: Proceedings of the 2018 CHI Conference on Human Factors in Computing Systems, pp. 1–12 (2018)
27. Marvin, G., Hellen, N., Jjingo, D., Nakatumba-Nabende, J.: Prompt engineering in large language models. In: International Conference on Data Intelligence and Cognitive Informatics, pp. 387–402. Springer Nature Singapore (2023)
28. Mesec, B.: The language model of artificial inteligence chatGPT – a tool of qualitative analysis of texts. Authorea Preprints (2023)
29. Perrin, A.J.: The CodeRead system: using natural language processing to automate coding of qualitative data. Soc. Sci. Comput. Rev. **19**(2), 213–220 (2001)
30. Reiss, M.V.: Testing the reliability of ChatGPT for text annotation and classification: a cautionary remark. arXiv preprint arXiv:2304.11085 (2023)
31. Saldaña, J., Omasta, M.: Qualitative Research: Analyzing Life. Sage Publications (2016)
32. Shaffer, D.W., Ruis, A.R.: How we code. In: Advances in Quantitative Ethnography: Second International Conference, ICQE 2020, Malibu, CA, USA, 1–3 Feb 2021, Proceedings 2, pp. 62–77. Springer International Publishing (2021)
33. Strauss, A., Corbin, J.: Basics of Qualitative Research. Sage Publications (1990)
34. Sutton, J., Austin, Z.: Qualitative research: data collection, analysis, and management. Can. J. Hosp. Pharm. **68**(3), 226 (2015)
35. Tai, R.H., et al.: An examination of the use of large language models to aid analysis of textual data. bioRxiv, 2023-07 (2023). https://doi.org/10.1101/2023.07.17.549361
36. Thomas, D.: A general inductive approach for qualitative data analysis. Am. J. Eval. **27**(2), 237–246 (2006). https://doi.org/10.1177/1098214005283748
37. Tierney, P.J.: A qualitative analysis framework using natural language processing and graph theory. Int'l. Rev. Res. Open Distrib. Learn. **13**(5), 173–189 (2012)
38. Törnberg, P.: How to Use Large-Language Models for Text Analysis (2023)
39. Tracy, S.J.: Qualitative quality: eight "big-tent" criteria for excellent qualitative research. Qual. Inq. **16**(10), 837–851 (2010)
40. Weston, C., Gandell, T., Beauchamp, J., McAlpine, L., Wiseman, C., Beauchamp, C.: Analyzing interview data: the development and evolution of a coding system. Qual. Sociol. **24**, 381–400 (2001). https://doi.org/10.1023/A:1010690908200
41. Xiao, Z., Yuan, X., Liao, Q.V., Abdelghani, R., Oudeyer, P.Y.: Supporting qualitative analysis with large language models: combining codebook with GPT-3 for deductive coding. In: Companion Proceedings of the 28th International Conference on Intelligent User Interfaces, pp. 75–78, Mar (2023). https://doi.org/10.1145/3581754.3584136
42. Yang, B., Nam, S., Huang, Y.: "Why my essay received a 4?": a natural language processing based argumentative essay structure analysis. In: International Conference on Artificial Intelligence in Education, pp. 279–290. Springer Nature Switzerland (2023)

43. Zambrano, A.F., Liu, X., Barany, A., Baker, R.S., Kim, J., Nasiar, N.: From nCoder to Chat-GPT: from automated coding to refining human coding. In: International Conference on Quantitative Ethnography, pp. 470–485. Springer Nature Switzerland (2023)

44. Zhang, H., Wu, C., Xie, J., Lyu, Y., Cai, J., Carroll, J.M.: Redefining qualitative analysis in the AI era: utilizing ChatGPT for efficient thematic analysis. arXiv preprint arXiv:2309.10771 (2023). https://doi.org/10.48550/arXiv.2309.10771

Aligning Tutor Discourse Supporting Rigorous Thinking with Tutee Content Mastery for Predicting Math Achievement

Mark Abdelshiheed(✉), Jennifer K. Jacobs, and Sidney K. D'Mello

University of Colorado Boulder, Boulder, CO 80309, USA
{mark.abdelshiheed,jennifer.jacobs,sidney.dmello}@colorado.edu

Abstract. This work investigates how tutoring discourse interacts with students' proximal knowledge to explain and predict students' learning outcomes. Our work is conducted in the context of high-dosage human tutoring where 9th-grade students ($N = 1080$) attended small group tutorials and individually practiced problems on an Intelligent Tutoring System (ITS). We analyzed whether tutors' talk moves and students' performance on the ITS predicted scores on math learning assessments. We trained Random Forest Classifiers (RFCs) to distinguish high and low assessment scores based on tutor talk moves, student's ITS performance metrics, and their combination. A decision tree was extracted from each RFC to yield an interpretable model. We found AUCs of 0.63 for talk moves, 0.66 for ITS, and 0.77 for their combination, suggesting interactivity among the two feature sources. Specifically, the best decision tree emerged from combining the tutor talk moves that encouraged rigorous thinking and students' ITS mastery. In essence, tutor talk that encouraged mathematical reasoning predicted achievement for students who demonstrated high mastery on the ITS, whereas tutors' revoicing of students' mathematical ideas and contributions was predictive for students with low ITS mastery. Implications for practice are discussed.

Keywords: Tutoring Discourse · Talk Moves · Math Tutoring · Decision Trees · Intelligent Tutoring Systems

1 Introduction

While "silence is one great art of conversation" is a popular quote in philosophy, a greater art in education is having productive conversations, as we learn to speak so we can speak to learn. Decades of research on classroom discourse have supported similar findings across a range of domains, including math, science, and English language arts [21,22,25]. Indeed, rich classroom talk has been highlighted as a key component in national educational standards in math [23,24]. Numerous theories purport that students benefit from collaborative interactions and dialogue, particularly in instructional settings [2,7,31]. There is emerging

A. M. Olney et al. (Eds.): AIED 2024, LNAI 14830, pp. 150–164, 2024.
https://doi.org/10.1007/978-3-031-64299-9_11

empirical evidence to back up these theories, suggesting that students learn more from discussion-based instruction compared to direct instruction [10,27,32].

However, as not every type of talk sustains learning, discourse frameworks such as transactive [19], exploratory [21], and accountable [22,29] talk have emerged to identify and define what type of talk is most consequential. Importantly, these frameworks enable the development of computational approaches to filtering and labeling relevant utterances.

In the context of human tutoring, which is our *present focus*, high-quality discourse has been shown to support students' engagement, critical thinking, and conceptual understanding [18,21]. However, many tutors lack training and experience in facilitating tutorials that incorporate rich discourse [30]. This work focuses on **talk moves** [16,26], which are a key component of accountable talk. While some studies on talk moves have demonstrated their association with improved discourse and student learning [10,27,32], prior work has yet to directly connect, *at scale*, instructional discourse models with learning outcomes, which is one goal of the present work.

To mitigate the high costs of human tutoring, researchers and practitioners are blending human tutoring with e-learning environments, specifically Intelligent Tutoring Systems (ITSs). There is a rich history of inferring learning outcomes based on students' behaviors in these environments, including time on task and accuracy in applying problem-specific principles [8,12] and the use of hints and problem-solving strategies [1,5,6,12]. Indeed, substantial work has leveraged artificial intelligence and machine learning tools to use students' logs in predicting performance in e-learning environments [5,14] and external assessments [8,12,17]. Despite the considerable success of these tools, prior research has yet to investigate how individual-based ITS performance interacts with group-based tutoring discourse in explaining and predicting students' learning outcomes.

In this work, we conduct a *large-scale evaluation* of whether the interaction of tutoring discourse (human tutor's talk moves) and students' content knowledge (performance on a math ITS) predicts students' math assessment scores. Our investigation is in the context of high-dosage, small group tutoring sessions with a human tutor, which have emerged as a key tool to address pandemic-related learning loss and help close achievement gaps [13]. Using data from 1080 9th-grade students from a large urban Midwestern district, we investigate whether an interpretable model can predict students' math achievement based on combinations of talk moves and ITS performance metrics. Our approach involves extracting a decision tree from a random forest classifier to get the benefits of high *interpretability* from the former and high *accuracy* from the latter.

1.1   Instructional Discourse

Productive discussions between teachers and students have long been a center of intensive study in educational research. For example, Webb et al. [32] showed that students' participation in classroom math conversations predicted their achievement. In particular, they found that teachers' encouragement of and follow-up on students' productive talk (i.e., talk moves) increased students' engagement, which cascaded to improve their learning outcomes.

Three popular perspectives to model meaningful teacher-student dialogues are transactive [19], exploratory [21], and accountable [22,29] talk. Transactive talk involves students transforming arguments they hear by building on each other's reasoning. This process involves refuting arguments till a final, winning argument is reached based on the group discussion. Exploratory talk involves dialogue in which students offer the relevant information they have such that everyone engages critically and constructively with others' ideas, all members try to periodically reach an agreement about major ideas, and all ideas are treated as worthy of attention and consideration.

Accountable talk theory identifies and defines an explicit set of discursive techniques that can promote rich, knowledge-building discussions in classrooms [22,29]. At the heart of accountable talk is the notion that teachers should organize discussions that promote students' equitable participation in a rigorous learning environment where their thinking is made explicit and publicly available to everyone in the classroom. Accountable talk outlines **three** general requirements of classroom discussions: accountability to the learning community, to content knowledge, and to rigorous thinking [26,29]. In this work, we focus on *talk moves*, which are linguistic acts that are intended to facilitate dialogue.

Prior studies have articulated that talk moves show potential in improving discourse and students' learning outcomes [10,27,32]. Chen et al. [10] developed a visualization tool to support teachers' reflections on classroom discourse, particularly the use of talk moves. They found that the tool significantly increased the teachers' use of productive talk moves compared to teachers who never used the tool. Additionally, students of the former set of teachers had significantly higher math achievement scores than their peers. O'Connor et al. [27] showed over two studies that teachers' use of talk moves to facilitate academically productive talk was associated with significantly higher standardized math tests in their students when compared to those who received direct instruction.

In essence, substantial research has shown the potential of rich discussions, as indicated by the use of talk moves, in improving students' learning outcomes. However, as far as we know, no attempts were made to connect group-based discourse during tutoring sessions with individual-based ITS performance.

1.2   Student Logs and Performance on Intelligent Tutoring Systems

Based on the considerable logistical challenges associated with offering frequent human tutoring, Intelligent Tutoring Systems (ITSs) are sought as interactive e-learning environments where students have the opportunity to individually learn in a personalized fashion [11]. Prior work has shown that tracing students' logs and overall performance on ITS is predictive of their final performance on ITSs [1,3–6,14,15] and on test scores [8,12,17]. The traced logs include time on task and the use of problem-solving strategies, amongst others.

To predict and influence learning outcomes on ITSs, Islam et al. [15] showed that an apprenticeship learning framework effectively modeled students' pedagogical decision-making strategies on a probability ITS and impacted students' learning gains positively. Hostetter et al. [14] revealed that students with specific personality traits benefited significantly more from personalized explanations individually tailored to their pedagogical decisions. Abdelshiheed et al. [1,3–6] found that the knowledge of how and when to use each problem-solving strategy predicted students' learning outcomes on a logic ITS and the transfer of metacognitive knowledge to a subsequent probability ITS.

In the context of predicting test scores, Baker et al. [8] found that individual student modeling frameworks performed significantly better than ensemble models on a genetics ITS for predicting students' paper test scores. Jensen et al. [17] showed that context-specific activity features extracted from interaction patterns on a math platform were significantly predictive of post-quiz performance. Feng et al. [12] tracked the students' number and average of hints received and requested to predict a standardized math assessment (i.e., the MCAS) of high school students. They found that the fitted regression models showed positive evidence of predicting the assessment scores.

In short, considerable work has leveraged students' performance on ITSs in predicting their achievement. However, as students practice individually on ITSs, it remains an open question whether their performance interacts with the instructional environment they experience when working with a human educator, such as the discourse occurring during small group tutorial sessions. In this study, we address this research question to explain and predict math assessment scores of students who receive tutoring from both an ITS and a human tutor.

1.3   Present Study

We investigate whether group-based human tutoring discourse interacts with the student's individual ITS performance to explain and predict the student's math achievement. We evaluate our research question in the *context* of high-dosage, small group tutorials with a human tutor on 1080 9th-grade students from a large urban Midwestern district. We focus on whether the *combination* of the tutor talk moves and the student's ITS performance can accurately predict the student's math assessment scores and changes in those scores over time.

We prioritize inducing an *interpretable* model that *accurately* explains and predicts students' math achievement in terms of the interaction between tutor talk moves and students' ITS performance. Specifically, we extract a Decision Tree (DT) from a Random Forest Classifier (RFC) to leverage the benefits of the high interpretability of a DT and the high accuracy of a RFC. We emphasize that the interpretability and accuracy of the model take higher *priority* than computational efficiency for the interest of the present work.

2   Methods

2.1   Participants

Students received tutoring during their regular school day (i.e., a second math period), where they attended small group tutorials with a human tutor every other day. Each group comprised at most six students (though most were two or three students) and was assigned to a tutor at the beginning of the year. Providing human tutors was part of a partnership between a large urban Midwestern district and *Saga Education*, a non-profit tutoring service provider. The frequency and duration of tutorials depended on many factors, such as the schedule, the tutor's pace, and the group's pace of processing information. A tutorial was typically 30 to 60 minutes long, and students received $2-3$ tutorials a week, yielding approximately $70-85$ tutorials in the school year.

Students alternated between working with the human tutor and individually working on the ***MATHia*** ITS from Carnegie Learning[1] to work on assigned math problems. Students practiced on the ITS without peer or human tutoring support. The ITS consists of 165 unique workspaces, each covering an algebraic or geometric topic. A workspace consists of topic-specific *skills* where a student aims to master each skill. On average, a workspace has four skills, yielding a total of 645 skills. Each workspace had a limited availability as it could be assigned only at a certain time of the year. Due to this limited availability and high content variability between workspaces, we collected the ***aggregate*** features of the ITS (Sect. 2.2). From now on, we refer to human tutors as '***tutors***' and MATHia ITS as '***ITS***' to distinguish the *human* factor from the *artificial* one.

The participants in this study are 9th-grade students who received math tutoring during the 2022-23 school year, where the majority of students ($> 80\%$) were low-income. During the school year, students completed five rounds of a math skills assessment. The assessments used or adapted items from existing measures with demonstrated reliability and validity and measured students' content knowledge at their current level as well as below-grade fundamentals. Each assessment consists of 30 questions that were graded in a binary manner, resulting in integer scores within the $[0, 30]$ range. The five assessments occurred roughly in August, October, January, March, and May.

Students occasionally skipped assessment rounds, tutorials, and ITS sessions. Out of 1521 students, we only analyzed data from 1080 students across 46 tutors who had **at least** one assessment and attended **at least** 50% of the tutorials and 50% of the ITS sessions during the school year. Table 1 shows the distribution of completed assessments over the final set of included participants.

Table 1. Completed Assessment Rounds Distribution Across Students

# Completed Assessments Per Student	# Students	# Completed Assessments
1	112	112
2	65	130
3	127	381
4	379	1516
5	397	1985
	Total: **1080**	Total: **4124**

---

[1] https://www.carnegielearning.com/solutions/math/mathia.

2.2 Input Features: Talk Moves and ITS Metrics

Tutorials were audio and video recorded as part of the standard protocol used by *Saga Education*, the tutoring service provider. To investigate how the *tutor talk* interacts with *students' proximal content knowledge*, we analyzed the tutors' talk moves from tutorials and the students' current performance on the ITS. In alignment with our goal of extracting an interpretable model, we selected a **limited** number of input features —talk moves and ITS metrics— to ensure exhausting as many possible feature combinations without computational limitations.

The talk moves framework [16, 26] categorizes an utterance made by a tutor or student into one of several labels that capture different dimensions of communication and learning. Table 2 lists the six tutor talk moves included in our study. For the scope of this work, we only focused on tutor talk moves within these categories. To automatically measure tutor talk moves, we leveraged a Robustly Optimized Bidirectional Encoder Representations from Transformers Pretraining Approach (RoBERTa) [20] model, which we fine-tuned on large data sets of talk in classrooms and a small number of Saga tutoring sessions. Details on model training and validation are discussed in Booth et al. [9]. Overall, the model was moderately accurate with a macro F1 of 0.765 for tutor talk moves.

For each student, on each completed assessment, we collected *six* features (the six tutor talk moves) for each tutorial session. For each talk move, we computed the **micro-average** *per session* during tutorials that occurred in between each assessment round (i.e., after the previous but before the current assessment date). We note that we had no way of identifying whether each tutor's talk move was directed to a specific student or a set of students in the tutorial.

Table 2. Description of Tutor Talk Moves

Category	Tutor Talk Move	Description
Learning Community	Keeping Students Together	Orienting students to each other and to be active listeners
	Getting Students to Relate to Each Other's Ideas	Prompting students to react to what another student said
	Restating	Verbatim repetition of all or part of what a student said
Content Knowledge	Press for Accuracy	Eliciting mathematical contributions and vocabulary
Rigorous Thinking	Press for Reasoning	Eliciting explanations, evidence, thinking aloud, and ideas' connection
	Revoicing	Repeating what a student said while adding on or changing the wording

Based on ITS data, we collected *five* aggregate features for each student to reflect the overall performance before each assessment round: 1) average number of mastered skills, 2) average number of opportunities needed before mastering a skill, 3) average time spent on a workspace, 4) average performance score on a workspace, and 5) average Adaptive Personalized Learning Score (APLS), which is based on a combination of the above.

2.3 Rationale of Decision Tree Extraction from Random Forest

Three popular, non-parametric, supervised learning algorithms are *decision trees, random forest classifiers,* and *random forest regressors*. A Decision Tree

(DT) produces a **single**, interpretable tree from *all* features of a dataset whose response variable is *discrete*. While a Random Forest Classifier (RFC) is also trained on a *discrete* response variable, it is an **ensemble** model based on the majority vote of many DTs that were each trained on a *random subset* of the input features and data points. A Random Forest Regressor (RFR) is similar to a RFC but rather works on a *continuous* response variable.

A main advantage of DTs is their high interpretability, as the final output is a single tree that is easy to understand. However, DTs are more prone to overfitting, as they can easily overlearn the patterns to produce a perfect split of the data, resulting in models without generalizability on unseen data. In contrast, RFCs and RFRs overcome the issue of overfitting by training many trees on random subsets of features and data points. However, RFCs and RFRs lack interpretability as the majority voting mechanism is difficult to explain, especially when the number of trees comprising the ensemble model increases.

Interestingly, DTs and RFCs have an advantage —that RFRs lack— from having a discrete response variable: the **no need** for **threshold exploration**. For example, assuming a response variable of an individual's income, the discrete version will have income labeled as *high* or *low* for each individual. Therefore, a DT or RFC would simply evaluate the Gini index or information gain for each potential split on high versus low incomes. However, a continuous version of the income will have various numbers, so a RFR would have first to determine (explore) the **threshold** of distinguishing high from low incomes before training the model. The issue with threshold exploration is that it is highly prone to local minima that evaluate suboptimal rather than optimal thresholds, despite many attempts to optimize such exploration [28].

In this work, although our assessment scores are within the $[0, 30]$ range and could be treated as continuous, we avoid the RFR's threshold exploration issue by following these three steps:

1- Attempting every reasonable threshold to binarize scores into high and low.
2- Training a Random Forest Classifier (RFC) on each reasonable threshold.
3- Avoiding the RFCs' lack of interpretability by extracting a Decision Tree (DT) from each RFC, which simultaneously avoids the DT's overfitting issue.

By exploring many possible thresholds for splitting, we are **not** hypothesizing that our approach is efficient. Rather, we prioritize **exhausting** all possible combinations of extracted decision trees to find one that best explains and predicts the assessment scores from the tutor's talk moves and the tutee's ITS metrics.

2.4 Procedure

Figure 1a shows the data analysis mechanism for a student with no missing data. For each assessment round, the preceding six tutor talk moves and five ITS metrics features were collected as described in Sect. 2.2. To account for missing data (as there are only 397(37%) out of 1080 students who had data on all five assessment rounds), Fig. 1b illustrates the **generic** format of collating data before each assessment round. Since assessment rounds happen on specific dates

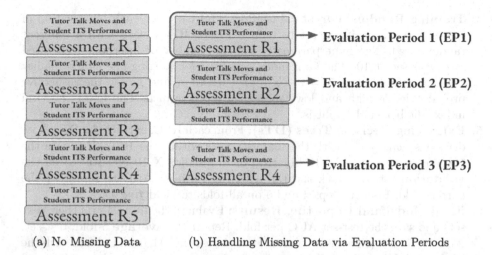

(a) No Missing Data   (b) Handling Missing Data via Evaluation Periods

Fig. 1. Data Collected for a Student in a School Year: Tutor Talk Moves, Student's ITS Performance, and Assessment Rounds Scores.

during the year, we leveraged the notion of an Evaluation Period (EP), where tutor talk moves and ITS metrics are only considered in the period that follows the previous assessment date. We had a total of 4124 EPs, one per completed assessment for each student, as suggested by the rightmost column of Table 1.

Algorithmic Procedure: Each EP is a record in our dataset with the input as 11 features (6 tutor talk moves + 5 student's ITS metrics) and the output as the assessment score. We extracted the decision tree using this **six-step** procedure:

1. **Binarizing Scores:** Pick a set of thresholds that distinguish high from low assessment scores. We explored thresholds from 15 to 24, as 2531(61%) out of 4124 assessment scores were within that range from the original $[0, 30]$ range. Convert scores into high and low, once per candidate threshold. Values greater than or equal to the threshold become 'high,' while those below it become 'low.'

2. **Stratified, Tutor-Based[2], Five Folds for Cross Validation:** Generate 1000 random partitions of the dataset, where each partition is a five-fold candidate, and each fold has unique tutors with their students. Pick the **best** five-fold candidate that minimizes the sum of squared error of this constraint: each fold having $\approx 20\%$ of the students being tutored by $\approx 20\%$ of the tutors. The selected five-fold partition of the 46 tutors (T) and 1080 students (S) was as follows: {(T:8, S:184), (T:8, S:194), (T:10, S:225), (T:10, S:238), (T:10, S:239)}.

3. **A 60-20-20 Nested Cross-Validation:** Each fold from Step 2 acts as the **test** set *exactly once*. For the remaining four folds, one is randomly chosen as the **validation** set, and the other three as the **training** set.

---

[2] Students belonging to the same tutor are in training or testing sets, but not both.

4. **Training Random Forest Classifiers (RFCs):** For each binarization of scores, use the **training** set to induce 10,000 RFCs, each with a different random seed. The hyperparameter for the number of internally generated trees was set at 10. The Gini index was used for judging the splits' quality and was computed as $G = 1 - \sum_{i=1}^{2} p_i^2$, where p_1 and p_2 are the respective probabilities of high and low labels in a given branch. The lower the Gini index, the better the split is.

5. **Extracting Decision Trees (DTs):** From each RFC, extract the DT whose decisions *agree most* with the majority vote within the RFC. Evaluate the 100,000 extracted DTs (10 binarization thresholds X 10,000 RFCs) on the **validation** set, and pick the DT with the **highest** AUC to represent the current fold. Repeat Steps 4 and 5 on all folds to yield *five* DTs.

6. **Final Model and Reporting Results:** Evaluate the five DTs on their **test** sets and save the test-set AUC per fold. Report the **average** 5-fold, test-set, AUC in Table 3 (rightmost column), but choose the DT from the fold with the **highest** AUC as the *final* model to evaluate on the whole dataset (Figs. 2 and 3). To aid visualization, convert the labels (high vs. low) back to original numeric values and show the mean(SD) of assessment scores for each branch.

To compare our approach to standard classifiers, we appropriately modified some steps of our procedure to yield the DTs, RFCs, and RFRs shown in the middle columns of Table 3. Importantly, to induce RFRs, we skipped score binarization (Step 1), then merged and modified Steps $4-6$ to train 10,000 RFRs based on the squared-error criterion, evaluate each on the validation set, and report the **average** 5-fold, test-set, performance. Since a RFR's default output is a coefficient of determination (R^2) between predictions and ground truth, we converted the R^2 to AUC via an effect size converter[3].

3   Results

We investigated *three* analyses via our algorithmic procedure: 1) how tutor talk moves and student's ITS metrics **interact** to predict assessment scores for each round, 2) whether the algorithmic procedure can accurately predict **changes** in

Table 3. Average 5-fold, test-set, AUCs of Talk Moves, ITS, and Their Combination

	Decision Tree (DT)	Random Forest Classifier (RFC)	Random Forest Regressor (RFR)	Extracted DT from RFC
		Predicting Assessment Scores		
Tutor Talk Moves	0.54	0.58	0.53	0.63
MATHia ITS	0.56	0.59	0.54	0.66
Combined (Talk Moves + ITS)	0.59	0.63	0.57	0.77
		Predicting the *Changes* in Assessment Scores		
Tutor Talk Moves	0.55	0.57	0.55	0.62
MATHia ITS	0.56	0.59	0.56	0.64
Combined (Talk Moves + ITS)	0.57	0.64	0.58	0.76

Unlike average 5-fold AUCs here, Figs. 2 and 3 evaluate final models on the *whole* dataset.

---

[3] https://www.escal.site.

assessment scores across rounds, and 3) how our decision tree extraction from a random forest classifier **compares** to other classifiers. Table 3 summarizes the average 5-fold AUCs of different predictors (columns) on different inputs (rows).

3.1 Interaction of Talk Moves and Intelligent Tutoring Metrics

Figure 2 shows the best-extracted decision trees from Table 3 (rightmost column, top half) but on the **whole** dataset (and hence the figures have *higher AUCs*[4]). Coincidentally, the three trees were generated by picking the threshold of 20 for splitting high and low scores. Each root node starts from 4124 assessment rounds, and the rounds are split as we progress downwards. The Mean (SD) is shown for high (in green) and low (in red) scores following the binary labeling. The AUC score ranges from 0 to 1 and reflects how well a decision tree separates high scores from low ones, where AUC of 0.5 denotes prediction by chance (luck).

Figure 2a shows that the decision tree resulting from tutor talk moves only ($AUC = 0.64$) is not very informative, despite the potentially promising right half of the tree. In particular, Press for Accuracy is the most discriminating talk move. The right half suggests that the use of Revoicing results in improved discrimination of 24% (from 17 to 21.1) for the High Press for Accuracy group. However, the left half of Fig. 2a shows no score discrimination for the Keep Together talk move among the Low Press for Accuracy group.

Figure 2b illustrates that the tree based on ITS metrics only ($AUC = 0.68$) is slightly better than Fig. 2a, as both halves of the tree are more stable. Specifically, for students who need more than four opportunities to master a skill —denoting low mastery— the assessment scores improve on average by 13% (from 18 to 20.4) when they have higher Adaptive Personalized Learning Score (APLS) scores. A similar 12% average improvement (from 18.2 to 20.3) occurs for high-mastery students when achieving higher performance scores.

The *best* decision tree ($AUC = 0.79$) emerges from combining all features as shown in Fig. 2c. The left half of the tree shows that Revoicing significantly distinguishes scores for low-mastery students. On the other hand, the right half suggests Press for Reasoning is a significant discriminator of scores —within one to two standard deviations— for high-mastery students. In brief, on average, the assessment scores witnessed a 50% improvement (from 15.2 to 22.8) for the high usage of Revoicing for low-mastery students and a 45% improvement (from 15.7 to 22.7) for the high usage of Press for Reasoning for high-mastery students.

3.2 Predicting the *Changes* in Assessment Scores

To assess whether our results would persist after including students' prior assessment scores, we repeated the algorithmic procedure (Sect. 2.4) using the *difference* between consecutive assessment rounds as the output. Accordingly, we excluded the 112 students in the first row of Table 1 as they only had one assessment, making it impossible to compute their score change. We adjusted the

---

[4] Final models yield respective AUCs of 0.64, 0.68, and 0.79 on the *whole* dataset.

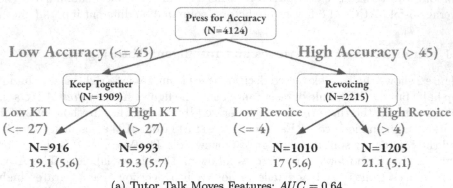

(a) Tutor Talk Moves Features: $AUC = 0.64$

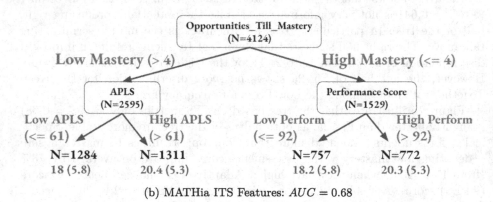

(b) MATHia ITS Features: $AUC = 0.68$

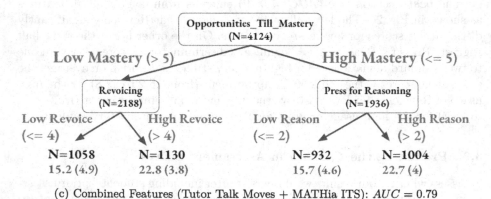

(c) Combined Features (Tutor Talk Moves + MATHia ITS): $AUC = 0.79$

Fig. 2. Extracted Decision Trees for Explaining Assessment Scores on *Whole Dataset*

candidate thresholds of splitting high versus low *changes* in assessment scores to be [2, 2.5, 3, 3.5], as 68% of the changes in scores were in that range.

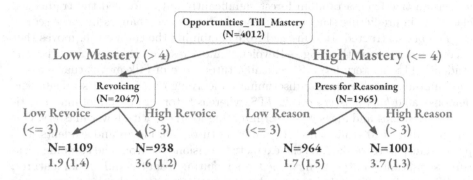

Fig. 3. Extracted Decision Tree for Predicting *Changes* in Scores ($AUC = 0.77$)

Figure 3 depicts the best-extracted decision tree from Table 3 (rightmost column, bottom half) but on the **whole** dataset. The tree comprised 968 students with 4012 assessments and was generated from a threshold of 2.5 for splitting high and low changes in scores. As the tree in Fig. 3 used the same nodes as the tree in Fig. 2c and had a comparable AUC of 0.77, this provides evidence of the robustness of the Revoicing and Press for Reasoning talk moves in discriminating assessment scores for low- and high-mastery students, respectively.

3.3 Comparison to Other Classifiers

To evaluate the accuracy of our procedure of extracting Decision Trees (DTs), we compared it against a traditional DT, a Random Forest Classifier (RFC), and a Random Forest Regressor (RFR), as shown in the middle columns of Table 3. None of these attempts came close to the results from the last column of Table 3. Specifically, the best traditional DT had an AUC of 0.59 and likely overfitted the data by picking suboptimal features, such as Press for Accuracy, at the root node due to its low Gini index score. The best RFC had an AUC of 0.64 and may have suffered from the majority vote mechanism, which minimized the role of meaningful trees' votes as they were a minority. Finally, the best RFR had an AUC of 0.58 (coefficient of determination: $R^2 = .018$), presumably due to the local minima resulting from the lack of picking the optimal threshold for splitting. This analysis suggests that our approach to generating a predictive model yields both robust and meaningful results.

4   Discussion and Conclusion

We investigated different methods of *interpreting* how a tutor's talk moves *interacts* with the student's content knowledge on a math Intelligent Tutoring System

(ITS) for predicting achievement on external math assessments. Our investigation occurred in the context of high-dosage, small group human tutoring sessions that were blended with individual ITS performance. We found that extracting a decision tree from a random forest significantly outperformed the traditional classifiers in predicting students' assessment scores and changes in those scores.

The best-extracted tree emerged from combining the tutor's talk moves that encouraged rigorous thinking —Revoicing and Press for Reasoning— with the student's ITS mastery level. Specifically, tutors' use of talk moves that encouraged mathematical *reasoning* discriminated learning outcomes for students who demonstrated *high mastery* on the ITS, whereas tutors' *revoicing* of mathematical contributions and ideas was discriminating for those with lower mastery. Our findings about the significance of combining tutor talk moves and students' ITS performance were verified, as the extracted decision tree from the combined features significantly outperformed the trees resulting from the individual features.

Whereas it is intriguing to consider the possibility that these different types of talk (revoicing vs. reasoning) might benefit students with different levels of mastery on the ITS, our results are **correlational** and not causal. In particular, tutors may have perceived higher-knowledge students (i.e., those with higher ITS mastery) as more confident and capable of showing their reasoning, being asked about their math ideas, or seeing other students in their group provide explanations. Meanwhile, tutors may have perceived lower-knowledge students (i.e., those who needed more time or effort to master ITS content) as feeling more comfortable with the tutor being the one to build on students' expressed math ideas. Although the tutors centered discussions around student thinking in both cases, pressing for reasoning was predictive of achievement for students with higher content mastery, while revoicing predicted achievement for students with lower content mastery. These findings speak to the significance of future interventions in raising human educators' awareness about these two talk moves and their differential impact on students with different levels of content mastery.

Limitations and Future Work. There are at least three caveats in our work. First, we did not analyze temporal dependencies between talk moves and ITS performance metrics. Future work should investigate possible causal and temporal relationships. Second, our approach of extracting decision trees from a random forest is not efficient as we prioritized more accurate models by exploring many candidate thresholds for splitting and by exhausting as many possible feature combinations without computational limitations. Future work should balance accuracy with efficiency. Finally, our analysis did not account for other student-specific factors, such as demographics, socio-economic status, and tutorial attendance, which are important directions for future work.

Acknowledgments:. This research was supported by the Learning Engineering Virtual Institute (LEVI) program and the National Science Foundation (grants #2222647 and #1920510). All opinions are those of the authors and do not necessarily reflect those of the funding agencies.

References

1. Abdelshiheed, M., Barnes, T., Chi, M.: How and when: The impact of metacognitive knowledge instruction and motivation on transfer across intelligent tutoring systems. IJAIED, pp. 1–34 (2023). https://doi.org/10.1007/s40593-023-00371-0
2. Abdelshiheed, M., Jacobs, J., D'Mello, S.: Not a team but learning as one: the impact of consistent attendance on discourse diversification in math group modeling. In: UMAP. Springer (2024)
3. Abdelshiheed, M., et al.: Metacognition and motivation: The role of time-awareness in preparation for future learning. In: CogSci, pp. 945–951 (2020)
4. Abdelshiheed, M., et al.: Bridging declarative, procedural, and conditional metacognitive knowledge gap using deep reinforcement learning. In: CogSci (2023)
5. Abdelshiheed, M., et al.: Leveraging deep reinforcement learning for metacognitive interventions across intelligent tutoring systems. In: Wang, N., Rebolledo-Mendez, G., Matsuda, N., Santos, O.C., Dimitrova, V. (eds.) AIED. Springer, Cham (2023). https://doi.org/10.1007/978-3-031-36272-9_24
6. Abdelshiheed, M., et al.: Example, nudge, or practice? Assessing metacognitive knowledge transfer of factual and procedural learners. UMUAI (2024)
7. Alexander, R.: A Dialogic Teaching Companion. Routledge (2020)
8. Baker, R.S.J., Pardos, Z.A., Gowda, S.M., Nooraei, B.B., Heffernan, N.T.: Ensembling predictions of student knowledge within intelligent tutoring systems. In: Konstan, J.A., Conejo, R., Marzo, J.L., Oliver, N. (eds.) UMAP 2011. LNCS, vol. 6787, pp. 13–24. Springer, Heidelberg (2011). https://doi.org/10.1007/978-3-642-22362-4_2
9. Booth, B.M., et al.: Human-tutor coaching technology (HTCT): automated discourse analytics in a coached tutoring model. In: LAK, pp. 725–735 (2024)
10. Chen, G., et al.: Efficacy of video-based teacher professional development for increasing classroom discourse and student learning. J. Learn. Sci. **29**(4–5), 642–680 (2020)
11. D'Mello, S.K., Graesser, A.: Intelligent tutoring systems: how computers achieve learning gains that rival human tutors. In: Handbook of Educational Psychology, pp. 603–629. Routledge (2023)
12. Feng, M., Heffernan, N.T., Koedinger, K.R.: Predicting state test scores better with intelligent tutoring systems: developing metrics to measure assistance required. In: Ikeda, M., Ashley, K.D., Chan, T.-W. (eds.) ITS 2006. LNCS, vol. 4053, pp. 31–40. Springer, Heidelberg (2006). https://doi.org/10.1007/11774303_4
13. Guryan, J., et al.: Not too late: improving academic outcomes among adolescents. Tech. Rep., National Bureau of Economic Research (2021)
14. Hostetter, J.W., et al.: XAI to increase the effectiveness of an intelligent pedagogical agent. In: IVA, pp. 1–9 (2023)
15. Islam, M.M., et al.: A generalized apprenticeship learning framework for modeling heterogeneous student pedagogical strategies. In: EDM (2024)
16. Jacobs, J., et al.: Promoting rich discussions in mathematics classrooms: using personalized, automated feedback to support reflection and instructional change. Teach. Teach. Educ. **112**, 103631 (2022)
17. Jensen, E., et al.: What you do predicts how you do: prospectively modeling student quiz performance using activity features in an online learning environment. In: LAK, pp. 121–131 (2021)
18. Kiemer, K., Gröschner, A., Pehmer, A.K., Seidel, T.: Effects of a classroom discourse intervention on teachers' practice and students' motivation to learn mathematics and science. Learn. Instr. **35**, 94–103 (2015)

19. Kruger, A.C.: Peer collaboration: conflict, cooperation, or both? Soc. Dev. **2**(3), 165–182 (1993)
20. Liu, Y., et al.: RoBERTa: a robustly optimized BERT pretraining approach. arXiv preprint arXiv:1907.11692 (2019)
21. Mercer, N., Hennessy, S., Warwick, P.: Dialogue, thinking together and digital technology in the classroom: some educational implications of a continuing line of inquiry. Int. J. Educ. Res. **97**, 187–199 (2019)
22. Michaels, S., et al.: Deliberative discourse idealized and realized: accountable talk in the classroom and in civic life. Stud. Philos. Educ. **27**, 283–297 (2008)
23. National Council of Teachers of Mathematics (NCTM): Principles and standards for school mathematics. Reston, VA (2000)
24. National Governors Association (NGA): Common core state standards. Washington, DC (2010)
25. Nystrand, M., Gamoran, A.: Instructional discourse, student engagement, and literature achievement. In: Research in the Teaching of English, pp. 261–290 (1991)
26. O'Connor, C., Michaels, S.: Supporting teachers in taking up productive talk moves: the long road to professional learning at scale. Int. J. Educ. Res. **97**, 166–175 (2019)
27. O'Connor, C., Michaels, S., Chapin, S.: "scaling down" to explore the role of talk in learning: from district intervention to controlled classroom study. In: Socializing Intelligence Through Academic Talk and Dialogue, pp. 111–126 (2015). https://www.jstor.org/stable/j.ctt1s474m1
28. Probst, P., et al.: Hyperparameters and tuning strategies for random forest. Wiley Interdiscip. Rev. Data Mining Knowl. Disc. **9**(3), e1301 (2019)
29. Resnick, L.B., Michaels, S., O'Connor, C.: How (well structured) talk builds the mind. Innov. Educ. Psychol. Perspect. Learn. Teach. Hum. Dev. **163**, 194 (2010)
30. Stein, C.A.: Let's talk: promoting mathematical discourse in the classroom. Math. Teach. **101**(4), 285–289 (2007)
31. Vygotsky, L., et al.: Interaction Between Learning and Development. Linköpings universitet, Linköping, Sweden (2011)
32. Webb, N.M., et al.: Engaging with others' mathematical ideas: Interrelationships among student participation, teachers' instructional practices, and learning. Int. J. Educ. Res. **63**, 79–93 (2014)

Automated Educational Question Generation at Different Bloom's Skill Levels Using Large Language Models: Strategies and Evaluation

Nicy Scaria[1]([⊠])(iD), Suma Dharani Chenna[1,2]([⊠])(iD),
and Deepak Subramani[1]([⊠])(iD)

[1] Computational and Data Sciences, Indian Institute of Science, Bengaluru, India
{nicyscaria,deepakns}@iisc.ac.in
[2] School of Computer Science and Engineering, VIT-AP University, Amaravati, India
sumadharanichenna@gmail.com

Abstract. Developing questions that are pedagogically sound, relevant, and promote learning is a challenging and time-consuming task for educators. Modern-day large language models (LLMs) generate high-quality content across multiple domains, potentially helping educators to develop high-quality questions. Automated educational question generation (AEQG) is important in scaling online education catering to a diverse student population. Past attempts at AEQG have shown limited abilities to generate questions at higher cognitive levels. In this study, we examine the ability of five state-of-the-art LLMs of different sizes to generate diverse and high-quality questions of different cognitive levels, as defined by Bloom's taxonomy. We use advanced prompting techniques with varying complexity for AEQG. We conducted expert and LLM-based evaluations to assess the linguistic and pedagogical relevance and quality of the questions. Our findings suggest that LLMs can generate relevant and high-quality educational questions of different cognitive levels when prompted with adequate information, although there is a significant variance in the performance of the five LLMs considered. We also show that automated evaluation is not on par with human evaluation.

Keywords: Large Language Models · Automated Educational Question Generation · Bloom's Taxonomy

1 Introduction

Transformer-based pre-trained large language models developed in recent years have drastically improved the quality of natural language generation (NLG) tasks [24]. With an exponential increase in training data and model size, these models can generate complex text with human expert-level quality. The release of OpenAI's ChatGPT made LLMs accessible to a wider audience who are not experts in natural language processing (NLP), allowing them to use them for their daily

© The Author(s), under exclusive license to Springer Nature Switzerland AG 2024
A. M. Olney et al. (Eds.): AIED 2024, LNAI 14830, pp. 165–179, 2024.
https://doi.org/10.1007/978-3-031-64299-9_12

tasks. The language models are tuned to follow the user instructions through instruction-tuning [24]. They have zero-shot capabilities [10], which means that if you prompt the LLM with detailed task descriptions, the model will create meaningful outputs. These LLMs have the potential to be used in different ways in education [9], including the creation of personalized content, assessments, and feedback.

High-quality assessments enable learners to deeply engage with the subject and relate their learning to the real world. Assessments that focus on different cognitive skills defined in Bloom's taxonomy levels [2] (described in Table 1) help educators identify gaps in student learning. This information allows them to adapt their teaching to better support students and also helps students understand their strengths and weaknesses. However, creating such assessments requires significant time and effort from educators [11]. Automated Educational Question Generation (AEQG) systems reduce the effort and cognitive load on teachers. For example, in a massive open online course (MOOC), a teacher typically engages with a large and diverse audience. Here, AEQG becomes critical to create assessments at different cognitive levels for this diverse audience at scale with minimal effort from the teacher. Students can participate in these assessments to analyze their cognitive skills and take remedial actions as necessary, thereby significantly improving their learning.

Related Work: In the pre-LLM era, AQG research focused mainly on generating questions using question-answer datasets such as SQuAD 2.0, and NQ, containing context, question, and answer [25]. However, the limited availability of public datasets impeded the progress of AQG systems capable of producing good quality questions. Recent research in question generation is focused on using pretrained or fine-tuned LLMs for the process. Models such as Text-To-Text Transfer Transformer (T5) and GPT3, along with context information, were used to generate questions [17]. Pre-training these models with educational text also improved the quality of the questions generated [3]. Recent research has shown also promising results in evaluating the quality of machine-generated content using LLMs using Chain-of-Thought (CoT) prompting on different evaluation criteria. G-Eval [13], an evaluator model based on GPT4, significantly outperformed previous models and aligned with human judgments on summarization tasks. However, the results of some studies that used fine-tuned GPT3 models to evaluate the pedagogical quality of machine-generated questions were unsatisfactory [3,16]. Human expert or crowd evaluations have been extensively used to analyze the pedagogical quality of machine-generated questions [6,19].

The questions generated by most AQG models generally test lower-order skills [20] or create questions that have answers directly mentioned in the text [25]. These questions are not enough to test the higher-order cognitive skills of students. Bloom's taxonomy [2] serves as a guide for educators to generate questions to test different cognitive skills. In a recent work [18], authors used GPT4 to automatically generate learning objectives based on Bloom's taxonomy for a university course on Artificial Intelligence.

1.1 Objective and Research Questions

Our approach utilizes the knowledge of the content inherently present in LLMs along with the addition of technical information on the question generation process in the prompt to generate educational questions. Although LLMs excel in various downstream tasks, they produce errors and inconsistencies [8], which can compromise the quality of the questions generated. This also varies significantly between different LLMs. Therefore, evaluating the quality of the questions generated by LLMs is essential. While metrics such as the BLEU score or perplexity can assess machine-generated questions, they typically only examine linguistic characteristics [22]. In the present work, we perform a manual expert evaluation using the services of two educators in the AEQG topic's domain and an automated LLM evaluation using an LLM that is not employed for AEQG.

Table 1. Revised Bloom's taxonomy [2] in ascending order in the cognitive dimension

Bloom's level	Description
Remember	Retrieve relevant knowledge from long-term memory
Understand	Construct meaning from instructional messages, including oral, written, and graphic communication
Apply	Carry out or use a procedure in a given situation
Analyze	Break material into foundational parts and determine how parts relate to one another and the overall structure or purpose
Evaluate	Make judgments based on criteria and standards
Create	Put elements together to form a coherent whole; reorganize into a new pattern or structure

We used zero-shot and few-shot techniques and CoT prompting to generate questions for a graduate-level data science course using LLMs of different sizes. Five different prompt strategies of varying complexity were used to create these questions. Then, we performed a manual expert evaluation using the services of two educators in the AEQG topic's domain and an automated LLM evaluation using an LLM that is not employed for AEQG. The evaluation was performed on a nine-item rubric to assess their linguistic and pedagogical quality [7] by experts and the LLM. The LLM evaluation is performed as a zero-shot classification task through a specially designed prompt.

Specifically, we investigated answers to the following research questions.

RQ1: Can instruction fine-tuned modern LLMs create high-quality and diverse educational questions at different cognitive levels based on Bloom's taxonomy?

RQ2: Does the size of the LLM significantly impact the model's performance in educational question generation?

RQ3: How does the amount of information provided in the prompt affect the quality of the questions generated?

RQ4: Can LLMs create questions that are relatable to a specific population or context?

RQ5: Can instruction fine-tuned LLMs evaluate generated educational questions effectively, similar to human evaluators, when given the same instructions?

In what follows, we first discuss the methodology. Then, we present and analyze the results before concluding with directions for future research.

2 Methodology

Our study consists of two parts. We use modern LLMs for AEQG in the first part through various prompting strategies. In the second part, we perform human evaluation and LLM evaluation.

2.1 Language Models

Training a language model from scratch or fine-tuning the available models is expensive due to constraints in the availability of educational data and the cost associated with training. Therefore, we used a mix of open-source and proprietary state-of-the-art LLMs for the study. The models used for question generation are Mistral (Mistral-7B-Instruct-v0.1), Llama2 (Llama-2-70b-chat-hf), Palm 2 (chat-bison-001), GPT-3.5 (gpt-3.5-turbo-0613), and GPT-4 (gpt-4-0613). Among these, Mistral has 7 billion parameters, and GPT models are rumored to have trillions of parameters[1]. For LLM-based evaluation of the questions, we used Gemini Pro (gemini-pro).

2.2 Question Generation

We used the five LLMs mentioned in Sect. 2.1 to generate questions of different cognitive levels. A higher temperature setting in LLMs results in a varied and unpredictable text, while a lower temperature setting makes the model output more deterministic and repetitive. Thus, we set the temperature of the LLMs at 0.9 to promote variety and diversity in the generated questions.

Content: The educational questions were generated for a graduate-level data science course comprising topics ranging from traditional machine learning algorithms, such as linear regression, to advanced topics in natural language processing, such as prompt engineering. We did not provide domain-specific information or context on these topics to the models for question generation. This approach was guided by the hypothesis that these models, trained using large amounts of recent Internet data, would possess inherent knowledge related to these contemporary course topics.

---

[1] The information about the number of parameters of the model is not publicly available.

Prompt Design: In the present study, we generated questions by instructing the models with five prompt styles/strategies (PS1 to PS5), each differing in complexity. These prompts followed specific techniques of pattern reframing, itemizing reframing, and assertions to make it easier for the instruction fine-tuned LLMs to follow the specific instructions [15]. Furthermore, the prompts encouraged the model to incorporate Indian-specific examples or context within the question to ensure relevance for Indian students. The first set of prompts (PS1) consisted solely of these core instructions. In contrast, subsequent sets progressively added more specific information and instructions in the prompt to further refine the quality and relevance of the generated questions.

There were mainly three significant additions to the prompts. First, the prompts were augmented with CoT instructions to make the LLM think sequentially about how to proceed with the task. CoT prompting has been shown to improve the quality of LLM-generated content [23]. In the prompt, the LLM was also given the persona of a graduate-level university course instructor creating questions for their students [15]. Second, the questions included definitions of the six cognitive levels of the revised Bloom's taxonomy. This approach was taken under the hypothesis that augmenting the LLM with explicit knowledge of the revised Bloom's taxonomy levels would enhance the quality of the generated questions. Third, an expert-crafted example was provided for each Bloom's taxonomy level. This few-shot approach, proven effective in different generative models [12], leverages human-crafted questions to guide the LLM's understanding of how the questions need to be framed and, in turn, enhance its question generation capabilities. Using these three, we created four different prompts: (PS2) CoT prompt with skill explanation, (PS3) CoT prompt with example questions, (PS4) CoT prompt with skill and example questions, and (PS5) CoT prompt with skill, skill explanation, and example questions.

We gave the same prompt to all LLMs in a specific combination, except for the topic-specific variables corresponding to each course topic, as given in the repository[2]. Each LLM generated six questions, one for each level of Bloom's taxonomy corresponding to the 17 course topics. Each model generated 102 questions, resulting in 510 questions for one combination, making it 2550 for the five prompt combinations.

2.3   Human Evaluation

Two experts evaluated the AEQG questions. Both experts deeply understand data science concepts and have experience teaching the subject to large graduate classes. The experts assessed the relevance and quality of the questions based on a nine-item rubric (Table 2; a modified version of [7]).

The experts were presented with LLM-generated questions in a random order along with the course topic associated with the questions. No further details about the prompt or its contents were shared. The experts hierarchically assessed

---

[2] https://github.com/nicyscaria/AEQG_Blooms_Evaluation_LLMs.

each question, starting from the top of the rubric to the bottom. In the evaluation, each group of the evaluation criteria, as indicated in Table 2, has a stopping point. This structured approach streamlines the evaluation process, minimizing the overall time and effort required for expert annotation. If *Understandable* is marked 'no', then none of the subsequent items are evaluated for that question and are automatically marked as 'NA', indicating not applicable. This design choice reflects the underlying principle that further evaluation of a question that is not understandable does not make sense. Similarly, within group 2, if the answer to *clear* is 'no', then the evaluation stops, and the remaining items are automatically marked as 'NA'. Additionally, if the response to *Clear* is 'more_or_less', then the *Rephrase* criteria of group 3 should be marked, and the evaluator should rephrase the question for clarity. The rephrased version of the question will be used for further evaluation. If the answer to *Answerable* in group 3 is 'no', the evaluation is stopped, and the remaining items of the rubric are marked 'NA'. In group 4, if either *Central* or *WouldYouUseIt* is marked 'no', then the evaluation is stopped, and the *Bloom'sLevel* criteria is labeled 'NA'; otherwise, the experts select the Bloom's level for the question and concludes the evaluation of one question. The process is repeated for all the questions.

Table 2. Hierarchical nine-item rubric used to evaluate questions generated by LLMs

Rubric item	Definition and response option
Understandable	Could you understand what the question is asking? *(yes/no)*
TopicRelated	Is the question related to the topic given in the prompt? *(yes/no)*
Grammatical	Is the question grammatically well-formed? *(yes/no)*
Clear	Is it clear what the question asks for? *(yes/more_or_less/no)*
Rephrase	Could you rephrase to make the question clearer? *(yes/no)*
Answerable	Can students answer the question with the information or context provided within? *(yes/no)*
Central	Do you think being able to answer the question is important to work on the topic given in the prompt? *(yes/no)*
WouldYouUseIt	If you were a teacher teaching the course topic would you use this question or the rephrased version in the course? *(yes/maybe/no)*
Bloom'sLevel	What is the Bloom's skill associated with the question? *(remember, understand, apply, analyze, evaluate, and create)*

The AEQG questions were considered high quality if they met the following criteria: (1) experts marked 'yes' for *Understandable*, *Grammatical*, *Clear*, and *Answerable*; (2) received a 'yes' or 'maybe' for *WouldYouUseIt*; or (3) being marked 'yes' for 'more_or_less' in the *Clear* criteria and subsequently marked 'yes' for *Rephrase*. Furthermore, we utilized *Bloom'sSkill* to understand whether the LLM adheres to the instructions provided in the prompt. The LLM adhered to the instructions provided if the *Bloom'sSkill* labels of the experts match the Bloom's skill level on the prompt.

Experts perceive the questions generated by LLMs in different ways. This perception is influenced by various factors, such as their preference for writing, personal assumptions, prior knowledge, and attention to detail [1]. Therefore, it is essential to have a measure to ensure the consistency of the expert evaluation. We measure inter-rater reliability using percentage agreement and Cohen's Kappa κ [14]. For the ordinal metrics, *Clear*, *WillYouUseIt* and *Bloom'sLevel*, we use quadratic weighted Cohen's κ [5] instead of simple Cohen's κ to penalize situations with a significant rating difference.

Along with the metrics discussed above, we explored the ability of LLMs to create questions relatable to a specific population or context, which, in this case, is India. To understand this, we curated and analyzed the contexts and themes specific to India that came up in the questions.

2.4 Automated Evaluation

Clearly, the above evaluation by a human expert is a laborious process. Automated evaluation offers a scalable and efficient alternative for assessing large-scale educational content. We use Gemini Pro for the LLM-based evaluation of the generated questions. To ensure deterministic behavior, we set the decoding temperature of the model to 0. In automated evaluation, we assess the quality of LLM-generated questions without reference questions using the same criteria outlined in Sect. 2.3. Recent studies have demonstrated the ability of LLMs to perform a reference-free evaluation for a variety of NLG tasks [13,21]. The prompt used in the prompt-based evaluator consisted of two components: (1) a detailed description of the evaluation criteria, evaluation instructions (instructions to evaluate the questions in a hierarchical manner as discussed in Sect. 2.3) along with the question and the course topic for which the question was generated, and (2) CoT instructions describing the evaluation steps, along with providing the evaluator LLM the persona of a graduate-level data science course instructor. The detailed prompt template can be found in the repository[3].

In addition to automated evaluation, we assessed the linguistic quality of the AEQG questions using a diversity measure based on the Paraphrase In N-gram Changes (PINC) score [4].

$$PINC(s, c) = \frac{1}{N} \sum_{n=1}^{N} \left(1 - \frac{|n\text{-}gram_s \cap n\text{-}gram_c|}{|n\text{-}gram_c|} \right) \tag{1}$$

PINC score is generally used to measure the novelty of n-grams in the automatically generated paraphrase of a sentence. In our case, we wanted to check if the questions generated by an LLM for a specific Bloom's skill use the same structure or words on different topics. For that, we considered every question generated for a specific skill by the model as the source and calculated the PINC score considering every other question of the same skill and by the same model as the candidate question. Finally, the average of these scores was calculated

---

[3] https://github.com/nicyscaria/AEQG_Blooms_Evaluation_LLMs.

for the AEQG questions generated by each LLM. A higher average PINC score indicates considerable diversity in the AEQG task.

3   Results and Analysis

We will be releasing the dataset, 'DataScienceQ'[3] containing 2550 questions generated for the present study. First, we present and analyze the results of the expert evaluation of these 2550 AEQG questions. We start by examining the agreement between experts on their evaluation using the percentage agreement and modified Cohen's κ values (Table 3). The percentage agreements and Cohen's κ values are calculated only for questions not labeled 'NA' as discussed in Sect. 2.3. The values in the table indicate that there is substantial agreement between experts on different evaluation criteria. The agreement is highest for the criteria *TopicRelated*, *Grammatical*, and *Central*. As expected, subjective criteria like *Rephrase* and *WouldYouUseIt* have the lowest agreement. In our analysis, an AEQG question is considered as "High Quality" only when both evaluators rate it as "High Quality". To assess alignment with Bloom's taxonomy, the subsequent analysis focused only on these "High Quality" questions. In what follows, we discuss the results for each research question stated in Sect. 1.1. Table 4 presents the key evaluation metrics (Sect. 2.3) for the analysis. Quality is measured as the percentage of AEQG questions that were selected as "High Quality" and skill is measured as the percentage of "High Quality" questions judged by experts to be at the same Bloom's taxonomy levels as the one given in the prompt to the LLM for the AEQG task.

RQ1: Can Instruction Fine-tuned Modern LLMs Create High-Quality and Diverse Educational Questions at Different Cognitive Levels Based on Bloom's Taxonomy? Among all AEQG questions from the different LLMs and prompting strategies, 78% were rated as "High Quality" and among these 65.56% were rated to match the intended skill level by both human raters (Table 4 'Overall' Columns and 'Overall' row). The temperature of all LLMs was set at 0.9 to promote textual diversity, resulting in a PINC score average of 0.92. These findings suggest that instruction fine-tuned LLMs demonstrate considerable potential to generate diverse and high-quality educational questions at different cognitive levels based on Bloom's taxonomy. For PS1-PS5, 72.55%, 70.58%, 71.56%, 86.27%, and 89.02% of the questions were identified as "High Quality" for Mistral 7B, Llama2 70B, Palm 2, GPT 3.5 and GPT 4 respectively. Similarly, 60%, 61.67%, 60%, 74.41%, and 70.04% of the questions followed adhered to Bloom's taxonomy level given by experts respectively.

RQ2: Does the Size of the LLM Significantly Impact the Model's Performance in Educational Question Generation? Table 4 presents the performance metrics of the five LLMs for the five sets of prompts. For quality and adherence to Bloom's taxonomy levels, GPT 4 and GPT 3.5 emerged as

Table 3. Expert inter-annotator agreement on the nine-item hierarchical rubric for AEQG questions.

Rubric item	Simple prompt		CoT & skill explanation		CoT & example	
	% agree	κ	% agree	κ	% agree	κ
Understandable	99.60%	0.67	99.61%	0.80	100.00%	1.00
TopicRelated	99.80%	0.95	99.80%	0.80	98.80%	0.93
Grammatical	100.00%	1.00	100.00%	1.00	100.00%	1.00
Clear	91.75%	0.67	97.42%	0.62	94.30%	0.61
Rephrase	90.38%	0.76	90.91%	0.62	80.77%	0.59
Answerable	93.41%	0.65	96.47%	0.69	94.75%	0.61
Central	97.42%	0.78	99.77%	0.98	99.77%	0.94
WouldYouUseIt	89.92%	0.58	94.82%	0.66	96.81%	0.69
Bloom'sLevel	82.07%	0.83	88.14%	0.90	90.00%	0.89

Rubric item	CoT, skill, and example		CoT, skill, skill explanation and example	
	% agree	κ	% agree	κ
Understandable	100.00%	1.00	100.00%	1.00
TopicRelated	99.80%	0.99	99.20%	0.96
Grammatical	100.00%	1.00	100.00%	1.00
Clear	93.30%	0.53	92.74%	0.66
Rephrase	92.31%	0.73	75.00%	0.53
Answerable	96.54%	0.81	93.89%	0.68
Central	100.00%	1.00	100.00%	1.00
WouldYouUseIt	95.64%	0.85	95.29%	0.82
Bloom's Level	90.36%	0.93	92.65%	0.93

the top performers. Palm 2, despite its larger size compared to Mistral 7B and LLama2 70B, demonstrated wide variance in the quality of the AEQG task for different prompt strategies. Palm 2 has only 36.99% Skill matching in a detailed and complex prompt (PS5), but it scores 70.51% in the PS3 prompt strategy. For PS5 prompts, the Mistral 7B model performs better than the Llama2 70B model, which is counterintuitive. The reason for this performance difference could be due to the way these models process a long prompt. Thus, no clear pattern exists between the model size and AEQG performance.

RQ3: How Does the Amount of Information Provided in the Prompt Affect the Quality of the Questions Generated? It is observed that the simple prompt (PS1) performed poorly in the quality of the questions generated (Table 4). Overall, the performance improved with the addition of more information to the prompt. However, the amount of improvement varied between the five LLMs. Figure 1 shows that for Mistral, Llama 2 and Palm 2, PS3 gave the highest quality questions, with PS4 being close behind. PS4 gave the highest

Table 4. Performance of the LLMs in the AEQG task. For each model and set of prompts PS1-PS5, the percentage of questions that are of high quality (Quality), adherence to Bloom's taxonomy level (Skill), and the PINC score are presented.

LLM	PS1: Simple prompt			PS2: CoT & skill explanation			PS3: CoT & example		
	Quality	Skill	PINC	Quality	Skill	PINC	Quality	Skill	PINC
Mistral 7B	70.59%	56.94%	0.94	75.49%	68.83%	0.95	78.43%	45.00%	0.94
Llama 2 70B	73.53%	57.33%	0.94	75.49%	71.43%	0.93	77.45%	58.23%	0.93
Palm 2	61.76%	55.56%	0.93	73.53%	65.33%	0.94	76.47%	**70.51%**	0.93
GPT 3.5	69.61%	**81.69%**	0.94	**94.12%**	**89.58%**	0.89	89.22%	64.84%	0.92
GPT 4	**75.49%**	74.03%	0.93	87.25%	82.02%	0.92	**91.18%**	65.59%	0.92
Overall	70.20%	66.36%	0.93	81.18%	76.32%	0.93	82.55%	61.04%	0.93
LLM	PS4: CoT, skill, & example			PS5: CoT, skill, skill explanation & example			Overall		
	Quality	Skill	PINC	Quality	Skill	PINC	Quality	Skill	PINC
Mistral 7B	71.57%	71.23%	0.93	66.67%	58.82%	0.94	72.55%	60.00%	0.94
Llama 2 70B	77.45%	73.42%	0.91	49.02%	40.00%	0.93	70.58%	61.67%	0.93
Palm 2	74.51%	69.74%	0.93	71.57%	36.99%	0.93	71.56%	60.00%	0.93
GPT 3.5	87.25%	73.03%	0.90	**91.18%**	**59.14%**	0.91	86.27%	**73.41%**	0.91
GPT 4	**96.08%**	**75.51%**	0.92	95.10%	56.64%	0.90	**89.02%**	70.04%	0.92
Overall	81.37%	72.77%	0.92	74.71%	51.18%	0.92	78.00%	65.56%	0.93

skill match for these three models, while the skill match for PS3 was low. Interestingly, PS5, which is the most complicated prompt used, reduced both quality and skill for these three models, indicating that while information enrichment improves AEQG, too much information in the prompt can be counterproductive. The GPT models also gave good performance for PS4, but their performance for PS2 to PS5 are more or less the same with respect to quality of questions. These two models did well in terms of skill match for PS2-PS4, and similar to the other three LLMs, the PS5 skill scores drop significantly. Our results indicate that a CoT prompt with a description of the skill and an example question performs best for AEQG.

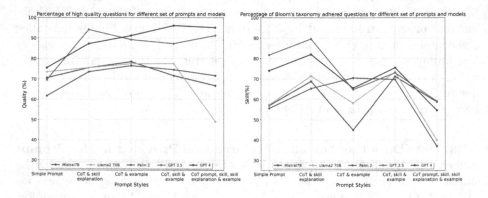

Fig. 1. Quality and Skill of prompting strategies for AEQG by different LLMs.

RQ4: Can LLMs Create Questions that are Relatable to a Specific Population or Context? In our study, LLMs were asked to create questions that are relatable to students in India. Nine recurring themes that are specific to India emerged (Fig. 2). These included questions on Bollywood movies, traffic challenges in Indian cities and their mitigation using computer vision techniques, and the Indian educational system. Given India's significant dependence on agriculture, many questions focused on climate, crop yield, crop diseases, and cropping patterns. In addition, numerous questions, specifically on the topic of natural language processing, incorporated examples in Indian languages such as Hindi, Tamil, Telugu, etc., presented both in their native scripts and transliterated forms (refer to 'DataScienceQ'). Interestingly, the Indian language text generated by open source models, Mistral 7B and Llama2 70B, was often inaccurate and poor quality. From Fig. 2, it is also clear that compared to GPT 4 and GPT 3.5, the recurring themes occur less in the questions generated by other models. Some questions exhibited a tendency to unnecessarily specify India when the context was general. The expert evaluators rephrased these questions. Furthermore, some questions reflect certain cultural generalizations about India that are not necessarily true, as evaluators have indicated. In the present paper, we did not implement any guardrails to reduce or eliminate biases in LLMs.

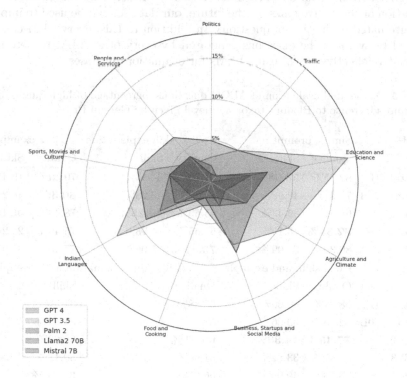

Fig. 2. Frequently repeated Indian contexts in the AEQG questions.

However, we plan to address and incorporate measures to mitigate such biases in future work.

RQ5: Can Instruction Fine-tuned LLMs Evaluate Generated Educational Questions Effectively, Similar to Human Evaluators, When Given the Same Instructions? We conducted an LLM-based evaluation to analyze the quality and adherence of machine-generated questions to different cognitive levels on the nine-item rubric in addition to the expert evaluation. We used Gemini Pro (gemini-pro), an LLM that is different from the five used for the AEQG task, for the evaluation (detailed methodology in Sect. 2.4). The results of the evaluation are given in Table 5. There is a significant discrepancy between LLM-based and expert evaluations. Interesting discrepancies emerged between Gemini Pro and expert evaluations, with Palm 2 excelling in automated evaluation, but underperforming in expert evaluation. In the Gemini Pro evaluation, even Llama 2 70B and Mistral 7B also performed better in some cases. Our automated evaluation using Gemini Pro revealed a tendency of the model to classify most machine-generated questions as belonging to the 'Apply' or 'Analyze' levels on *Bloom'sLevel*. The observed performance dip of the LLMs in adhering to the evaluator-LLM's Bloom's level can be attributed to this fact. This result indicates that extreme caution must be exercised when using LLMs for automated evaluation of generative tasks. In the future, our data set can be used to improve the automated evaluation of questions. In addition to this, every LLM can also be used to evaluate the questions generated by every other LLM to determine the relative effectiveness of using LLMs for evaluation purposes.

Table 5. Automated evaluation of AEQG questions: percentage of high-quality questions and adherence to Bloom's taxonomy level given by Gemini Pro.

LLM	Simple prompt		CoT & skill explanation		CoT & example	
	Quality	Skill	Quality	Skill	Quality	Skill
Mistral 7B	82.35%	**53.98%**	61.76%	44.44%	70.59%	19.44%
Llama 2 70B	79.41%	40.02%	65.69%	43.28%	**80.39%**	40.24%
Palm 2	69.61%	39.43%	65.69%	**49.25%**	78.43%	**40.00%**
GPT 3.5	**82.35%**	48.15%	67.65%	34.78%	75.49%	28.24%
GPT 4	80.39%	38.09%	**75.49%**	35.06%	77.45%	37.97%
LLM	CoT, skill, and example		CoT, skill, skill explanation and example			
	Quality	Skill	Quality		Skill	
Mistral 7B	58.82%	40.00%	55.88%		31.57%	
Llama 2 70B	62.75%	39.06%	53.92%		34.54%	
Palm 2	**77.45%**	**44.30%**	**69.61%**		36.62%	
GPT 3.5	66.67%	33.82%	51.96%		30.18%	
GPT 4	73.53%	40.00%	66.67%		**38.24%**	

4   Discussion and Conclusion

Our study demonstrates that LLMs can produce high-quality and diverse educational questions aligned with Bloom's taxonomy, requiring minimal input from educators, but the performance varies based on the size of the model and the prompt used to generate these questions. Larger proprietary models like GPT 4 and GPT 3.5 outperform smaller open source models across all the metrics in expert evaluation, but the same does not hold for the Palm 2 model. While adding a lot of information (skill explanation, example questions, and CoT instructions) significantly reduced the performance of the LLMs, particularly for open-source models, optimal results were achieved with prompts including CoT instructions paired with either skill, skill explanation, or example questions. CoT instructions, with examples, resulted in more high-quality questions while compromising on the adherence to Bloom's skill. On the other hand, Bloom's skill explanations with CoT instructions slightly reduced the number of high-quality questions but significantly boosted the performance of adherence to Bloom's skill. The questions generated often incorporated contextually relevant Indian contexts, although some instances exhibited generalizations about India. Teachers can employ our PS3 strategy (which is the most efficient among the five strategies) to craft questions at different Bloom's levels. Additionally, this approach can be applied to generate questions for national examinations in different subjects, ensuring an unbiased and evenly distributed assessment across different topic areas.

The evaluation of 2550 questions took a considerable amount of time and effort from expert evaluators. Although attention was paid to making the evaluation process objective, experts' decisions can still be subjective depending on who is evaluating them. However, the LLM-based evaluation proved to be less effective in our case. There was a considerable difference across all metrics in the case of the Gemini Pro evaluation compared to the expert evaluation. Interestingly, in the Gemini Pro evaluation, Palm 2 outperformed other models on different evaluation metrics, even though expert evaluation suggested otherwise. This discrepancy might stem from the fact that Palm 2 and Gemini Pro are both Google's models, potentially sharing similar training data or methodologies. We found that our evaluator model is suboptimal and does not align with the expert evaluation. This could be due to the lack of such examples that these models would have seen during their training. This requires the training of the LLM on evaluation datasets on specific subjects and evaluation metrics to make it robust for evaluation. This is a potential future direction of research. In the current work, we have only used Gemini Pro for automated evaluation. We plan to use every LLM to evaluate the questions generated by every other LLM in the future to understand the relative effectiveness of utilizing LLMs for evaluating machine generated questions in general.

Our approach used the inherent knowledge of the content possessed by the LLMs on the topic for AEQG. This revealed limitations in their understanding of specific domains. For example, Mistral 7B and Llama2 70B struggled on the topics of "prompt engineering", producing questions related to general engineering. This limitation can be addressed using the Retrieval-Augmented Generation (RAG) technique, which allows retrieving relevant information associated with the subject or the course topic from an external database for creating questions. Another interesting observation in the study was the inability of most models to generate high-quality questions at the 'Create' level of Bloom's taxonomy. Thus, the present paper can be extended to use existing resources from the Internet or course material through databases to improve the performance of the model in question generation. We did not study the generation of questions for topics other than data science content, and such an exploration using the methodology showcased here could be a potential future direction for research.

References

1. Amidei, J., Piwek, P., Willis, A.: Rethinking the agreement in human evaluation tasks. In: Proceedings of the 27th International COLING, pp. 3318–3329 (2018)
2. Anderson, L.W., Krathwohl, D.R.: A Taxonomy for Learning, Teaching, and Assessing: A Revision of Bloom's Taxonomy of Educational Objectives, complete Addison Wesley Longman Inc., Boston (2001)
3. Bulathwela, S., Muse, H., Yilmaz, E.: Scalable educational question generation with pre-trained language models. In: Wang, N., Rebolledo-Mendez, G., Matsuda, N., Santos, O.C., Dimitrova, V. (eds.) AIED 2023. LNCS, vol. 13916, pp. 327–339. Springer, Heidelberg (2023). https://doi.org/10.1007/978-3-031-36272-9_27
4. Chen, D., Dolan, W.B.: Collecting highly parallel data for paraphrase evaluation. In: Proceedings of ACL 2021, pp. 190–200 (2011)
5. Cohen, J.: Weighted kappa: nominal scale agreement provision for scaled disagreement or partial credit. Psychol. Bull. **70**(4), 213 (1968)
6. Horbach, A., Aldabe, I., Bexte, M., de Lacalle, O.L., Maritxalar, M.: Linguistic appropriateness and pedagogic usefulness of reading comprehension questions. In: Proceedings of LREC 2020, pp. 1753–1762 (2020)
7. Horbach, A., Aldabe, I., Bexte, M., de Lacalle, O.L., Maritxalar, M.: Linguistic appropriateness and pedagogic usefulness of reading comprehension questions. In: Proceedings of the Twelfth Language Resources and Evaluation Conference, pp. 1753–1762 (2020)
8. Ji, Z., et al.: Survey of hallucination in natural language generation. ACM Comput. Surv. **55**(12), 1–38 (2023)
9. Kasneci, E., et al.: ChatGPT for good? on opportunities and challenges of large language models for education. Learn. Individ. Differ. **103**, 102274 (2023)
10. Kojima, T., Gu, S.S., Reid, M., Matsuo, Y., Iwasawa, Y.: Large language models are zero-shot reasoners. Adv. NeurIPS **35**, 22199–22213 (2022)
11. Kurdi, G., Leo, J., Parsia, B., Sattler, U., Al-Emari, S.: A systematic review of automatic question generation for educational purposes. IJAIED **30**, 121–204 (2020)
12. Liu, P., Yuan, W., Fu, J., Jiang, Z., Hayashi, H., Neubig, G.: Pre-train, prompt, and predict: a systematic survey of prompting methods in natural language processing. ACM Comput. Surv. **55**(9), 1–35 (2023)
13. Liu, Y., Iter, D., Xu, Y., Wang, S., Xu, R., Zhu, C.: G-eval: NLG evaluation using gpt-4 with better human alignment. In: Bouamor, H., Pino, J., Bali, K. (eds.) Proceedings of the 2023 Conference on EMNLP, pp. 2511–2522 (2023)
14. McHugh, M.L.: Interrater reliability: the kappa statistic. Biochemia medica **22**(3), 276–282 (2012)

15. Mishra, S., Khashabi, D., Baral, C., Choi, Y., Hajishirzi, H.: Reframing instructional prompts to gptk's language. In: Findings of ACL 2022, pp. 589–612 (2022)
16. Moore, S., Nguyen, H.A., Bier, N., Domadia, T., Stamper, J.: Assessing the quality of student-generated short answer questions using gpt-3. In: Hilliger, I., Munoz-Merino, P.J., De Laet, T., Ortega-Arranz, A., Farrell, T. (eds.) EC-TEL. LNCS, vol. 13450, pp. 243–257. Springer, Heidelberg (2022). https://doi.org/10.1007/978-3-031-16290-9_18
17. Nguyen, H.A., Bhat, S., Moore, S., Bier, N., Stamper, J.: Towards generalized methods for automatic question generation in educational domains. In: Hilliger, I., Munoz-Merino, P.J., De Laet, T., Ortega-Arranz, A., Farrell, T. (eds.) EC-TEL. LNCS, vol. 13450, pp. 272–284. Springer, Heidelberg (2022). https://doi.org/10.1007/978-3-031-16290-9_20
18. Sridhar, P., Doyle, A., Agarwal, A., Bogart, C., Savelka, J., Sakr, M.: Harnessing llms in curricular design: Using gpt-4 to support authoring of learning objectives. arXiv preprint arXiv:2306.17459 (2023)
19. Steuer, T., Bongard, L., Uhlig, J., Zimmer, G.: On the linguistic and pedagogical quality of automatic question generation via neural machine translation. In: De Laet, T., Klemke, R., Alario-Hoyos, C., Hilliger, I., Ortega-Arranz, A. (eds.) EC-TEL 2021, Proceedings, vol. 12884, pp. 289–294. Springer, Heidelberg (2021). https://doi.org/10.1007/978-3-030-86436-1_22
20. Ushio, A., Alva-Manchego, F., Camacho-Collados, J.: Generative language models for paragraph-level question generation. In: Proceedings of EMNLP 2022 (2022)
21. Wang, J., Liang, Y., et al.: Is ChatGPT a good NLG evaluator? a preliminary study. In: Dong, Y., Xiao, W., Wang, L., Liu, F., Carenini, G. (eds.) Proceedings of the 4th New Frontiers in Summarization Workshop. ACL (2023)
22. Wang, Z., Valdez, J., Basu Mallick, D., Baraniuk, R.G.: Towards human-like educational question generation with large language models. In: Rodrigo, M.M., Matsuda, N., Cristea, A.I., Dimitrova, V. (eds.) AIED 2022. LNCS, vol. 13355, pp. 153–166. Springer, Heidelberg (2022). https://doi.org/10.1007/978-3-031-11644-5_13
23. Wei, J., et al.: Chain-of-thought prompting elicits reasoning in large language models. Adv. NeurIPS **35**, 24824–24837 (2022)
24. Zhang, H., Song, H., Li, S., Zhou, M., Song, D.: A survey of controllable text generation using transformer-based pre-trained language models. ACM Comput. Surv. **56**, 1–37 (2022)
25. Zhang, R., Guo, J., Chen, L., Fan, Y., Cheng, X.: A review on question generation from natural language text. ACM TOIS **40**(1), 1–43 (2021)

VizChat: Enhancing Learning Analytics Dashboards with Contextualised Explanations Using Multimodal Generative AI Chatbots

Lixiang Yan[✉], Linxuan Zhao, Vanessa Echeverria, Yueqiao Jin,
Riordan Alfredo, Xinyu Li, Dragan Gaševi'c,
and Roberto Martinez-Maldonado

Monash University, Clayton, VIC 3108, Australia
lixiang.yan@monash.edu

Abstract. Learning analytics dashboards (LADs) serve as pivotal tools in transforming complex learner data into actionable insights for educational stakeholders. Despite their potential, the effectiveness of LADs, particularly the visualisations they utilise, has been under scrutiny. Concerns have been raised about their potential to cause cognitive overload, especially for users with limited data visualisation literacy, thus questioning their practical utility in supporting decision-making and reflective practices. This tool paper tackles these concerns by introducing VizChat, an open-sourced, prototype chatbot designed to augment LADs by providing contextualised, AI-generated explanations for visualisations. Developed on multimodal generative AI (GPT-4V) and retrieval-augmented generation (Langchain), VizChat offers on-demand, contextually relevant explanations that aim to improve user comprehension without overwhelming them with excessive information. Through a case study, we demonstrated VizChat's diverse capabilities, including actively seeking clarifications on ambiguous queries, personalising responses based on previous user interactions, providing contextually relevant explanations of specific visualisations, integrating information from multiple visualisations for a comprehensive response, and offering detailed insights into the data collection and analysis processes behind each visualisation. Such efforts support the paradigm shift from exploratory to explanatory approaches in LADs, highlighting the potential of integrating generative AI and chatbots to enhance the educational value of learning analytics.

Keywords: Generative Artificial Intelligence · Learning Analytics ·
Visualisation · Chatbots · Multimodal · GPT · LLM

1 Introduction

Learning analytics dashboards (LADs) have become an instrumental part of educational technology, transforming intricate learner data into comprehensible

A. M. Olney et al. (Eds.): AIED 2024, LNAI 14830, pp. 180–193, 2024.
https://doi.org/10.1007/978-3-031-64299-9_13

and actionable insights for educators, learners, and other stakeholders [33,40]. Within these dashboards, visualisations are frequently utilised to provide a clear, visual representation of educational data for supporting self-regulated learning [25], reflective practices [10], pedagogical interventions [19], and adviser-student dialogues [5]. However, the effectiveness of these LADs in supporting learning and teaching practices remains an ongoing debate [17]. Specifically, concerns have been raised regarding whether the complexity of the visualisations may lead to cognitive overload for learners and educators, inhibiting their practical utilities [14,30], especially those with a low level of data visualisation literacy [8,31]. This is a growing concern as the visualisations may become increasingly complex due to the advancements in multimodal learning analytics and educational data mining, which have enhanced data collection and processing capacity [28,32]. Consequently, it is essential to ensure LADs can effectively communicate various insights to educational stakeholders without overwhelming them.

A paradigm shift from exploratory to explanatory visualisations in LADs is critical to ensure that these visual insights remain practical and meaningful to educational stakeholders [10,33,40]. The problem of expecting learners and educators to explore and comprehend the visualisations in LADs without any additional support has already been evident in prior literature. That is, most of these are end users without extensive data visualisation literacy, who would struggle with basic chart interpretations [23] and may exhibit interpretation biases [3]. Consequently, it is essential to augment visualisations in LADs with explanatory features that could facilitate user comprehension. Incorporating narratives and data storytelling elements into visualisations has been identified as a potent approach to improving user comprehension [35]. Such integrations have proven more effective in communicating educational insights to learners and educators than conventional graphs and figures [12,24]. However, the design and integration of these elements into LADs are often complex and highly variable as data stories are strongly dependent on the context and domain, posing challenges in terms of scalability and applicability across different LADs or learning environments.

An alternative approach to support user comprehension of the visualisations in LADs involves embedding interactive features that provide on-demand explanations [6,34]. This approach has become increasingly feasible with the advancements in multimodal generative AI technology, such as GPT-4V, which is capable of generating explanations from textual and visual inputs [1]. The conversational capabilities of such AI technologies could enable the development of adaptable chatbots within LADs aimed at improving user understanding via interactive dialogues. These chatbots can offer AI-generated explanations without cluttering the visualisations with excessive text or labels, thus maintaining a balance between information richness and usability [27,38]. The growing maturity of retrieval-augmented generation (RAG) methodologies could also enhance the accuracy and contextual relevance of these AI-generated explanations [36,37], making them grounded in learning design and educational theories. Additionally, autonomous AI agents developed based on generative AI technologies could monitor and facilitate the communication between educational stakeholders and

LADs, ensuring a more accurate understanding of users' queries by actively seeking clarifications [42]. Together, these technological advancements in generative AI could empower the development of adaptive chatbots that support learners and educators to better comprehend the educational insights communicated through the visualisations in LADs.

In this tool paper, we take the initiative in this direction by developing a prototype chatbot, VizChat, for LADs. VizChat is an open-source, adaptive chatbot designed to supplement visualisations in LADs with interactive, AI-generated explanations. It utilises the capabilities of multimodal generative AI, specifically GPT-4V, to provide explanations that are accurate and contextually relevant. These explanations draw upon a foundation of customisable, predefined knowledge from the learning design and data procedures, ensuring that they are both informative and relevant to the learning context. Additionally, we present a case study illustrating how VizChat could generate accurate and context-sensitive explanations of various visualisations within an established LAD that has been incorporated into practice to aid student reflection.

2   Background and Related Work

2.1   Learning Analytics Dashboards and Explanatory Visualisations

LADs aim to support educational stakeholders in monitoring teaching and learning practices [33]. LADs are often composed of visual representations and analyses of educational data to engage users to search for trends and answer questions using these data [40]. However, LADs face limitations, such as presenting complex visualisations, ineffective communication of insights [7], and misalignment with teachers' pedagogical needs [18]. To address these issues, research has shifted towards *explanatory LADs*, which are designed to visually guide users towards key insights that are contextualised with information derived from the learning design and are aligned with stakeholders' pedagogical needs [9].

Using information visualisation (InfoVis) techniques and data storytelling principles, LADs include clear titles that convey the main message, narratives with text explanations, and the use of arrows and colour-coded information to draw attention to specific data points [9,12]. Research has shown that these explanatory LADs can efficiently direct attention to key pedagogical insights [9,24], foster deep reflections and aid teachers and students in understanding their learning process [9,12] and aid teachers with low visual literacy skills in an efficient interpretation of the information [31].

Despite these benefits, a significant challenge remains: automatically translating contextual information from the learning activity into visual elements. This process is often treated as a co-design process, where designers, researchers, and educators define pedagogical intentions and align them with visual elements [10,12,24,31]. Recent solutions have aimed to streamline this process by asking educators to define pedagogical rules, but translating these rules into visual elements remains the task of designers and researchers [13]. On the other hand, other solutions have begun to use unsupervised learning and natural language processing (NLP) to automatically generate students' narratives to enable academic advisors to understand student success and risk behaviours [2].

2.2 Generative AI and Chatbots in Education

Recent advancements in Generative AI have unlocked new possibilities for developing learning content and engaging with educational data [39,43]. For instance, Large Language Models (LLMs) such as GPT and Llama, capable of generating textual content in response to user prompts, have shown promise in creating instructional materials and practice quizzes with minimal human intervention [45]. Additionally, speech-to-text models, notably OpenAI's Whisper [15], have facilitated the automatic transcription of lecture recordings, thereby broadening the range and modality of accessible learning materials [15]. Image-generating diffusion models like Midjounery and DALL-E have demonstrated their capability to foster creative practices in art-centric STEAM education and support multimedia learning by producing images from text descriptions [21]. The emergence of multimodal models, such as GPT-4V, has expanded generative AI's abilities to interpret and synthesize both textual and visual information, creating avenues for more comprehensible explanations [1]. These breakthroughs in foundational models suggest a promising future where conveying insights to learners and educators could become increasingly interactive and diversified, enhancing the modalities through which these insights are communicated [43].

The ability of multimodal generative AI to simultaneously understand natural and visual language at a high-quality, zero shot level is critical for the development of explanatory LADs with interactive dialogues. While chatbots have been extensively used in various educational settings, including intelligent tutoring systems and feedback mechanisms [29], their integration into LADs has been limited. This limitation stems from the challenges associated with achieving high-quality, zero-shot understanding of visual language [20]. Moreover, simply generating descriptions of visualisations falls short of providing meaningful explanations. For explanations to benefit learners and educators, they must be contextually grounded in the specific learning design. This contextual grounding is essential for supporting decision-making processes and provoking in-depth reflection [33,40]. Therefore, to construct chatbots for LADs that meet these requirements, it is essential to employ multimodal generative AI models capable of understanding both natural and visual languages.

The maturity of supporting infrastructures could enhance the ability of generative AI to provide accurate and contextualised explanations for the different visualisations in LADs. Generative AI models are facing criticism for their tendency to hallucinate well-articulated but inaccurate content [16], which poses a risk of misleading educational stakeholders. An increasingly popular method to mitigate model hallucination is using retrieval augmented generation (RAG). This technique confines content generation to material that is contextually relevant [36]. For instance, pertinent information can be transformed into vector embeddings—representations of objects like words or entities as vectors within a continuous space that encapsulates semantic or relational meanings [26]. This information can then be dynamically retrieved during interactions with generative AI models by conducting a semantic search. A common method is calculating the cosine similarity between the stored embeddings and the embeddings generated from user queries. This strategy has proven to be effective in reducing the occurrence of model hallucination and improving the domain-specific accuracy of the generated outputs [37].

Generative AI can also return irrelevant responses when users supply inadequate information in their prompts [41]. Expecting all learners and educators to master the art of crafting effective prompts is unrealistic. However, the development of AI agents—autonomous and adaptive entities that function independently to achieve set goals without needing constant user input—presents a solution to enhance the relevance of user prompts. These agents can actively facilitate the refinement of prompts, ensuring that interactions with generative AI yield more pertinent and useful outcomes [42]. Therefore, integrating multimodal generative AI with RAG and autonomous AI agents could enable the creation of context-aware chatbots for LADs. Such chatbots would be capable of providing learners and educators with precise and contextualised explanations of various visualisations.

3   VizChat: System Architecture

In this section, we outline the system architecture of an open-source[1] prototype of VizChat (refer to Fig. 1). This architecture integrates the capabilities of multimodal generative AI (e.g., GPT-4V) in understanding both natural and visual languages with recent advancements in RAG. This combination aims to deliver accurate and contextually relevant explanations for visualisations within LADs. The prototype was developed as a Chrome extension (based on ChatGPTBox[2]) that can be activated and overlaid on any web-based LADs with the functionality of capturing screenshots and chatting with users. Further details are provided below.

A: Knowledge Database. Creating a knowledge database filled with all relevant contextual information is crucial to minimise the risk of generative AI delivering hallucinated or irrelevant explanations that could mislead learners and educators [36]. The process of building this database is straightforward, relying largely on existing documentation. As illustrated in Fig. 1 (Section A), LAD administrators and educators can upload documents pertinent to the learning design, such as course and task descriptions. These documents should align with the objectives of the LADs, whether they aim to support students' overall progression in a course or their performance in specific tasks. Data and learning analysts can also contribute by uploading records or descriptions relevant to data analysis and visualisation, offering more depth for each visualisation explanation. For example, detailing the axes of a scatter plot helps elucidate the pattern being visualised.

---

[1] https://github.com/LinxZhao/VizChat-pub.
[2] https://github.com/josStorer/chatGPTBox.

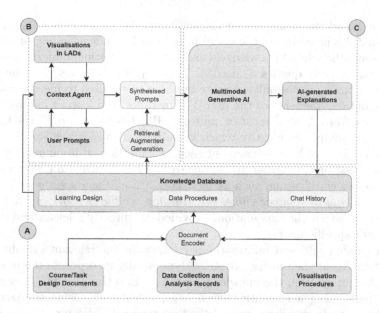

Fig. 1. VizChat: a prototype system architecture for augmenting learning analytics dashboards with contextualised explanations using multimodal generative AI. The sections correspond to the three essential components of the system, including A) knowledge database, B) prompt synthesis, and C) contextualised explanation.

These textual materials were encoded into a knowledge database of vector embeddings using Chroma[3] and LangChain[4], a Python library designed for creating context-aware applications with large language models (LLMs). Specifically, we utilised LangChain's document loaders and text splitters to parse the documents. We then used OpenAI's embedding model (text-embedding-ada-002) to convert the text into vector embeddings. This step may not be necessary for simpler LADs that contain minimal contextual information, as it is possible to incorporate all relevant content into a single prompt. However, for more complex LADs that feature intricate learning scenarios, multiple data modalities, and various types of visualisations, converting text to embeddings is essential. This approach helps to mitigate the performance drop of LLMs when dealing with lengthy contexts [22] and enables the retrieval of the most relevant information for a user's query through semantic search.

B: Prompt Synthesis. After setting up the knowledge database, the subsequent key element in VizChat is the Context Agent. Its role is to accurately interpret users' queries and utilise RAG to create a context-specific prompt for generating explanations. The Context Agent is enhanced with GPT-4V's capabilities and has access to the knowledge database. It begins by capturing screenshots of the LADs and generating descriptions for each visualisation based on

---

[3] https://www.trychroma.com/.
[4] https://python.langchain.com/.

pre-stored knowledge about those visualisations. When users pose a question, the Context Agent evaluates the query in light of its understanding of the visualisations and the related knowledge. It may request additional information from users to clarify their queries as needed. This approach ensures that the generated explanations are not only accurate but also customised to meet the specific inquiries and requirements of the user. For instance, if a learner inquires, *"The figure shows that I talk the most in my team. What does it mean?"*, the Context Agent first attempts to identify which visualisation the learner is referring to, using its database of knowledge about each visualisation. If it cannot determine which visualisation the learner means due to a lack of information, the agent will ask for further clarification, such as by asking, *"Could you specify which figure you are referring to (the bar chart, the heatmap, or the network graph)?"* This process ensures that the explanations provided are directly relevant and tailored to the user's specific query.

After gaining sufficient information to determine the relevant visualisations, the Context Agent then utilises RAG to dynamically incorporate the most pertinent information from the knowledge database into the prompt. This process involves performing a semantic search (e.g., calculating cosine similarity [4]) across the vector embeddings to find the text segments (default at 200 OpenAI tokens or approximately 150 English words each but can be adjusted by users based on their document contents) most relevant to the user's query (default at top 10 segments). Taking the earlier example, the agent would seek out information about the visualisations concerning students' verbal communication and the anticipated verbal interactions according to the learning design from the knowledge database. By doing this, the Context Agent can craft a prompt that encapsulates the user's intention and integrates the most relevant contextual information, ensuring the explanation is grounded in the specific context of the query. This tailored prompt is then passed on to the generative component of the system for creating the explanation.

C: Contextualised Explanation. The final component in the VizChat system, as shown in Fig. 1 (Section C), focuses on generating explanations derived from the contextually enhanced prompt crafted by the Context Agent. This prompt, which includes both a screenshot of the LAD and a contextually enriched query, is forwarded to a multimodal generative AI model, specifically GPT-4V in our implementation, for generating explanations that interpret the visualisations within the given context. The system is designed to memorise each user interaction. Every exchange between a user and VizChat is re-incorporated into the knowledge database by transforming these conversations into vector embeddings, which can be retrieved to enhance the personalisation and relevance of subsequent explanations. This mechanism prevents the Context Agent from asking for similar clarifications repetitively, which might diminish user experience. For LADs that offer interactive capabilities, allowing users to customise views to reflect different aspects of learning data (for example, various stages of learning activities), this dynamically evolving knowledge database is crucial. It enables the generation of explanations that address specific queries and also connects disparate segments of learning data, leveraging insights from previous interactions to provide more comprehensive and contextually grounded explanations.

4   Illustrative Case Study

In this section, we present a case study with three authentic dialogue examples, illustrating how VizChat could generate accurate and contextualised explanations of various visualisations within a LAD (Fig. 2 - A) that has been used to support student reflection. We first provided an overview of the LAD and its visualisations. As shown in Fig. 2 (B–D), three examples were then given to demonstrate VizChat's capabilities in seeking and memorising user clarification for generating context-specific responses, integrating multiple visualisations to synthesise cohesive explanations, and elaborating data analysis and collection process to enhance transparency. These examples were chosen based on the findings of a prior evaluation study of the LAD with students, where they reported challenges in navigating complex visualisations and comprehending how the visualisations were formed based on their data [11]. Due to restricted access to a variety of LADs and their associated contextual information, the examples provided should not be considered a formal evaluation of VizChat. Instead, they are intended to showcase the system's potential capabilities. Future research is needed to conduct a thorough assessment of VizChat's reliability and informativeness across a diverse range of LADs, in order to build a solid empirical foundation for its effectiveness. Our decision to make VizChat open-source is driven by the goal of facilitating such evaluative studies, enabling the wider research community to easily access and test the system. In practice, after activating the VizChat Chrome extension, a pop-up chat box will appear on the bottom right of any LADs web interface. The examples were snapshots of the chat history with VizChat for the following LAD.

Multimodal Learning Analytics Dashboard. The multimodal LAD, designed for a clinical simulation unit in undergraduate nursing education, aims to facilitate teacher-guided reflection during debriefing sessions. It focuses on cultivating skills such as prioritisation, teamwork, and communication. The LAD and its visualisations are the result of collaborative efforts involving researchers and educators across various co-design iterations, with effective application in real learning environments (further details in [11]). For instance, educators recommended a social network visualisation to quickly discern communication patterns among team members and other simulation actors. Figure 2 (A) illustrates that the LAD comprises five distinct visualisations. The timeline visualisation marks the four stages of the learning activity and key activities logged by educators during simulations. The user can select different stages on the timeline, which will also influence the visualised data for the other four visualisations. The prioritisation bar chart reflects students' strategies, such as focusing on the primary patient or attending to other tasks, derived from their spatial behaviour tracked via wearable devices (additional information in [44]). The ward map shows students' verbal and spatial distribution during simulations, with colour

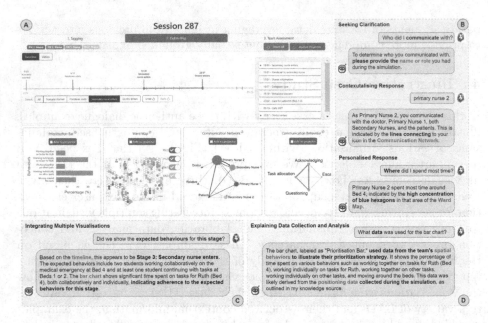

Fig. 2. VizChat in a LAD for student reflection. A) LAD with four visualisations. B) VizChat clarifies queries, contextualises responses, and switches visualisations for accurate responses. C) VizChat integrates multiple visualisations for explanations. D) VizChat explains data analysis and collection methods behind the prioritisation bar.

saturation indicating frequent verbal interactions (recorded by microphones) and hexagon placement showing spatial distribution. The social network visualisation maps communication between students and with patient manikins and a doctor. Lastly, the undirected and simplified epistemic network represents the co-occurrence of team communication codes, automatically generated from verbal data [46].

Constructing Knowledge Database. We infused VizChat's knowledge database with essential information regarding the task background of the simulation scenario, learning activity design (e.g., different phases of the learning activity), data collection procedure, and data analysis related to each visualisa-

Knowledge Database		
Learning Design	**Data Collection**	**Data Analysis**
Text: The main focus for students will be the post-op patient in Bed 4 (Ruth Jenkins), however the requirements the other patients will mean students will need to prioritise care... **Embeddings**: [-0.022295473143458366, 0.043492648750543594, ...]	**Text**: A range of sensing technologies was used to collect multimodal data. These included wireless headsets with a unidirectional microphone for capturing... **Embeddings**: [0.02358199842274189, 0.011037209071218967, ...]	**Text**: The timeline shows the specific stage of the simulation that has been visualised. Right to the timeline is logged activities that the teachers created to document key... **Embeddings**: [-0.005268605425953865, 0.04427655041217804, ...]

Fig. 3. Snapshot of VizChat's knowledge database for the case study LAD.

tion. This information was from existing learning material created by educators, prior publications of the research team that designed the LAD and its analytics, and documented expert domain knowledge during our prior studies. Figure 3 shows a snapshot of the constructed knowledge database.

Seeking and Memorising Clarifications. In the first example, we demonstrated VizChat's capability to request clarifications from users, followed by providing contextually relevant responses based on the most pertinent visualisations in the LAD. As depicted in Fig. 2 (B), when a user's query is ambiguous or lacks sufficient details to identify their representation in the LAD, VizChat would actively ask the user to provide more clarifications (e.g., *"...please provide the name or role you had during the simulation."*). This information is essential for providing personalised responses that address the user's needs. For example, VizChat could accurately pinpoint the most relevant visualisation to the user's query (e.g., *"...Communication Network"*) and articulate the response with some concise explanations of how to interpret the corresponding visualisation (e.g., *"...lines connecting..."*). To avoid asking for similar clarifications repetitively, this conversation was stored in the knowledge database, retrieved during subsequent interactions, and incorporated as part of the integrated prompt to contextualise VizChat's ongoing replies. This ability to memorise the user's prior responses was evident in the following turn of conversation, where the user asked a different question and VizChat was able to provide an explanation based on the memorised representation from a different visualisation that was most relevant to their current question (e.g., *"...high concentration of blue hexagons in the area of the Ward Map"*).

Integrating Multiple Visualisations. Apart from providing explanations based on one specific visualisation, VizChat can also synthesise analyses from multiple visualisations to create a comprehensive response. As shown in Fig. 2 (C), even when the user does not specify the stage being referred to in their question, VizChat can use the timeline visualisation to deduce the relevant stage (e.g., *"Based on the timeline, this appears to be Stage 3: Secondary nurse enters."*). Utilising this insight, VizChat retrieves information pertinent to the query (e.g., *"...expected behaviours..."*) and provides a detailed explanation (e.g., *"The expected behaviours include..."*). By interpreting the semantic meaning of the query and the retrieved knowledge, VizChat also selects the most relevant visualisation (e.g., the prioritisation bar chart) and analyses it to offer a thorough response that describes the team's prioritisation behaviours, comparing them with the expected behaviours for that stage (e.g., *"The bar chart shows..., indicating adherence to the expected behaviours for this stage"*).

Explaining Data Collection and Analysis. VizChat, through its knowledge database, can also provide detailed explanations about the underlying data collection and analysis processes, thereby enhancing the transparency of each visualisation in the LAD. For example, as illustrated in Fig. 2 (D), when queried about the data underpinning the bar chart, VizChat accurately described the data source (e.g., *"...used data from the team's spatial behaviours to illustrate* *prioritisation strategy."*), offered a descriptive explanation of the axes (e.g., *"It shows the percentage of time spent..."*), identified the specific data used (e.g., *"...**positioning data** collected during the simulation."*), and clarified the source of this information (e.g., *"...as outlined in my **knowledge source**."*).

5   Discussion and Conclusion

In this tool paper, we introduced VizChat, a novel prototype chatbot system leveraging multimodal generative AI and RAG to provide context-sensitive explanations of visualisations in LADs. VizChat aims to facilitate a shift from traditional exploratory methods–where learners and educators independently explore and interpret insights within LADs–to an explanatory approach. This approach enhances LADs with features that support user comprehension through customised explanations [10, 40]. Our case study showcased VizChat's diverse capabilities: it actively seeks clarifications on ambiguous queries, personalises responses based on previous user interactions, provides contextually relevant explanations of specific visualisations, integrates information from multiple visualisations for a comprehensive response, and offers detailed insights into the data collection and analysis processes behind each visualisation. These functionalities can be beneficial for educational stakeholders, particularly those with limited data visualisation literacy, to derive meaningful insights from LADs through interactive conversations [8, 23].

This work has several limitations that future studies should address to further validate and enhance the utility of VizChat. First, the cases in Sect. 4 are illustrative and do not constitute a comprehensive evaluation of VizChat's effectiveness. It is essential for future research to assess the accuracy, reliability, and informativeness of VizChat through practical use cases involving learners and teachers within LADs. Second, there is a need to compare VizChat's explanatory capabilities with other data storytelling methods [12, 24] across varying levels of data visualisation literacy. Exploring VizChat's potential in enhancing data visualisation literacy among users warrants attention. Making VizChat open-source aims to stimulate such evaluative research across AI in education, educational technology, and human-computer interaction domains. Third, the integration of RAG for response improvement [36] requires validation, highlighting the importance of the knowledge database's quality. Lastly, the use of the proprietary GPT-4V model [1] raises concerns about cost, given its usage fees. Future research should explore the cost-benefit balance of using such advanced models versus more affordable, possibly lower-performing alternatives for broad implementation. Addressing these areas is crucial for progressing from exploratory to explanatory visualisations in LADs and enhancing their educational value.

Acknowledgement. This research was in part supported by the Australian Research Council (DP220101209, DP240100069) and Jacobs Foundation. The research of Lixiang Yan and Dragan was also sponsored by Defense Advanced Research Projects Agency (DARPA) under agreement number HR0011-22-2-0047. The U.S. Government is authorised to reproduce and distribute reprints for Governmental purposes notwithstanding any copyright notation thereon. The views and conclusions contained herein are those of the authors and should not be interpreted as necessarily representing the official policies or endorsements, either expressed or implied, of DARPA or the U.S. Government.

References

1. Achiam, J., et al.: GPT-4 technical report. arXiv preprint arXiv:2303.08774 (2023)
2. Al-Doulat, A., et al.: Making sense of student success and risk through unsupervised machine learning and interactive storytelling. In: Bittencourt, I.I., Cukurova, M., Muldner, K., Luckin, R., Millán, E. (eds.) AIED 2020. LNCS (LNAI), vol. 12163, pp. 3–15. Springer, Cham (2020). https://doi.org/10.1007/978-3-030-52237-7_1
3. Alhadad, S.S.: Visualizing data to support judgement, inference, and decision making in learning analytics: insights from cognitive psychology and visualization science. J. Learn. Anal. 5(2), 60–85 (2018)
4. Banerjee, D., Singh, P., Avadhanam, A., Srivastava, S.: Benchmarking LLM powered chatbots: methods and metrics. arXiv preprint arXiv:2308.04624 (2023)
5. Charleer, S., Moere, A.V., Klerkx, J., Verbert, K., De Laet, T.: Learning analytics dashboards to support adviser-student dialogue. IEEE Trans. Learn. Technol. 11(3), 389–399 (2017)
6. Chen, C.M., Wang, J.Y., Hsu, L.C.: An interactive test dashboard with diagnosis and feedback mechanisms to facilitate learning performance. Comput. Educ.: Artif. Intell. 2, 100015 (2021)
7. Corrin, L.: Evaluating students' interpretation of feedback in interactive dashboards. Score Reporting Research and Applications, pp. 145–159 (2018)
8. Donohoe, D., Costello, E.: Data visualisation literacy in higher education: an exploratory study of understanding of a learning dashboard tool. Int. J. Emerging Technol. Learn. (iJET) 15(17), 115–126 (2020)
9. Echeverria, V., Martinez-Maldonado, R., Granda, R., Chiluiza, K., Conati, C., Buckingham Shum, S.: Driving data storytelling from learning design. In: Proceedings of the 8th International Conference on Learning Analytics and Knowledge, pp. 131–140 (2018)
10. Echeverria, V., Martinez-Maldonado, R., Shum, S.B., Chiluiza, K., Granda, R., Conati, C.: Exploratory versus explanatory visual learning analytics: driving teachers' attention through educational data storytelling. J. Learn. Anal. 5(3), 73–97 (2018)
11. Echeverria, V., et al.: TeamSlides: a multimodal teamwork analytics dashboard for teacher-guided reflection in a physical learning space. In: Proceedings of the 14th Learning Analytics and Knowledge Conference, pp. 112–122 (2024)
12. Fernandez Nieto, G.M., Kitto, K., Buckingham Shum, S., Martínez-Maldonado, R.: Beyond the learning analytics dashboard: alternative ways to communicate student data insights combining visualisation, narrative and storytelling. In: Proceedings of the 12th Learning Analytics and Knowledge Conference, pp. 219–229 (2022)
13. Fernandez-Nieto, G.M., Martinez-Maldonado, R., Echeverria, V., Kitto, K., Gašević, D., Buckingham Shum, S.: Data storytelling editor: a teacher-centred tool for customising learning analytics dashboard narratives. In: Proceedings of the 14th Learning Analytics and Knowledge Conference, pp. 678–689 (2024)

14. Gibson, A., Martinez-Maldonado, R.: That dashboard looks nice, but what does it mean? towards making meaning explicit in learning analytics design. In: Proceedings of the 29th Australian Conference on Computer-Human Interaction, pp. 528–532 (2017)

15. Gris, L.R.S., Marcacini, R., Junior, A.C., Casanova, E., Soares, A., Aluísio, S.M.: Evaluating OpenAI's whisper ASR for punctuation prediction and topic modeling of life histories of the museum of the person. arXiv preprint arXiv:2305.14580 (2023)

16. Ji, Z., Lee, N., Frieske, R., Yu, T., Su, D., Xu, Y., Ishii, E., Bang, Y.J., Madotto, A., Fung, P.: Survey of hallucination in natural language generation. ACM Comput. Surv. **55**(12), 1–38 (2023)

17. Jivet, I., Scheffel, M., Drachsler, H., Specht, M.: Awareness Is Not Enough: pitfalls of learning analytics dashboards in the educational practice. In: Lavoué, É., Drachsler, H., Verbert, K., Broisin, J., Pérez-Sanagustín, M. (eds.) EC-TEL 2017. LNCS, vol. 10474, pp. 82–96. Springer, Cham (2017). https://doi.org/10.1007/978-3-319-66610-5_7

18. Kaliisa, R., Mørch, A., Kluge, A.: 'My point of departure for analytics is extreme Skepticism': implications derived from an investigation of university teachers' learning analytics perspectives and design practices. Technol. Knowl. Learn. **27** (2022). https://doi.org/10.1007/s10758-020-09488-w

19. Kim, J., Jo, I.H., Park, Y.: Effects of learning analytics dashboard: analyzing the relations among dashboard utilization, satisfaction, and learning achievement. Asia Pac. Educ. Rev. **17**, 13–24 (2016)

20. Lee, K., et al.: Pix2Struct: screenshot parsing as pretraining for visual language understanding. In: International Conference on Machine Learning, pp. 18893–18912. PMLR (2023)

21. Lee, U., et al.: Prompt aloud!: incorporating image-generative AI into steam class with learning analytics using prompt data. Educ. Inf. Technol., 1–31 (2023)

22. Liu, N.F., et al.: Lost in the middle: How language models use long contexts. arXiv preprint arXiv:2307.03172 (2023)

23. Maltese, A.V., Harsh, J.A., Svetina, D.: Data visualization literacy: investigating data interpretation along the novice-expert continuum. J. Coll. Sci. Teach. **45**(1), 84–90 (2015)

24. Martinez-Maldonado, R., Echeverria, V., Fernandez Nieto, G., Buckingham Shum, S.: From data to insights: a layered storytelling approach for multimodal learning analytics. In: Proceedings of the 2020 Chi Conference on Human Factors in Computing Systems, pp. 1–15 (2020)

25. Matcha, W., Gašević, D., Pardo, A., et al.: A systematic review of empirical studies on learning analytics dashboards: a self-regulated learning perspective. IEEE Trans. Learn. Technol. **13**(2), 226–245 (2019)

26. Mikolov, T., Sutskever, I., Chen, K., Corrado, G.S., Dean, J.: Distributed representations of words and phrases and their compositionality. In: Advances in Neural Information Processing Systems , vol. 26 (2013)

27. Noroozi, O., Alikhani, I., Järvelä, S., Kirschner, P.A., Juuso, I., Seppänen, T.: Multimodal data to design visual learning analytics for understanding regulation of learning. Comput. Hum. Behav. **100**, 298–304 (2019)

28. Ochoa, X.: Multimodal learning analytics - rationale, process, examples, and direction. In: Lang, C., Siemens, G., Wise, A.F., Gašević, D., Merceron, A. (eds.) The Handbook of Learning Analytics, pp. 54–65. SoLAR, 2 edn. (2022)

29. Okonkwo, C.W., Ade-Ibijola, A.: Chatbots applications in education: a systematic review. Comput. Educ.: Artif. Intell. **2**, 100033 (2021)

30. Park, Y., Jo, I.H.: Factors that affect the success of learning analytics dashboards. Educ. Tech. Res. Dev. **67**, 1547–1571 (2019)
31. Pozdniakov, S., Martinez-Maldonado, R., Tsai, Y.S., Echeverria, V., Srivastava, N., Gasevic, D.: How do teachers use dashboards enhanced with data storytelling elements according to their data visualisation literacy skills? In: LAK23: 13th International Learning Analytics and Knowledge Conference, pp. 89–99 (2023)
32. Romero, C., Ventura, S.: Educational data mining and learning analytics: an updated survey. Wiley Interdisc. Rev.: Data Min. Knowl. Discov. **10**(3), e1355 (2020)
33. Sahin, M., Ifenthaler, D.: Visualizations and dashboards for learning analytics: a systematic literature review. Vis. Dashboards Learn. Anal., 3–22 (2021)
34. Scheers, H., De Laet, T.: Interactive and explainable advising dashboard opens the black box of student success prediction. In: Technology-Enhanced Learning for a Free, Safe, and Sustainable World: 16th European Conference on Technology Enhanced Learning, EC-TEL 2021, Bolzano, Italy, September 20-24, 2021, Proceedings 16, pp. 52–66. Springer (2021). https://doi.org/10.1007/978-3-030-86436-1_5
35. Shao, H., Martinez-Maldonado, R., Echeverria, V., Yan, L., Gašević, D.: Data storytelling in data visualisation: does it enhance the efficiency and effectiveness of information retrieval and insights comprehension? In: CHI'24. In press. ACM, ACM, Honolulu, HI, USA (2024)
36. Shuster, K., Poff, S., Chen, M., Kiela, D., Weston, J.: Retrieval augmentation reduces hallucination in conversation. arXiv preprint arXiv:2104.07567 (2021)
37. Siriwardhana, S., Weerasekera, R., Wen, E., Kaluarachchi, T., Rana, R., Nanayakkara, S.: Improving the domain adaptation of retrieval augmented generation (rag) models for open domain question answering. Trans. Assoc. Comput. Linguistics **11**, 1–17 (2023)
38. Therón, R.: Visual learning analytics for a better impact of big data. In: Burgos, D. (ed.) Radical Solutions and Learning Analytics. LNET, pp. 99–113. Springer, Singapore (2020). https://doi.org/10.1007/978-981-15-4526-9_7
39. UNESCO: Generative artificial intelligence in education: what are the opportunities and challenges? (2023). https://www.unesco.org/en/articles/generative-artificial-intelligence-education-what-are-opportunities-and-challenges
40. Verbert, K., Ochoa, X., De Croon, R., Dourado, R.A., De Laet, T.: Learning analytics dashboards: the past, the present and the future. In: Proceedings of the 10th Learning Analytics and Knowledge Conference, pp. 35–40 (2020)
41. White, J., et al.: A prompt pattern catalog to enhance prompt engineering with chatgpt. arXiv preprint arXiv:2302.11382 (2023)
42. Wu, Q., et al.: AutoGen: enabling next-gen LLM applications via multi-agent conversation framework. arXiv preprint arXiv:2308.08155 (2023)
43. Yan, L., Martinez-Maldonado, R., Gasevic, D.: Generative artificial intelligence in learning analytics: contextualising opportunities and challenges through the learning analytics cycle. In: Proceedings of the 14th Learning Analytics and Knowledge Conference, pp. 101–111 (2024)
44. Yan, L., et al.: The role of indoor positioning analytics in assessment of simulation-based learning. Br. J. Edu. Technol. **54**(1), 267–292 (2023)
45. Yan, L., et al.: Practical and ethical challenges of large language models in education: a systematic scoping review. Br. J. Edu. Technol. **55**(1), 90–112 (2024)
46. Zhao, L., et al.: METS: Multimodal learning analytics of embodied teamwork learning. In: Proceedings of the 13th International Conference on Learning Analytics and Knowledge, pp. 186–196. LAK2023 (2023)

Affect Behavior Prediction: Using Transformers and Timing Information to Make Early Predictions of Student Exercise Outcome

Hao Yu[1], Danielle A. Allessio[2], William Rebelsky[2], Tom Murray[2], John J. Magee[3], Ivon Arroyo[2], Beverly P. Woolf[2], Sarah Adel Bargal[4], and Margrit Betke[1(✉)]

[1] Boston University, Boston, MA 02215, USA
{haoyu,betke}@bu.edu
[2] University of Massachusetts Amherst, Amherst, MA 01003, USA
{allessio,wrebelsky}@umass.edu, {tmurray,ivon,bev}@cs.umass.edu
[3] Clark University, Worcester, MA 01610, USA
jmagee@clarku.edu
[4] Georgetown University, Washington, D.C. 20057, USA
sarah.bargal@georgetown.edu

Abstract. Early prediction of student outcomes, as they practice with an intelligent tutoring system is crucial for providing timely and effective interventions to students, potentially improving their engagement and other productive behaviors, and ultimately enhancing their learning. In this work, we propose a novel approach for predicting the outcome of a student solving a mathematics problem in an intelligent tutor using early visual and tabular cues. Our approach analyzes only the first several seconds of a student's problem-solving process captured in a video feed, along with timing information obtained from their learning log. Our model, EPATT (Early Prediction using an Affect-aware Transformer and Timing information), extracts facial affective embeddings from video frames using transfer learning and analyzes their temporal dependencies using a Transformer. The timing information about when students take certain actions (e.g., requesting a hint) is combined with the video representation, enhancing the model's ability to predict student performance quickly. Experimental results show that EPATT achieves superior performance over baselines and state-of-the-art on a student dataset for exercise outcome prediction, demonstrating the efficacy of our approach and the potential impact of early outcome prediction for the development of better intelligent tutors.

Keywords: Affect Behavior Prediction · Exercise Outcome Prediction · Early Prediction · Visual Data · Timing Log Data

A. M. Olney et al. (Eds.): AIED 2024, LNAI 14830, pp. 194–208, 2024.
https://doi.org/10.1007/978-3-031-64299-9_14

1   Introduction

A fundamental task of personalized digital learning environments, or intelligent tutoring systems (ITS), is to interpret student behavior and provide appropriate interventions. Researchers have worked on designing and analyzing systems that can recognize and respond to users' emotional states, e.g., developing technologies that detect facial expressions, gestures, and other physiological signals to infer emotional states [12, 25]. Intelligent tutors can leverage real-time affective feedback from students to improve their learning experience [2, 14, 15].

Methods of predicting student affective and/or cognitive outcomes have focused on analyzing an entire stream of student video as students engage with a learning activity –using facial analysis techniques [18] or affect transfer learning [26, 27]. However, these methods require considerable computational overhead by processing a lengthy video as students are exposed to an entire activity, spanning several minutes, which makes their use challenging in real-time intelligent tutors, where inferences of how students are doing should ideally happen almost instantaneously. Moreover, due to the complex nature of student behaviors, achieving high prediction performance has proven to be extremely challenging for these approaches [18, 26, 27]. Even human observers demonstrate low accuracy predicting exercise solving outcomes solely from video [27], indicating the need for accessing additional information to improve prediction accuracy.

To tackle these two main challenges, this paper proposes a multimodal approach for predicting student exercise outcomes at an *early* stage by analyzing the first k seconds of student video, where k can be set to be as small as 5 s or as large as 20 s, and augmenting this with student learning log data available in the same k-second timeframe. This is the first work to combine visual affective analysis with student log data in the context of prediction and *early* prediction of student exercise outcomes.

We follow the student learning outcomes classification for the MathSpring intelligent tutor platform [3] that was proposed to support and encourage students as they solved math exercises [18]. Each exercise attempted by a student had an outcome label automatically annotated by the intelligent tutor (e.g., skipped, solved on first try, solved with hint, ...). Prior work [18, 26, 27] predicts such outcome labels using as input the student's entire video feed of solving the exercise. This requires both the time spent solving the problem and the time needed for the model to compute the outcome prediction. In contrast, we propose *early* prediction of students' exercise outcomes. Such early prediction increases the efficacy of an intelligent tutor by enabling personalized support in the form of real-time interventions for individual learning needs. An intelligent tutor that reacts quickly and provides an early intervention such as hints or encouragement in the first several seconds when a student starts working on an exercise could reduce the likelihood of a student deciding to skip this exercise.

The recent success of Transformers [30] in computer vision [13, 22] has demonstrated the potential of applying Transformers in behavior prediction tasks. We employ a Transformer for analyzing student visual information, especially due to its ability and efficiency in capturing temporal dependencies in student videos, which is important for understanding student behavior and affective states from a short video segment. We design an affect-aware Transformer for efficiently analyzing student visual information, particularly when we only have access to the first 5 s (or 20 s) of video instead of the entire sequence. We leverage transfer learning to learn a per-frame affect embedding by pre-training an affect network for facial expression recognition. We also extract relevant facial embeddings such as gaze direction and head poses using a facial analysis network. These embeddings capture valuable cues and insights into students' emotions and behavior.

We further propose augmenting the visual representation with timing information from student log data such as the exact timestamp a student takes certain actions, e.g., ask for a hint, attempt to answer the exercise, and correctly solve the exercise. Such information provides complementary insights into a student's learning process and can be used to better understand their behavior and affective states. Visual information remains instrumental for *early* prediction as visual information is available before timing information becomes available. Finally, we combine the timing information (whenever available at the early stage of attempting an exercise) with state-of-the-art visual affect analysis techniques (Vision Transformers) through a multimodal fusion module to generate student exercise outcome prediction. Our resulting model, EPATT (Early Prediction using an Affect-aware Transformer with Timing information), is able to predict the exercise outcome in the first several seconds of a student's attempt and achieves superior performance over the previous state-of-the-art on a dataset for exercise outcome prediction. We summarize our contributions as follows:

- We propose the first affect model, called EPATT, that can predict student exercise outcomes at an *early* stage by analyzing only the first 5 (or up to 20) seconds of student data.
- We propose incorporating state-of-the-art facial affective embeddings from video frames using transfer learning and analyzing their temporal dependencies using an affect-aware Transformer for exercise outcome prediction.
- We propose a multimodal system that augments video representation with timing information obtained from students' learning log data. Incorporating timing information about when students take certain actions (e.g., requesting a hint), enhances the ability of our model to perform early prediction of student outcomes.
- We present a system performance trade-off that demonstrates the importance of visual data for *early* prediction in intelligent tutors as they become available, before other data do, such as action timing information.
- We present extensive experimental results demonstrating EPATT's superior performance over state-of-the-art and show each of the proposed components (visual information; log data) contributes to improving the performance.

2 Related Work

Exercise Outcome Prediction. In prior research, the problem of predicting student exercise outcomes in intelligent tutors has been studied [18,26,27]. Joshi et al. [18] first introduced a labeled video dataset of student interactions with MathSpring and predicted excercise outcomes using traditional facial affect signals, such as head pose, gaze, and facial action units (AUs). Ruiz et al. [26] augmented the video dataset and developed a transfer learning approach to leverage a deep affect representation for exercise outcome prediction, achieving state-of-the-art performance. Ruiz et al. [27] extended that analysis by including student engagement prediction, enabling exploration of how engagement and learning outcomes correlate. Unlike the previous work that considered only video information and predicted exercise outcomes based on the entire video, this current research combines state-of-the-art visual representation with timing information to predict students' exercise outcomes in an early stage, i.e., during the first 5–20 s of each video, supporting more timely tutor interventions for students.

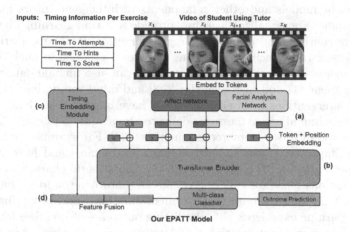

Fig. 1. Our proposed model, EPATT, for affect-aware early prediction of student problem solving outcomes. It consists of a Video Embedding Module for extracting affect and facial embeddings (a), a Transformer Encoder for modeling temporal dependencies in the videos (b), a Timing Embedding Module to encode the timing information (c), and a Fusion Module that combines the video information with timing information (d). (Colour figure online)

Early Action Recognition. Early action recognition is the identification of actions in streaming video sequences before they are fully observed. It is a challenging task due to the limited information a model has available. Ryoo [28] proposed to use dynamic bag-of-words to efficiently represent ongoing activities with spatio-temporal features probabilistically. Hoai et al. [17] presented a maximum-margin framework based on a structured output SVM for early detection of temporal events. Ma et al. [23] improved LSTM training with novel ranking losses for activity detection and early detection tasks.

Early action recognition is instrumental for many applications, e.g., predicting *accident-about-to-happen* as opposed to *accident-happened* in autonomous vehicles. For the task of real-time gesture recognition, accurately classifying gestures early, when they are only partially observed, is instrumental because it minimizes latency and improves user experience. Early gesture recognition has been explored for interactive games [1], showing that predicting which of twelve gestures a person performs can be done in less than two seconds (50 frames) with an accuracy of 45% using a sequence-to-sequence motion forecasting model. Compared to early action recognition, the goal of our work is to make early predictions of student problem-solving outcomes based on videos of students working on these exercises. The motivation here is to provide the opportunity for future online tutors to deliver interventions in a timely manner, e.g., hints or encouragement when the student is first struggling with the exercise.

Timing Information and Interventions in Intelligent Tutors. Some benefits of student models and other components in intelligent tutors include the ability to promote personalized learning, provide real-time learning analysis, use self-adaptive content, and designate targeted practice. Real-time learning data in an intelligent tutor includes how long it takes a student to click on a hint, choose an answer choice or solve an exercise. It can also include data that indicates engagement [33], off-task behavior [6,8] and mind-wandering [21].

Several interventions in intelligent tutors have improved student affect and outcome, including affective messages delivered by avatars and empathetic messages that respond to students' recent emotions [32]. For example, interventions have led to improved grades in state standardized exams and have influenced students' perceptions of themselves [9,19,32]. Empathetic characters that provide interventions generate superior results by improving student interactions with the system, addressing negative student emotions, and providing gains in the overall learning experience [20]. Predicting outcomes of exercises for students is a valuable source of information for planning and executing interventions in intelligent tutors [10,33] since they enable the tutor to provide hints when the system predicts that a student will not successfully complete an exercise.

3   Methods

The proposed model for early prediction of student exercise outcomes consists of a Video Embedding Module, a Transformer Encoder, a Timing Embedding Module, and a Fusion Module (Fig. 1).

The input video $\mathbf{X}$ of a student solving an exercise is defined to be of size $N \times H \times W \times 3$, where N is the number of frames in the video and each frame is an RGB image of size $H \times W$. We retrieve the timing information $\mathbf{T} \in \mathbb{R}^M$ from student learning log data as additional input to the model. In this work, we use $M = 7$ types of timing information: from the beginning of the exercise, the time until the student asks for the first/second/third hint, the time until the student makes the first/second/third attempt and the time until the student correctly solves the exercise (in seconds). Given the input video $\mathbf{X}$ and timing information $\mathbf{T}$, our task can be formulated as a multi-class classification problem, which predicts the outcome class $y \in \{1, \ldots, C\}$ of the exercise.

3.1 Frame-Wise Video Embedding and Tokenization

Given the raw video frames $\mathbf{X} = (x_1, x_2, \ldots, x_N)$, we adopt multiple deep learning networks, as was first proposed by [27], to embed the videos of students' faces and gestures while they solve exercises, Fig. 1a. As facial expressions and gestures are important cues for inferring exercise outcomes, an affect network was trained using in-the-wild images for facial expression recognition. Transfer learning was leveraged to learn an affect representation for each frame of student videos $\rho(x_i), i = 1, \ldots, N$ (Fig. 1, green). We also extract per-frame facial Action Unit (AU), gaze direction, and head pose using a facial analysis network, Open-Face [7], which we denote as $\psi(x_i), i = 1, \ldots, N$ (Fig. 1, yellow). We concatenate the outputs of the two networks as the final affective embedding for each frame (green-yellow boxes in Fig. 1).

Our methodological contribution here is to use the affective embeddings as the frame-wise tokens $(z_1, z_2, \ldots, z_N) \in \mathbb{R}^{N \times d}$, where $z_i = \rho(x_i) \oplus \psi(x_i)$ and d is the dimension of the concatenated embedding, for the Transformer Encoder (Fig. 1b). Similar to BERT's [class] token [11], we add a learnable embedding to the sequence of embedded tokens, denoted as $z_0 \in \mathbb{R}^d$ ([CLS] in Fig. 1). The state of this added embedding at the output of the Transformer Encoder ($z_0^{(L)} \in \mathbb{R}^d$) summarizes the video information and serves as the video representation, which is denoted as $\mathbf{E}_v$ (Fig. 1, purple, and Eq. 4 below).

We add position embeddings $\mathbf{E}_{pos} \in \mathbb{R}^{(N+1) \times d}$ to the frame tokens to retain positional information (Fig. 1, vermilion), which is essential for understanding the sequential order of student behavior. To achieve this, we add standard learnable 1D position embeddings to each patch embedding, which are learned during training to encode the temporal order of the frame tokens.

3.2 Transformer Encoder Module

The Transformer Encoder [30] consists of L identical layers. Each layer ℓ comprises multi-headed self-attention (MSA) [30], layer normalization (LN) [5], and MLP blocks as follows:

$$\mathbf{z}^{(0)} = [z_0; z_1; z_2; \ldots; z_N] + \mathbf{E}_{pos}, \tag{1}$$

$$\hat{\mathbf{z}}^{(\ell)} = \mathrm{MSA}(\mathrm{LN}(\mathbf{z}^{(\ell-1)})) + \mathbf{z}^{(\ell-1)}, \; \ell = 1, \ldots, L, \tag{2}$$

$$\mathbf{z}^{(\ell)} = \mathrm{MLP}(\mathrm{LN}(\hat{\mathbf{z}}^{(\ell)})) + \hat{\mathbf{z}}^{(\ell)}, \; \ell = 1, \ldots, L, \tag{3}$$

$$\mathbf{E}_v = \mathrm{LN}(z_0^{(\ell)}). \tag{4}$$

The MLP block consists of two linear projections separated by a GELU activation function [16]. The token dimension d remains fixed across all layers, and the final video embedding is obtained as the output of the [CLS] token $\mathbf{E}_v \in \mathbb{R}^d$.

3.3   Timing Embedding Module and Fusion Module

The timing information $\mathbf{T} \in \mathrm{R}^M$ includes $M = 7$ real numbers representing the duration from the start of the exercise to the occurrence of an action, measured in milliseconds. In some exercise samples, certain types of timing information may be unavailable due to actions not occurring within the first 5–20 s. To handle such cases where we have missing values in $\mathbf{T}$, we replace all the missing timing information with the out-of-domain value -1, which is not found naturally in the input data (time durations are always positive). In-domain values range between 0 and 20 s. While there could be alternative options for this value, experimental evaluation demonstrated that -1 yields marginally better results compared to e.g. a large value of 40 s. Before feeding $\mathbf{T}$ into an embedding layer, we standardize the values by subtracting the mean and scaling to unit variance, which improves the performance and training stability of the model. We encode the timing information using a fully connected embedding layer α and obtain the timing embedding $\mathbf{E}_t = \alpha(\mathbf{T}) \in \mathrm{R}^d$, which has the same dimension d as the video representation (Fig. 1, blue). We maintain an equal dimensionality for timing representation and video representation, in order to encourage them to contribute relatively equally to the classification task.

The timing and video embeddings are fused through concatenation. By directly concatenating both representations, we emphasize the complementary nature of timing and video data, rather than their cross-modality relationships, as we assume that different modalities affect the outcome prediction relatively independently. This approach reduces the complexity of the model, lowering the risk of overfitting, which is beneficial given the small size of the student dataset used in this work (see Section Dataset). Finally, we use a linear multi-class classifier $\mathcal{C}$ to predict exercise outcome $y = \mathcal{C}(\mathbf{E}_v \oplus \mathbf{E}_t)$ based on the multi-modal representation (Fig. 1, orange).

4   Experiments

4.1   Dataset

We used the MathSpring Children Dataset [27]. It consists of 968 videos of fifty-one sixth-grade male and female students of different ethnicities recorded with consent while they were working on a single exercise, as well as the exercise outcome and student learning log data. We extracted the timing information for each exercise from the log data.

The outcome of a student problem-solving interaction with the tutor is labeled according to a series of cognitive/affective states, which characterize the correctness of the solution and the degree of effort excerpted in math practice. These effort levels are predictors of longer-term outcomes such as emotions, affective gains, achievement and learning [31]. They depend on the success rate (mistakes made), hints requested, and time excerpted during problem-solving [4], and consist of: SOF (solved on the first attempt), SHINT (solved correctly after seeing hints), GUESS (guessed the answer –several incorrect answers before correct, and no hints seen), ATT (solved after one incorrect attempt –self-corrected), NOTR (responded so quickly that there was no time to read the exercise), SKIP (skipped the exercise, no action), and GIVEUP (made mistakes or asked for hints, yet the student eventually gave up and requested a new exercise).

We trained models on the original seven classes and also restructured the effort labels into 2 behavior classes, which we believe will enhance our understanding of student behavior and facilitate targeted interventions. The two new behavior outcome classes are positive (SOF, SHINT) and negative (GUESS, ATT, NOTR, GIVEUP). SOF and SHINT show that the student is working actively and positively on the exercise. We consider all other effort outcomes as outcomes for which the tutor needs to interrupt the student and provide an intervention. Results for both the 7-Effort-class and 2-Behavior-class prediction tasks are reported in Section Results.

The distribution of video length in the MathSpring Children Dataset for 7 student effort outcomes is illustrated in Fig. 2. The box plot shows the wide variability in the length of videos present in the dataset (we use a logarithmic scale for better visualization). The recorded clips range from brief 1–2 s sequences to extended recordings lasting up to 20 min, and the length of videos also varies across different efforts, indicating different student behaviors and complexity of the learning activities captured in the dataset.

We consider only the first 5, 10, 15, and 20 s of each video for exercise outcome prediction. Figure 2 shows that on average, the length of videos largely exceeds the thresholds we use for the prediction, especially for SHINT and GIVEUP. That is, a prediction can be typically made by the tutor before a student responds within the tutor. This provides the intelligent tutor with ample time to deliver appropriate interventions. For example, an intelligent tutor can identify a struggling student who may be on the verge of giving up on an exercise in the early stages, and can provide timely interventions with appropriate support. When training and evaluating the model using different thresholds, we use subsets of the dataset by removing videos that are shorter than the threshold for a fair comparison, which also prevents the model from relying solely on video length to make predictions rather than understanding the underlying visual patterns. The subsets for 5, 10, 15, and 20 s contain 880 samples, 826 samples, 776 samples, and 714 samples, respectively.

4.2 Experimental Setup

Model Variants. To evaluate the performance of our model, we implemented multiple baseline methods for comparison. The majority vote classifier selects the majority class in the dataset, which serves as a random baseline. The timing-only model trains a classifier on top of the timing embedding, considering timing information only. We also reproduce the previous state-of-the-art method for exercise outcome prediction, ATL-BP [27], and report its performance. The visual-only model, EPAT (EPATT without Timing), trains the Transformer without the timing embedding. Finally, our full model, EPATT, combines the affective visual

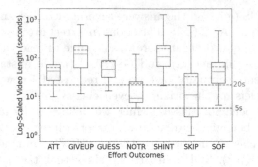

Fig. 2. Distribution of video length (log scale) for 7 outcomes in the MathSpring Children Dataset. The box is drawn from the 25th percentile to the 75th percentile with an orange horizontal line to denote the median exercise length and a green dashed line to denote the average. The boundary of the lower/upper whisker is the minimum/maximum video length of this class in the dataset. The blue and red horizontal dashed lines represent the decision-time thresholds of our model from 5 s to 20 s, respectively. The plot shows that our models attempt to make an outcome prediction much earlier than the average time a student spends on an exercise for 5 of the 7 student effort outcome classes. (Color figure online)

features with timing embedding to predict outcomes. The performance of the models listed here is presented in Tables 1 and 2.

Implementation Details. We implemented our models in PyTorch and conducted experiments on one NVIDIA TITAN Xp GPU. For the Affect Network, we pre-trained a ResNet-50 network on a subset of 50,000 randomly sampled images from the AffectNet dataset [24] and validated the network on 5,000 randomly selected images. Following [27], we downsampled the videos to three frames per second from the original 30 frames per second, to reduce both processing time and storage requirements. For the Transformer Encoder, we employed 8-heads self-attention, and $\ell = 4$ layers. The token dimension and hidden size d is set to 248. We applied dropout [29] to the output of each sub-layer, before it is added to the sub-layer input and normalized. We trained the full model using the Adam optimizer. We used a learning rate of 3×10^{-4} for 100 epochs, and a batch size of 1 following the previous work [27].

Evaluation Protocol and Metrics. Following the experimental setup by [27], we performed five-fold cross-validation, where the training set contains 80% data and the testing set contains the remaining 20%. To test generalization to new users, we adopted a leave-users-out experimental setup where users are exclusively split into either the training or test set. In other words, we enforce the rule that no video clips of the same user can be in both the test and training sets. In this manner we can measure how the system performs when applied to an unseen user. The balanced accuracy, mean F_1-score, and Cohen's Kappa coefficient are reported. The mean F-score is a commonly used metric that combines precision and recall, while balanced accuracy measures the average accuracy of each class, taking into account class imbalance. Cohen's kappa measures the agreement between the predicted and actual class labels, taking into account the possibility of agreement occurring by chance. The range of Cohen's kappa is [-1,+1], and a Cohen's kappa of 0 or lower indicates random predictions.

5   Results

Seven-Effort Outcome Prediction. We trained models for predicting seven student effort outcomes using information from the first five, ten, fifteen, and twenty seconds of each video, see Table 1.

First 5-second results. When considering only the first five seconds, our proposed model demonstrates superior performance compared to all baselines.

Table 1. Results for Early Prediction of 7-Effort Outcomes: Five-fold cross-validation averages and standard deviations are reported. The full model EPATT outperforms all the baselines and previous state-of-the-art [27].

Data	Method	F-Score	Balanced Acc.	κ
5 s	Majority Vote	0.07±0.01	0.15±0.01	0.00±0.00
	Timing Module (timing info only)	0.17±0.05	0.25±0.05	0.03±0.02
	ATL-BP [27]	0.19±0.03	0.20±0.02	0.08±0.04
	EPAT (*Ours* visual info only)	0.24±0.04	0.28±0.05	**0.14±0.05**
	EPATT (*Ours* visual + timing info)	**0.25±0.05**	**0.29±0.05**	0.11±0.04
10 s	Majority Vote	0.08±0.02	0.15±0.01	0.00±0.00
	Timing Module (timing info only)	0.18±0.03	0.22±0.03	0.08±0.03
	ATL-BP [27]	0.19±0.02	0.20±0.02	0.06±0.02
	EPAT (*Ours* visual info only)	0.23±0.03	0.24±0.03	0.14±0.03
	EPATT (*Ours* visual + timing info)	**0.28±0.05**	**0.29±0.04**	**0.17±0.07**
15 s	Majority Vote	0.07±0.01	0.15±0.01	0.00±0.00
	Timing Module (timing info only)	0.20±0.05	0.22±0.04	0.12±0.04
	ATL-BP [27]	0.18±0.01	0.19±0.02	0.06±0.04
	EPAT (*Ours* visual info only)	0.22±0.03	0.23±0.02	0.15±0.02
	EPATT (*Ours* visual + timing info)	**0.31±0.04**	**0.31±0.04**	**0.19±0.03**
20 s	Majority Vote	0.08±0.01	0.14±0.00	0.00±0.00
	Timing Module (timing info only)	0.26±0.04	0.27±0.03	0.19±0.06
	ATL-BP [27]	0.18±0.04	0.20±0.04	0.07±0.05
	EPAT (*Ours* visual info only)	0.26±0.03	0.28±0.05	0.15±0.02
	EPATT (*Ours* visual + timing info)	**0.31±0.02**	**0.32±0.04**	**0.26±0.06**

Notably, both, the proposed full model and the model with visual information only, outperform the previous state-of-the-art method ATL-BP [27] by large

Table 2. Results for Early Prediction of 2-Behavior Outcomes. Positive and negative are predicted using the first 5, 10, 15, and 20 s of the data.

Data	Method	F-Score	Balanced Acc.	κ
5 s	Majority Vote	0.34±0.04	0.50±0.00	0.00±0.00
	Timing Module (timing info only)	0.39±0.01	0.52±0.01	0.03±0.02
	ATL-BP [27]	0.59±0.04	0.60±0.03	0.19±0.07
	EPAT (*Ours* visual info only)	**0.59±0.04**	**0.60±0.05**	**0.19±0.09**
	EPATT (*Ours* visual + timing info)	0.55±0.08	0.58±0.04	0.15±0.08
10 s	Majority Vote	0.35±0.04	0.50±0.00	0.00±0.00
	Timing Module (timing info only)	0.44±0.05	0.55±0.03	0.09±0.05
	ATL-BP [27]	0.61±0.04	0.61±0.04	0.22±0.08
	EPAT (*Ours* visual info only)	0.60±0.03	0.60±0.03	0.21±0.06
	EPATT (*Ours* visual + timing info)	**0.61±0.04**	**0.62±0.03**	**0.24±0.07**
15 s	Majority Vote	0.35±0.02	0.50±0.00	0.00±0.00
	Timing Module (timing info only)	0.48±0.04	0.56±0.03	0.13±0.05
	ATL-BP [27]	0.59±0.03	0.59±0.04	0.18±0.07
	EPAT (*Ours* visual info only)	0.57±0.02	0.57±0.02	0.15±0.03
	EPATT (*Ours* visual + timing info)	**0.61±0.06**	**0.63±0.04**	**0.26±0.07**
20 s	Majority Vote	0.35±0.01	0.50±0.00	0.00±0.00
	Timing Module (timing info only)	0.52±0.03	0.59±0.02	0.19±0.04
	ATL-BP [27]	0.62±0.05	0.63±0.04	0.25±0.09
	EPAT (*Ours* visual info only)	0.61±0.02	0.62±0.02	0.23±0.04
	EPATT (*Ours* visual + timing info)	**0.67±0.04**	**0.68±0.04**	**0.35±0.09**

margins of 6 and 5 pp in F-Score, and 9 and 8 pp in balanced accuracy, respectively (Table 1, 5 s). By incorporating timing information, the performance of our model slightly improves in terms of F-Score (1 pp) and balanced accuracy (1 pp) while the visual-only model achieves a higher Cohen's Kappa coefficient.

First 10-second results. The proposed model consistently outperforms all the baselines on all the metrics (Table 1, 10 s). We can observe that the full model achieves a large increase over the visual-only model in this setting (5 pp in F-score, 5 pp in balanced accuracy, and 3 pp in Cohen's Kappa). It demonstrates that timing information provides valuable supplementary information to the visual model and can enhance the overall performance of the model.

First 15-second results. The proposed model achieves the highest F-score (0.31), and balanced accuracy (0.31) (Table 1, 15 s). The proposed full model and the model with visual information only consistently outperform baselines and the previous state-of-the-art method ATL-BP [27].

First 20-second results. The proposed full model achieves a large performance increase over all baselines and the visual-only model (Table 1, 20 s). The average performance gain on all metrics compared with the previous state-of-the-art

ATL-BP [27] is 14.7 pp, and incorporating timing information also improves the performance by an average increase of 6.7 pp on all metrics.

Two-Behavior Outcome Prediction. To better understand student behavior and provide the opportunity for interventions, we also trained models for 2 restructured behavior classes (positive & negative), see Table 2.

First 5-second results. The proposed model with visual information only achieves the highest F-score (0.59), balanced accuracy (0.60), and Cohen's Kappa (0.19) (Table 2, 5 s). The performance of the proposed full model, EPATT, is slightly lower compared to the visual-only model, suggesting that facial features contribute more remarkably in the very early stage of predicting student outcomes. Moreover, the relatively low scores for the timing-only model (F-score 0.39, Cohen's Kappa 0.03) show that timing information alone is not sufficient for effective classification when considering only the first 5 s of information. This could be due to the fact that in the first 5 s, most students have just finished reading the statement of the exercise, without having the chance to take any action. Therefore, the visual information, mainly consisting of facial expressions, is more informative in predicting early outcomes during this period.

First 10-second results. We observe that EPATT now outperforms the proposed model with visual information only (Table 2, 10 s). Also, the model with timing information only achieves reasonable performance (0.44 in F-score, 0.55 in balanced accuracy, and 0.09 in Cohens Kappa). This further shows that as time passes, students start to take actions, making the timing information more informative and useful in predicting student outcomes.

First 15-second results. EPATT achieves the highest F-score (0.61), balanced accuracy (0.63) and Cohen's Kappa (0.26) (Table 2, 15 s). The performance gain by considering timing information further increases (4 pp in F-score, 6 pp in balanced accuracy, and 11 pp in Cohen's Kappa).

First 20-second results. (Table 2, 20 s). EPATT outperforms both the previous state-of-the-art [27] and the visual-only model by large margins (5 pp and 6 pp in F-score, 5 pp and 6 pp in balanced accuracy, and 10 pp and 12 pp in Cohen's Kappa, respectively). The average performance gain compared with the previous state-of-art [27] is 6.7 pp. This result confirms that our model, which combines timing information with visual affect analysis, can accurately and effectively predict student exercise outcomes in the early stages.

6   Discussion and Conclusions

By accurately predicting student exercise outcomes during the early stages of problem-solving, intelligent tutors can provide timely interventions and support, leading to more effective learning outcomes. Combining the analysis of the timing information about a student's actions with visual affect analysis provides a mechanism for such early prediction. Visual information is instrumental when outcome data is not yet available, and action timing information greatly boosts visual performance as it becomes more available.

The observed trends for early student exercise outcome prediction demonstrate that both visual and logged timing data are important. Visual information has the benefit of being available as early as the student starts solving the problem. Log data has the benefit of having a more direct mapping of performance through metrics, but becomes available with time. Our model trades off the use of direct performance indicators in return for being able to make early predictions.

We designed an affect-aware Transformer to model the temporal dependencies of student video sequences. Our experiments show that the proposed model achieves a significant improvement over the state-of-the-art model on all metrics, with our best model achieving an average performance increase of 14.7 percent points for the 7-Effort problem and 6.7 percent points for the 2-Behavior class problem when using 20 s of data. Overall, our study demonstrates the potential of combining visual affect analysis with timing information to predict student exercise outcomes in the early stages. Our proposed model can assist intelligent tutoring systems in providing timely interventions and support for students, ultimately targeting better learning outcomes.

Acknowledgement. This work has been partially funded by the U.S. National Science Foundation, grants # 1551572, # 1551590, # 1551589, and # 1551594.

References

1. Agrawal, R., Joshi, A., Betke, M.: Enabling early gesture recognition by motion augmentation. In: The 11th International Conference on Pervasive Technologies Related to Assistive Environments, Corfu, Greece, June 26–29, pp. 98–101 (2018)
2. Arroyo, I., Cooper, D.G., Burleson, W., Woolf, B.P., Muldner, K., Christopherson, R.: Emotion sensors go to school. In: Proceedings of the 2009 Conference on Artificial Intelligence in Education, pp. 17–24. IOS Press (2009)
3. Arroyo, I., Mehranian, H., Woolf, B.P.: Effort-based tutoring: an empirical approach to intelligent tutoring. In: Educational Data Mining, CiteSeerX (2010)
4. Arroyo, I., Woolf, B.P., Burelson, W., Muldner, K., Rai, D., Tai, M.: A multimedia adaptive tutoring system for mathematics that addresses cognition, metacognition and affect. Int. J. Artif. Intell. Educ. **24**, pp. 387–426 (2014)
5. Ba, J.L., Kiros, J.R., Hinton, G.E.: Layer normalization. arXiv preprint arXiv:1607.06450 (2016)
6. Baker, R.S.: Modeling and understanding students' off-task behavior in intelligent tutoring systems. In: Proceedings of the SIGCHI Conference on Human Factors in Computing Systems, pp. 1059–1068 (2007)
7. Baltrusaitis, T., Zadeh, A., Lim, Y.C., Morency, L.: OpenFace 2.0: facial behavior analysis toolkit. In: 13th IEEE International Conference on Automatic Face and Gesture Recognition, pp. 59–66 (2018)

8. Cetintas, S., Si, L., Xin, Y.P.P., Hord, C.: Automatic detection of off-task behaviors in intelligent tutoring systems with machine learning techniques. IEEE Trans. Learn. Technol. **3**(3), 228–236 (2009)

9. Craig, S.D., Graesser, A.C., Perez, R.S.: Advances from the office of naval research STEM Grand Challenge: expanding the boundaries of intelligent tutoring systems. IJ STEM Ed. **5**(1), 1–4 (2018). https://doi.org/10.1186/s40594-018-0111-x

10. Delgado, K., et al.: Student engagement dataset. In: Proceedings of the IEEE/CVF International Conference on Computer Vision, pp. 3628–3636 (2021)

11. Devlin, J., Chang, M., Lee, K., Toutanova, K.: BERT: pre-training of deep bidirectional transformers for language understanding. In: Proceedings of the 2019 Conference of the North American Chapter of the Association for Computational Linguistics: Human Language Technologies, NAACL-HLT, pp. 4171–4186 (2019)

12. D'Mello, S.K., Bosch, N., Chen, H.: Multimodal-multisensor affect detection. In: The Handbook of Multimodal-Multisensor Interfaces: Signal Processing, Architectures, and Detection of Emotion and Cognition-Volume 2, pp. 167–202, Association for Computing Machinery and Morgan and Claypool (2018)

13. Dosovitskiy, A., et al.: An image is worth 16x16 words: transformers for image recognition at scale. In: International Conference on Learning Representations (2021)

14. D'Mello, S., et al.: A time for emoting: When affect-sensitivity is and isn't effective at promoting deep learning. In: International Conference on Intelligent Tutoring Systems, pp. 245–254 (2010)

15. Gordon, G., et al.: Affective personalization of a social robot tutor for children's second language skills. In: 30th AAAI Conference on Artificial Intelligence (2016)

16. Hendrycks, D., Gimpel, K.: Gaussian error linear units (GELUs). arXiv preprint arXiv:1606.08415 (2016)

17. Hoai, M., De la Torre, F.: Max-margin early event detectors. Int. J. Comput. Vision **107**, 191–202 (2014)

18. Joshi, A., et al.: Affect-driven learning outcomes prediction in intelligent tutoring systems. In: IEEE International Conference on Automatic Face and Gesture Recognition (2019)

19. Karumbaiah, S., Lizarralde, R., Allessio, D., Woolf, B.P., Arroyo, I., Wixon, N.: Addressing student behavior and affect with empathy and growth mindset. In: Proceedings of 10th International Conference on Educational Data Mining (2017)

20. Kim, Y.: Empathetic virtual peers enhanced learner interest and self-efficacy. In: Workshop on Motivation and Affect in Educational Software, 12th International Conference on Artificial Intelligence in Education, pp. 9–16 (2005)

21. Lee, W., et al.: Measurements and interventions to improve student engagement through facial expression recognition. In: 24th HCI International Conference, pp. 286–301 (2022)

22. Liu, Z., et al.: Swin transformer: hierarchical vision transformer using shifted windows. In: Proceedings of the IEEE/CVF International Conference on Computer Vision (2021)

23. Ma, S., Sigal, L., Sclaroff, S.: Learning activity progression in LSTMs for activity detection and early detection. In: Proceedings of the IEEE Conference on Computer Vision and Pattern Recognition, pp. 1942–1950 (2016)

24. Mollahosseini, A., Hassani, B., Mahoor, M.H.: AffectNet: a database for facial expression, valence, and arousal computing in the wild. IEEE Trans. Affect. Comput. **10**, 18–31 (2019)

208   H. Yu et al.

25. Monkaresi, H., Bosch, N., Calvo, R.A., D'Mello, S.K.: Automated detection of engagement using video-based estimation of facial expressions and heart rate. IEEE Trans. Affect. Comput. **8**(1), 15–28 (2016)
26. Ruiz, N., et al.: Leveraging affect transfer learning for behavior prediction in an intelligent tutoring system. In: IEEE International Conference on Automatic Face and Gesture Recognition (2021)
27. Ruiz, N., et al.: ATL-BP: a student engagement dataset and model for affect transfer learning for behavior prediction. IEEE Trans. Biometrics Behav. Identity Sci. **5**(3), 411–424 (2023). https://doi.org/10.1109/TBIOM.2022.3210479
28. Ryoo, M.S.: Human activity prediction: early recognition of ongoing activities from streaming videos. In: International Conference on Computer Vision (2011)
29. Srivastava, N., Hinton, G., Krizhevsky, A., Sutskever, I., Salakhutdinov, R.: Dropout: a simple way to prevent neural networks from overfitting. J. Mach. Learn. Res. **15**(1), 1929–1958 (2014)
30. Vaswani, A., et al.: Attention is all you need. In: NeurIPS (2017)
31. Wixon, M., Arroyo, I., Muldner, K., Burleson, W., Rai, D., Woolf, B.: The opportunities and limitations of scaling up sensor-free affect detection. In: Educational Data Mining 2014 (2014)
32. Woolf, B.P., et al.: The effect of motivational learning companions on low achieving students and students with disabilities. In: Aleven, V., Kay, J., Mostow, J. (eds.) ITS 2010. LNCS, vol. 6094, pp. 327–337. Springer, Heidelberg (2010). https://doi.org/10.1007/978-3-642-13388-6_37
33. Yu, H., et al.: Measuring and integrating facial expressions and head pose as indicators of engagement and affect in tutoring systems. In: Sottilare, R.A., Schwarz, J. (eds.) HCII 2021. LNCS, vol. 12793, pp. 219–233. Springer, Cham (2021). https://doi.org/10.1007/978-3-030-77873-6_16

To Read or Not to Read: Predicting Student Engagement in Interactive Reading

Beata Beigman Klebanov$^{(\boxtimes)}$, Jonathan Weeks, and Sandip Sinharay

ETS, Princeton, USA
{bbeigmanklebanov,jweeks,ssinharay}@ets.org

Abstract. In this study, upper-elementary-age students used an interactive reading app to read from a classic children's novel during a summer program. Students took turns reading with an adult virtual narrator (audiobook). We use process and background data to explore factors that could predict whether a reader will read their next turn or skip it. We find that skipping quickly becomes self-perpetuating, underscoring the need to support the teacher in providing just-in-time personalized intervention to help students avoid the disengagement trap.

Keywords: children's reading · summer reading · engagement

1 Introduction

The 2022 NAEP results show that 37% of U.S. fourth graders read below the Basic level, an increase since 2019 (34%).[1] There is thus an urgent need to help students recover from the learning loss induced by the pandemic, including a recognition that school-based learning might need to be supplemented after-school and in the summer.[2] Summer enrichment programs can supplement instruction in a more relaxed atmosphere where literacy or math activities are mixed with field trips, games, and other fun activities. However, due to the relaxation of the school discipline and expectations, sustained engagement with learning activities could be a challenge. We analyze data from a summer program where students engaged in independent reading of a children's novel using an electronic shared-reading platform. The research question that we focus on is: What process and demographic factors can help predict reader disengagement?

One of the major pedagogical challenges is classroom orchestration, which refers to design and implementation of classroom activities by the teacher to optimize outcomes. Reviewing the main aspects of orchestration in technology-enhanced learning, [17] pointed out the importance of (a) planning and design, (b) management of the time and workflow, (c) awareness of what is happening

---

[1] https://nces.ed.gov/nationsreportcard/reading/.
[2] https://njtutoringcorps.org/budget-statement/.

A. M. Olney et al. (Eds.): AIED 2024, LNAI 14830, pp. 209–222, 2024.
https://doi.org/10.1007/978-3-031-64299-9_15

in the classroom as a whole and with individual students, (d) adaptation to the emergent occurrences during the learning activity, and (e) interplay of learner- and teacher-driven elements of orchestration. Specific students becoming disengaged may be an instance of an emergent occurrence requiring teacher action, as that student's learning may become undermined and the attention of the disengaged student could turn to activities disruptive to others in the classroom.

To help forestall student disengagement, we explore the engagement dynamics in order to understand what characteristics of a student's ongoing and prior activity in the reading app could predict disengagement. If such prediction is possible, it opens up a possibility of supporting effective classroom orchestration by alerting the teacher in real time to allow for a well-timed adaptation action.

2   Related Work

Reader disengagement from a reading activity is often described as 'mind wandering', or following some internal train of thought instead of paying attention to the text; whether or not mind wandering is necessarily unintentional is subject to debate [6]. It is a common phenomenon and has been observed in both reading and listening to a narration of a text, with the second engendering more mind-wandering [10]. While the exact phenomenon in question and methods for detecting it vary substantially [21,27], it is a robust finding in the literature that mind wandering often leads to compromised comprehension of the text [6]: Missing important information early on may lead to failure to build an adequate mental model of the text [23] and cascade further into failure to make relevant inferences later in the reading process [22]. This process may result in a 'vicious cycle', where an initial failure of attention results in impaired comprehension which in turn promotes further inattention [6]. This cyclical view may also suggest some dynamism, where the initial onset of mind wandering might be unintentional but its further maintenance, upon realization of incomprehension, may be a conscious decision.

Both strong and struggling readers may exhibit mind wondering. For the strong readers, mental resources may not be fully engaged with a relatively easy reading activity and start 'working' on something else in parallel. For the struggling readers, the difficulty of the reading may result in the reader disconnecting attention from the reading. In addition to task difficulty, fatigue and lack of interest in the topic have been linked to mind wandering during reading [6]. Individual characteristics such as larger working memory capacity and stronger ability to execute attention control were linked to less mind wandering [25].

Researchers of technology-supported reading investigated detection of mind wandering during literacy activities through monitoring of reading speed [8], eye tracking [7], tracking of scrolling behavior [3], etc. The large bulk of the research is focused on detecting mind wandering as it unfolds or right after it has occurred; however, recent studies also started looking at predicting mind wandering using physiological signs, finding that certain types of arm movements tend to occur about 5 min ahead of mind wandering episodes [24]. Prediction of possible mind wandering may help forestall its occurrence, by, for example, helping the reader take a timely break or changing the activity.

While readers can sometimes apply self-initiated recovery strategies, such as re-reading a passage when they realized their minds were wandering, this does not always happen [26]. At the moment, interventions to mitigate mind wandering are generally either reactive (following a detection of mind wandering that is either unfolding or has already occurred) or proactive, focused on techniques such as practicing mindfulness and/or physical activity breaks [14,15], that are most effective when practiced regularly, without a direct relation to any measurements during the reading activity. In a reactive intervention, when eye-tracking detects mind wandering, the reading activity is stopped and a comprehension question is shown; if the response is incorrect, the correct answer is shown to help re-build comprehension; the reader is also prompted to re-read the text [7]. In a different study, eye-tracking-based detection of mind wandering resulted in the reader being prompted for a written self-explanation related to the text content, in order to promote deeper engagement with the content [13].

The psychology literature on mind wandering is generally more concerned with the onset of mind-wandering than with its maintenance (under what conditions it persists [19]), and focuses on brief episodes of disconnect – 10 to 15 s. To the best of our knowledge, little is known about the tendency of such episodes to recur, or to develop into a more sustained disengagement. Finally, the reviewed studies were laboratory studies; experts have called for research that would translate laboratory findings into the real world, commenting that it is "an endeavour fraught with complexity and risk, but it is an essential step for research to remain relevant and to contribute to broad societal good" [6].

In this study, we extend the prior research in three ways. First, we investigate a **real-life context** where children read during summer camp, with the attendant lack of control over the experimental environment, including acoustic, technical, and behavioral 'noise', such as construction outside the window, occasionally poor WiFi signal, and anticipation of a subsequent sports or game camp activity, respectively. In such contexts, it may not yet be realistic to use sensitive and expensive equipment with advanced capabilities, although the technology is advancing rapidly [9]. Second, the goal of the reading in our context is not strictly learning and comprehension to a high standard of coherence [4] and the text is not in the expository or informational genres typically used in mind wandering studies [6]. Since this is a relatively **informal reading activity** outside of the regular school context, the target standards of coherence are generally lower and more similar to reading for recreational and entertainment purposes. Third, we investigate a **long-term** engagement with the reading activity, across multiple reading episodes per day and across multiple days. Thus, while we are able to measure aspects of reader activity at a resolution only somewhat inferior to that of the lab studies reviewed above – minutes rather than seconds — we can investigate a much longer-term pattern of engagement.

3   The Interactive Reading App

We built Relay Reader™ [12], a reading and listening app[3], to help developing readers improve fluency while enjoying a good story. Figure 1 shows a screenshot.

Fig. 1. A screenshot of the Relay Reader app.

The user takes turns reading aloud with a pre-recorded narrator. A turn can be set by the user to between 70 and 200 words on average, separately for the narrator and the user. While the narrator is reading, the text is highlighted for the reader to follow along. The unit of highlight is a **span** – a phrase or a short sentence read by the narrator on one breath. Spans are detected semi-automatically [12]. For example, Fig. 1 shows a highlighted 15-word span.

When it is the user's turn to read, the user clicks on a button to start the audio recording and on another button to indicate that they are done. The audio recording is sent for processing to measure performance metrics such as reading accuracy and fluency [2,11]; the measurements are reported to the teacher in a periodic report but are currently not communicated to the user of the app.

After every four turns (two narrator's and two reader's), the reader is asked two multiple-choice **comprehension questions**. The questions are plot-oriented and ask about characters, their appearance, feelings, relationships; about locations and events, or other aspects of the plot. The questions were created manually to ensure effective coverage of important plot elements. The

---

[3] https://relayreader.org.

questions are not meant to trigger deep inference – they probe attention as much as comprehension, and serve a dual purpose of continuously assessing the reader and helping the reader recall an important story element though instant feedback showing the correct answer. In the context of mind wandering research, the app incorporates frequent changes of activity (listening, reading, answering questions) and questions as a mechanism to recover some information that may have been lost if a turn was not read fully or not listened to with attention.

The questions usually ask about information that was explicitly stated (or paraphrased) in the most recent user or narrator turn. We call the latest span in the text that states a piece of information necessary to answer the question the **anchor** of the question; once the bookmark passes the anchor, the question can be asked. After every four turns, the two questions with the nearest preceding anchors are asked. To achieve this local nature of the questions, we created a question for about every 100 words of running text. For example, a question that could be anchored to "I am the ghost of the Talking Cricket" is "Who did Pinocchio meet in the forest?". Note that the anchor does not contain all the information, as it says nothing about the forest, but that information was stated before and so the reader would have been exposed to it prior to this point.

Apart from the recordings of the user's oral reading, the app logs time-stamped events that follow the reader's activity, such as beginning and end of each narrator and student turns, the content of the turn that was read (in narrator's case) or was supposed to be read (in the user's case), responses to the questions. To alleviate any privacy and ethical concerns regarding data collection and usage, we followed our institutions's privacy policy[4] and IRB guidelines.

4   Data

4.1   Data Collection Context and Participants

The data collection occurred in summer programs in 5 locations belonging to the same national organization in the North-East of USA. The 133 participating students were mostly 3–5 graders, split into multi-age groups of 10 per instructor, on average (range: 7–14 students per instructor). While all sites received the same guidelines of 20 min of reading 3 times a week, different sites, instructors, and weeks varied – some sessions were longer than others and fewer sessions occurred during some weeks at some sites due to other summer activities. Of the 133 students, 65 were male, 60 female; gender information was not available for 8 students. In terms of grade, 47% of the students finished 4th grade, 27% finished 3rd grade, 17% finished 5th grade, and 5 students (3%) finished 2nd (4) and 1st (1) grades. Grade information was not available for 7 students. All students read *The Adventures of Pinocchio*, a 130-page (39K-word) classic novel by Carlo Collodi (translated from the Italian by Carol Della Chiesa), and listened to the narration by Mark Smith obtained from LibriVox[5] during narrator turns.

---

[4] https://www.ets.org//legal/privacy.html.
[5] https://librivox.org/the-adventures-of-pinocchio-by-carlo-collodi/.

4.2 Definitions

The following types of data derived from the logs will be used to compute the variables for our models. Descriptive statistics are shown in Table 1.

Span A phrase or short sentence that can be read on one breath; see Sect. 3 for more details. A span averages 8 words, with a standard deviation of 4.4.

Turn The passage that a participant (narrator or student) is to read out loud.

Turn Length The number of spans in a given turn.

Turn Duration The number of seconds that elapsed while the narrator was reading (for narrator turns); the number of seconds that elapsed between the timestamp of "Start Reading" button pressed by the student and "Done Reading" button pressed by the student (for student turns).

RCQ A comprehension question asked during the activity; see Sect. 3.

Skipped Turn If the duration of a turn is such that it would have taken less than one second per span had the reader actually read it, the turn is extremely unlikely to have been fully read, as it would entail a reading rate of about 480 words per minute (60 sec × 8 words) – an unrealistic reading rate, even for silent reading.[6] We consider such turns skipped. Note that it is not possible to skip a narrator's turn, only student turns can be skipped.

Table 1. Descriptive statistics of the dataset, per reader, after the pre-processing described in Sect. 5.2. The data come from $n = 133$ readers.

Variable	Mean	Stdev	Min	Max
# Reading turns	54.6	27.5	2	104
# Reading turns per day	6.5	4.3	2	20
# Skipped reading turns	18.6	18.8	0	83
# Spans per turn	15.9	5.5	0	40
RCQ % correct	65.3	22.6	0	100
Duration of a reading turn (in seconds)	62.3	49.5	2.9	281.1

5 Predicting Disengagement

5.1 Independent Variables (predictors)

Our research question relates to finding process and demographic factors that may predict disengagement. We operationalize the prediction problem as one of estimating the probability of a student skipping their upcoming reading turn given the characteristics of the immediate task, of the student's reading behavior so far and demographic variables. The problems that process variables are trying to capture are described below. Table 2 shows turn-level statistics.

---

[6] Most adults read English fiction silently at a rate of 200–320 words per minute [5].

1. *I am already disengaged*: A variable that counts the number of skips in the three most recent reading turns: **SkipsIn3**.
2. *I am getting fatigued by this activity*: The number of turns already read today is counted by the **UnitDay** variable; every two turns form a unit.
3. *I am doing too big a share of the reading*: The ratio of the length of the most recent narrator turn to the length of the upcoming student's turn: **NSRatio**.
4. *I am not quite following the story*: Proportion of RCQs answered incorrectly so far: **PIncorrect**.

Note that UnitDay, NSRatio, and PIncorrect correspond to variables found to be related to mind wandering in the literature – fatigue, an aspect of task difficulty (in a student's perception), and reading comprehension. The SkipsIn3 variable addresses the dynamic of the reading engagement in time and has not been considered in mind wandering research so far, to the best of our knowledge.

Table 2. Descriptive statistics per user reading turn; $n = 7{,}208$ turns.

Variable	Mean	Stdev	Min	Max
SkipsIn	1.03	1.27	0	3
NSRatio	1.80	1.52	0.33	21
UnitDay	3.23	2.15	1	10
PIncorrect	0.34	0.23	0	1

We also consider grade and gender. If grade has predictive value, for example, if it is harder for younger students to sustain attention, instructors in the multi-grade classrooms might stop the activity earlier for younger readers. Some prior research and public discourse suggest that boys may have more difficulty engaging with reading, especially of fiction; see [20] for a critique.

5.2   Data Pre-processing for Analysis

The data were structured using a person-period format where all data associated with a particular student reading turn appear in a row. That is, the number of rows per student corresponds to the number of student reading turns. Indicator variables were included for day and unit within day. Lagged variables for three previous turns were merged to the records for each current turn. For instance, the row with data for turn 4 includes the data for turns 3, 2, and 1. We also created cumulative variables; their values are accumulated up to the given time point. Lastly, since students seldom read more than 20 passages in a day, the data were truncated to only include up to the 10th UnitDay (20th reading turn).

5.3 Models

The outcome of interest is the odds of a student skipping during the upcoming turn. Given that there are multiple observations for each student, it is important to take into account within-student clustering. As a first step, we fit a two-level mixed-effects model with random intercepts and no additional predictors. The intraclass correlation was 0.7, which suggests that clustering at the student level will lead to more accurate statistical inference. As a next step, we examined whether there was meaningful variation in the effect of time. We considered random effects for observations nested within days and observations nested within time points. The former relates to potential clustering over shorter time intervals, whereas the latter – over the full duration of participation. In both cases, the intraclass correlations were very small (around 0.015), suggesting that the probability of skipping is not expected to vary much from day to day. We therefore opted for a two-level model with fixed effects for various predictors and random intercepts for the students.

In the subsequent steps, we examined several models. For each model, we considered the significance of the predictors, the practical significance of the coefficients, the increase in pseudo R^2 [16], the impact on model fit (reduction in AIC and BIC), and the correlation between the fixed effects (an indicator of multicollinearity).

We fitted generalized linear mixed models to the data. The models were implemented using a logit link function; the parameters were estimated via maximum likelihood (Laplace approximation) with the lme4 [1] package in R [18]. The dependent variable is a binary indicator of whether the current reader's turn is skipped. The odds of student i skipping are:

$$odds = P(Skip_i = 1)/(1 - P(Skip_i = 1)) \tag{1}$$

The baseline model was specified as in Eq. 2, where β_{00} is the average log-odds and u_{0i} is the deviation for the cluster-specific log-odds:

$$logit(odds) = \beta_{00} + u_{0i} \tag{2}$$

In the next model we specified fixed effects for the six variables listed in Sect. 5.1 and a random effect for the intercept with student at level-2. In this model, all of the process variables were significant predictors of skipping while neither of the demographic variables was a significant predictor. We therefore removed the demographic variables from the model and refitted the following final model specified in Eq. 3:

$$logit(odds) = \beta_{00} + \beta_{10}SkipsIn3_i + \beta_{20}UnitDay_i + \\ + \beta_{30}NSRatio_i + \beta_{40}PIncorrect_i + u_{0i} \tag{3}$$

6 Results and Discussion

The baseline model with the random effect only (Eq. 2) fits with AIC = 5,486, BIC = 5,500, and Log Likelihood = -2,741. Table 3 shows summary statistics

Table 3. Model estimates and fit statistics for the model specified in Eq. 3 for predicting the probability of a reader skipping their current reading turn. $n = 7{,}208$.

	Variance	St. Dev	
Random Effects			
Student	1.92	1.387	
	Coefficient	Signif	St. Error
Fixed Effects			
Intercept	-2.98	$p < 0.001$	(0.192)
SkipsIn3	1.01	$p < 0.001$	(0.041)
UnitDay	0.16	$p < 0.001$	(0.020)
NSRatio	-0.27	$p < 0.001$	(0.030)
PIncorrect	1.69	$p < 0.001$	(0.260)
Fit Statistics			
AIC		4,528	
BIC		4,570	
Log Likelihood		$-2{,}258$	
Pseudo R^2 (Fixed+Random)		0.82	
Pseudo R^2 (Fixed)		0.49	

for the model in Eq. 3. The log likelihood of the model is larger than that of the baseline model by 18%. The fixed effects explain 49% of the variance in the observations, as measured by pseudo R^2 for the fixed effects [16].

The model coefficients for each fixed effect show the odds ratio of skipping to not-skipping the current turn controlling for all the other variables. For example, each unit read within a day adds 0.16 to the skipping odds; thus, all else being equal, skips are likelier later in the daily reading session. If the narrator's latest turn is of about the same length as the student's (NSRatio close to 1), the odds of skipping are reduced by 0.27. Skipping each of the preceding three turns increases the odds of skipping the next one by 1. Having low story comprehension increases the likelihood of skipping (note that the variable codes for percent *incorrect*).

We observe that the signs of the coefficients of the variables that implement literature-based constructs align with results reported in prior work – low comprehenders and students who are becoming fatigued from the activity are likelier to skip; if the upcoming turn is substantially shorter than the preceding narrator's turn, which may be perceived by the student as having a much easier job to do than the narrator, the reader is less likely to skip the turn. These results provide evidence for generalization of the findings in the literature to a real-life extended reading context in an informal educational settings.

Our most sobering finding is that disengagement is something that a student can get trapped in very quickly. Table 4 shows the probabilities of skipping given different number of skips within the last three turns. A reader who has not skipped any of the preceding three turns is highly unlikely to skip the next turn. In contrast, having skipped the three immediately preceding turns, a reader with average comprehension (65%) has a 40% chance of skipping the next turn – even if fatigue is low and the reading is balanced between reader and narrator.

Table 4. Predicted probability of skipping the current turn for a 65%-comprehension reader as a function of skipping none, one, two, or three of the three preceding turns and of the unit in the day. X means the model predicts no skipping (negative odds).

Unit Day	Probability of skipping for an average-comprehension reader			
	SkipIn3=0	SkipIn3=1	SkipIn3=2	SkipIn3=3
2	X	X	X	0.40
3	X	X	X	0.45
4	X	X	X	0.49
5	X	X	0.11	0.53
6	X	X	0.22	0.56
7	X	X	0.30	0.59
8	X	X	0.37	0.61
9	X	X	0.43	0.64
10	X	X	0.47	0.66

The model is not successful in predicting a reader's first skip. Without prior skips, all reasonable constellations of other variables predict no skipping (due to the strong negative intercept).[7] Given that there is no useful predictor of the first skip and given the high odds of further skipping following three consecutive skips, our findings underscore the importance of a good design of the activity based on the model and of real-time adaptation of the activity in a personalized manner if some students do show a snowballing disengagement pattern.

To exemplify the way the model informs activity design, let us consider Table 5 that shows the importance of the fatigue and story comprehension factors. Two skips early in the session might not set the reader up for subsequent skipping, but the chances of skipping increase substantially later in the session for readers with low and average comprehension. Recall that one unit corresponds to two reading turns; due to the interleaved nature of the activity, for there to be two reading turns, there also must have been two narrator turns. Assuming a 150-word average narrator turn and 100-word average student turn, these correspond, roughly, to the amount of text that could be read in about one minute (According to Table 1, average student turn lasted 62.3 s). Thus, a

---

[7] The intercept of -2.98 is not compensated even if it is the 10th unit of the day ($10*0.16 = 1.6$) and the reader is guessing RCQs (PIncorrect=0.75, the odds increase is 1.27), $1.27+1.60 = 2.87 < |-2.98|$, so the model would still predict no skipping.

unit of the activity is expected to take about 4 min of net reading and listening, plus short breaks, responses to questions, re-plays if necessary – about 5 min of activity. According to Table 5, on the 6th unit of the day, that is, after about 25 min of the activity, students who skipped two preceding turns and have low story comprehension have 38% chance of skipping the next turn. Perhaps 20–25 minutes would be a good target duration of this activity for the class. If the scheduling can be personalized, students with stronger story comprehension could continue for a few more turns.

Table 5. Impact of fatigue (UnitDay) and story comprehension (RCQ percent correct) on predicted probabilities of skipping after skipping two of the preceding three turns.

| Unit Day | Probability of skipping the next turn given 2 skips | | |
	45% RCQ	65% RCQ	85% RCQ
2	X	X	X
3	0.13	X	X
4	0.23	X	X
5	0.32	0.11	X
6	0.38	0.22	X
7	0.44	0.30	0.09
8	0.48	0.37	0.20
9	0.52	0.43	0.29
10	0.55	0.47	0.36

To exemplify opportunities for a personalized adjustment of the activity, Table 6 illustrates the impact of the variable that captures the ratio of the length of the narrator's turn to the student's. According to the model, a high ratio predicts lower probability of skipping. For example, a student with low story comprehension has a 23% chance of skipping on unit 4 if the narrator and the student have turns of approximately equal length, which would go down to 3% if the next student turn were only about half as long as the narrator's (see columns 2,3 row 2 in Table 6). Likewise, a student with average story comprehension who skipped two turns would have a 22% chance of skipping the next turn in the equal turn length situation on unit 5, but the probability would go down to only 1% if the next reading turn were much shorter than that narrator's turn (see columns 4,5 row 3 in the Table). Since narrator and student turn length are separately adjustable in the app, a shorter reader turn and a longer narrator turn can be arranged to help keep the student on track. It is also possible that other teacher actions, such as asking the student to pause and conversing with them, would provide a break in the activity and help redirect the trajectory of subsequent reading away from further skipping.

Finally, we observe that the background variables we explored – grade and gender – were not significant predictors of skipping. This result suggests that all

Table 6. Illustration of the impact of the ratio of the narrator to student turn length (NSRatio). N = S corresponds to a case where narrator and student turns are approximately of the same length; in N = 2S the narrator is reading twice the amount.

	Probability of skipping the next turn given 2 skips			
Unit Day	45% RCQ		65% RCQ	
	N = S	N = 2S	N = S	N = 2S
2	X	X	X	X
3	0.13	X	X	X
4	0.23	0.03	0.11	X
5	0.32	0.16	0.22	0.01

the students in the grade range we explored – mostly 3–5 graders – were equally able to engage with the activity, which may make it easier to administer in a multi-age group of readers. We also found no evidence that boys are less likely to be engaged in reading fiction – contrary to a common belief.

7   Conclusion

In this study, upper-elementary-age students used an interactive reading app to read from a children's novel during a literacy enrichment part of a summer program. Students took turns reading with an adult virtual narrator (audiobook). We used process data to explore factors that could predict whether a reader will read their upcoming reading turn or skip it. We found that prior skipping is the strongest predictor of future skipping, suggesting that disengagement is likely to snowball if left unattended.

We illustrated a way the model can be used to estimate a reasonable duration of the activity and to plan for a personalized activity adjustment action in case continued disengagement hazard is flagged for a particular student. Our results point towards the existence of a window of opportunity for a teacher's corrective action – between the first observed skip and the next one or two, since the first cannot be predicted and the third is predictive of a high likelihood of a reader getting trapped in a sustained pattern of disengagement. After a flagged first skip, the teacher may have only about 5 min to act before the student gets to the third consecutive skip. Identifying possible effective teacher actions for various types of students is an urgent area for future research.

Our results suggest that factors reported in the literature as conducive to mind wandering during reading, namely, fatigue, task difficulty, and low comprehension, generalize to the new context examined in this study – a relatively informal reading of a novel by upper elementary students during summer camp. Our findings also show a quick transition from a one-off to sustained disengagement, thus providing empirical evidence to the 'vicious cycle' hypothesis discussed in the mind wandering literature [6]. Collecting further empirical evidence, with readers of different ages reading in a different context, is also an important avenue for future work in order to deepen the understanding of the fine-grained dynamics of the process of reader disengagement, which, in turn, could lead to improved prevention, detection, and recovery solutions.

8   Limitations

A number of limitations of the current work are recognized; they all point to directions for future work to examine the robustness of the findings. One limitation is the use of one book; other reading materials may generate different engagement patterns. Secondly, the activity took place in a multi-age group; a more agewise homogeneous group could show a different engagement pattern.

Acknowledgment. This material is based upon work partially supported by the National Science Foundation and the Institute of Education Sciences under Grant #2229612. Any opinions, findings, and conclusions or recommendations expressed in this material are those of the author(s) and do not necessarily reflect the views of the National Science Foundation or the U.S. Department of Education.

References

1. Bates, D., Mächler, M., Bolker, B., Walker, S.: Fitting linear mixed-effects models using lme4. J. Stat. Softw. **67**(1), 1–48 (2015)
2. Beigman Klebanov, B., Loukina, A.: Exploiting structured error to improve automated scoring of oral reading fluency. In: Proceedings of AIED), pp. 76–81 (2021)
3. Biedermann, D., et al.: Detecting the disengaged reader-using scrolling data to predict disengagement during reading. In: LAK23: 13th International Learning Analytics and Knowledge Conference, pp. 585–591 (2023)
4. Van den Broek, P., Lorch, R., Linderholm, T., Gustafson, M.: The effects of readers' goals on inference generation and memory for texts. Memory Cogn. **29**(8), 1081–1087 (2001)
5. Brysbaert, M.: How many words do we read per minute? A review and meta-analysis of reading rate. J. Memory Language **109**(104047) (2019)
6. D'Mello, S., Mills, C.: Mind wandering during reading: an interdisciplinary and integrative review of psychological, computing, and intervention research and theory. Language Linguist. Compass **15**(4), e12412 (2021)
7. Faber, M., Bixler, R., D'Mello, S.: An automated behavioral measure of mind wandering during computerized reading. Behav. Res. Methods **50**, 134–150 (2018)
8. Franklin, M.S., Smallwood, J., Schooler, J.W.: Catching the mind in flight: using behavioral indices to detect mindless reading in real time. Psychon. Bull. Rev. **18**, 992–997 (2011)
9. Hutt, S., Wong, A., Papoutsaki, A., Baker, R., Gold, J., Mills, C.: Webcam-based eye tracking to detect mind wandering and comprehension errors. Behav. Res. Methods, 1–17 (2023)
10. Kopp, K., D'Mello, S.: The impact of modality on mind wandering during comprehension. Appl. Cogn. Psychol. **30**(1), 29–40 (2016)
11. Loukina, A., et al.: Automated estimation of oral reading fluency during summer camp e-book reading with My Turn To Read. In: Proceedings of INTERSPEECH, pp. 21–25 (2019)

12. Madnani, N., et al.: My Turn to Read: an interleaved e-book reading tool for developing and struggling readers. In: Proceedings of the ACL: System Demonstrations, pp. 141–146 (2019)
13. Mills, C., Gregg, J., Bixler, R., D'Mello, S.: Eye-Mind reader: an intelligent reading interface that promotes long-term comprehension by detecting and responding to mind wandering. Hum.-Comput. Interact. **36**(4), 306–332 (2021)
14. Mrazek, M., Zedelius, C., Gross, M., Mrazek, A., Phillips, D., Schooler, J.: Mindfulness in education: enhancing academic achievement and student well-being by reducing mind-wandering. In: Mindfulness in social psychology, pp. 139–152. Routledge (2017)
15. Müller, C., Otto, B., Sawitzki, V., Kanagalingam, P., Scherer, J.S., Lindberg, S.: Short breaks at school: effects of a physical activity and a mindfulness intervention on children's attention, reading comprehension, and self-esteem. Trends Neuroscience Educ. **25**, 100160 (2021)
16. Nakagawa, S., Schielzeth, H.: A general and simple method for obtaining R2 from generalized linear mixed-effects models. Methods Ecol. Evol. **4**(2), 133–142 (2013)
17. Prieto, L., Holenko Dlab, M., Gutiérrez, I., Abdulwahed, M., Balid, W.: Orchestrating technology enhanced learning: a literature review and a conceptual framework. Int. J. Technol. Enhanced Learn. **3**(6), 583–598 (2011)
18. R Core Team: R: A Language and Environment for Statistical Computing. R Foundation for Statistical Computing, Vienna, Austria (2019). https://www.R-project.org
19. Randall, J., Oswald, F., Beier, M.: Mind-wandering, cognition, and performance: a theory-driven meta-analysis of attention regulation. Psychol. Bull. **140**(6), 1411 (2014)
20. Scholes, L., Spina, N., Comber, B.: Disrupting the 'boys don't read' discourse: primary school boys who love reading fiction. Br. Edu. Res. J. **47**(1), 163–180 (2021)
21. Schooler, J., Smallwood, J., Christoff, K., Handy, T., Reichle, E., Sayette, M.: Meta-awareness, perceptual decoupling and the wandering mind. Trends Cogn. Sci. **15**(7), 319–326 (2011)
22. Smallwood, J.: Mind-wandering while reading: attentional decoupling, mindless reading and the cascade model of inattention. Language Linguist. Compass **5**(2), 63–77 (2011)
23. Smallwood, J., McSpadden, M., Schooler, J.: When attention matters: the curious incident of the wandering mind. Memory Cogn. **36**, 1144–1150 (2008)
24. Southwell, R., Peacock, C., D'Mello, S.: Getting the wiggles out: movement between tasks predicts future mind wandering during learning activities. In: Proceedings of AIED, pp. 489–501 (2023)
25. Unsworth, N., McMillan, B.: Mind wandering and reading comprehension: examining the roles of working memory capacity, interest, motivation, and topic experience. J. Exp. Psychol. Learn. Mem. Cogn. **39**(3), 832 (2013)
26. Varao-Sousa, T., Solman, G., Kingstone, A.: Re-reading after mind wandering. Can. J. Exper. Psychol./Revue Canadienne de psychologie expérimentale **71**(3), 203 (2017)
27. Weinstein, Y., De Lima, H., Van Der Zee, T.: Are you mind-wandering, or is your mind on task? The effect of probe framing on mind-wandering reports. Psychon. Bull. Rev. **25**, 754–760 (2018)

Short Papers

Short Papers

How Well Can You Articulate that Idea? Insights from Automated Formative Assessment

Mahsa Sheikhi Karizaki[1]([✉]), Dana Gnesdilow[2], Sadhana Puntambekar[2], and Rebecca J. Passonneau[1]([✉])[iD]

[1] Pennsylvania State University, State College, PA 16801, USA
{mfs6614,rjp49}@psu.edu
[2] University of Wisconsin-Madison, Madison, WI, USA
gnesdilow@wisc.edu, puntambekar@education.wisc.edu

Abstract. Automated methods are becoming increasingly used to support formative feedback on students' science explanation writing. Most of this work addresses students' responses to short answer questions. We investigate automated feedback on students' science explanation essays, which discuss multiple ideas. Feedback is based on a rubric that identifies the main ideas students are prompted to include in explanatory essays about the physics of energy and mass. We have found that students revisions generally improve their essays. Here, we focus on two factors that affect the accuracy of the automated feedback. First, learned representations of the six main ideas in the rubric differ with respect to their distinctiveness from each other, and therefore the ability of automated methods to identify them in student essays. Second, sometimes a student's statement lacks sufficient clarity for the automated tool to associate it more strongly with one of the main ideas above all others.

Keywords: Automated Essay Feedback · Student Writing Clarity

1 Introduction

Science writing has been found to enhance students' inquiry and reasoning skills [5,8]. Artificial Intelligence has been used to support formative assessment of short answer responses to guide revision [3,16]. However, using AI tools for revision feedback of science essays is still novel, therefore little is known about accuracy of AI feedback on essays. In a project that provides a web-delivered short curriculum on middle school roller coaster physics, students are prompted to write essays, then revise them based on automated feedback. We find that the feedback accuracy depends on the inherent distinctiveness of propositions that express the main ideas and on how clearly the students express themselves.

The essay feedback comes from PyrEval [2,13], software that detects main ideas in short passages written to the same prompt. PyrEval can use any pre-trained model to convert spans of text to semantic vectors. Before classroom

A. M. Olney et al. (Eds.): AIED 2024, LNAI 14830, pp. 225–233, 2024.
https://doi.org/10.1007/978-3-031-64299-9_16

deployment, to optimize accuracy of the feedback, we tested multiple semantic vector methods on a set of manually labelled student data. After classroom use, we manually labeled a new set of essays to assess accuracy on the new classroom sample. By examining patterns of cosine similarities of main idea vectors used as exemplars versus students' main idea vectors, we find that some main ideas are more distinctive than others, and that some student statements have similar cosine similarities to multiple ideas. Both factors that affect PyrEval accuracy.

2   Related Work

Formative feedback, meaning feedback during a unit or course to support further learning, has been found to be most beneficial when it focuses on the *what, how and why* of a problem rather than on verification of results [12]. A series of papers from a group at UC Berkeley have investigated the use of automated guidance in support of short answer explanations from middle school students. They have compared automated feedback alone and in combination with information about the personalized nature of the feedback [14], alone or in combination with students providing feedback on a sample essay [4], and finally alone or in combination with an interface that models the revision process [3]. In all three cases, automated guidance was from the C-rater-ML tool [7], reported to have a 0.72 Pearson correlation with expert humans [14].

Similar investigations by the Concord Consortium [9,17], mostly with high school students, aimed at improving students' understanding of uncertainty in science [10]. All studies relied on C-rater-ML, achieving QWK scores with humans between 0.78 and 0.93, depending on the study. One study compared generic argumentation feedback to student-specific feedback through use of C-rater-ML, with the latter leading to greater improvements in revisions [17]. Another study compared feedback on argumentation writing alone or in combination with feedback on students' use of science simulations and data [9].

There is relatively little work on automated formative feedback for essay revision. Zhang et al. [16] presented eRevise, which provides rubric-based feedback on students' use of evidence for source-based essays. The authors found that reliance on word embeddings had the best combination of performance accuracy and ability to provide student-specific feedback. Tests with middle school students showed that eRevise led to improved scores on revisions. For our middle school essays, we found that PyrEval performed better using word embeddings rather than contextualized embeddings (cf. Sect. 6), with accuracies from 0.74 to 0.80 across datasets.

3   Roller Coaster Physics Curriculum

During a 2–3 week design-based physics unit, middle school students learned about the physics of energy and energy transfer. They conducted virtual experiments in a web-based environment using a roller coaster simulation, recorded

ID	Sim	Text Description
1	0.69	The greater the height, the greater the potential energy (PE)
2	0.77	As the cart moves downhill, PE decreases and kinetic energy increases
3	0.60	The total energy of the system is always the sum of PE and KE
4	0.75	The law of conservation of energy states that energy cannot be created or destroyed, only transformed
5	0.75	The initial drop should be higher than the hill
6	0.70	Higher mass of the cart corresponds to greater total energy of the system

(a) The Six Main Ideas.

Feedback		My Confidence
Height and Potential Energy	✓	Medium
Relation between Potential Energy and Kinetic Energy	?	High
Total energy	?	Low
Energy transformation and Law of Conservation of Energy	?	High
Relation between initial drop and hill height	✓	Medium
Mass and energy	✓	High

(b) Sample Feedback Checklist: a green check mark means PyrEval detected the idea; a gold question mark indicates PyrEval did not. The 'My Confidence' column reflects PyrEval's average accuracy for a given idea.

Fig. 1. Main Idea content units with average cosine similarities (Sim).

their data, and answered multiple choice and open-ended questions before submitting explanation essays and essay revisions. An essay prompt guided them to explain six ideas, such as the influence of height on potential energy. The six ideas are shown in Fig. 1a, with the type of checklist feedback they might receive in Fig. 1b. Elsewhere, we reported that students' revised essays improved based on the automated feedback [11].

4   PyrEval

PyrEval is designed as a lightweight content assessment tool, and is easily adapted to new datasets due to its modular design. Its use of pre-trained semantic vectors means that it requires no training data. Below we explain how we tuned it to our data, including hyperparameter selection. Its three modules are an essay preprocessor, a module to build the content model, and one to assess essays.

The preprocessor converts essays into lists of semantic vectors corresponding to main clauses. This module supports user selection of different semantic vector methods, as we demonstrate in experiments in Sect. 6.

The second module automatically constructs a content model known as a pyramid from five reference essays (exemplars). A pyramid is a list of weighted content units where each content unit represents different ways of expressing the same content extracted from the reference essays. It groups the clause vectors from different exemplar essays into content units (CUs) of at most five vectors. CUs with fewer than five vectors have lower importance. The typical pyramid has a few CUs with the maximum weight of 5, and increasingly more content units for each lower weight. Aligning a pyramid to our rubric is described below.

PyrEval's assessment module [13] constructs a hypergraph graph for each student essay. Each hypernode is a sentence with one internal node per clause, where CUs are associated with internal nodes they are similar to. Edges connect clauses in different sentences that are associated to the same CU. An adaption of a greedy maximal independent set algorithm finds the set of matches (nodes) that give highest sum of CU weights.

Table 1. Comparison of Six Semantic Vector Methods on GT1

Method	$topk$	t	Accuracy
WTMF	3	0.55	0.795
WTMF with MidPhys	3	−0.01	0.675
WTMF Refinement	3	−0.01	0.705
BERT	3	0.85	0.752
Fine Tuned BERT	3	0.83	0.752
BERT + WTMF	3	0.83	0.756

5 Data

Two datasets are used here, one to tune PyrEval to the middle school essays, and one to analyze PyrEval accuracy. In year 2 of the project, we selected 7 high quality student essays to construct 21 pyramids to choose from. We labeled 39 additional essays of varying quality for presence of each main idea, which we refer to as Ground Truth 1 (GT1). Three annotators from the project worked independently, then arrived at a consensus labeling, which was updated several times while testing alternative pyramids. We aimed for a pyramid with exactly six content units of weight 5 (the maximum weight) corresponding one-to-one to the six main ideas in the curriculum. After selecting the pyramid with the best performance on GT1, we manually edited the 5 corresponding reference essays to further improve the pyramid. The *Sim* column of Fig. 1a shows the average pairwise cosine similarity of the five vectors within each of the six main idea content units in our final pyramid.

In year 3 of the project, original and revised essays from 60 students were labeled, which we refer to as Ground Truth 2 (GT2). Raters examined the PyrEval feedback, and labeled it as correct or incorrect. Inter-rater reliability was measured on 20% of the essays from two researchers working independently. Substantial agreement of Cohen's Kappa = 0.768 was achieved. Then one of the researchers labeled the remainder of the data.

6 Experiments and Results

Our previous work found WTMF, a matrix factorization vector method, to outperform other word embedding methods [6]. Here we compare six additional vector methods: 1) WTMF with its original corpus; 2) WTMF on the original corpus augmented with MidPhys, a dataset consisting of 11,245 constructed responses from middle school students to 55 physics questions; 3) refinement of the WTMF vector space; 4) BERT contextualized vectors [1]; 5) BERT fine-tuned on MidPhys; 6) concatenation of vectors 4 and 5. For each method, we performed grid search over two PyrEval hyperparameters: t, the threshold cosine similarity value of a student essay vector to a pyramid content unit to be added to the assessment hypergraph, and *topk*, the number of different student essay vectors that can be associated with the same content unit.

Table 2. PyrEval accuracies, recall, precision and F1.

Dataset	PAcc.	NAcc.	Acc.	Rec.	Pre.	F1
GT1	80.64	76.56	79.50	92.77	80.20	86.03
GT2-O	73.73	77.14	74.72	88.67	73.72	80.51
GT2-R	77.00	55.32	74.17	91.98	76.99	83.82
GT2	75.53	70.39	74.44	90.50	75.52	82.34
All	76.78	70.05	75.47	91.09	76.71	83.28

The third method adapts an approach to refine vectors for opposite sentiment words to have lower cosine similarity [15], that relied on a human ranking of sentiment words. We refined the cosine similarities of a set of key physics terms to be more distant, for word pairs like "potential" and "kinetic," using tf-idf scores computed on the MidPhys corpus to rank words.

Table 1 compares the six semantic vector methods on the original ground truth dataset GT1. Because method 1 had the highest accuracy, we used this method in our project.

Table 2 reports accuracy on the GT2 dataset. That it is somewhat lower than on GT1 is to be expected, given that GT1 was relatively small in size. Accuracies are broken down into positive accuracy (or sensitivity) and negative accuracy (or specificity). PyrEval's use of a greedy maximal independent set approach optimizes for the highest sum of matched CU weights, thus inherently favors positive over negative accuracy.

Table 3 shows varied accuracy across the six main ideas. (Accuracy "bins" in the Fig. 1b checklist are based on GT1 results.) In GT2, the ideas PyrEval identifies most accurately are, in descending order, statements that: define the law of conservation of energy (main idea 4), explain the roller coaster initial drop must be higher than the hill that follows (main idea 5), and that greater mass of the cart results in greater total energy (main idea 6). Main idea three accuracy is modest (71.66%), and accuracy is lower for main ideas one and two.

7   Distinctiveness of Ideas and Student Writing Clarity

PyrEval has higher accuracy on the fourth and fifth main ideas (MIs), which we attribute to greater distinctiveness of lexical items used to express them, such as *transformed, transferred* for MI4 and *initial, drop, hill* for MI5. All the other ideas mention energy, potential energy and kinetic energy, which are relatively close in vector space. While the term *mass* is unique to MI6, its embedding is close to energy terms. The distinctiveness of the main ideas can be quantified by average cosine similarities of all pairs of vectors from the main idea content units, as shown in Table 4. Averaging across pairs gives the lowest similarities (greatest distinctiveness) for MI4 (0.27) and MI5 (0.28), moderate for MI3 (0.37), and around 0.43 for MIs 6, 2 and 1. See below for the *Count* column.

Table 3. Accuracy on Main Ideas 1–6 (as percentages).

Dataset	MI 1	MI 2	MI 3	MI 4	MI 5	MI 6
GT1	76.92	82.05	69.23	89.74	71.79	84.62
GT2 O	63.33	56.66	66.66	91.66	86.66	83.33
GT2 R	63.33	61.66	76.66	86.66	86.66	70.00
GT2	63.33	59.16	71.66	89.16	86.66	76.66
All	66.66	64.77	71.06	89.30	83.01	78.61

Figure 2 illustrate cases of clauses with $\geq t$ similarity to multiple MIs. The top of the figure shows two phrases that are more poorly written, and that are candidate matches to three main ideas. The lower half of the figure shows two well articulated statements, with $\geq t$ similarity to exactly one main idea. Column 2 of Table 4 shows how often each MI in a given pair is a candidate match for multiple clauses in a student essay. Pairs of main ideas are shown in ascending order of the number of clauses that have a similarity to both MIs above the threshold t. Main ideas 1 and 5 are the most "confusable" for PyrEval, with 1,152 clauses having similarity $\geq t$ to both.

We plotted distributions of cosine similarities of student vectors to main ideas in a random selection of 117 GT2 essays (out of 159), then verified the consistency of our observations on the remaining 42. We selected one plot to show here. We binned essays by number of PyrEval errors into High, (N = 58; 0–1 errors), Mid (N = 45; 2 errors), and Low (N = 14; ≥ 3 errors) accuracy. Clauses from the High and Mid bins had similarities above t for 1.63 main ideas (sd = 0.84). Clauses from the Low bin exceeded t for 1.73 main ideas (sd = 0.78). When a clause is a candidate match for up to 3 (*topk*) content units, the algorithm is more likely to err. Figure 3 plots the cosine similarity (x-axis) by number of clause-main idea pairs at that cosine similarity (y-axis) in an accurately assessed essay of average length versus an inaccurately assessed long essay. The accurate essay (darker bars) has a lower count of clauses overall, but more importantly, very few that

ID	1	2	3	4	5	6
			Low Clarity Examples			
1	0.63	-	0.52	0.51	-	0.57
2	-	0.58	-	0.53	-	0.52
			High Clarity Examples			
3	-	-	-	0.71	-	-
4	-	-	-	-	-	0.69

Fig. 2. Clauses with low versus high clarity, and main ideas they are similar to.

Table 4. Average cosine similarity of all pairs of main ideas.

Pair	Count	Avg. Sim.
4–5	5	0.06
1–4	6	0.21
3–4	16	0.38
3–5	25	0.16
2–4	57	0.30
4-6	69	0.40
1–3	158	0.38
5–6	170	0.27
2–3	214	0.39
2–5	472	0.40
1–6	532	0.44
3–6	534	0.54
2–6	802	0.48
1–2	986	0.59
1–5	1,152	0.53

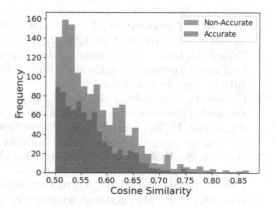

Fig. 3. Cosine similarity distributions of clauses in the full assessment hypergraph for an accurate short essay, and a long inaccurate essay.

have a cosine similarity of 0.70 and above. In contrast, the inaccurate essay has about ten times as many at that cosine similarity and above, which increases the chances that the assessment algorithm would select the wrong node.

8   Conclusion

Through error analysis of a software tool that provides formative feedback on students' science explanation essays, we presented two perspectives on distinctiveness of ideas. First, science explanation statements converted to semantic vectors have different degrees of distinctiveness. Second, students' statements of an idea can be more or less clearly articulated. Both factors affect the accuracy of a software tool we employed to provide formative feedback on students' essays.

Acknowledgements. This work was supported by NSF DRK award 2010351.

References

1. Devlin, J., Chang, M.W., Lee, K., Toutanova, K.: BERT: pre-training of deep bidirectional transformers for language understanding. In: Proceedings of the 2019 NAACL Conference, pp. 4171–4186. ACL (2019). https://doi.org/10.18653/v1/N19-1423
2. Gao, Y., Sun, C., Passonneau, R.J.: Automated pyramid summarization evaluation. In: Bansal, M., Villavicencio, A. (eds.) Proceedings of the 23rd CoNLL, Hong Kong, China, pp. 404–418. Association for Computational Linguistics (2019). https://doi.org/10.18653/v1/K19-1038
3. Gerard, L., Linn, M.C.: Computer-based guidance to support students' revision of their science explanations. Comput. Educ. **176**, 104351 (2022). https://doi.org/10.1016/j.compedu.2021.104351
4. Gerard, L., Linn, M.C., Madhok, J.: Examining the impacts of annotation and automated guidance on essay revision and science learning. In: Looi, C.K., Polman, J.L., Cress, U., Reimann, P. (eds.) Transforming Learning, Empowering Learners: The International Conference of the Learning Sciences (ICLS) (2016)
5. Graham, S., Kiuhara, S.A., MacKay, M.: The effects of writing on learning in science, social studies, and mathematics: a meta-analysis. Rev. Educ. Res. **90**(2), 179–226 (2020). https://doi.org/10.3102/0034654320914744
6. Guo, W., Diab, M.: Modeling sentences in the latent space. In: Proceedings of the 50th Annual Meeting of the ACL, pp. 864–872. Association for Computational Linguistics (2012). https://aclanthology.org/P12-1091
7. Heilman, M., Madnani, N.: ETS: Domain adaptation and stacking for short answer scoring. In: Manandhar, S., Yuret, D. (eds.) Second Joint Conference on Lexical and Computational Semantics (*SEM), SemEval 2013, Atlanta, GA, pp. 275–279. Assoc. for Computational Linguistics (2013). https://aclanthology.org/S13-2046
8. Klein, P.D., Boscolo, P.: Trends in research on writing as a learning activity. J. Writ. Res. **7**(3), 311–350 (2016). https://doi.org/10.17239/jowr-2016.07.03.01
9. Lee, H.S., Gweon, G.H., Lord, T., Paessel, N., Pallant, A., Pryputniewicz, S.: Machine learning-enabled automated feedback: supporting students' revision of scientific arguments based on data drawn from simulation. J. Sci. Educ. Technol. 168–192 (2021). https://doi.org/10.1007/s10956-020-09889-7
10. Pallant, A., Lee, H.S., Pryputniewicz, S.: How to support secondary school students' consideration of uncertainty in scientific argument writing. J. Geosci. Educ. **68**(1), 8–19 (2020)
11. Puntambekar, S., Dey, I., Gnesdilow, D., Passonneau, R.J., Kim, C.: Examining the effect of automated assessments and feedback on students' written science explanations. In: Proceedings of the 17th ICLS, pp. 1866–1867 (2023)
12. Shute, V.J.: Focus on formative feedback. Rev. Educ. Res. **78**(1), 153–189 (2008). https://doi.org/10.3102/0034654307313795
13. Singh, P., Passonneau, R.J., Wasih, M., Cang, X., Kim, C., Puntambekar, S.: Automated support to scaffold students' written explanations in science. In: Rodrigo, M., et al. (eds.) Artificial Intelligence in Education, vol. 13355, pp. 660–665. Springer, Cham (2022). https://doi.org/10.1007/978-3-031-11644-5_64

14. Tansomboon, C., Gerard, L.F., Vitale, J.M., Linn, M.C.: Designing automated guidance to promote productive revision of science explanations. Int. J. Artif. Intell. Educ. **17**, 729–757 (2017). https://doi.org/10.1007/s40593-017-0145-0

15. Yu, L.C., Wang, J., Lai, K.R., Zhang, X.: Refining word embeddings for sentiment analysis. In: Proceedings of the 2017 EMNLP, pp. 534–539. Association for Computational Linguistics (2017). https://doi.org/10.18653/v1/D17-1056

16. Zhang, H., et al.: eRevise: using natural language processing to provide formative feedback on text evidence usage in student writing. In: IAAI-19 (2019). https://doi.org/10.1609/aaai.v33i01.33019619

17. Zhu, M., Liu, O.L., Lee, H.S.: The effect of automated feedback on revision behavior and learning gains in formative assessment of scientific argument writing. Comput. Educ. **143**, 103668 (2020). https://doi.org/10.1016/j.compedu.2019.103668

Educational Content Personalization for Neurodiversity: A Survey of Technologies Supporting Linguistic Development in Individuals with Autism Spectrum Disorder (ASD)

Keylla Ramos Saes[1]([✉]) [iD], Anarosa Alves Franco Brandão[1]([✉]) [iD],
Elthon Manhas de Freitas[1]([✉]) [iD], and Fraulein Vidigal de Paula[2]([✉]) [iD]

[1] Laboratório de Técnicas Inteligentes - Escola Politécnica - USP - BRA,
São Paulo, Brazil
{keylla.saes,elthon}@alumni.usp.br, anarosa.brandao@usp.br
[2] Instituto de Psicologia, USP, BRA, São Paulo, Brazil
fraulein@usp.br
https://www.poli.usp.br/ , https://www.ip.usp.br/

Abstract. The surge of online education via e-learning platforms offers flexibility but raises concerns about the uniform presentation of content, lacking the personalized touch of traditional teaching. These concerns become particularly significant when considering neurodiversity, with Autism Spectrum Disorder (ASD) being a noteworthy example. ASD, a neurodevelopmental disorder, significantly impacts social interaction, behavior, and communication, especially pragmatic skills. Given the autistic inclination for logical and consistent thinking, coupled with the challenge of intuitive understanding compromising communication effectiveness, we argue that computational solutions serve as ideal tools for developing language and social skills. This survey delves into existing literature on intelligent technologies designed to support personalized education in linguistic development for children and adolescents with ASD level 1. Our findings reveal a gap in intelligent technologies that specifically target linguistic development, focusing on pragmatic language and the social use of the language for individuals with ASD level 1. The growing population of these students emphasizes the need for incorporating intelligent technologies to personalize education and address specific challenges they face. This serves as a crucial contribution to effective inclusion, tackling unique hurdles in linguistic and social learning.

Keywords: Autism · Neurodiversity · Artificial intelligence · Personalization · E-learning · Language pragmatic

1   Introduction

The internet's widespread availability and declining hardware costs have spurred digitalization across sectors, notably in education [12]. E-learning platforms democratize content access, transcending physical barriers of traditional education [14]. However, this shift poses challenges, particularly for students. Dynamic adaptations by instructors, vital for comprehension, are lacking, potentially leading to cognitive gaps [14].

This concern is further accentuated in the case of neurodiverse students, necessitating specialized support and tailored levels of guidance throughout the learning process. This engenders critical questions pertaining to the effective inclusion of students with special needs. Against this backdrop, there arises a compelling need to explore strategies for advancing within the transitioning educational landscape, evolving the online learning paradigm through intelligent e-learning platforms, and expanding the boundaries of the classroom inclusively for neurodiverse students [10].

This study addresses the educational needs of children and adolescents with Autism Spectrum Disorder (ASD), a population that has grown significantly in public schools, increasing by 280% according to the Brazilian census [20]. ASD is characterized by atypical development, behavioral manifestations, and deficits in communication and social interaction [2,3]. Individuals may also exhibit repetitive behaviors and restricted interests, with severity ranging from mild to high support needs: level 1 (mild), level 2 (moderate), and level 3 (high) [2,3].

According to the Diagnostic and Statistical Manual of Mental Disorders - DSM-5 [2], individuals with ASD exhibit variations in the domain of structural aspects of language, including syntax, morphology, and phonology. These variations range from the absence of spoken language development to intact structural language skills, characterized by fluent and complex sentences. However, even in cases of preserved structural skills, deficits in the pragmatic use of language may still be present, as highlighted in Fig. 1 [2,6].

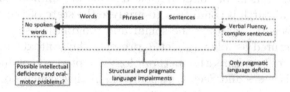

Fig. 1. Language structural scheme - source: [6]

Da Silva Junior and Rodrigues [17] highlight that current technological solutions for communication in Autism Spectrum Disorder (ASD) primarily target individuals needing high support, focusing on basic communication. However, individuals with ASD, often struggle with interpreting information intuitively, leading to social discomfort. This limitation affects comprehension of contexts, irony, and metaphors, hindering social interaction [11]. Ibanos et al. [7] further note that children with ASD encounter difficulties in understanding linguistic facts and communicative intentionality, impacting social interaction. In the context of early treatment and stimulation for ASD, as advocated by the Brazilian Ministry of Health [15], personalizing content for the learning of children with ASD is a significant challenge for professionals such as psychologists, teachers, and therapists. The diversity of profiles requires individual approaches, often manual, creating an additional workload for professionals. The limitation of available digital tools restricts development opportunities for individuals with ASD level 1, who, although verbal and able to communicate, face significant social challenges due to literal interpretation in communication [3]. Individuals on the autism spectrum exhibit differentiated neural processing, favoring deliberative reasoning over social interactions. This contributes to characteristics such as high neuroticism, low extroversion, and low agreeableness, aligning with stereotypes of Information Technology (IT) professionals. The study indicates that computers, being logical and consistent, are of particular interest to individuals with autism. Therefore, we believe that the adoption of technological support, including artificial intelligence (AI), can enhance the development of pragmatic language, providing personalized and engaging therapies, expanding therapeutic reach to more individuals on the autism spectrum [8,18]. In fact, in a recent systematic literature review, Chiu and colleagues [4] underscore AI's diverse applications in education, including task assignment, personalized feedback, adaptive learning environments, enhanced interactivity, teacher support through content creation, identifying learning issues, and customized teaching. Additionally, AI aids in analyzing student work for immediate, personalized feedback and adaptive assessments in the assessment domain. The definitions of AI usage align with the identified needs of the autistic population, especially in the perspective of personalized interactions. Developments in communication techniques and artificial intelligence can provide new potential methods for autism treatment. To offer contributions on this field, we carried out a systematic review. In this paper we search the literature related to existing intelligent technological solutions to support the development of communication skills of children and adolescents with ASD. Such a search is conducted to analyze the state-of-the-art of research in this issue, specially regarding the social and pragmatic use of the language. The text is structured as follows: Sect. 1 presents the context and the problem introduction, Sect. 2 presents the methodology adopted to conduct the review, 3 presents the discusses and Sect. 4 presents the review conclusions.

2 Systematic Review Methodology

In order to search for studies that could answer the research question "What are the existing intelligent and personalized technological solutions to support development of communication skills, in their context of social use and pragmatic language for children and adolescents with Autism Spectrum Disorder?", a systematic literature review was conducted following the protocol described next:

2.1 Review Protocol

According to Kitchenham [9] we had defined the following protocol:

1. The scientific online libraries ACM and IEEE Xplore, as well as the indexed database Scopus (limited to computer science articles) and the AIED proceedings' series, were chosen as the repositories to conduct the search;
2. The time lapse adopted is 2012 to 2024, because of the Brazilian autism law (No. 12.764/2012) establishment;
3. The search string was built considering the following combination of keywords, technically adapted to the repository search engine: (((autism) OR (asd) OR (autistic)) AND ((communic*) OR (language) OR (semantics) OR (pragmatic) OR (prosody) OR (morphosyntax) OR (metalanguage) OR (phonetics) OR (narrative) OR (lexical) OR (syntactic)) AND ((technology) OR (e-learning) OR (artificial intelligence) OR ((e-learning content personalization)) AND ((educa*) OR (learning) OR (teach*)))
4. The documents considered for the search are journal or conference papers written in English or Spanish and describing research related to intelligent solutions to pragmatic problems in ASD people.
5. As inclusion criteria we considered papers addressing intelligent or personalized computational solutions focused on therapies in the field of communication and pragmatic language.
6. As exclusion criteria we considered papers not related to therapy using intelligent technology with a focus on personalization, communication, and pragmatic language

2.2 Conducting the Review

We started the review by executing the search string adapted according to each search engine only considering the results from 2012 onwards. Such choice is based on the year the law that guarantees educational rights for individuals with ASD is established in Brazil (No.12,764/2012). The Fig. 2 illustrates the review process: after running the search string in the aforementioned repositories, there were 587 papers from ACM, 611 from IEEExplore and 635 from Scopus, yielding 1833 papers. Then, we exclude 185 duplicate papers before applying the inclusion/exclusion criteria. The search results were imported into the Start tool [21], and the inclusion/exclusion criteria were subsequently manually applied within the tool, following a careful review of the titles and abstracts of the articles and 48 articles were selected for full reading.

As established in the review protocol, exclusion criteria discarded papers not related to therapy using intelligent technology with a focus on personalization, communication, and pragmatic language. All papers selected to be read are referenced and available at this link[1]. Table 1 presents some bibliometric information about the discarded papers. Nevertheless, selected papers were categorized by subject and some bibliometric information is presented in table 2.

---

[1] https://acesse.one/OYJMC.

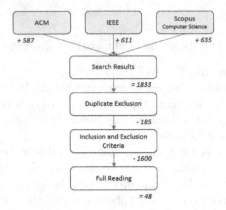

Fig. 2. Search selection diagram

For the sake of information regarding this review within the AIED conference proceedings series, no results were returned when the search string was executed considering publications from 2012 on. Nevertheless, the keyword "autism" returned nine papers, among them only one being related to adaptive learning [19]. In addition for the string "e-learning personalized content", 22 papers were returned, two of them [1,16] offering some insights for future research.

In the next section we discuss some findings after fully reading the selected papers.

3   Discussion

This systematic review identifies some gaps in intelligent solutions aimed at providing educational support to foster communication and pragmatic skills in individuals with autism spectrum disorder (ASD). While research exists on diagnostics, communication interfaces, and emotion recognition for individuals at levels 2 and 3, there is a scarcity of intelligent educational solutions to support reading and comprehension focusing on those at level 1. This subgroup, consisting of individuals with mild impairments and low support needs, encounters specific challenges in such tasks involving the language pragmatics.

Within the spectrum of technical solutions, such as video technology, serious games, augmented reality, and social robots, there is extensive exploration to enhance fundamental daily skills, especially in high-support autistics. However, there is considerable potential for the evolution of these technologies, aiming to specifically address pragmatic deficits and social skills challenges present in low-support autistics at level 1. While some studies have predominantly focused on emotion recognition, a significant contribution emerges from research related to social stories. However, current solutions in this category exhibit a notable limitation in generalization, being highly specific and focused in early stages.

Content personalization challenges exist in traditional and digital education [13], particularly for students with ASD [5]. The convergence of cited studies

Table 1. Distribution by exclusion criteria

Criteria	Description	#	%
(E01)	Autism diagnosis	276	17.3%
(E02)	Social Robots	200	12.5%
(E03)	Research on other topics in autism	209	13.1%
(E04)	Non-language-related Systematic Review	208	6.9%
(E05)	Non-verbal autistic solution	120	7.5%
(E06)	Not related to autism	114	7.1%
(E07)	Virtual reality solution	162	10.1%
(E08)	Serious Game	131	8.2%
(E09)	Emotion's Identification	118	7.4%
(E10)	Mobile Learning solution	91	5.7%
(E11)	Augmented reality solution	56	3.5%
(E12)	Literacy	10	0.6%
(E13)	Food therapies	2	0.1%

Table 2. Distribution by inclusion criteria

Criteria	Description	#	%
(I01)	Video Technology to pragmatic development	2	4.2%
(I02)	Social Robots to pragmatic development	2	4.2%
(I03)	Language Literature Review	3	6.3%
(I04)	Social Story to pragmatic development	13	25.0%
(I05)	Serious Game to learn pragmatic	2	4.2%
(I06)	Personalized Content Recommendation	5	10.4%
(I07)	Pragmatic in ASD Diagnosis	5	10.4%
(I08)	E-Learning Ecosystems to pragmatic development	4	8.3%
(I09)	Intelligent software to pragmatic development	11	27.1%

in this review highlights the diversity of profiles in the autism spectrum, emphasizing this need. Despite the observed advancements in educational platforms and e-learning systems in incorporating personalization features, considerable challenges persist in adapting content to language accessible to this specific audience, considering their aversions, preferences and hyperfocus.

4   Conclusion

This study highlights the gaps present in the field of e-learning personalization directed towards neurodiverse individuals, with a specific emphasis on the audience with Autism Spectrum Disorder (ASD) level 1, characterized by a reduced need for support in daily activities. Despite possessing a degree of independence, these individuals face challenges in the domain of social interaction and understanding pragmatic language. In the perspective of distance learning through e-learning, the identified difficulties are primarily centered on customizing educational content, considering the unique characteristics and interests of the students.

The presented challenges underscore the importance of future research aimed at enhancing the educational landscape for neurodiverse individuals, especially those positioned on the autism spectrum level 1. The systematic review provides a valuable contribution by offering insights into the need for personalized and targeted educational strategies, considering the cognitive and social specificities of these students, to effectively promote the development of pragmatic language and facilitate their engagement in the educational environment. Moreover, this review searches only within AI in education literature, and further research must be conducted in psychology databases such as APA.

Acknowledgments. This study was financed in part by the Coordenação de Aperfeiçoamento de Pessoal de Nível Superior - Finance code 01.

References

1. Aleven, V., et al.: The beginning of a beautiful friendship? Intelligent tutoring systems and MOOCs. In: Conati, C., Heffernan, N., Mitrovic, A., Verdejo, M.F. (eds.) AIED 2015. LNCS (LNAI), vol. 9112, pp. 525–528. Springer, Cham (2015). https://doi.org/10.1007/978-3-319-19773-9_53
2. Association, A.P., et al.: DSM-5: Manual diagnóstico e estatístico de transtornos mentais. Artmed Editora (2014)
3. Caetano, S.C., Lima-Hernandes, M.C.P., de Paula, F.V., Resende, B.D., Módolo, M.: Autismo. Linguagem e Cognição, Paco Editorial (2015)
4. Chiu, T.K., Xia, Q., Zhou, X., Chai, C.S., Cheng, M.: Systematic literature review on opportunities, challenges, and future research recommendations of artificial intelligence in education. Computers and Education: Artificial Intelligence 4, 100118 (2023). https://doi.org/10.1016/j.caeai.2022.100118
5. Costa, E., Aguiar, J., Magalhães, J.: Sistemas de recomendação de recursos educacionais: conceitos, técnicas e aplicações. Jornada de Atualização em Informática na Educação 1(1) (2013)
6. Grzadzinski, R., Huerta, M., Lord, C.: DSM-5 and autism spectrum disorders (ASDS): an opportunity for identifying ASD subtypes. Molecular autism (2013)
7. Ibaños, A.M.T., Costa, J.C.d.: A natureza da pragmática: percurso teórico em um piscar de olhos 1, 286–293 (2017). https://doi.org/10.15448/1984-7726.2017.3.29360
8. Jia, R., Jia, H.H.: What makes us it people? Autistic tendency and intrinsic interests in it. In: Proceedings of the 2019 on Computers and People Research Conference, pp. 153–156 (2019). https://doi.org/10.1145/3322385.3322408
9. Kitchenham, B.: Procedures for performing systematic reviews, Technical report. Keele University and Empirical Software Engineering National ICT Australia (2004)

10. Klašnja-Milićević, A., Ivanović, M.: E-learning personalization systems and sustainable education (2021). https://doi.org/10.3390/su13126713
11. L.E.R.: Desenvolvimento linguístico (2023). https://encr.pw/8H8Rf
12. Matthew, U.O., Kazaure, J.S., Okafor, N.U.: Contemporary development in e-learning education, cloud computing technology & IoT. EAI Endorsed Trans. Cloud Syst. (2021). https://doi.org/10.4108/eai.31-3-2021.169173
13. Moura, J.G., Brandao, L.O., Brandao, A.A.F.: A web-based learning management system with automatic assessment resources. In: 2007 37th Annual Frontiers In Education Conference - Global Engineering: Knowledge Without Borders, Opportunities Without Passports, pp. F2D-1–F2D-6 (2007). https://doi.org/10.1109/FIE.2007.4418100
14. Petretto, D.R., et al.: The use of distance learning and e-learning in students with learning disabilities: A review on the effects and some hint of analysis on the use during covid-19 outbreak. Clin. Pract. Epidemiol. Ment. Health CP & EMH **17**, 92'(2021). https://doi.org/1745017902117010092
15. da Saúde, M.: Definição - transtorno do espectro autista (tea) na criança (2023). https://linhasdecuidado.saude.gov.br/portal/transtorno-do-espectro-autista/definicao-tea/
16. Shi, L., Gkotsis, G., Stepanyan, K., Al Qudah, D., Cristea, A.I.: Social personalized adaptive e-learning environment: topolor - implementation and evaluation. In: Lane, H.C., Yacef, K., Mostow, J., Pavlik, P. (eds.) AIED 2013. LNCS (LNAI), vol. 7926, pp. 708–711. Springer, Heidelberg (2013). https://doi.org/10.1007/978-3-642-39112-5_94
17. da Silva Junior, E.F., da Hora Rodrigues, K.R.: Ferramentas computacionais como soluções viáveis para alfabetização e comunicação alternativa de crianças autistas: Um mapeamento sistemático sobre as tecnologias assistivas existentes. In: Anais do X Workshop sobre Aspectos da Interação Humano-Computador para a Web Social, pp. 71–80. SBC (2019)
18. Sousa, T.A., Ferreira, V.D., dos S. Marques, A.B.: How do software technologies impact the daily of people with autism in Brazil: a survey. In: Proceedings of the XV Brazilian Symposium on Information Systems, pp. 1–8 (2019)
19. Tsatsou, D., et al.: Adaptive learning based on affect sensing. In: Penstein Rosé, C., et al. (eds.) AIED 2018, Part II. LNCS (LNAI), vol. 10948, pp. 475–479. Springer, Cham (2018). https://doi.org/10.1007/978-3-319-93846-2_89
20. da Unesp, J.: Com número de diagnósticos em crescimento, transtorno do espectro autista ainda é desafio para pesquisa neurológica (2023). https://encr.pw/wyjtT
21. Zamboni, A., Thommazo, A., Hernandes, E.C.M., Fabbri, S.: Start uma ferramenta computacional de apoio à revisão sistemática. In: Proc.: Congresso Brasileiro de Software (CBSoft'10), Salvador, Brazil, pp. 91–96. UFBA (2010)

Beyond the Obvious Multi-choice Options: Introducing a Toolkit for Distractor Generation Enhanced with NLI Filtering

Andreea Dutulescu[1(✉)], Stefan Ruseti[1], Denis Iorga[1], Mihai Dascalu[1,2], and Danielle S. McNamara[3]

[1] National University of Science and Technology Politehnica,
313 Splaiul Independentei, 060042 Bucharest, Romania
andreea.dutulescu@stud.acs.upb.ro,
{stefan.ruseti,denis_nicolae.iorga,mihai.dascalu}@upb.ro
[2] Academy of Romanian Scientists, Str. Ilfov, Nr. 3, 050044 Bucharest, Romania
[3] Learning Engineering Institute, Arizona State University, PO Box 871104,
Tempe, AZ 85287, USA
danielle.mcnamara@asu.edu

Abstract. The process of generating challenging and appropriate distractors for multiple-choice questions is a complex and time-consuming task. Existing methods for an automated generation have limitations in proposing challenging distractors, or they fail to effectively filter out incorrect choices that closely resemble the correct answer, share synonymous meanings, or imply the same information. To overcome these challenges, we propose a comprehensive toolkit that integrates various approaches for generating distractors, including leveraging a general knowledge base and employing a T5 LLM. Additionally, we introduce a novel strategy that utilizes natural language inference to increase the accuracy of the generated distractors by removing confusing options. Our models have zero-shot capabilities and achieve good results on the DGen dataset; moreover, the models were fine-tuned and outperformed state-of-the-art methods on the considered dataset. To further extend the analysis, we introduce human annotations with scores for 100 test questions with 1085 distractors in total. The evaluations indicated that our generated options are of high quality, surpass all previous automated methods, and are on par with the ground truth of human-defined alternatives.

Keywords: Multiple-choice questions · Distractor generation · Challenging distractors · Natural language inference

1 Introduction

In the domain of education, the rise of AI technologies has catalyzed a paradigm shift in pedagogical methodologies, offering automation opportunities for

A. M. Olney et al. (Eds.): AIED 2024, LNAI 14830, pp. 242–250, 2024.
https://doi.org/10.1007/978-3-031-64299-9_18

educators and aiding them in creative tasks. One area of research has been quiz generation, where advanced language models became appropriate for real-life use [4]. While considerable scholarly attention has been devoted to the domain of question generation in the context of educational content development, there exists a noticeable gap in research about distractor generation (i.e., the generation of incorrect answers), specifically in the construction of multiple-choice options. Addressing this limitation is imperative for the refinement of educational tools, ensuring a more thorough examination of students' comprehension and analytical abilities within a pedagogical framework.

Multiple-choice questions are widely used in classroom quizzes to test students' general knowledge. However, manually designing such tests and choosing the most appropriate distractors to serve as possible choices along with the correct answer is a tedious task. The problem of automatically generating foils has been studied before but is far from solved. A good set of distractors in a multiple-choice setting must fulfill two main conditions in the literature regarding educational tests [11], namely: (1) *Validity*: they must not be synonyms or imply the same information as the correct answer; (2) *Difficulty*: they must test a deep understanding of the subject, they have to appear as valid possible answers and not be easily discarded as incorrect ones.

Multiple approaches and directions of study have been taken for distractor generation - i.e., creating incorrect or misleading options in the context of multiple-choice questions to assess a learner's knowledge. The majority of these approaches have the same two main components: (1) candidate generation and (2) candidate selection. [9] experimented with two semantic networks to generate distractors for general knowledge questions and used a probabilistic topic model to choose words related to a certain concept. In order to further take into account the question and its semantics, [1] used pre-trained language models (e.g., SciBERT) to generate possible distractors for fill-in-the-blanks questions. The models were fine-tuned on the CLOTH [13] and DGen [9] datasets to generate the specific distractors.

Ensuring the validity of the distractors and filtering out generated samples that resemble the correct answer has yet to be thoroughly approached. This is important because there is a high risk in the attempt to create the most challenging distractors of selecting a foil that would be the synonym or implication of the correct answer, making the question invalid for a student. [9] and [1] did not specifically filter out the invalid distractors and relied on the fine-tuning task to generate and select good candidates. [8] only filtered out WordNet synonyms [7] of the answer. However, these methods are limited to concrete repetitions, context-unaware synonyms of the answer, or syntax inconsistencies.

In this paper, we propose a toolkit that automatically generates distractors for general knowledge science questions, using general knowledge bases and Pre-trained Language Models. Our approach pays special attention to distractor validity, as the overall correctness of a question's multiple choices is of utmost importance in a classroom quiz. We leverage a novel approach of using natural language inference to filter distractors that would be an implication, synonym, or the correct answer. Moreover, we employ a ranking mechanism to propose the most appropriate and challenging distractors. Our main contributions are as follows:

- Introduce and open-source a comprehensive toolkit that surpasses current state-of-the-art models for distractor generation on both automatic and human evaluation;
- Propose a novel filtering method based on natural language inference that ensures the relevance and validity of the proposed distractors;
- Leverage human annotations in a qualitative evaluation that argues for comparable quality between our generated and human-curated distractors.

2   Method

Figure 1 showcases an overview of the pipeline employed for this task, with each component detailed in its corresponding sub-section.

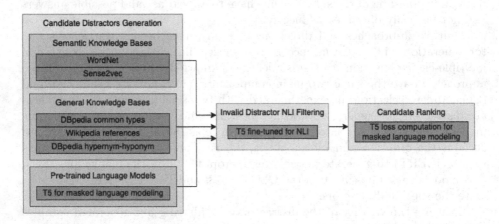

Fig. 1. Overall architecture of our method.

2.1   Dataset

The dataset considered for both automatic and human evaluation is DGen [9]. DGen is a curated dataset with multiple-choice, fill-in-the-blank science questions. Each item contains the sentence, the correct answer, and three distractors. We used the default partitions of the dataset comprised of 2,322 entries for train and 259 for testing.

2.2   Candidate Generation

We generate multiple-choice candidates with various methods, as we want to provide a comprehensive toolkit with multiple complementary methods readily integrated that cover and generalize well for diverse school subjects. Each method independently generates distractors, followed by a semantic filtering and candidate ranking procedure. In the case of DGen, the generated candidates are split into individual words to fulfill the restriction of single-word answers.

Semantic Knowledge Bases. Semantic databases are used to compute possible distractors for a given word by employing existing links between concepts. Our considered alternatives are WordNet [7] and Sense2Vec [12]. From WordNet, we use the hypernym-hyponym relation in which different items of the same class as the correct answer (i.e., siblings in the hypernym taxonomy with the answer) are selected to participate in the candidate set. Sense2Vec ensures a broader and more permissive search to select the most similar concepts regarding the embedding representation of the correct answer.

General Knowledge Bases. As the main focus of this task is to generate foils for science, general knowledge, and school subjects, a large multi-purpose semantic repository is used, namely DBpedia [6]. In order to start searching for appropriate distractors, the DBpedia resource corresponding to the answer must be located. In this regard, the lookup API service is used to gather the URIs of the entities related to the answer. The proposed entities are ranked with a pretrained language model that computes the likelihood of that entity label being generated by a language model instead of the actual answer stem. The entity with the highest probability is used as the corresponding answer resource.

Three approaches for generating candidates with DBpedia are employed: (1) selecting the entities of the same class as the answer with the highest number of common types with the answer; (2) extracting the entities referred to by links in the Wikipedia page corresponding to the answer; and (3) picking the entities with the same hypernym as the correct answer.

Pre-trained Language Models. A pre-trained language model, namely T5 [10], is used to generate candidate foils for quiz questions. The general approach is to generate the most likely stems that would either fill in or answer a certain question without additional context. A list of possible candidates for the answer is sampled from the model's distribution. T5 was chosen for its masked language prediction pre-training objective, which perfectly matches the fill-in-the-blanks scenario in the dataset.

In the case of fill-in-the-blank items, the question comes as a sequence of text containing a marked blank space. That space should be completed with a choice that maintains the sequence's fluency, context, and factual correctness. For this sub-task, the blank space is replaced with a special masking token near which a reference to the correct answer is appended, thus ensuring that the model is answer-aware, guiding the generation towards a different answer. This prompt is forwarded to a T5 model, more specifically, in the form of "<mask_token> (or gravity) causes rocks to roll downhill.", for the item: "____ causes rocks to roll downhill.", and the correct answer: "gravity".

Invalid Distractors Filtering. We consider it highly important not to propose distractors that may invalidate the test item, such as distractors containing, implicating, or having the same meaning as the correct answer. These invalid distractors are filtered out by leveraging natural language inference models to detect whether the correct answer implies the proposed foils. More formally, two sequences are computed: one that contains the correct answer (e.g., "igneous rocks form from cooled magma or lava."), and one with the generated candidate (e.g., "granite rocks form from cooled magma or lava."). If the first sentence entails the second one, it means that the candidate is not suitable, as it can be inferred from the correct answer, making the test item invalid since the question would have two correct answers. A T5 model fine-tuned for natural language inference is used for this part. If the model's output for a (correct answer sentence, candidate sentence) pair is an entailment, then the candidate is not part of the final proposal.

Along with this filtering, we also discarded distractors that do not have the same part of speech as the correct answer for the dataset format.

2.3 Candidate Ranking

At this stage, the remaining candidates are valid choices to be proposed. Their ranking is established as the likelihood of being the correct answer since the correct answers should have been filtered out in the previous stage. The score is computed as the negative log-likelihood of the candidate being generated instead of a special token. A T5 model is used for this, and the prompt forwarded to the model has the same format as the one described in Subsect. 2.2, except this time the model is not used for inference, but rather to compute the Cross-Entropy loss for a given label (the candidate). The loss value is the score computed for that candidate, and a lower score implies a better candidate for the distractor.

We experimented with two different approaches: (1) a zero-shot setting in which a vanilla T5 model is used to compute the score and (2) a fine-tuned T5 model on the train partition of the dataset. The fine-tuning was done by forwarding the prompt with the ground-truth distractors replacing the special tokens.

2.4 Evaluation Metrics

We replicated the setup for the automatic evaluation on which [1] evaluated their performance. We considered relevant the following metrics, calculated for the X proposed distractors in regard to the ground-truth set: P@X (Precision), F1@X, MRR@X (Mean Reciprocal Rank [3]), and NDCG@X (Normalized Discounted Cumulative Gain [5]).

3   Results

The evaluation was performed on the test partition of the dataset with the metrics described above and is presented in Table 1. We report our best approaches (i.e., with ranking and fine-tuning, and with or without NLI filtering) alongside the current state-of-the-art on this dataset, CDGP [1].

Our method achieves higher or comparable results on all metrics, especially on P@1, meaning that our first proposed distractor is often found in the ground-truth set.

Table 1. Evaluation of our methods versus the current state-of-the-art

Method	P@1	F1@3	F1@5	MRR@5	MRR@10	NDCG@5	NDCG@10
CDGP [1]	12.40	12.74	**12.93**	23.22	25.00	28.49	**34.42**
Ours w/ NLI	18.53	12.22	11.38	24.51	25.40	28.17	31.02
Ours w/o NLI	**20.07**	**13.25**	11.58	**26.15**	**27.06**	**30.14**	32.97

4   Summary and Discussion

As we argue that the comparison with three ground-truth distractors is far from suitable for assessment and for comparing the effectiveness of an automated model, we experimented with human annotators to decide the best distractors generated with different methods. The first 100 entries of the DGen's test partition were used for human annotation. For the annotation data, we employed and generated distractors with the five automated methods targeted for comparison besides the DGen dataset [9] initial human distractors (Dataset GT):

- Distractors proposed by [1] (CDGP);
- Our proposed distractors:
  - without NLI filtering and without ranking fine-tuning (w/o NLI, w/o FT);
  - without NLI filtering and with ranking fine-tuning (w/o NLI, w/ FT);
  - with NLI filtering and without ranking fine-tuning (w/ NLI, w/o FT);
  - with NLI filtering and ranking fine-tuning (w/ NLI, w/ FT).

We selected the top three candidates from each approach and all distractors were combined in a set of candidates for each item. Three annotators were asked to assign a score to each candidate in the set based on the question and correct answer provided. The scores were assigned as follows: 0 (Invalid), 1 (Poor), 2 (OK), 3 (Good). More information and details about the annotation process can be found in the detailed annotation procedure[1]. We computed the inter-rater

---

[1] https://github.com/readerbench/distractor-generation.

agreement as the ICC(2,k) correlation [2]. We obtained a score of 0.67, which is rated by [2] as a good agreement.

The scores for each method are plotted in Fig. 2. Based on the human annotations that take into account the quality of a distractor, our method with fine-tuning yields scores comparable to the dataset's ground truth. This argues the high performance of our method, which matches the quality of the manually curated dataset and emphasizes the necessity of relying on human annotation rather than relying solely on ground-truth comparisons in the context of foil generation.

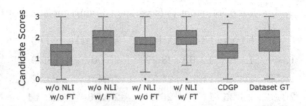

Fig. 2. Human annotation scores distribution per method

Moreover, we observe the importance of employing NLI filtering to ensure a sound approach. Our corresponding variants significantly surpass the current state-of-the-art and obtain a similar quality as the ground truth. The fact that our method employing just NLI filtering manages to have high-quality distractors is especially important since it proves zero-shot capabilities and impressive adaptability for future domains and items. Our method that employs NLI filtering, without any fine-tuning, receives higher results than the current state-of-the-art fine-tuned on the dataset. Moreover, with no fine-tuning, the NLI filtering has a high impact on the performance, as can be observed from the w/o NLI w/o FT versus w/ NLI w/o FT comparison.

Table 2 highlights the distribution of distractor scores for each method and the improvements introduced by different variations of our approach. In terms of invalid distractors (i.e., a score of 0), our NLI filtering manages to reduce them by almost 50% in regards to the current state-of-the-art. Moreover, our best method has the highest frequency of distractors rated as high-quality (i.e., a score of 3).

Table 2. Distribution of distractors scores per method.

Method	Candidate Scores			
	0 (Invalid)	1 (Poor)	2 (OK)	3 (Good)
w/o NLI, w/o FT	107	71	49	73
w/o NLI, w/ FT	66	63	60	111
w/ NLI, w/o FT	43	96	72	89
w/ NLI, w/ FT	41	72	78	109
CDGP	82	102	49	67
Dataset GT	28	93	86	93

5   Conclusions

In conclusion, this paper introduces an open-source comprehensive toolkit for generating fill-in-the-blank question distractors, leveraging knowledge bases and pre-trained language models. Additionally, we include a filtering method based on natural language inference that substantially reduces the generation of invalid distractors. By leveraging NLI, we can robustly eliminate confusing options that share similar meanings or represent alternative correct answers, ensuring that the generated distractors engage learners in a sound way. This feature sets our toolkit apart from existing methods that often fail to effectively filter out incorrect choices. The proposed method surpasses current standards in both automatic and human evaluation and is comparable with the quality of human-curated distractors.

More importantly, we go beyond relying solely on automatic evaluation measures and highlight the value of manual assessment. Involving human annotators in the evaluation process reveals that automated measures alone are inadequate for accurately assessing the quality of generated distractors. As many types of multiple-choice questions are used in quizzes and knowledge tests, a future study direction can lead towards adapting the current method to consider various types of questions and going beyond fill-in-the-blank sentences. The T5 model can be easily adapted to generate possible answers to such questions.

Acknowledgment. The research reported here was supported by the Institute of Education Sciences, U.S. Department of Education, through Grant R305T240035 to Arizona State University and by the Learning Engineering Tools Competition. The opinions expressed are those of the authors and do not represent the views of the Institute or the U.S. Department of Education.

References

1. Chiang, S.H., Wang, S.C., Fan, Y.C.: Cdgp: Automatic cloze distractor generation based on pre-trained language model. In: Findings of EMNLP 2022, ACL, Abu Dhabi, United Arab Emirates (2022)

2. Cicchetti, D.V.: Guidelines, criteria, and rules of thumb for evaluating normed and standardized assessment instruments in psychology. Psychol. Assess. **6**(4) (1994)
3. Craswell, N.: Mean reciprocal rank. Encyclopedia of database systems **1703** (2009)
4. Elkins, S., Kochmar, E., Serban, I., Cheung, J.C.: How useful are educational questions generated by large language models? In: Wang, N., Rebolledo-Mendez, G., Dimitrova, V., Matsuda, N., Santos, O.C. (eds.) AIED 2023. CCIS, vol. 1831, pp. 536–542. Springer, Cham (2023). https://doi.org/10.1007/978-3-031-36336-8_83
5. Järvelin, K., Kekäläinen, J.: Cumulated gain-based evaluation of IR techniques. ACM Trans. Inf. Syst. (TOIS) **20**(4) (2002)
6. Lehmann, J., et al.: Dbpedia–a large-scale, multilingual knowledge base extracted from Wkipedia. Semantic web **6**(2) (2015)
7. Miller, G.A.: WordNet: a lexical database for English. In: Human Language Technology: Proceedings of a Workshop held at Plainsboro, New Jersey, March 8-11, 1994 (1994)
8. Panda, S., Palma Gomez, F., Flor, M., Rozovskaya, A.: Automatic generation of distractors for fill-in-the-blank exercises with round-trip neural machine translation. In: Proceedings of ACL: Student Research Workshop, Dublin, Ireland. ACL (2022)
9. Ren, S., Zhu, K.Q.: Knowledge-driven distractor generation for cloze-style multiple choice questions. In: Proceedings of AAAI, vol. 35 (2021)
10. Roberts, A., et al.: Exploring the limits of transfer learning with a unified text-to-text transformer. Technical report, Google (2019)
11. Towns, M.H.: Guide to developing high-quality, reliable, and valid multiple-choice assessments. J. Chem. Educ. **91**(9), 1426–1431 (2014)
12. Trask, A., Michalak, P., Liu, J.: sense2vec-a fast and accurate method for word sense disambiguation in neural word embeddings. arXiv preprint arXiv:1511.06388 (2015)
13. Xie, Q., Lai, G., Dai, Z., Hovy, E.: Large-scale cloze test dataset created by teachers. In: Proceedings of EMNLP 2018, Brussels, Belgium. ACL (2018)

Exploring the Potential of Automated and Personalized Feedback to Support Science Teacher Learning

Jamie N. Mikeska[1]([email]) [iD], Beata Beigman Klebanov[1], Alessia Marigo[1], Jessica Tierney[1], Tricia Maxwell[1], and Tanya Nazaretsky[2]

[1] ETS, Princeton, NJ 08541, USA
jmikeska@ets.org
[2] École Polytechnique Fédérale de Lausanne, 1015 Lausanne, Switzerland

Abstract. This study investigates how automated and personalized feedback can support elementary teachers' learning within digital teaching simulations. We interviewed 15 participants to elicit their understanding and evaluation of feedback about a teacher's facilitation of a science discussion, which was generated using natural language processing (NLP). Findings indicate that participants: (a) perceived strong alignment between the feedback and the discussion facilitated, (b) used the feedback to identify future instructional moves, and (c) noted how the feedback can support reflection.

Keywords: Natural Language Processing · Teaching Simulations · Feedback

1 Introduction

Interactive technologies, such as digital teaching simulations (**DTSims**), are increasingly used to support teacher learning of key teaching competencies [9, 18, 19]. While prior studies suggest that personalized feedback is critical for learning [15], limited research has examined feedback in the context of DTSims. Yet, recent advances in NLP have shown promise for automating the evaluation of multi-speaker discourse [22, 24], which can be used to provide feedback to learners. In this study, we used interviews to examine how 15 teachers and teacher educators understood and evaluated an automated personalized feedback report about a teacher's ability to facilitate a science discussion. The study's three research questions (RQs) examine participants' perceptions about the alignment between the feedback and two aspects -- the discussion facilitated (RQ1) and key improvement areas (RQ2) and the feedback's usefulness (RQ3).

2 Related Work

2.1 Using Feedback and Digital Simulations to Support Teacher Learning

Feedback is "information provided by an agent (e.g., teacher, peer) regarding aspects of one's performance or understanding…" [12, p. 81] and is used to help learners identify their strengths and areas for growth [5, 10, 15]. Studies have shown that teachers tend

A. M. Olney et al. (Eds.): AIED 2024, LNAI 14830, pp. 251–258, 2024.
https://doi.org/10.1007/978-3-031-64299-9_19

to view feedback positively and value feedback they can implement in the classroom [3, 11]. One context for providing feedback is within DTSims where teachers can receive feedback and reflect as they learn how to engage in critical teaching practices, such as facilitating discussions or communicating with parents [8, 10, 18].

While DTSims can take varied forms, TeachLivE or Mursion's simulated classrooms are some of the most popular DTSims, in use across over 100 teacher education programs worldwide [18, 19]. In these simulations a teacher interacts in real-time with five student avatars on a computer. A human-in-the-loop (a simulation specialist) is behind the scenes responding as all five student avatars with the help of voice modulation and other software to power the avatars' physical movements and verbal responses, although the teacher never sees the simulation specialist [2]. The simulation specialist undergoes extensive training to learn how to respond as students at specific grade levels and engage in the content-specific aspects of the teaching interaction.

Research suggests that productively using DTSims involves structured implementation cycles where teachers prepare for, engage in, and reflect on their simulated teaching [18, 19]. A critical component is personalized feedback to help teachers identify their instructional strengths and determine areas for growth and how to improve [4, 6, 18]. Yet, scaling such feedback requires NLP to evaluate the multi-speaker discourse.

2.2 Using NLP to Evaluate Classroom Discourse

NLP models have been built in recent research to identify utterances where teachers implement high-quality discourse strategies, such as providing conversational uptake, asking open-ended questions, pressing students for reasoning, or getting students to relate to each other's ideas [1, 7, 14, 21, 26]. These annotated utterances provide a basis for feedback to teachers [6, 13]. Traditionally, the models are built using supervised machine learning requiring a substantial number of labeled examples to support generalization, such as the over 20K utterance-level annotations in the TalkMoves dataset [1, 7, 14, 25]. However, successful learning using less annotated data with transformer models has been demonstrated on other datasets [14]. In our research, we followed a similar approach by labeling fewer than 1K of almost 10K utterances across 157 discussions [6]. This annotated data are relatively clear-cut cases, those that raters picked as examples to justify their scores. Selection of a small number of clear-cut examples is an easier and faster task for a human to perform than a more comprehensive annotation of a random sample that would entail making decisions on not-so-clear-cut cases. Successful learning from such data can result in better utilizing human time to develop and evaluate machine learning and NLP systems for analyzing classroom discourse.

3   Data and Methods

3.1   Study Context and Previous Work

This study draws upon previous research. First, across three earlier projects, research teams collected video and transcript data of elementary preservice teachers (PSTs) facilitating a Mystery Powder (**MP**) science discussion in Mursion's upper elementary simulated classroom. In this discussion, the teacher's goal is to have five student avatars come to consensus on the mystery powder's identity and the most useful properties to identify the mystery powder [16]. Human raters also used a three or four level rubric to score each discussion on five dimensions of facilitating high-quality discussions (one dimension was attending to student ideas) and generated written justifications to identify evidence from teacher and student utterances. The Qualitative Data Repository has example MP videos, transcripts, scores and justifications [17].

Second, in a more recent project, NLP was used for automated evaluation of the teacher and student utterances in 157 MP discussion transcripts for two key aspects of attending to student ideas: eliciting meaningful student contributions and using student ideas. For each of these aspects, an automated utterance-level binary classifier was built [20] that predicted, for each utterance in each discussion transcript, whether the utterance was (class 1) or was not (class 0) an example of the target behavior. To train the classifier, we utilized the utterances picked by raters as justifications as class 1 examples and all utterances from transcripts for which human raters indicated that there were no positive examples of the target phenomenon as class 0 examples. Findings showed that current NLP technology, specifically the transformer-based models used to build the classifiers [23], were sufficient to enable good generalization to utterances in transcripts unseen during training. Details about the models, parameters, outputs, and evaluation results can be found in Nazaretsky et al. [20].

Third, we used these models to classify all student utterances as providing meaningful contributions or not and all teacher utterances for encouraging the use of student ideas or not. Figure 1 shows a short snippet from one MP discussion between a teacher (Mr. W) and two student avatars (Mina and Carlos). The underlined text shows meaningful student contributions. The italicized text shows the teacher using student ideas.

Mr. W: Mina and Will, why did you choose weight as an important property?
Mina: Because it falls under some of the things that we can see and measure.
Mr. W: *Carlos, do you want to explain to them about why you thought that weight wasn't important?*
Carlos: Sure. Well, actually I don't think weight is really that important, because the weight of the object doesn't really change what the object is. If you were to add more powder, it would change the weight, but that doesn't change what the powder is.
Mina: I guess I see what you mean by that, but I still think that we found the correct thing.

Fig. 1. Example of Teacher and Student Utterances in a MP Science Discussion

3.2 Feedback Report Development

For the current study, our team translated these automated utterance-level predictions into personalized feedback reports. Each report contained seven components: (a) Eliciting Student Contributions, (b) Meaningful Contributions, (c) Flow of Discussion: Student Contributions, (d) Individual Student Participation During the Discussion, (e) Making Use of Students' Ideas, (f) Flow of Discussion: Teacher Contributions, and (g) Student vs. Teacher Contributions. Figure 2 shows a screenshot of component (b), which illustrates the extent to which each student provided meaningful contributions. Each component included a title and guiding question, a graphical representation of the automated evaluation results from one discussion, a "Summary of Performance" highlighting the main takeaways about how well the teacher did on that component, and ideas for how the teacher could improve under "Recommended Next Steps."

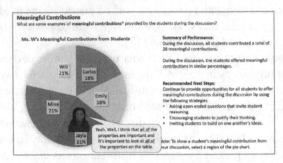

Fig. 2. Example of Feedback Component about Meaningful Contributions

3.3 Participant Sample

The study had 15 participants – five PSTs, six in-service teachers (ISTs), and four teacher educators (TEs). The PSTs had completed a university-based elementary science methods course, the ISTs were currently teaching elementary science, and the TEs were professors who had taught an elementary science methods course. The 11 teachers had previously facilitated the MP discussion and each TE had used the MP task with PSTs on an earlier project. Participants identified as Female (87%), Male (13%), White/Caucasian (73%), Hispanic/Latino (20%), or Asian/Asian American (7%).

3.4 Data Collection and Analysis

Each participant reviewed a video of either their own (for the PSTs or ISTs) or a teacher's (for the TEs, as they did not facilitate a discussion) MP science discussion and then reviewed a personalized feedback report that identified the teacher's strengths and areas of growth in attending to student ideas across the seven feedback components. Next, each participant completed an online survey about their perceptions of each component, including noting whether it was useful, easy to understand and aligned with the discussion facilitated. Finally, one team member facilitated and audio recorded an interview with each participant to elicit their understanding and evaluation of the components. This study focused on analyzing participants' responses to interview questions about the extent to which they felt like the information provided with each component aligned with their review of the discussion (RQ1) and what the teacher could do to improve in a specific component in future discussions (RQ2). RQ3 examined whether and why they thought the report was useful to support teachers' learning.

Analysis proceeded in four steps. First, each interviewer used the interview transcript and their notes to record the main ideas the participant gave to each question. Second, for RQ1 and RQ3, the interviewer used the participant's response to determine if the participant perceived the feedback as aligned, somewhat aligned, or not aligned with the

MP discussion facilitated (RQ1) and as useful, somewhat useful, or not useful to support teacher learning (RQ3). For RQ2, the interviewer used the interview data to determine if their answer about how they could improve in the component was fully, somewhat, or not aligned with the information provided in the feedback report. Third, we calculated the percentage of responses per category. In the final step, we used qualitative content analysis to identify key patterns in the reasons participants gave.

4   Results

4.1   RQ1: Alignment Between Feedback Components and the MP Discussion

Across the seven components, 92.4% of the responses (97 out of 105) indicated that the feedback shown was aligned with the MP science discussion that they reviewed. The strong alignment was primarily due to the similarities they noted between the feedback and their observations of: (a) the teacher's instructional moves or (b) how the student avatars engaged in the discussion. For example, PST01 explained how the feedback about the extent to which they elicited meaningful student contributions (feedback component b) "correlates well" with the discussion they facilitated because "I knew I was asking Carlos more questions…now I see it visually…[and] see how this choice affected the participation of the other four students." Similarly, TE22 explained how the feedback was aligned with the discussion they observed because "there was a low level of meaningful contributions…there were so few it was hard to identify them." Only a few participants noted that the feedback was only somewhat aligned (7 responses, 6.7%) or not aligned (1 response, 1.0%) with the discussion facilitated.

4.2   RQ2: Using Personalized Feedback to Identify Improvement Areas

When describing how they could improve in each of the feedback components, 99% of the responses were strongly aligned with the feedback. Only one participant's improvement ideas were partially aligned with the feedback provided. The most common pattern was the use of the ideas in the "Recommended Next Steps" part of each component to identify how the teacher could improve. Most participants used those ideas to describe what they needed to do to improve their or the teacher's instructional practice. For example, IST11 explained: "I would make sure I kept a summary of who and what they [student contributions] were and have it present when they were in the classroom."

4.3   RQ3: Usefulness of Personalized Feedback to Support Teacher Learning

While all participants agreed that the personalized feedback report provides useful infor-mation to help support teachers' growth, they did so for varied reasons. First, participants thought the reports were useful because they can provide teachers with recommended next steps to help them improve. PST05 said the feedback report "does a good job of laying out next steps…even though my performance was fairly effective there are still things I could do to…have more equal discussion so everyone can share" while TE20 appreciated "the comparative feedback and concrete next steps" so teachers know what to focus on to strengthen their instruction. Second, participants thought the feedback was useful because it helps teachers identify their strengths and areas of growth. As PST03 explained: "…it helps with being able to visually see where they are lacking and strengths…I originally thought my questions would have my students talk a lot, but…I need to work on how to elicit more questioning." Finally, some participants valued the potential of personalized feedback for helping them to identify changes in their instruction over time and for providing an opportunity for them to analyze and reflect on their instruction. For example, PST01 remarked how the feedback report allowed them to notice how much they had "grown as an educator" and that it "allows you to analyze your own teaching from a more professional standpoint."

5   Discussion

This study examined PSTs', ISTs', and TEs' perceptions of automated personalized feedback in the context of DTSims. This feedback is one component within an AI in Education (AIED) system. Evaluating the extent to which the feedback aligns with the discussion facilitated and is perceived as useful for supporting teacher learning is critical to ensure the accuracy of the feedback and that varied stakeholders value this aspect of the AIED system. If the automated feedback fails to align with the discussion or stakeholders do not perceive it as useful, then they will be less likely to use it or worse, they will be aiming to modify their instruction in unproductive ways. Like previous research [3, 12], study participants perceived the feedback positively and valued it for helping teachers to identify strategies they could implement in their classrooms.

One key contribution of this study is illustrating how automated evaluation of multi-speaker discussions can be translated into personalized feedback that has the potential to support teacher learning. This feedback addressed specific and varied components of how teachers attended to student ideas, providing teachers with an in-depth view of how they engaged in this complex instructional skill. Study findings provide strong evidence that participants used the feedback in intended ways – consistently identifying areas for future growth and learning about teaching strategies they (or the teacher) could use to strengthen their discussion facilitation. These results align with previous findings of the importance of feedback identifying learners' strengths and improvement areas [3, 12]. Overall, findings suggest that NLP can be used to generate automated personalized feedback to help develop teachers' ability to engage in key teaching practices.

6   Summary and Conclusion

Study findings indicated that participants: (a) perceived strong alignment between the feedback and the discussion facilitated, (b) were consistently able to use the feedback to identify specific teaching moves the teacher could use to strengthen their discussion facilitation, and (c) noted the usefulness of feedback to support teacher learning. These findings suggest that personalized feedback generated from automated evaluation of teacher and student discussion is an important mechanism to support teachers in learning from their engagement within DTSims. To address some of the study's limitations, namely the limited sample size, the use of participants' self-report and one feedback report, the participants' familiarity with the discussion tasks, and the focus on one teaching practice within one content area, it will be important to examine how NLP can be used to generate personalized feedback across content areas and teaching practices, how teachers apply what they learn from such feedback to their practice in classrooms, and how varying aspects of feedback reports and participants' familiarity with the simulation task relate to changes in teachers' instructional practice.

Acknowledgments. The MP data used in this study was collected in previous grants funded by the National Science Foundation (grant #1621344, #2032179, and #2037983). Any opinions, findings, and conclusions or recommendations expressed in these materials are those of the author(s) and do not necessarily reflect the views of the National Science Foundation.

Disclosure of Interests. The authors have no competing interests to declare that are relevant to the content of this article.

References

1. Alic, S., Demszky, D., Mancenido, Z., Liu, J., Hill, H., Jurafsky, D. Computationally identifying funneling and focusing questions in classroom discourse. In: Proceedings of the 17th Workshop on Innovative Use of NLP for Building Educational Applications (BEA 2022), pp. 224–233, Jul (2022)
2. Bondie, R., Mancenido, Z., Dede, C.: Interaction principles for digital puppeteering to promote teacher learning. J. Res. Technol. Educ. **53**(1), 107–123 (2021)
3. Chizhik, E., Chizhik, A.: Value of annotated video-recorded lessons as feedback to teacher-candidates. J. Technol. Teach. Educ. **26**(4), 527–552 (2018)
4. Cohen, J., Wong, V., Krishnamachari, A., Berlin, R.: Teacher coaching in simulated environment. Educ. Eval. Policy Anal. **42**(2), 208–231 (2020)
5. de Kleijn, R.A.: Supporting student and teacher feedback literacy: an instructional model for student feedback processes. Assess. Eval. High. Educ. **48**, 1–15 (2021). https://doi.org/10.1080/02602938.2021.1967283
6. Demszky, D., Liu, J., Hill, H.C., Jurafsky, D., Piech, C.: Can automated feedback improve teachers' uptake of student ideas? Evidence from a randomized controlled trial in a large-scale online course. Educ. Eval. Pol. Anal. (2023)
7. Demszky, D., Liu, J., Mancenido, Z., Cohen, J., Hill, H., Jurafsky, D., Hashimoto, T.B.: Measuring conversational uptake: a case study on student-teacher interactions. In: Proceedings of the 59th Annual Meeting of the Association for Computational Linguistics. pp. 1638–1653 (2021)
8. Dieker, L., Rodriguez, J., Lignugaris, K., Kraft, B., Hynes, M., Hughes, C.: The potential of simulated environments in teacher education: current and future possibilities. Teach. Educ. Spec. Educ. **37**(1), 21–33 (2014)
9. Dieker, L., Hughes, C., Hynes, M., Straub, C.: Using simulated virtua environments to improve teacher performance. School-Univ. Partners. **10**(3), 62–81 (2017)
10. Gamlem, S.: Feedback to support learning: changes in teachers' practice and beliefs. Teach. Dev. **19**(4), 461–482 (2015)
11. Garet, M., Wayne, A., Brown, S., Rickles, J., Song, M., Manzeske, D.: The impact of providing performance feedback to teachers and principals (NCEE 2018-4001). National Center for Education Evaluation and Regional Assistance (2017)

12. Hattie, J., Timperley, H.: The power of feedback. Rev. Educ. Res. **77**(1), 81–112 (2007). https://doi.org/10.3102/003465430298487
13. Jacobs, J., Scornavacco, K., Harty, C., Suresh, A., Lai, V., Sumner, T.: Promoting rich discussions in mathematics classrooms: using personalized, automated feedback to support reflection and instructional change. Teach. Teach. Educ. **112**, 103631 (2022)
14. Jensen, E., et al.: Toward automated feedback on teacher discourse to enhance teacher learning. In: Proceedings of the CHI Conference on Human Factors in Computing Systems, pp. 1–13 (2020)
15. Lipnevich, A.A., Smith, J.K., (eds.): The Cambridge Handbook of Instructional Feedback. Cambridge University Press (2018)
16. Mikeska, J.N., et al.: Conceptualization and development of a performance task for assessing and building elementary preservice teachers' ability to facilitate argumentation-focused discussions in science: the mystery powder task. (Research Memorandum No. RM-21-06). ETS (2021)
17. Mikeska, J.N., et al.: S1 mystery powder science elementary task. Qual. Data Reposit. (2021). https://doi.org/10.5064/F6FLYLN6
18. Mikeska, J.N., Howell, H., Dieker, L., Hynes, M.: Understanding the role of simulations in K-12 mathematics and science teacher education: outcomes from a teacher education simulation conference. Contemp. Issues Technol. Teach. Educ. **21**(3), 781–812 (2021)
19. Mikeska, J.N., Howell, H., Kinsey, D.: Do simulated teaching experiences impact preservice teachers' ability to facilitate argumentation-focused discussions in mathematics and science? J. Teach. Educ. **74**(5), 422–436 (2023)
20. Nazaretsky, T., Mikeska, J.N., Beigman Klebanov, B.: Empowering teacher learning with AI: automated evaluation of teacher attention to student ideas during argumentation-focused discussion. In: Proceedings of Learning Analytics and Knowledge Conference, Arlington, TX (2023)
21. O'Connor, C., Michaels, S.: Supporting teachers in taking up productive talk moves: the long road to professional learning at scale. Int. J. Educ. Res. **97**, 166–175 (2019)
22. Razumovskaia, E., Glavas, G., Majewska, O., Ponti, E., Korhonen, A., Vulic, I.: Crossing the conversational chasm: a primer on natural language processing for multilingual task-oriented dialogue systems. J. Artif. Intell. Res. **74**, 1351–1402 (2022)
23. Sanh, V., Debut, L., Chaumond, J., Wolf, T.: DistillBERT, a distilled version of BERT: smaller, faster, cheaper and lighter. arXiv abs/1910.01108 (2019)
24. Southwell, R., et al.: Challenges and feasibility of automatic speech recognition for modeling student collaborative discourse in classrooms. Int'l. Educ. Data Min. Soc. (2022)
25. Suresh, A., Jacobs, J., Harty, C., Perkoff, M., Martin, J., Sumner, T.: The TalkMoves dataset: K-12 mathematics lesson transcripts annotated for teacher and student discursive moves. In: Proceedings of the 13th Language Resources and Evaluation Conference, pp. 4654–4662 (2022)
26. Tran, N., Pierce, B., Litman, D., Correnti, R., Matsumura, L.: Utilizing natural language processing for automated assessment of classroom discussion. In: Proceedings International Conference on Artificial Intelligence in Education, pp. 490–496 (2023)

Coding with AI: How Are Tools Like ChatGPT Being Used by Students in Foundational Programming Courses

Aashish Ghimire[✉] and John Edwards

Utah State University, Logan, UT 84322, USA
{a.ghimire,john.edwards}@usu.edu

Abstract. Tools based on generative artificial intelligence (AI), such as ChatGPT, have quickly become commonplace in education, particularly in tasks like programming. We report on a study exploring how students use a tool similar to ChatGPT, powered by GPT-4, while working on Introductory Computer Programming (CS1) assignments, addressing a gap in empirical research on AI tools in education. Utilizing participants from two CS1 class sections, our research employed a custom GPT-4 tool for assignment assistance and the ShowYourWork plugin for keystroke logging. Prompts, AI replies, and keystrokes during assignment completion were analyzed to understand the state of students' programs when they prompt the AI, the types of prompts they create, and whether and how students incorporate the AI responses into their code. The results indicate distinct usage patterns of ChatGPT among students, including the finding that students ask the AI for help on debugging and conceptual questions more often than they ask the AI to write code snippets or complete solutions for them. We hypothesized that students ask conceptual questions near the beginning and debugging help near the end of program development do not find statistical evidence to support it. We find that large numbers of AI responses are immediately followed by the student copying and pasting the response into their code. The study also showed that tools like these are widely accepted and appreciated by students and deemed useful according to a student survey - suggesting that the integration of AI tools can enhance learning outcomes and positively impact student engagement and interest in programming assignments.

Keywords: LLM · Chatbot · ChatGPT · AI in Education · Keystrokes

1 Introduction

The advent of generative artificial intelligence (AI) has ushered in a new era across various sectors, including education. Among these AI advancements, tools like ChatGPT, particularly those powered by Generative Pre-trained Transformers (GPT), have garnered increasing attention. Their integration into educational

A. M. Olney et al. (Eds.): AIED 2024, LNAI 14830, pp. 259–267, 2024.
https://doi.org/10.1007/978-3-031-64299-9_20

practices, particularly in programming and computer science education, signifies a notable shift in instructional methodologies. This shift raises questions about the role and effectiveness of these tools in enhancing student learning outcomes, particularly in foundational courses such as Introductory Computer Science. This research aims to delve into the burgeoning field of AI application in education, focusing on the usage and impact of a ChatGPT-like tool in introductory programming class(CS1) coding assignments. Specifically, the study addresses these three research questions:

RQ1 *How do students employ generative AI-based tools, such as ChatGPT, while completing their CS1 coding assignments?*

RQ2 *What discernible patterns emerge from students' usage of this tool during assignments?*

RQ3 *Does a tool like ChatGPT make programming classes more accessible, improve students' efficiency, or help new programmers learn programming?*

To investigate these questions, the study utilized participants from two sections of a CS1 class, incorporating a custom GPT-4 tool designed for assignment assistance along with the ShowYourWork [3] plugin to the PyCharm integrated development environment(IDE) for recording keystrokes. Additionally, a post usage survey was conducted to collect students' feedback. This approach allowed for a comprehensive analysis of student interactions with the AI tool and a comparison of their performance in assignments completed with and without the aid of AI. The subsequent sections of this paper will detail the methodology, present the findings, and discuss the implications of these results in the context of modern computer science education.

2 Related Works

Recent advances in generative AI and natural language processing have enabled the development of sophisticated large language models (LLMs) like GPT-4, Codex, GitHub Copilot, and ChatGPT. These models are not just technical marvels but have profound implications for computing education research and practice [2]. Similarly, GitHub Copilot demonstrated its efficacy by generating solutions that met the requirements of introductory programming assignments [6]. There's a growing consensus on shifting focus from basic coding skills to higher-order thinking and problem-solving abilities [1]. Other studies explore the use of LLMs by students in a computing context [4,5,9], including over-reliance, which may hinder learning [7]. At least one study examines students' views on using ChatGPT [8].

3   Methodology

3.1   Participants

Our study was done in compliance with a protocol approved by our university's institutional review board (IRB). The participants of this study were students enrolled in two sections of a Computer Science 1 (CS1) course at our institution, a mid-sized research university in the United States. The study commenced during the final two weeks of the Fall 2023 semester. Students were given two programming assignments and given the option of using a tool based on LLMs for assistance. Students did not need to participate in the study in order to use the AI tool.

Programming assignments were to be completed in Python. The first programming assignment involves writing a graphical car racing game. Starter code provides the graphical structure and render loop. Students are asked to be creative in designing the game play. The second programming assignment provides starter code that provides a menu to allow the user to sort a deck of cards and search for specific cards. There are logic errors in the starter code which make the program give the wrong results. The student is asked to identify and fix the errors. The assignments were designed to reinforce learning objectives related to methods, classes, objects, and operator overloading, involving work with multiple files and starter code.

A total of 48 students from both sections, out of 246, participated in the study. However, not all participants contributed to the dataset equally; some did not submit their keystroke data, and others did not engage with the LLM-based tools sufficiently to be included in the full data analysis. Ultimately, the keystroke data and AI tool usage data from 25 students were used for in-depth analysis. No demographic data were collected from the participants. The only background information gathered was regarding their prior programming experience, through a single question on the subject. Of the participants, 21 completed the post-assignment survey, providing valuable insights into their experiences and perceptions of using AI tools in their assignments. This selective participation and data contribution highlights the varied engagement levels with both the study and the LLM-based tool, underscoring the need for further investigation into factors influencing students' willingness to utilize such technologies in educational settings.

3.2   Tools

We used three tools for data collection. The first is a custom GPT-4 wrapper. It enabled the logging of prompts submitted by students, responses generated by GPT-4, and additional data such as timestamps, context, and follow-up counts. The second tool is the *ShowYourWork* IDE plugin. All students in the course were required to install ShowYourWork into the PyCharm IDE. ShowYourWork logs all keystrokes made within the PyCharm IDE during assignment completion. And finally, we asked students to complete a survey on their programming experience and perceptions of the usefulness of the AI tool after the study assignments.

3.3 Data Collection and Analysis

In collaboration with the course instructor (not an investigator in this study), students were instructed to complete their coding assignments with the option of freely using the custom GPT-4 powered tool. During this process, the ShowYour-Work plugin continuously recorded their keystrokes, while the AI tool archived all student prompts and corresponding LLM responses in a structured database. This comprehensive dataset was pivotal for our analysis.

4   Results

In this section, we attempt to answer our research questions by analyzing the data collected from prompts, keystrokes, and surveys.

4.1   RQ1: How Do Students Employ Generative AI-Based Tools, Such as ChatGPT, While Completing Their CS1 Coding Assignments?

A. Prompt Type. The custom tool was programmed to not answer any questions that were not related to these topics, using meta prompting and system instructions. Students could ask any questions about computer science and mathematics to the AI tool. We first categorized the prompt types into 4 categories according to the following definitions:

1. **Debugging Help**: Prompts that seek help to identify or fix errors in the provided code snippet.
2. **Code Snippet**: Prompts that ask for a specific part of the code, like a function or a segment.
3. **Conceptual Questions**: Prompts that are more about understanding concepts or algorithms rather than specific code.
4. **Complete Solution**: Prompts that request an entire solution or a complete code snippet.

We leveraged OpenAI GPT-4 Turbo for categorizing the prompts using meta-prompting. We then calculated the inter-rater reliability between human raters and GPT-4. For percentage agreement metrics, we observed a percentage agreement of $\frac{17}{20} = 85\%$ with Cohen's Kappa $\kappa = 0.76$.

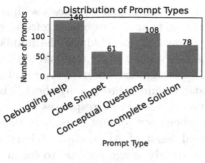

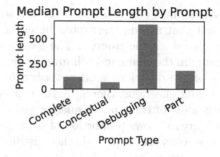

(a) Type of prompts

(b) Bar chart showing median prompt length to AI tool in no. of characters

Fig. 1. Prompt type and prompt length

Figure 1a shows the count of various types of prompts. Asking for help with debugging code and asking conceptual questions were the most common types of prompts, as opposed to asking for full or partial code directly. Figure 1b shows the bar chart of the median length of prompts sent to the AI tool. The median prompt length for Debugging prompts was over 500, whereas the median prompt length for each of the other three prompt types was under 250. This makes sense, since when asking for help debugging the student will include the code in the prompt. The plot indicates that most of the prompts are under 200 characters, meaning the most common prompts did not contain starter code but were rather more conceptual in nature or had a smaller code snippet.

By examining the students' keystrokes before and after their engagement with the AI tool, this question seeks to understand the nature of engagement and the kind of support provided by the AI tool. We first examined when in the coding process students use the AI assistance. Students tend to start slow on their assignments and don't immediately use the AI tool. However, students who use the AI tool use it at least once before they complete one-fifth of their assignment.

We hypothesized that the type of prompts students made (e.g., Debugging Help) would change as students progressed toward completion of the assignment. For example, we expected that students would ask more conceptual and/or complete solution types of questions near the beginning and debugging questions in the middle and at the end of development. We found no statistical support for this hypothesis. Median percentages of assignment completed for each query type were relatively close to each other (Debugging Help: 46%, Code Snippet: 57%, Conceptual Questions: 39%, Complete Solution: 44%). We performed two Mann-Whitney U tests between Conceptual Questions and Complete Solution with no statistical significance ($U = 407, p = 0.55$) and between Conceptual Questions and Debugging Help ($U = 1037, p = 0.19$). This could mean that, indeed, students ask varied types of questions throughout development, or that our sample size is insufficient to statistically detect the patterns.

264   A. Ghimire and J. Edwards

We define a conversation chain as a series of prompts and responses with the AI tool that are uninterrupted by closing the webpage, a webpage refresh, or by a refresh of the context. The length of a conversation chain is the number prompts in the chain and is limited to seven, after which the context is refreshed. Most of the questions students asked for CS1 assignments were conceptual or debugging questions; therefore, the conversation chains were usually short. This means most of the student queries were solved in one or two responses.

Figure 2a shows proportion of work done when each prompt is made in terms of keystrokes. As expected, there is an initial burst of AI prompt activity as students are starting their assignments. Interestingly, usage appears to continue throughout program development. In addition, there appear to be roughly 10 prompts right near the end of development.

Figure 2b shows the time and keystrokes between two consecutive prompts by the same student. Most of the consecutive prompts occur within the first 5 to 15 min. This indicates that students are not spending a lot of time between prompts but rather trying the solution, varying their prompt, and asking again within a short time period. As expected, and as we see in the discussion below, many prompts are separated by few keystrokes but a big paste, indicating students are copying the AI response into their code (this was allowed). However, a surprise is the number of cases where, in the short time between prompts (1–30 minutes), students typed 500 or even 1000 characters. These students engage in a flurry of programming activity between prompts, possibly trying out ideas from the AI response, or possibly typing in the AI-generated code instead of pasting it.

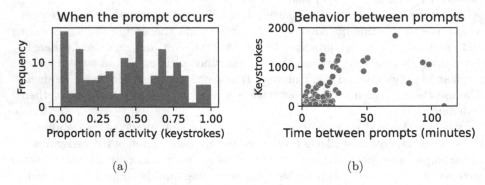

(a)   (b)

Fig. 2. (a) Histogram of proportion of activity when AI is prompted. (b) Scatter plot of the number of keystrokes vs time (in minutes) between prompts. Prompt pairs with greater than 120 min between them (there are 21 such pairs) are not shown.

Next, we explored the activity that occurs immediately following a prompt to the AI. Figure 3a shows the histogram of the proportion of LLM calls that were followed by a large paste event of more than 20 characters for each student. For example, the figure shows that six students pasted the output from the AI tool into their code about half the time. All students pasted response text at least once. For this analysis, we only looked at prompts that were not classified as asking conceptual questions (which is the second most common type of prompt). We confirmed that the paste text came from the last response of the AI by testing if the pasted text was a substring of the AI response. Figure 3b shows that half of the pastes were exactly the AI response. For prompts asking for code or help with debugging, students often end up copying and pasting the response.

4.2 RQ3: Does a Tool Like ChatGPT Make Programming Classes More Accessible, Improve Students' Efficiency, or Help New Programmers Learn Programming? How Do Students Feel About Such a Tool?

We conducted a post-assignment survey, and students expressed that the tool was useful and helped them complete assignments more quickly. Our survey reveled that students are already using similar tools in their programming classes. About one-third student indicated using such tools most of the time and about half 42% students using sometimes or about half of the time. Less than one quarter student used it less than 25% of time. Over 92% agreed these tool helped the complete assignment faster.

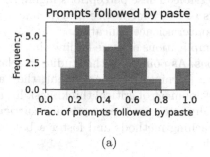
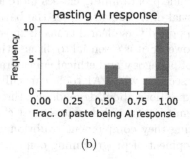

(a) (b)

Fig. 3. Paste activity following prompts. (a) Histogram of percentage of AI calls followed by a big 'paste' event (over 20 characters) for each student. (b) Percentage of paste events that are direct substrings of the AI response.

Programming can be an intimidating subject for some students, and tools like these have been touted for their potential as a personalized tutor. Over 90% of respondents agreed with the statement, "Tools like these help increase the accessibility of programming classes or encourage me to take programming classes". 10% were neutral and no respondent disagreed. Finally, over 85% of students mostly agreed with the statement, "If offered, I would use tools like this one in future classes or assignments".

5   Conclusion and Discussion

This study aimed to explore the impact and usage patterns of generative AI-based tools, like ChatGPT, on student performance and engagement in CS1 coding assignments. Through detailed analysis of interactions between students and the AI tool, as well as students' keystrokes and survey responses, several key findings emerged.

Firstly, the integration of a custom GPT-4 powered tool in programming assignments revealed significant usage among students, particularly for debugging and conceptual understanding. This suggests that such tools can serve as effective aids in the learning process, potentially reducing the time students spend stuck on particular problems and enhancing their overall learning experience.

The analysis of keystroke data and AI tool interactions indicated that students primarily used the AI tool for assistance with debugging and conceptual questions, with most interactions resulting in short conversation chains. This finding points to the efficiency of AI tools in providing targeted, immediate assistance, which, in turn, may contribute to improved problem-solving skills and deeper conceptual understanding.

Survey responses further supported the utility of the AI tool, with a vast majority of students reporting that it helped them complete assignments faster and made programming classes more accessible. These perceptions highlight the potential of AI tools to lower the barriers to entry for novice programmers and to support diverse learning needs in computer science education.

However, it is essential to discuss the implications of these findings in the context of pedagogy and ethical considerations. As pointed in the studies in related work section, while AI tools can enhance learning and engagement, they also raise questions about dependency, the development of critical thinking skills, and academic integrity. Educators must carefully integrate these tools into curricula, ensuring they complement traditional teaching methods and foster a balanced development of programming competencies.

5.1   Threats to Validity and Future Works

Our study was conducted at a single institution with a relatively small sample size, limiting generalizability. A threat to internal validity is the fact that we did not control which assignment the student was working on (due to small sample size) and behavior may have been different for different assignments.

Future studies could explore the long-term impact of AI tool usage on learning outcomes, investigate its effects across diverse educational contexts, and examine strategies to mitigate potential drawbacks. As AI technology continues to evolve, ongoing research and dialogue among educators, researchers, and policymakers will be crucial in harnessing its potential to enrich learning experiences while maintaining academic integrity and fostering comprehensive skill development.

References

1. Chen, M., et al.: Evaluating large language models trained on code. arXiv preprint arXiv:2107.03374 (2021)
2. Denny, P., et al.: Computing education in the era of generative AI. arXiv preprint arXiv:2306.02608 (2023)
3. Edwards, J., Hart, K., Shrestha, R., et al.: Review of CSEDM data and introduction of two public CS1 keystroke datasets. J. Educ. Data Min. **15**(1), 1–31 (2023)
4. Hedberg Segeholm, L., Gustafsson, E.: Generative language models for automated programming feedback (2023)
5. Kazemitabaar, M., Hou, X., Henley, A., Ericson, B.J., Weintrop, D., Grossman, T.: How novices use LLM-based code generators to solve CS1 coding tasks in a self-paced learning environment. arXiv preprint arXiv:2309.14049 (2023)
6. Puryear, B., Sprint, G.: Github copilot in the classroom: learning to code with ai assistance. J. Comput. Sci. Coll. **38**(1), 37–47 (2022)
7. Vaithilingam, P., Zhang, T., Glassman, E.L.: Expectation vs Experience: evaluating the usability of code generation tools powered by large language models. In: Chi Conference on Human Factors in Computing Systems Extended Abstracts, pp. 1–7 (2022)
8. Yilmaz, R., Yilmaz, F.G.K.: Augmented intelligence in programming learning: examining student views on the use of ChatGPT for programming learning. Comput. Hum. Behav. Artif. Hum. **1**(2), 100005 (2023)
9. Zastudil, C., Rogalska, M., Kapp, C., Vaughn, J., MacNeil, S.: Generative AI in computing education: perspectives of students and instructors. In: 2023 IEEE Frontiers in Education Conference (FIE). pp. 1–9. IEEE (2023)

Identifying Gaps in Students' Explanations of Code Using LLMs

Rabin Banjade(✉)⬤, Priti Oli⬤, Mahmudul Islam Sajib⬤, and Vasile Rus⬤

University of Memphis, Memphis, USA
{rbnjade1,poli,msajib,vrus}@memphis.edu

Abstract. This study investigates methods based on Large Language Models (LLMs) to identify gaps or missing parts in learners' self-explanations. This work is part of our broader effort to automate the evaluation of students' freely generated responses, which in this work are learners' self-explanations of code examples during code comprehension activities. We experimented with two methods and four distinct LLMs in two distinct settings. One method prompts LLMs to identify gaps in learners' self-explanations, whereas the other method relies on LLMs performing a sentence-level semantic similarity task to identify gaps. We evaluated these methods in two settings: (i) simulated data generated using LLMs and (ii) actual student data. Results revealed the semantic similarity method significantly improves task performance over the zero-shot prompting for gap identification (the holistic method), i.e., over the standard method of prompting LLMs to directly address the gap identification task.

Keywords: Self Explanation · Large Language Models · Automated Assessment · Scaffolding · Code Comprehension

1 Introduction

This paper explores several methods based on large language models to detect gaps or missing parts in learners' self-explanations of targeted instructional content, e.g., explanations of code examples. Self-explanation, i.e., elucidating learning material to oneself through speech or writing [8], has a positive impact on learning [1]. Students who self-explain while engaging in various instructional activities, such as reading code examples or solving problems, tend to learn more and develop a more profound understanding of the subject matter. However, as Renkl and colleagues [12] point out, the effectiveness of self-explanation might not be effective in instances where learners either neglect pertinent explanations or engage passively with the content. Building upon these insights, Intelligent Tutoring Systems (ITS) that rely on instructional strategies such as scaffolded self-explanation have been developed to enhance learning outcomes as demonstrated by various studies [6].

Scaffolding students' self-explanation relies on accurately assessing students' explanations in terms of correctness or completeness. This entails addressing two

A. M. Olney et al. (Eds.): AIED 2024, LNAI 14830, pp. 268–275, 2024.
https://doi.org/10.1007/978-3-031-64299-9_21

crucial tasks: identifying incorrect student responses and identifying incomplete responses, the latter being the focus of our study. For instance, in the case of incorrect responses, the system may correct potential misconceptions articulated by the student, while for incomplete responses, it may offer appropriate hints to encourage learners to think about the missing components-such as steps in a problem solution. In this study, we explore using LLMs to identify missing parts in student self-explanations of code examples.

LLMs have emerged as state-of-the-art systems that can generate coherent content in response to input prompts [2], including computer code and accompanying explanations, which is relevant to our work here. Indeed, this generative feature can be very useful for our target task of assessing student responses because, for instance, in current state-of-the-art ITSs, this task relies on expert-authored benchmark explanations that are considered correct and complete (unless they are well-known misconceptions). The benchmark explanations play a crucial role in automated methods that rely on semantic similarity approaches to evaluate the correctness and completeness of student responses [3,15]. In such approaches, if a student's answer is semantically equivalent to a corresponding benchmark response, the student response is deemed as having the same correctness and completeness value as the expert-generated benchmark response. It should be noted that semantic similarity has been the dominant approach to assessing self-explanations.

Using experts to author the benchmark responses is tedious and expensive. Therefore, such approaches do not scale well across topics and domains. LLMs have been shown to help significantly with code explanation generation for typical code examples used in intro-to-programming courses [7,10,14]. Furthermore, there is evidence that LLMs can competitively address the semantic similarity task between two texts [4,9].

Therefore, we present in this paper the result of our investigation on the role of LLMs in addressing two major challenges in the automated assessment of students' self-explanation: (i) the generation of benchmark/reference explanations and (ii) auto-assessment of student-generated free responses during instructional tasks. For the latter task, auto-assessment, we explore two approaches: (1) a direct approach in which we prompt LLMs to identify gaps in self-explanations and (2) a semantic similarity approach in which we prompt LLMs to assess the similarity of a student self-explanation with respect to a benchmark/reference explanation.

As research questions, this paper addresses the following key questions.

RQ1: Can LLMs detect gaps in code explanations, indicating incomplete explanations?

RQ2: Can LLMs-generated explanations be used with semantic-similarity approaches to identify gaps in students' code explanations?

RQ3: Can LLMs accurately determine the semantic similarity of self-explanations at the sentence level?

2 Methodology

To meet our research objectives, we developed a comprehensive methodology. First, we generate code explanations to be used as benchmark explanations. Second, we used two approaches to identify missing details in student explanations. The two approaches are (1) a holistic approach in which LLMs are prompted to identify the missing parts given a student explanation and the corresponding code example and (2) a point-wise, semantic similarity approach at sentence level in which student explanations are broken down into units of analysis (sentences) and then LLMs are prompted to identify which such sentences match corresponding sentences in the benchmark explanations (a sentence in the benchmark/reference explanation is called an expectation). The methodology involves a selection of relevant models and prompts and the evaluation of LLMs in identifying missing expectations in student explanations.

Code Explanation Generation. Our first step involved generating code explanations using four different models and different prompts to evaluate different types of code explanations. We aimed to understand the diversity and quality of explanations produced by these models. We employed four different LLMs to generate benchmark code explanations: GPT-4-0613 [11], GPT-3.5 Turbo-0613, Mixtral-8x7b-Instruct-v0.1 [16], and llama-2-70b-chat [17]. These LLMs have achieved state-of-the-art results in various tasks while differing in training data and algorithms, although the model size and training data for OpenAI models are not disclosed. We queried all the models using the unified interface provided by *LiteLLM*[1], ChatGPT-4 and ChatGPT-3.5 were queried through OpenAI API, whereas Mixtral and LLama-2 were prompted via the *Together API*[2]. For all these models, we used a temperature parameter value of 0 for consistency and reproducible results. We generated explanations for 10 Java code examples, sourced from the DeepCode codeset [13], which were diverse in complexity and concept coverage. These examples were previously used in a human subject experiment where self-explanations were collected from students.

P1: Provide a line-by-line explanation of the given java code {code}
P2: Explain the code to the student in a way that they understand necessary concepts. Focus on conceptual understanding of the code.{code}
P3: Summarize the given java code {code}.
P4: In this code example, {code}, we will focus on the concept of loops. We do that with the help of a program whose goal is to find the smallest divisor of a positive number. Your task is to read the code shown and understand what it does. Once you are done reading the code, type your explanation of what the code does. Try to identify the major blocks of code and their goals and how those goals are implemented. Please go on and do your best

---

[1] https://litellm.ai/.
[2] https://docs.together.ai/docs/quickstart.

to explain your understanding of the code and its output in as much detail as you can.

The prompts (P1,P2, P3,P4) were chosen to understand how LLMs generate explanations and which explanation to use as a benchmark explanation for our task of identifying gaps in student explanations. Prompt **P4** is identical to the prompt given to students when prompted to self-explain the code whereas prompts P1, P2, and P3 were used to generate different types of code explanations based on previous studies [10]. We evaluated the quality of explanations generated from each prompt to be used in our main task.

2.1 Data

We evaluated our methods using two data settings: student data and simulated data. From 60 students at a large US public university, we collected self-explanations prompted by **P4**. Sampling was based on word length to ensure detailed explanations, yielding 50 diverse explanations annotated with missing expectations relative to DeepCode codeset benchmarks [13]. For simulated data, we derived a dataset from LLM-generated explanations by randomly removing one expectation/sentence per explanation and repeated three times to create three samples each with one missing expectation. This simulated gaps in student explanations, with removed expectations serving as ground truth for evaluating LLMs' detection of missing components. Initial observations indicated that sentences from the generated explanations aligned with expectations, representing comprehension of specific code parts/concepts.

2.2 Prompting to Identify Gaps in Explanations of Code

In order to prompt LLMs to identify gaps or missing expectations in student or simulated explanations of code, we went through a prompt selection process that involved several hits and trials. We used two different settings to prompt for identifying missing parts in code explanations: (1) the holistic approach in which we prompt to identify what is missing given the student explanation and the corresponding code, and (2) the pointwise, sentence-level semantic similarity-based approach. For the holistic approach, we designed two different types of prompts with providing reference or benchmark explanations (see below prompt P6) and without (P7 - see below). For consistent evaluation between student data and simulated missing data, we prompted to generate a single sentence for each prompt.

P6: Given the following code:{code} and the following reference explanation: {reference explanation}, your task is to identify what is missing in the following student explanation:{student explanation} of the code. Generate the missing part as a single sentence.
P7: Given the following code:{code}, your task is to identify what is miss-

> ing in the following student explanation:{student explanation} of the code.
> Generate the missing part as a single sentence.

As the holistic prompting didn't yield very promising results, we have introduced a novel approach to guide LLMs in identifying and generating the absent expectation: the pointwise, sentence-level semantic similarity approach. Rather than directly requesting the missing expectation, our method evaluates pairwise resemblances between the explanations produced by LLMs and those produced by students. When dealing with the simulated data, the pairwise similarity consistently yields completely accurate outcomes, given that we compare the same pair of sentences except for the missing one. As a result, we only report results here for the pairwise similarity technique when applied to the student explanations. The LLM-generated explanations function as expert/benchmark explanations. To identify the missing expectation, in this case, we pinpoint the sentence within the expert explanation (LLMs' explanation) that exhibits the lowest pairwise semantic similarity with any sentence in the simulated student explanation. This particular sentence is considered to result in missing expectations.

For implementing the pairwise similarity approach, we adopt the subsequent prompt format:

> **P8**: Provide a semantic similarity score on a scale of 0 to 1, 0 being least similar and 1 being most similar, for the following two sentences {reference sentence} and {student sentence}.

We chose a scale of 0 to 1 for similarity prompting as Gatto et al. [5] showed that prompting for a similarity value between 0 to 1 has a better performance compared to using a 1–5 scale. This has been confirmed by our experiments as well.

Evaluation. We evaluated code explanation generation using two metrics: *Correctness* and *Completeness*. Correctness assessed the accuracy of explanations, while Completeness measured coverage of code concepts. For missing expectation generation, we focused solely on Correctness, which indicates the proportion of accurate responses in identifying missing parts.

3   Results and Discussion

We compared the completeness and correctness metric of code explanations generated from four different LLMs using four prompting strategies, as mentioned in the methodology section. We obtained 100% correct explanations for all the prompts from all the models. However, we obtained 88%, 94%, 71%, and 80% completeness scores for prompts P1, P2, P3, and P4, respectively, averaged across all the models (n = 40) for each prompt. This evaluation aimed to understand which prompt generated the most accurate explanations (correct and complete), which we can use to generate reference/benchmark explanations instead of using experts. Prompt (**P2**) leads to the best explanations. The other prompts led to explanations that showed specific characteristics that made them inconsiderable as reference explanations. For instance, prompting for line-by-line explanations (**P1**) generates explanations focusing on code's syntactic elements, whereas code summarization (**P3**) prompts miss out on important details necessary for understanding and learning. Prompt **P4** generates explanations mostly on the block level and misses important syntactic knowledge components. In sum, explanations generated by **P2** are balanced regarding functional and syntactic level explanations. Therefore, we obtained reference explanations using the prompt (**P2**). Explanations from all the models for **P2** were similar in completeness and correctness, so we randomly sampled explanations from different models as benchmark explanations. One observation was that LLAMA2 explanations consisted of conversation-style filler sentences such as, "Sure, here is an explanation of the given code," which we removed. For each explanation, we sampled a reference explanation for the same code such that the two differed.

3.1   Prompting for Missing Expectations

We prompted LLMs in zero-shot settings to identify missing parts (expected unit answers where the unit is a sentence) in student explanations. As mentioned earlier in the *Methodology* section, we evaluated the performance of LLMs on this task using two prompt settings, (1) with and (2) without the use of reference explanations, for two sources of explanations, (1) simulated data and (2) student data.

Table 1 shows our results for prompting for missing expectations under different settings using simulated and student data. As seen in the table, prompting to identify (and generate) missing expectations with or without reference explanations does not improve performance in terms of correctness. It also shows that additional context from reference explanation does not provide an additional boost in both simulated data and student data. Comparing among models, Mixtral and GPT-4 provide better performance in terms of correctness. Also, when the same prompts were given to student data, the performance of LLMs decreased, which can be attributed to the diverse nature of student explanations, which often adopted conversational styles that deviated from conventional grammar and writing standards in English. One of the observations worth noting is that generated missing expectations were not specific. For example, even though the goal was to understand the program that solved a specific task, most of the generated missing expectations were about complexity analysis; this was mostly prevalent in code examples such as binary search and insertion sort. One of the major challenges, as we observed, is to guide LLMs to focus on missing details compared to reference explanations, which would set the scope of learning. One could probably achieve this with various advanced prompting techniques, which can be further explored; we resorted to a prevalent approach, i.e., semantic similarity-based assessment of student answers. Towards this direction, As discussed in the methodology section, we prompted for pairwise similarity between sentences as shown in Table 1 indicated by **P8**. Our evaluation is based on a single missing expectation; our results indicate that we can vastly increase the correctness of our generated explanations. Our results also indicate that language models can perform well for semantic similarity tasks, as indicated by other works [5].

Table 1. Correctness comparison (n = 50) for results obtained by prompting LLMs for missing expectation generation in simulated and student data.

Model	Simulated Data		Student Data		
	P6	P7	P6	**P7**	P8
GPT-3.5	47%	44%	37%	48%	82%
GPT-4	77%	66%	40%	41%	94%
LLAMA2	66%	60%	34%	42%	84%
MIXTRAL	77%	77%	52%	38%	90%

4   Conclusion and Future Work

In summary, we studied if we can identify missing gaps in student explanations using four state-of-the-art LLMs. Our experiments under different settings using explanations generated from LLMs as reference explanations showed that prompting for similarity can yield better results for finding missing expectations than zero-shot prompting of LLMs. This indicates that LLMs can determine semantic similarity in sentence level for student explanation in code comprehension tasks. One of the limitations of our work is maintaining temporal validity due to LLMs' evolving landscape. LLMs offer new research opportunities in generating programming exercises, unit tests, code explanations, and providing automated feedback on student code submissions, but effectively guiding novice programmers remains a challenge. Our efforts to utilize LLMs for scaffolding students' code understanding represent progress in this direction. We plan to explore various prompting techniques and leverage open-source models like LLAMA2 to enhance transparency and scalability in education.

Acknowledgments. This work has been supported by the following grants awarded to Dr. Vasile Rus: the Learner Data Institute (NSF award 1934745); CSEdPad (NSF award 1822816); iCODE (IES award R305A220385). The opinions, findings, and results are solely those of the authors and do not reflect those of NSF or IES.

References

1. Aleven, V.A., Koedinger, K.R.: An effective metacognitive strategy: learning by doing and explaining with a computer-based cognitive tutor. Cogn. Sci. **2**(26), 147–179 (2002)
2. y Arcas, B.A.: Do large language models understand us? Daedalus **151**(2), 183–197 (2022)

3. Bexte, M., Horbach, A., Zesch, T.: Similarity-based content scoring - a more classroom-suitable alternative to instance-based scoring? In: Rogers, A., Boyd-Graber, J., Okazaki, N. (eds.) Findings of the Association for Computational Linguistics: ACL 2023, pp. 1892–1903. Association for Computational Linguistics, Toronto, Canada (2023). https://doi.org/10.18653/v1/2023.findings-acl.119, https://aclanthology.org/2023.findings-acl.119

4. Brown, T., et al.: Language models are few-shot learners. In: Advances in Neural Information Processing Systems, vol. 33, pp. 1877–1901 (2020)

5. Gatto, J., Sharif, O., Seegmiller, P., Bohlman, P., Preum, S.M.: Text encoders lack knowledge: leveraging generative LLMs for domain-specific semantic textual similarity. arXiv preprint arXiv:2309.06541 (2023)

6. Graesser, A.C., McNamara, D.S., VanLehn, K.: Scaffolding deep comprehension strategies through point&query, autotutor, and iStart. Educ. Psychol. **40**(4), 225–234 (2005)

7. MacNeil, S., Tran, A., Mogil, D., Bernstein, S., Ross, E., Huang, Z.: Generating diverse code explanations using the GPT-3 large language model. In: Proceedings of the 2022 ACM Conference on International Computing Education Research-Volume 2, pp. 37–39 (2022)

8. McNamara, D.S., Magliano, J.P.: Self-explanation and metacognition: the dynamics of reading. In: Handbook of Metacognition in Education, pp. 60–81. Routledge (2009)

9. Oli, P., Banjade, R., Chapagain, J., Rus, V.: Automated assessment of students' code comprehension using LLMs. arXiv preprint arXiv:2401.05399 (2023)

10. Oli, P., Banjade, R., Chapagain, J., Rus, V.: The behavior of large language models when prompted to generate code explanations. In: Proceedings of the workshop on Generative AI for Education (GAIED) at the Thirty-Seventh Conference on Neural Information Processing Systems (NeurIPS 2023). arXiv (2023)

11. OpenAI, R.: GPT-4 technical report. arxiv 2303.08774. View in Article **2**, 13 (2023)

12. Renkl, A.: Learning mathematics from worked-out examples: analyzing and fostering self-explanations. Eur. J. Psychol. Educ. **14**(4), 477–488 (1999)

13. Rus, V., Brusilovsky, P., Tamang, L.J., Akhuseyinoglu, K., Fleming, S.: Deepcode: an annotated set of instructional code examples to foster deep code comprehension and learning. In: Crossley, S., Popescu, E. (eds.) ITS 2022. LNCS, vol. 13284, pp. 36–50. Springer, Cham (2022). https://doi.org/10.1007/978-3-031-09680-8_4

14. Sarsa, S., Denny, P., Hellas, A., Leinonen, J.: Automatic generation of programming exercises and code explanations using large language models. In: Proceedings of the 2022 ACM Conference on International Computing Education Research-Volume 1, pp. 27–43 (2022)

15. Sung, C., Dhamecha, T., Saha, S., Ma, T., Reddy, V., Arora, R.: Pre-training BERT on domain resources for short answer grading. In: Proceedings of the 2019 Conference on Empirical Methods in Natural Language Processing and the 9th International Joint Conference on Natural Language Processing (EMNLP-IJCNLP), pp. 6071–6075 (2019)

16. team, M.A.: Mixtral of experts (2022). https://mistral.ai/news/mixtral-of-experts/

17. Touvron, H., et al.: Llama 2: open foundation and fine-tuned chat models. arXiv preprint arXiv:2307.09288 (2023)

A Multi-task Automated Assessment System for Essay Scoring

Shigeng Chen[1]([✉])[iD], Yunshi Lan[2][iD], and Zheng Yuan[1][iD]

[1] Department of Informatics, King's College London, London, UK
{shigeng.chen,zheng.yuan}@kcl.ac.uk
[2] School of Data Science and Engineering, East China Normal University,
Shanghai, China
yslan@dase.ecnu.edu.cn

Abstract. Most existing automated assessment (AA) systems focus on holistic scoring, falling short in providing learners with comprehensive feedback. In this paper, we propose a Multi-Task Automated Assessment (MTAA) system that can output detailed scores along multiple dimensions of essay quality to provide instructional feedback. This system is built on multi-task learning and incorporates Orthogonality Constraints (OC) to learn distinct information from different tasks. To achieve better training convergence, we develop a training strategy, Dynamic Learning Rate Decay (DLRD), to adapt the learning rates for tasks based on their loss descending rates. The results show that our proposed system achieves state-of-the-art performance on two benchmark datasets: ELLIPSE and ASAP++. Furthermore, we utilize ChatGPT to assess essays in both zero-shot and few-shot contexts using an ELLIPSE subset. The findings suggest that ChatGPT has not yet achieved a level of scoring consistency equivalent to our developed MTAA system and that of human raters.

Keywords: Automated Essay Scoring · Multi-Task Learning · ChatGPT Automated Assessment · Zero-Shot Learning · Few-Shot Learning

1 Introduction

Automated assessment (AA), mimicking the judgment of examiners evaluating the quality of student writing, is one of the most important educational Natural Language Processing (NLP) applications. Originally used for summative purposes in standardised testing such as the TOEFL[1] and GRE[2], these systems are now frequently found in classrooms [4,5].

Traditional ML-based AA systems typically rely on hand-crafted features and models like SVM and linear regression have been proposed [6–9], while neural-based AA systems often employ word embeddings like GloVe [10] and incorporate

---

[1] https://www.ets.org/toefl.html.
[2] https://www.ets.org/gre.html.

© The Author(s), under exclusive license to Springer Nature Switzerland AG 2024
A. M. Olney et al. (Eds.): AIED 2024, LNAI 14830, pp. 276–283, 2024.
https://doi.org/10.1007/978-3-031-64299-9_22

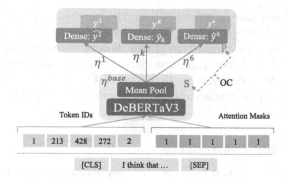

Fig. 1. The architecture of MTAA.

deep neural networks such as CNN [11] and LSTM [12] to achieve better system performance [13–16]. In contrast, Transformer-based AA systems which leverage large pre-trained language models like DistilBERT [22] and XLNet [21] outperform their counterparts by handling long-distance relationships and exhibiting strong generalization abilities [17–19]. Most prior AA systems focusing on holistic scoring are unable to provide in-depth feedback to learners. A few studies [28–30] have explored the development of an AA system to support multi-dimensional essay evaluation. However, they often neglect the inter-connectedness among various assessment measures.

In this paper, we propose a Multi-Task Automated Assessment (MTAA) system that eliminates the need for feature engineering and evaluates essays across various dimensions of essay quality. Specifically, a multi-task learning (MTL) framework has been designed where Dynamic Learning Rate Decay (DLRD) has been employed to promote balanced training across different tasks, and Orthogonality Constraints (OC) [23] have been employed to facilitate the encoding of various facets of the inputs from the shared and task-specific networks. We evaluate our system on two public benchmarks, ELLIPSE [1] and ASAP++ [2], and new state-of-the-art results have been achieved. In addition, we engage Chat-GPT, which has recently been used for automatic scoring [26,27], in a comparative evaluation under both zero-shot and few-shot conditions. The results show that ChatGPT has not yet reached the scoring consistency of our developed MTAA system and that of human raters.

2 Multi-task Automated Assessment

2.1 Architecture Design

MTAA, as shown in Fig. 1, is a model with hard parameter sharing [24] that utilizes a backbone as a shared encoder to optimize multiple tasks and task-specific decoders to perform predictions. The shared encoder of the model consists of a pre-trained base version of DeBERTaV3 [25] and a mean pooling layer with the former capturing intricate information from the inputs and the latter extracting compressed shared representations. The task-specific encoder is structured with various branches, each one consisting of two densely connected layers for a task. These layers are utilized to extract knowledge from shared representations and to evaluate an essay across various dimensions. OC and DLRD are integrated into the MTL model to extract specific task-related information and promote faster training convergence.

2.2 Orthogonality Constraints

To encourage the shared and task-specific encoders to encode diverse information facets of tasks within the MTL framework, we incorporate Orthogonality Constraints (OC), as introduced in [20,23], into our MTAA system:

$$\ell_{oc}^k = \sum_{k=1}^{m} \left\| \mathbf{S}^\top \mathbf{P}^k \right\|_F^2 \tag{1}$$

where $\| \cdot \|_F^2$ refers to the squared Frobenius norm. $\mathbf{S}$ and $\mathbf{P}^k$ are two matrices, whose rows are the shared and task-specific representations, as shown in Fig. 1.

2.3 Dynamic Learning Rate Decay

To ensure effective and balanced MTL learning, as well as preventing certain tasks from overpowering others during model optimization, we propose Dynamic Learning Rate Decay (DLRD) to adapt different learning rates for different tasks. Specifically, DLRD keeps a moderate learning rate for the shared encoder and assigns smaller learning rates to task-specific encoders exhibiting high learning speeds. The DLRD involves two steps:

1. Calculating task weights:

$$\omega^k(t) = \frac{r^k(t-1)^\alpha}{\bar{r}}, r^k(t-1) = \frac{\ell^k(t-2)}{\ell^k(t-1)} \tag{2}$$

where t is the index of training iteration. Exponent α serves as a factor for adjusting the magnitude of differences in task weights and a greater value of α (>1) amplifies the disparities among task weights. The average value $\bar{r}$ scales the weights to prevent dominance by tasks with higher loss descent rates.

2. Dynamical learning rate decay:

$$\eta^k = \eta^{\text{base}} \, \omega^k(t), \eta^{\text{base}} = \eta_0 \gamma^{\lfloor \frac{t}{n} \rfloor} \tag{3}$$

where γ represents the decay factor. Floor function $\lfloor \frac{t}{n} \rfloor$ represents the frequency of learning rate decay, occurring every certain number of epochs. η^{base} is the learning rate for the shared network, and learning rate η^k is calculated for the task-specific network based on its task weight. In our implementation, we set $\eta_0 = 1e-5$, $\alpha = 10$, and $\gamma = 0.3$. Additionally, we set $n = 1$ to perform learning rate decay in every epoch.

3   Experiments

3.1   Datasets

Our system is developed using two multi-dimensional AA benchmark datasets. We further divide both datasets into 80% for training and 20% for testing.[3]

ELLIPSE [1] consists of 3,911 argumentative essays written by English language learners in grades 8–12, with each sample comprising a text and a corresponding score set representing Cohesion, Grammar, Vocabulary, Phraseology, Syntax, and Conventions levels.[4] Each essay was independently rated by two expert annotators using a five-point scoring rubric. Score discrepancies of two or more points were resolved through annotating team discussion. The average word count of the essays is 430, with most falling within 250 to 500 words. The scores range from 1.0 to 5.0 in increments of 0.5, where each of the six dimensions demonstrates an approximate normal distribution, with the mean at about 3.0. We also observe that the Pearson correlation coefficients among all six evaluation dimensions exceed 0.6, suggesting a substantial positive linear relationship.

ASAP++ [2] has been developed on top of ASAP [3], offering multi-dimensional scores for first six prompts. Prompts 1–2 assess argumentative essays on Content, Organization, Word Choice, Sentence Fluency, and Conventions. Prompts 3-6 evaluate source-dependent essays on Content, Prompt Adherence, Language, and Narrativity. For details on attribute definitions and essay statistics, please refer to the original paper.

3.2   Metrics

The performance of the models is assessed by a broad range of AA metrics, including the root mean square error (RMSE), Pearson correlation coefficient (PCC) and Spearman's rank correlation coefficient (SCC), and the Quadratic Weighted Kappa (QWK), all of which are widely adopted in AA [6,8,9,13–17,28–31]. We evaluate the model performance using a column-wise mean technique. Specifically, for each task, we compute metric scores based on actual-predicted pairs and then average these scores across all tasks/dimensions to obtain an overall assessment metric value.

3.3   Results

We compare our proposed MTAA system with several robust baselines, including BERT and RoBERTa, as well as MTL_vanilla, which maintains the same network structure as the MTAA but does not incorporate the OC and DLRD mechanisms.

---

[3] The train/test split can be found at https://github.com/Aries-chen/MTAA/blob/main/README.md.

[4] The ELLIPSE rubric is available at: https://docs.google.com/document/d/1OSbRELoWKlq8chYmujAaHJqMwFZnwt2PnnbSXfOJkIY/edit.

Table 1. Performance on ELLIPSE. The abbreviations are: Coh. for Cohesion, Syn. for Syntax, Voc. for Vocabulary, Phr. for Phraseology, Gra. for Grammar, and Con. for Conventions. Avg. represents the average scores across all dimensions. $ChatGPT_0$ and $ChatGPT_3$ denote the use of ChatGPT in zero-shot and few-shot settings, respectively. The best scores for each metric are highlighted in bold.

Datasets	Models	RMSE ↓							PCC ↑						
		Coh.	Syn.	Voc.	Phr.	Gra.	Con.	Avg.	Coh.	Syn.	Voc.	Phr.	Gra.	Con.	Avg.
ELLIPSE	BERT	.54	.47	.48	.47	.52	.48	.49	.60	.69	.65	.68	.65	.68	.66
	RoBERTa	.51	.46	.43	.45	.49	.46	.47	.65	.70	.68	.71	.72	.73	.70
	$MTL_{vanilla}$	.51	.45	.43	.44	.48	.46	.46	.66	.72	.69	.73	.73	.73	.71
	MTAA	.51	.45	.42	.44	.46	.44	**.45**	.66	.72	.70	.73	.74	.75	**.72**
$ELLIPSE_{Subset}$	$ChatGPT_0$	.89	.67	.76	.89	.88	.74	.80	.26	.64	.63	.50	.52	.62	.53
	$ChatGPT_3$	.81	.59	.58	.69	.81	.62	.68	.26	.63	.62	.58	.46	.56	.52
	MTAA	.58	.38	.36	.44	.50	.46	**.45**	.54	.72	.74	.70	.72	.68	**.68**

Datasets	Models	SCC ↑							QWK ↑						
		Coh.	Syn.	Voc.	Phr.	Gra.	Con.	Avg.	Coh.	Syn.	Voc.	Phr.	Gra.	Con.	Avg.
ELLIPSE	BERT	.57	.66	.63	.65	.63	.66	.63	.56	.64	.61	.65	.61	.63	.62
	RoBERTa	.62	.67	.65	.69	.71	.71	.67	.61	.66	.65	.67	.69	.71	.66
	$MTL_{vanilla}$	.63	.69	.67	.72	.72	.71	**.69**	.62	.69	.64	.69	.70	.70	.67
	MTAA	.63	.69	.67	.71	.73	.72	**.69**	.63	.69	.67	.69	.72	.71	**.68**
$ELLIPSE_{Subset}$	$ChatGPT_0$	.21	.59	.64	.49	.48	.61	.50	.12	.33	.28	.22	.32	.42	.29
	$ChatGPT_3$	.19	.58	.63	.55	.41	.53	.48	.25	.56	.56	.51	.43	.55	.48
	MTAA	.40	.59	.66	.70	.74	.64	**.62**	.49	.67	.68	.69	.69	.65	**.64**

Furthermore, we compare our proposed system with ChatGPT in both zero-shot and few-shot (i.e. 3-shot) manners.[5] Due to budget constraints, we utilized a representative subset of ELLIPSE that maintains the percentage of samples at each score level from the test set.[6]

Results on ELLIPSE are presented in Table 1. We can see that the proposed MTAA demonstrates superior performance compared to the baselines (i.e., BERT, RoBERTa, and $MTL_{vanilla}$) on the evaluated ELLIPSE dataset. Specifically, it achieves scores of 0.45 for RMSE, 0.72 for PCC, 0.69 for SCC, and 0.68 for QWK, setting new state-of-the-art performance. When evaluated on the subset, our MTAA model significantly outperforms ChatGPT in both zero-shot and few-shot settings. While these methods show similar performance in terms of PCC and SCC, the few-shot approach significantly excels over the zero-shot one for RMSE and QWK. We also notice that all models yield the worst performance on Cohesion compared to other dimensions.

---

[5] We used the GPT-4-0613 API. The prompts used in our experiments are available at https://github.com/Aries-chen/MTAA/blob/main/Few-shot_prompt.txt.

[6] This comparison is excluded from ASAP++ due to its lack of a clear evaluation rubric, making it difficult to provide precise prompts for ChatGPT inputs.

Table 2. Performance on ASAP++. The abbreviations are: Cont. for Content, Org. for Organization, WoCh. for Word Choice, SeFl. for Sentence Fluency, Conv. for Conventions, PrAd. for Prompt Adherence, Lang. for Language, and Narr. for Narrativity. Avg. represents the average scores across all dimensions. The best scores for each metric are highlighted in bold.

Metrics	Models	Argumentative essays						Source-dependent essays				
		Cont.	Org.	WoCh.	SeFl.	Conv.	Avg.	Cont.	PrAd.	Lang.	Narr.	Avg.
RMSE ↓	MTAA	.76	.73	.73	.68	.71	**.72**	.56	.57	.62	.58	**.58**
PCC ↑	MTAA	.76	.76	.75	.75	.74	**.75**	.84	.83	.80	.81	**.82**
SCC ↑	MTAA	.75	.75	.73	.74	.73	**.74**	.84	.83	.79	.80	**.82**
QWK ↑	MTAA	.72	.70	.70	.72	.70	**.71**	.80	.80	.74	.76	**.77**
	Ridley et al. (2020) [29]	.54	.41	.53	.54	.36	.48	.54	.57	.53	.61	.56
	Ridley et al. (2021) [30]	.56	.46	.56	.55	.41	.51	.56	.57	.54	.61	.57
	Chen & Li. (2023) [31]	.57	.48	.58	.58	.42	.53	.57	.58	.55	.61	.58

Results on **ASAP ++** are presented in Table 2. Again, our proposed MTAA system outperforms all its competitors [29–31] by a large margin when evaluated on QWK and yields new state of the art.[7]

4   Discussion

MTL is particularly well-suited for multi-dimensional AA tasks due to their ability to first extract shared information before exploiting task-specific information. This advantage stems from the fact that the assessment measures in these tasks are often related, but not necessarily identical. The benefits of our approach are further enhanced by the integration of OC and DLRD, which promote task-specific representations while effectively balancing learning speeds across tasks. Furthermore, the proposed design eliminates the need for manual task weight adjustment, thereby making the model more robust and generalizable.

Regarding the low performance in Cohesion compared to all the other dimensions, we speculate that the abstract and complex scoring measure poses significant challenges for AA systems. Making use of detailed and concrete features, e.g. 'reference and transitional words and phrases' (as outlined in the ELLIPSE Cohesion Rubric), might be beneficial.

The performance of ChatGPT in our multi-dimensional AA task is much lower than those reported in other NLP tasks. Upon analyzing the zero-shot outputs, we discovered that the scores across all measures were consistently lower than those provided by human raters. This observation suggests that ChatGPT acts as a more "stringent" assessor. However, when we provided it with three essay-score pairs for few-shot evaluation, ChatGPT became more "lenient" and the scores aligned more closely with those given by human raters. This is evidenced by the improvements in RMSE and QWK as shown in Table 1. Despite ChatGPT's underperformance in our task, it possesses unique strengths, such as

---

[7] Previous work has only reported QWK.

providing more specific feedback [26], including grammar corrections and word suggestions.

5   Conclusion

We proposed a MTAA system that supports multi-dimensional essay scoring. Specifically, we introduced OC to obtain more task-specific representations and designed DLRD to dynamically adjust the learning rates for the tasks to achieve balanced training. Our system achieves state of the art on ELLIPSE and ASAP++ public benchmarks. Additionally, we explored the potential of Chat-GPT in multi-dimensional AA and found that ChatGPT has not yet matched the consistency of our MTAA system or that of human raters. Our future research interests lie in investigating the performance of ChatGPT in providing multi-dimensional feedback, such as offering detailed and constructive suggestions in addition to scoring. We aim to develop a more comprehensive and effective AA system that not only assigns scores but also guides students towards improving their writing skills across multiple dimensions.

References

1. Franklin, A., et al.: Feedback prize - English language learning. Kaggle (2022). https://kaggle.com/competitions/feedback-prize-english-language-learning
2. Mathias, S., Bhattacharyya, P.: ASAP++: enriching the ASAP automated essay grading dataset with essay attribute scores. In: LREC (2018)
3. Hamner, B., Morgan, J., Vandev, L., Shermis, M., Vander Ark, T.: The Hewlett foundation: automated essay scoring. Kaggle (2012). https://kaggle.com/competitions/asap-aes
4. Ramineni, C., Trapani, C., Williamson, D., Davey, T., Bridgeman, B.: Evaluation of the e-rater® scoring engine for the TOEFL® independent and integrated prompts. ETS Research Report Series, Wiley Online Library, vol. 2012, no. 1, pp. i–51 (2012)
5. Ramineni, C., Trapani, C., Williamson, D., Davey, T., Bridgeman, B.: Evaluation of e-rater for the GRE issue and argument prompts. Educational Testing Service Princeton, NJ (2012)
6. Yannakoudakis, H., Briscoe, T., Medlock, B.: A new dataset and method for automatically grading ESOL texts. In: ACL-HLT, pp. 180–189 (2011)
7. Contreras, J.O., Hilles, S., Abubakar, Z.B.: Automated essay scoring with ontology based on text mining and NLTK tools. In: ICSCEE, pp. 1–6. IEEE (2018)
8. Kumar, Y., Aggarwal, S., Mahata, D., Shah, R.R., Kumaraguru, P., Zimmermann, R.: Get it scored using AutoSAS - an automated system for scoring short answers. CoRR, abs/2012.11243 (2020)
9. Phandi, P., Chai, K.M.A., Ng, H.T.: Flexible domain adaptation for automated essay scoring using correlated linear regression. In: EMNLP, pp. 431–439 (2015)
10. Pennington, J., Socher, R., Manning, C.: GloVe: global vectors for word representation. In: EMNLP, pp. 1532–1543 (2014)
11. Kim, Y.: Convolutional neural networks for sentence classification. CoRR, abs/1408.5882 (2014)

12. Hochreiter, S., Schmidhuber, J.: Long short-term memory. Neural Comput. **9**(8), 1735–1780 (1997)
13. Taghipour, K., Ng, H.T.: A neural approach to automated essay scoring. In: EMNLP, pp. 1882–1891 (2016)
14. Dong, F., Zhang, Y., Yang, J.: Attention-based recurrent convolutional neural network for automatic essay scoring. In: CoNLL, pp. 153–162 (2017)
15. Wang, Y., Wei, Z., Zhou, Y., Huang, X.: Automatic essay scoring incorporating rating schema via reinforcement learning. In: EMNLP, pp. 791–797 (2018)
16. Tay, Y., Phan, M., Tuan, L.A., Hui, S.C.: SkipFlow: incorporating neural coherence features for end-to-end automatic text scoring. In: AAAI, vol. 32, no. 1 (2018)
17. Rodriguez, P.U., Jafari, A., Ormerod, C.M.: Language models and automated essay scoring. CoRR, abs/1909.09482 (2019)
18. Vaswani, A., et al.: Attention is all you need. In: Advances in Neural Information Processing Systems, vol. 30. Curran Associates, Inc. (2017)
19. Andersen, Ø.E., Yuan, Z., Watson, R., Cheung, K.Y.F.: Benefits of alternative evaluation methods for automated essay scoring. Int. Educ. Data Mining Soc. (2021)
20. Zhang, A., et al.: On orthogonality constraints for transformers. In: ACL-IJCNLP 2021 (Vol. 2: Short Papers), pp. 375–382 (2021)
21. Yang, Z., Dai, Z., Yang, Y., Carbonell, J.G., Salakhutdinov, R., Le, Q.V.: XLNet: generalized autoregressive pretraining for language understanding. CoRR, abs/1906.08237 (2019)
22. Sanh, V., Debut, L., Chaumond, J., Wolf, T.: DistilBERT, a distilled version of BERT: smaller, faster, cheaper and lighter. CoRR, abs/1910.01108 (2019)
23. Bousmalis, K., Trigeorgis, G., Silberman, N., Krishnan, D., Erhan, D.: Domain separation networks. CoRR, abs/1608.06019 (2016)
24. Caruana, R.: Multitask learning. Mach. Learn. **28**, 41–75 (1997)
25. He, P., Gao, J., Chen, W.: DeBERTaV3: improving DeBERTa using ELECTRA-style pre-training with gradient-disentangled embedding sharing. CoRR, abs/2111.09543 (2021)
26. Yoon, S., Miszoglad, E., Pierce, L.R.: Evaluation of ChatGPT feedback on ELL writers' coherence and cohesion. arXiv preprint arXiv:2310.06505 (2023)
27. Wu, X., He, X., Liu, T., Liu, N., Zhai, X.: Matching exemplar as next sentence prediction (MeNSP): zero-shot prompt learning for automatic scoring in science education. In: Wang, N., Rebolledo-Mendez, G., Matsuda, N., Santos, O.C., Dimitrova, V. (eds.) AIED 2023. LNCS, vol. 13916, pp. 401–413. Springer, Cham (2023)
28. Ke, Z., Inamdar, H., Lin, H., Ng, V.: Give me more feedback II: annotating thesis strength and related attributes in student essays. In: ACL, pp. 3994–4004 (2019)
29. Ridley, R., He, L., Dai, X., Huang, S., Chen, J.: Prompt agnostic essay scorer: a domain generalization approach to cross-prompt automated essay scoring. CoRR, abs/2008.01441 (2020)
30. Ridley, R., He, L., Dai, X., Huang, S., Chen, J.: Automated cross-prompt scoring of essay traits. In: AAAI, vol. 35, no. 15, pp. 13745–13753 (2021)
31. Chen, Y., Li, X.: PMAES: prompt-mapping contrastive learning for cross-prompt automated essay scoring. In: ACL, pp. 1489–1503 (2023)

Content Knowledge Identification with Multi-agent Large Language Models (LLMs)

Kaiqi Yang[1], Yucheng Chu[1], Taylor Darwin[2], Ahreum Han[2], Hang Li[1], Hongzhi Wen[1], Yasemin Copur-Gencturk[2], Jiliang Tang[1], and Hui Liu[1(✉)]

[1] Michigan State University, East Lansing, MI 48824, USA
{kqyang,chuyuch2,lihang4,wenhongz,tangjili,liuhui7}@msu.edu
[2] University of Southern California, Los Angeles, CA 90089, USA
{tdarwin,ahreumha,copurgen}@usc.edu

Abstract. Teachers' mathematical content knowledge (CK) is of vital importance and need in teacher professional development (PD) programs. Computer-aided asynchronous PD systems are the most recent proposed PD techniques. However, current automatic CK identification methods face challenges such as diversity of user responses and scarcity of high-quality annotated data. To tackle these challenges, we propose a Multi-Agent LLMs-based framework, LLMAgent-CK, to assess the user responses' coverage of identified CK learning goals without human annotations. Leveraging multi-agent LLMs with strong generalization ability and human-like discussions, our proposed LLMAgent-CK presents promising CK identifying performance on a real-world mathematical CK dataset MaCKT.

Keywords: Math Knowledge Development · Large Language Models · Multi-Agent Systems

1 Introduction

Professional Development (PD) keeps educators informed about current education trends and provides them with the necessary knowledge [5]. Teachers may enter the profession with different levels of preparation of subject-specific Content Knowledge (CK) [4]. Historically, PD for educators is primarily conducted with human experts via in-person or synchronous online meetings, which is costly and not available for *all* teachers such as those in rural areas [9]. In addition, the need for timely evaluation and feedback exacerbates the challenge, especially for the programs lacking real-time PD facilitators. These challenges have motivated computer-aided asynchronous PD as an alternative to traditional human-driven PD, with which teachers improve their PD equally with fewer concerns about costs and limitations of time or location [1].

For asynchronous PD programs, it's challenging to inspect teachers' mastered CK. Prior systems commonly use free-text questions to test teachers' CK and

A. M. Olney et al. (Eds.): AIED 2024, LNAI 14830, pp. 284–292, 2024.
https://doi.org/10.1007/978-3-031-64299-9_23

utilize rule-based systems and traditional machine learning methods to automatically identify teachers' CK[1] [4]. However, these methods have some limitations: static matching rules fail to discern the semantic meaning of the diversified text, while machine learning methods heavily rely on annotated training datasets, which are domain-specific and costly. This impedes their wide usage in education scenarios.

In this paper, we propose a multi-agent LLMs-based framework, LLMAgent-CK, to identify the user responses' coverage of CK learning goals. Leveraging the Large Language Model's (LLM) broad knowledge learned during pre-training, LLMAgent-CK has strength in generating reliable results without labeled data. By introducing discussions between multiple LLMs, LLMAgent-CK eliminates the potential bias of results from a single model and provides a better alignment with human expertise feedback. At last, since LLMs are generative models, LLMAgent-CK provides both identified results and the reasons for them. We conduct comprehensive experiments with a real-world mathematical CK dataset MaCKT. The experiment showcases LLMAgent-CK can achieve up to 95.83% precision, comparable with the human annotators.

2 Related Works

Automatic Short Answer Grading (ASAG) [22] is a well-established research topic focusing on scoring student answers. Early studies [11,19] perform word- and phrase-level pattern matching to grade. [20] represents both answer and reference text with TF-IDF features. With the rise of deep learning, text embedding techniques including word2vec [14] and BERT [7] have been widely adopted in studies to enhance the semantic representations of text [10]. At last, with the emergence of LLMs, recent works [15,18] have started to adopt LLM with prompt-tuning tricks for ASAG tasks.

Multi-agent LLMs [2,21] framework involves the fusion of multiple LLM-driven agents working cooperatively. The collaborative synergy results in more nuanced and effective solutions. [12] proposes a cooperative agent framework of role-playing, enabling agents to cooperate on complex tasks. [16] creates a sandbox environment consisting of virtual entities, every intelligent agent being capable of autonomously interacting in the environment. Besides collaborations, the multi-agent debate framework has also been explored to enhance the outputs. [8,13] use it for translation and arithmetic problems; [2] applies it to evaluate the quality of generated responses on open-ended questions and traditional natural language generation (NLG) tasks. In this work, we design a multi-agent framework to simulate the discussion and consensus between humans, generating expert-like feedback to the teacher's responses.

3 Problem Statement

Given question text Q consisting of the main question and hints, teachers are asked to provide CK-related answers with free text T. To identify the learning

---

[1] A formal definition of content knowledge identification problem is provided in Sect. 3.

goals reached by $\mathcal{T}$, the system compares $\mathcal{T}$ with expert-designed learning goal list $\mathcal{E} = [e_1, e_2, \dots, e_M]$, where e_m denotes the m-th goal. The output of the identification system will be binary responses $O = [o_1, o_2, \dots, o_M]$, where $o_m \in \{0, 1\}$ indicates whether e_m is reached by teacher's response.

4 Method

In this section, we introduce our multi-agent LLMs conversation framework LLMAgent-CK which can identify the CK learning goals in teachers' responses. We first give overview of the framework. Then, we detail the major designs of the framework. At last, we present three implementations of the framework simulating the common discussion of humans.

4.1 An Overview

An overview of LLMAgent-CK is shown in Fig. 1. It consists of three types of LLM-powered agents (i.e. *Administrator, Judger, Critic*) and two controlling strategies for these agents (i.e. discussion strategy and decision strategy). *Administrator* agent pre-processes and passes the input data into the framework; a group of *Judger* agents (Num. of *Judgers* = N) makes their answers; the *Critic* agent makes up the final outputs based on their answers or proposes new conversations if needed. For the controlling strategies, discussion strategies can control the ways of message sharing between agents, while decision strategies guide the *Critic* agent to make final decisions and manage the new conversation.

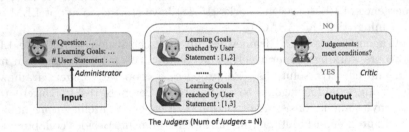

Fig. 1. An overview of LLMAgent-CK. The blue arrows indicate the **discussion strategy** for message-passing, and the red arrows indicate the **decision strategy** for decision-making. (Color figure online)

4.2 Role Design

Administrator Agent. A_A is built on LLMs that preprocess the input data and broadcast it to other agents. The input learning goals $\mathcal{E}$ are provided by human experts, which could be sub-optimal for LLMs to understand. Therefore, we assign the ability of rephrasing [6] to A_A. Then it broadcasts all the input data (the question $\mathcal{Q}$, rephrased learning goals $\mathcal{E}$, and the user's responses $\mathcal{T}$) to the other agents.

Judger Agent. A_J identifies the learning goals reached by user responses. Given information from A_A, A_J examines the user responses and identifies which goal is correctly reached. Finally, it outputs the set of learning goals reached by the user responses, as well as explanatory evidence for prediction. It is optional to broadcast the replies of one A_J to other A_Js dependent on the discussion strategy. If the broadcast is allowed, the input prompt goes with outputs from other A_J.

Critic Agent A_C. Given the input Q and $\mathcal{E}$, A_C checks if the outputs from all A_J reach an agreement with decision strategies; otherwise, A_C summarizes the reason for disagreement and proposes new discussions.

4.3 Controlling Strategy Design

Discussion Strategy determines how the intermediate outputs of agents are shared with other agents. We define a strategy *Open Judge*, where A_J are aware of the outputs from other A_Js. It simulates the cooperation scenarios where there are open discussions between people. Conversely, the strategy *Closed Judge* turns off the answers sharing between A_Js, which imitates scenarios where people make decisions independently.

Decision Strategy leverages and ensemble group thinking of diverse agents. We design *Total Agreement* where consensus is made when all the A_J have the same outputs. The second strategy is *Majority Voting*: A_C makes final decisions based on the majority consensus of A_Js. If there is no consensus, A_C summarizes the outputs and proposes new conversations. Decision-making strategies simulate the discussion process.

4.4 Practical Implementations

We demonstrate 3 practical implementations deriving from the LLMAgent-CK framework as Fig. 2. With different numbers of agents and controlling strategies, we can design new implementations to simulate different cooperation styles of humans.

LLMAgent-CK-Single is a naive implementation simulating the scenario where people make decisions individually, thus the number of A_J is $N = 1$. Without discussion and decision, this implementation checks the ability of a single LLM and serves as a baseline model for other LLM-based implementations.

LLMAgent-CK-Discuss includes two A_Js and allows them to communicate when making decisions. It simulates a public discussion where all the participants are aware of others' opinions. The controlling strategies are *Open Judge* and *Total Agreement*.

LLMAgent-CK-Vote deploys controlling strategies *Closed Judge* and *Majority Voting*. Above five independent A_J, A_C decides on agreements through majority voting. It simulates real-world majority voting activities based on independent individual decisions.

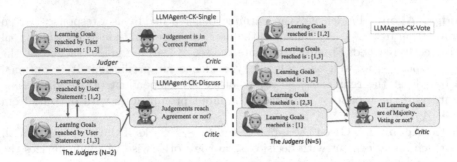

Fig. 2. Illustration of 3 implementations: LLMAgent-CK-Single (left-top), LLMAgent-CK-Discuss(left-bottom) and LLMAgent-CK-Vote(right). In the figure, the discussion strategies are indicated by blue arrows. (Color figure online)

5   Experiment

In this section, we illustrate the experiments of the proposed framework. We introduce the experimental settings and the results of three implementations.

5.1   Experimental Setting

Dataset: The data in this study were collected during an AI-based professional development program for middle school mathematics teachers [3], which was open-text responses from participating teachers when they were completing activities. The majority of teachers were female (81%) and white (69%), similar to the demographic profile of the U.S. teaching workforce. We construct a dataset MaCKT based on the data with selected topics of math CK. There are domain experts to discuss and annotate the ground-truth labels of reached goals. Below is one example:

– **Questions** *"6 workers can paint one house in 7 h. Is the relationship between the number of workers and time to paint the house proportional?"*.
– **Learning Goals**: *"(1)as the number of workers changes, the time it takes also changes"*, *"(2)the amount of time it takes is not constant..."*, *"(3)the relationship is not proportional"*.
– **User Response**: *"...for 6 workers it would take 7 h, for 12 workers it would take 3.5 h."*.
– **Expert Annotation**: *[1,2]*.

We sample 9 questions from MaCKT (including 452 user responses) denoted as *Q1, Q2, ..., Q9*. The ground-truth labels are only utilized for evaluation purposes rather than model training.

Metrics: With binary responses $O = [o_1, o_2, \ldots, o_M]$, we define metrics as below. Given a question having J user responses $\mathcal{T}$ with the true output O_j and prediction $\hat{O}_j$:

$$\text{Question-level Recall} = \frac{\sum_j^J |O_j \cap \hat{O}_j|}{\sum_j^J |O_j|}; \text{Question-level Precision} = \frac{\sum_j^J |O_j \cap \hat{O}_j|}{\sum_j^J |\hat{O}_j|};$$

$$\text{Question-level F1 Score} = 2 * \frac{\text{Question-level Recall} * \text{Question-level Precision}}{\text{Question-level Recall} + \text{Question-level Precision}};$$

where $|O_j \cap \hat{O}_j|$ is the number of goals correctly identified by models.

Baselines: Note that our framework follows the zero-shot setting without using ground truth labels for training, in which most machine learning methods are not applicable. We define two semantic matching baselines: with embeddings of questions and responses from pre-trained language models, we calculate the cosine similarities. A match is claimed when the similarity is larger than a pre-defined threshold obtained by grid-search with the best performance. We select the BERT [7] and Sentence BERT (S_BERT) [17] models.

Implementation Details: For LLMAgent-CK, we choose GPT-3.5 Turbo and GPT-4 Turbo as the backbone LLMs. To set the agent roles, we select the temperature from 0 to 0.7. We follow the zero-shot (no labeled data) setting, and no annotations are provided to the models.

5.2 Results

Single-Agent Baseline Comparison. We first show the performance of the baselines and LLMAgent-CK-Single. As shown in Table 1, both two state-of-the-art GPT models achieve superior performances for all 9 questions, especially given the agreement across education experts is around 85% to 90%. The performances of semantic matching baselines are consistently poorer than GPT models. Even though the Question-level Recall of baselines appears high, there is a significant discrepancy between high recall scores and lower precision scores. This suggests the baselines tend to indiscriminately classify learning goals as being reached, suggesting a limited capacity due to the complexity of the input data.

Multi-agent Comparison. Based on our prior experiment, we select GPT-4 Turbo as the LLM backbone. Without loss of generality, we select 4 of the 9 questions[2] from the dataset, i.e. $Q1$, $Q4$, $Q7$, and $Q8$. With the comparison in this subsection, we find that the multi-agent frameworks further improve the performance of LLMs, even when the individual model works poorly. Besides, the voting mechanism leads to even better results than the discussion.

In Fig. 3, we show the results of LLMAgent-CK-Discuss and LLMAgent-CK-Vote. With multi-agent discussion, most of the performances improved compared with the single-agent setting. For example, for the question $Q4$, the three metrics increase from 54.35%, 75.75%, 63.29% (LLMAgent-CK-Single with GPT-4

---

[2] Note that the selected question has the lowest Question-level Recall scores with GPT-4 Turbo under LLMAgent-CK-Single. We made such a selection to check the effects of multi-agent frameworks on the hard problems for LLMs.

Table 1. Performance of the baselines and LLMAgent-CK-Single. GPT v3.5 and GPT v4 stand for the backbones of LLMAgent-CK-Single. Best performances are marked with **bold font**.

Metric	Model	Q1	Q2	Q3	Q4	Q5	Q6	Q7	Q8	Q9	Ave
Question-level Recall	BERT	**95.06**	**100**	**100**	**95.65**	**94.37**	**100**	**92.31**	**97.22**	**98.15**	**96.97**
	S_BERT	87.65	87.50	88.31	97.83	87.32	95.65	100	87.50	90.74	91.39
	GPT v3.5	75.31	87.01	78.57	80.43	84.51	39.13	67.31	63.89	75.93	72.45
	GPT v4	64.20	87.01	80.36	54.35	64.79	86.96	65.38	61.11	62.96	69.68
Question-level Precision	BERT	47.53	29.63	53.47	45.83	44.97	52.27	48.98	52.63	24.65	44.44
	S_BERT	48.30	32.67	55.28	46.88	46.97	52.38	48.15	52.07	25.79	45.39
	GPT v3.5	57.55	70.53	41.12	63.79	61.86	40.91	67.31	69.70	41.00	57.09
	GPT v4	**81.25**	**80.72**	**86.53**	**75.75**	**95.83**	**90.9**	**89.47**	**89.79**	**62.96**	**83.69**
Question-level F1	BERT	63.37	45.71	69.68	61.97	60.91	68.66	64.00	68.29	39.41	60.22
	S_BERT	62.28	47.57	68.00	**63.38**	61.08	67.69	65.00	65.28	40.16	60.05
	GPT v3.5	65.24	77.91	53.99	71.15	71.43	40.00	67.31	66.67	53.25	62.99
	GPT v4	**71.73**	**83.75**	**83.33**	63.29	**77.31**	**88.89**	**75.55**	**72.72**	**62.96**	**75.50**

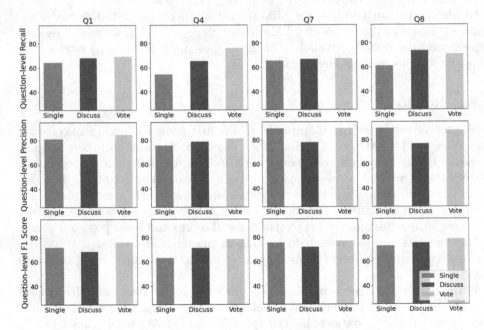

Fig. 3. Performance of LLMAgent-CK-Discuss (green) and LLMAgent-CK-Vote (blue). We also show the results of LLMAgent-CK-Single (red) as the reference. (Color figure online)

Turbo) to 65.22%, 78.95%, and 71.43%. In most cases, the multi-agent frameworks help to remedy the limitation of the single LLM. The LLMAgent-CK-Vote achieves higher scores on almost all four selected questions. Especially, the Question-level F1 Score of four questions improves by +7.87%, +7.23%, +5.01%, +3.28% compared with LLMAgent-CK-Discuss. As Question-level F1

Score indicates a more balanced view of the performance, this indicates that the LLMAgent-CK-Vote design is more powerful than LLMAgent-CK-Discuss.

6 Conclusion

We propose a Multi-Agent LLMs-based framework, LLMAgent-CK, to solve the challenging CK identification problem. With comprehensive experiments on a mathematical CK dataset MaCKT, we demonstrate its capabilities in generating promising identification results without the need for annotations. For future research, we aim to explore a broader range of communication strategies to improve its overall performance and efficiency. Moreover, we plan to employ advanced fine-tuning techniques to elevate the framework's effectiveness.

Acknowledgments. This material is based upon work supported by the National Science Foundation under Grant No. (NSF 2234015) and Institute of Education Sciences (R305A180392). Any opinions, findings, conclusions, or recommendations expressed in this material are those of the author(s) and do not necessarily reflect the views of the National Science Foundation and Institute of Education Sciences.

References

1. Burns, M., et al.: Barriers and supports for technology integration: views from teachers. UNESCO Global Monitoring Report (2023)
2. Chan, C.M., et al.: ChatEval: towards better LLM-based evaluators through multi-agent debate. arXiv preprint arXiv:2308.07201 (2023)
3. Copur-Gencturk, Y., Li, J., Cohen, A.S., Orrill, C.H.: The impact of an interactive, personalized computer-based teacher professional development program on student performance: a randomized controlled trial. Comput. Educ. (2024)
4. Copur-Gencturk, Y., Orrill, C.H.: A promising approach to scaling up professional development: intelligent, interactive, virtual professional development with just-in-time feedback. J. Math. Teach. Educ. (2023)
5. Copur-Gencturk, Y., Papakonstantinou, A.: Sustainable changes in teacher practices: a longitudinal analysis of the classroom practices of high school mathematics teachers. J. Math. Teach. Educ. (2016)
6. Deng, Y., et al.: Rephrase and respond: let large language models ask better questions for themselves. arXiv preprint arXiv:2311.04205 (2023)
7. Devlin, J., et al.: BERT: pre-training of deep bidirectional transformers for language understanding. arXiv preprint arXiv:1810.04805 (2018)
8. Du, Y., et al.: Improving factuality and reasoning in language models through multiagent debate. arXiv preprint arXiv:2305.14325 (2023)
9. Esquibel, J.S., Darwin, T.: The teacher talent pipelines: a systematic literature review of rural teacher education in the virtual age. In: Handbook of Research on Advancing Teaching and Teacher Education in the Context of a Virtual Age (2023)
10. Gomaa, W.H., Fahmy, A.A.: Ans2vec: a scoring system for short answers. In: Hassanien, A.E., Azar, A.T., Gaber, T., Bhatnagar, R., F. Tolba, M. (eds.) AMLTA 2019. AISC, vol. 921, pp. 586–595. Springer, Cham (2020). https://doi.org/10.1007/978-3-030-14118-9_59

11. Jordan, S.: Short-answer e-assessment questions: five years on. In: International Computer Assisted Assessment Conference (2012)
12. Li, G., et al.: CAMEL: communicative agents for "mind" exploration of large scale language model society. arXiv preprint arXiv:2303.17760 (2023)
13. Liang, T., et al.: Encouraging divergent thinking in large language models through multi-agent debate. arXiv preprint arXiv:2305.19118 (2023)
14. Mikolov, T., et al.: Efficient estimation of word representations in vector space. arXiv preprint arXiv:1301.3781 (2013)
15. Nicula, B., et al.: Automated assessment of comprehension strategies from self-explanations using LLMs. Information (2023)
16. Park, J.S., et al.: Generative agents: interactive simulacra of human behavior. In: ACM UIST (2023)
17. Reimers, N., Gurevych, I.: Sentence-BERT: sentence embeddings using Siamese BERT-networks. arXiv preprint arXiv:1908.10084 (2019)
18. Schneider, J., et al.: Towards LLM-based autograding for short textual answers. arXiv preprint arXiv:2309.11508 (2023)
19. Siddiqi, R., Harrison, C.: A systematic approach to the automated marking of short-answer questions. In: IEEE INMIC. IEEE (2008)
20. Wang, H.C., Chang, C.Y., Li, T.Y.: Assessing creative problem-solving with automated text grading. Comput. Educ. (2008)
21. Wu, Q., et al.: AutoGen: enabling next-gen LLM applications via multi-agent conversation framework. arXiv preprint arXiv:2308.08155 (2023)
22. Zhang, Y., Lin, C., Chi, M.: Going deeper: automatic short-answer grading by combining student and question models. UMUAI **30** (2020)

Using Knowledge Graphs to Improve Question Difficulty Estimation from Text

Enrico Gherardi[1], Luca Benedetto[2][(✉)] (iD), Maristella Matera[1] (iD),
and Paula Buttery[2] (iD)

[1] Politecnico di Milano, Milan, Italy
{enrico.gherardi,maristella.matera}@polimi.it
[2] ALTA Institute, University of Cambridge, Cambridge, UK
{luca.benedetto,paula.buttery}@cl.cam.ac.uk

Abstract. Question Difficulty Estimation (QDE) is a crucial task in many educational settings. Previous research focused on Natural Language Processing (NLP) to overcome the limitations of traditional QDE methods, but no work experimented the use of Knowledge Graphs (KGs) to provide a taxonomy of the topics assessed in exams. We propose two ways of incorporating KG information into existing models for QDE from text and, by experimenting on a publicly available dataset, show that they outperform the models that use text information exclusively, with a decrease in MAE of up to 8% with respect to the best-performing baseline (BERT-based QDE). We study how the models generalise to topics different from those used for training, and observe that while in most cases KGs are still capable of outperforming the baselines, a simpler model such as DistilBERT is more robust to previously unseen topics.

Keywords: Natural Language Processing · Question Difficulty Estimation · Knowledge Graph

1 Introduction

The estimation of the difficulty of exam questions is important for any educational setup: for instance, it enables to identify low quality questions or the ones that are unsuitable for a certain group of learners, as well as to perform custom recommendations. Traditionally, it is done with either manual calibration or pretesting [6]. In manual calibration, one (or more) subject expert assigns a numerical value representing the difficulty to each question, and this is inherently subjective. With pretesting, the new questions are administered to students in an actual test scenario and their difficulty is estimated based on the correctness of the students' answers (e.g., with Item Response Theory [10]), which accurately estimates the difficulty but introduces a significant delay between question generation and its usability for scoring students. Recent research addressed these issues by leveraging Natural Language Processing (NLP) techniques to perform QDE from Text (QDET), thus estimating question difficulty solely based on textual information [1,6]. However, none of these approaches leveraged the potential

A. M. Olney et al. (Eds.): AIED 2024, LNAI 14830, pp. 293–301, 2024.
https://doi.org/10.1007/978-3-031-64299-9_24

of Knowledge Graphs (KGs), which model the relationships between the various knowledge entities (i.e., topics) of the exam, and can be available during question creation. In this work, we fill this gap by enhancing previously proposed QDET models to use KGs for modelling question topics[1]. By experimenting on a publicly available dataset, we observe that this provides valuable additional information that improves the accuracy of existing text-based models. We also study the generalisation capabilities of the QDET models to questions of different topics[2] and observe that, in this case, the KGs do not provide a clear advantage with respect to previous text-only models. The contributions of this work[3] are twofold: i) we show that KGs can enhance previously proposed QDET models, and ii) we study the capabilities of QDET models (both with and without KGs) to generalise to questions about different topics.

2   Related Work

Whilst research on QDE has a fairly long history, it was only in recent years that NLP approaches gained popularity in this task. Recent surveys [1,6] on QDET highlighted a shift from the usage of features such as readability and word-complexity measures towards approaches that rely upon modern NLP techniques based on deep learning. This was confirmed in a recent study [2], which quantitatively compared the accuracy of several QDET models on two different datasets. The study found that the approach based on Transformers [17] proposed in [3] proved to be the most accurate both on maths questions and English reading comprehension questions (with BERT [8] slightly outperforming Distil-BERT [14]). As for other proposed approaches, the linguistic features (e.g., used in [4,7,11,16]) perform fairly well on reading comprehension questions, but are not effective on maths questions. On the contrary, word embeddings (such as word2vec [13]) used in [9], and features based on information retrieval (such as TF-IDF [12]) used in [5,15], are quite accurate on content knowledge assessment (e.g., maths and science), but not on reading comprehension questions. Still, all these models are outperformed by the Transformer-based models.

3   Proposed Models

We build upon the state of the art that uses only textual information, which are the Transformer-based approaches proposed in [3], and experiment with two approaches to leverage KGs. With **Stacking** (Sect. 3.1), we build one regressor based on the KG and another one based on the text of the questions. Then, we train a higher-level model that learns to weigh the outputs of the two regressors. With **Embedding Concatenation** (Sect. 3.2), we embed the information from the KGs into a vector, concatenate it with the textual embedding of the question, and train a regressor estimating the difficulty.

---

[1] Please note that we focus on difficulty as defined in Item Response Theory and not on question complexity as defined in other frameworks (e.g., Bloom's taxonomy).

[2] E.g., we train the model on *Geometry* questions and evaluate it on *Algebra* questions.

[3] Code is available at https://github.com/enricogherardi/BERT-KG-for-QDE.

3.1   Stacking

The proposed stacking technique, named **Average & BERT**, combines three key components to enhance predictive accuracy (see Fig. 1.a).

1) *Knowledge Graph Regressor*: it predicts question difficulty solely based on KG information. Given the specific structure of our KG (Subsect. 4.1), in which the topic-related information for a question can be summarized by its leaf node, we experimented with two approaches - i) embedding the leaf node using Node2Vec and ii) utilizing the average difficulty of leaf nodes. The latter yielded superior results, thus becoming our chosen method.
2) *Text Regressor*: it leverages a pretrained BERT model to derive a fixed-dimensional representation of the text of the question, and a linear layer to predict the difficulty from the final hidden state of the [CLS] token.
3) *Meta-Learner*: the two regressors specialise in handling distinct information types and the meta-learner – which consists of a single output neuron that operates with a linear activation function – acts as the final prediction layer. Thus, it synthesize the outputs from the *Text Regressor* and the *KG Regressor* and optimises the fusion of the predictions from the two base models, refining the final estimation of question difficulty.

3.2   Embedding Concatenation

The proposed approach, named **Node2Vec & BERT**, integrates textual information and KG representations by employing embedding concatenation. It encompasses four components (see Fig. 1.b).

1) *Text Contextual Embedding*: it generates contextual embeddings from the question text using a modified BERT model, which extracts the fixed-dimensional representation of the input sequence by capturing the final hidden state of the [CLS] token, thus capturing overall sequence information.
2) *Knowledge Graph Embedding*: leveraging a simplified KG representation (Subsect. 4.1) – where each question is associated with a leaf node – this component captures the graph information via the Node2Vec algorithm whose embedding represents the specific node (i.e., topic) associated with the question.
3) *Concatenation Layer*: it concatenates the two embeddings, creating a unified feature vector that preserves the information from both text and KG.
4) *Regression Layer*: it yields the final prediction by processing the concatenated embeddings to perform the regression task. The fusion of BERT-based textual embeddings with KG embeddings derived through Node2Vec provides the model with a comprehensive understanding of the question context and its associated topic.

4   Experimental Setup

4.1   Experimental Dataset

We experiment on the publicly available dataset released for the NeurIPS 2020 Education Challenge [18]; which is, to the best of our knowledge, the only public

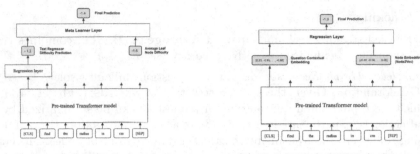

(a) Average & BERT Architecture.   (b) Node2Vec & BERT Architecture.

Fig. 1. Proposed Architectures.

dataset providing both the text of the questions and the graph information. This dataset is sourced from the Eedi platform (https://eedi.com/), and consists of responses to mathematics questions from students ranging from 7 to 18 years of age. The students' responses were sourced from September 2018 to May 2020. We work on three datasets collected from the publicly released data: i) the *Interactions*, ii) the *Questions*, and iii) the *Knowledge Graph* datasets.

Interactions contains the records of the students' answers to the questions, providing information such as a unique user ID, question ID, and the correctness of the answer. In total, it contains around 1,400,000 students' responses, with an average correctness of 53.7%. In total, there are 948 different questions and 4,918 different students; the average number of answers per student is 282, while the average number of answer per question is 1,459. This dataset is used to obtain, with an Item Response Theory (IRT) [10] model, the "true" difficulty of the questions, which will later be employed as the ground truth when training and evaluating the QDET models. Specifically, the estimated difficulty is distributed as a Gaussian in the range $[-5; 5]$, with mean close to 0.0.

Questions contains the set of questions which were administered to students. For each question, it provides an image containing the text of the question, the text of the answer options (all questions are Multiple-Choice Questions), and – if needed – a figure required to answer the question. To incorporate the textual information of the questions into our analysis, we apply Optical Character Recognition (OCR) techniques to extract from the images the question text, which is associated with the IRT difficulty estimated from *Interactions*.

Knowledge Graph provides the information about the topics which are assessed by the questions. Each question is associated with multiple nodes in the graph, with each node representing a distinct topic. Specifically, the KG of this dataset exhibits some peculiar characteristics: i) there is a single type of relationship between the nodes (child nodes represent sub-topics of their parent nodes), ii) the graph is undirected, connected, and acyclic, thus forming a tree structure, and iii) each question is associated with one and only one leaf node (and some non-leaf nodes). Therefore, the information pertaining to a specific question within the given topic can be summarized by its leaf node- its most

Table 1. MAE comparison for different models on the test set.

Model	MAE
Average	0.696
DistilBERT	0.519
BERT	0.515
Node2Vec & BERT	0.490
Average & BERT	**0.472**

specific topic. The tree has a root node (*Math*) and extends up to four levels of depth: the three main subjects (children nodes of *Math*) are *Numbers* (associated with 509 questions), *Geometry* (295 questions), and *Algebra* (144 questions); at level 3 there are 33 nodes and at level four there are 56 nodes.

4.2 Baselines

We consider two baselines: i) *average*, which assigns to each question the average difficulty of the training set questions, and ii) the two supervised transformer-based approaches proposed in [3], which consist in fine-tuning BERT and Distil-BERT to estimate question difficulty from question text[4]. To implement BERT and DistilBERT, we use the Hugging Face transformers library [19].

5 Results

5.1 Evaluation of Error in Difficulty Estimation

As the QDET is a regression task estimating a continuous numerical value representing question difficulty, we evaluate the models using Mean Absolute Error (MAE), a metric commonly used in previous research [6]. Table 1 compares the errors obtained for the proposed models and the baselines. All results are obtained on the test set (20% of the original dataset).

As expected, all transformer-based models significantly improve the average baseline, providing at least a 25% decrease in MAE. The fine-tuned models based on BERT and DistilBERT have comparable performance, confirming the results from [3] and the original DistilBERT paper [14]. Both models leveraging the KG outperform all the baselines: the reduction compared to the best performing baseline is 5% for Node2Vec & BERT and 8% for Average & BERT. This confirms that the information contained in the KG about the topics assessed by the questions is indeed useful to improve the accuracy of QDET models.

---

[4] In [3], the authors experiment with an additional Masked Language Modelling pre-training on a corpus of documents related to the same topics assessed by the questions, and observe that this improves the QDET accuracy of both BERT and DistilBERT. Unfortunately, a similar corpus is not available for the Eedi dataset.

Table 2. Evaluation of the generalisation capabilities of the QDET models, showing the MAE on both the topics they were trained on and the two other topics.

Training node	Model	Evaluation node (MAE)		
		Geometry	Numbers	Algebra
Geometry	BERT	0.499	**0.652**	0.501
	DistilBERT	0.526	0.662	**0.496**
	Node2Vec & BERT	0.496	0.655	0.503
	Average & BERT	**0.479**	0.656	0.507
Numbers	BERT	0.625	0.564	0.596
	DistilBERT	**0.575**	0.580	0.550
	Node2Vec & BERT	0.609	**0.559**	0.547
	Average & BERT	0.579	0.560	**0.497**
Algebra	BERT	0.556	0.697	0.483
	DistilBERT	**0.538**	0.694	0.492
	Node2Vec & BERT	0.553	0.680	0.502
	Average & BERT	0.548	**0.662**	**0.472**

Also, Average & BERT performs slightly better than Node2Vec & BERT, most likely due to the fact that the (internal) *Knowledge Graph Regressor* component used in Average & BERT, which estimates the difficulty by using the average difficulty of the specific leaf node, performs slightly better than the *Knowledge Graph Embedding* component used in Node2Vec & BERT (0.567 vs 0.608 MAE).

5.2   Evaluation of Generalisation Capabilities to New Topics

To investigate the capabilities of the proposed models to generalise to new topics, we train the QDE models on one topic at a time, and evaluate them on the other topics. For example, we train the models on *Numbers* and test them on *Geometry* and *Algebra*[5]. The results are shown in Table 2.

Considering the MAE on the same topic used for training, we can see the same hierarchy of performance as in the previous section, with both models leveraging KG information generally outperforming the others, and Average & BERT being overall the better performing. The difference between the MAE of the text-only Transformer-based models and two proposed models leveraging KG information is generally smaller than before, when considering the same topic for training and evaluation. Most likely, this is because by splitting the dataset into three smaller training datasets the available information is not enough to leverage to a full extent the additional information provided by the KG. In other words, the KG provides useful information, but this requires more training data.

---

[5] We evaluate the generalisation capabilities of the model only on topics which are relatively close to each other (i.e., three areas in maths), and we would expect a greater performance decrease in the case of completely unrelated topics.

Lastly, considering the generalisation capabilities to other topics, we can see that DistilBERT exhibits the best generalisation capabilities across the considered models (it is the best performing in three out of six topic combinations), and Average & BERT seems the second-best model in terms of generalisation capabilities (being the best performing in two out of six combinations).

6 Conclusions and Future Works

This work has proposed two approaches to incorporate information from Knowledge Graphs (KGs) into existing text-based QDET models. We integrate topic information from the leaf node of the KG either i) by using Node2Vec, a graph embedding algorithm, or ii) by averaging the difficulty of training questions associated with the same node. Then, we merge these outputs with a text-based model using BERT through i) Embedding Concatenation (Node2Vec) or ii) Stacking (Average). By experimenting on a publicly available dataset of maths questions, we show that both approaches outperform previous models, thus proving the usefulness of KG for QDET. Specifically, the Node2Vec & BERT approach leads to a reduction in MAE of 5%, while Average & BERT to a reduction in MAE of 8%. We also studied the capabilities of QDET models and the two proposed approaches leveraging KG information to generalise to questions of other topics, different from the ones used for training, and showed that although the models we propose are overall capable of more accurate predictions, the advantage is reduced when considering the generalisation capabilities to new topics, suggesting that they need to be separately trained on each topic. On the other hand, DistilBERT proved capable of impressive capabilities when evaluating its generalisation capabilities to new topics: although it is slightly worse than BERT (and significantly worse than the KG-based models) on seen topics, it exhibits better generalisation capabilities, and might therefore be preferred when model size needs to be taken into account on where enough data for training the more accurate KG models is not available.

Being this the first work experimenting with KGs for QDET, there are several avenues open for future research. One direction could explore graph neural networks for complex KGs or other more advanced KG embedding algorithms, as well as experimenting with other datasets to study whether our findings are consistent across datasets. Also, more work will be needed to understand whether there is a threshold value for the dataset size, such that for smaller datasets one should use text-only methods such as DistilBERT, and for larger datasets the more advanced (and capable) approaches proposed in this paper using KGs.

Acknowledgments. This research was partially funded by Cambridge University Press & Assessment.

References

1. AlKhuzaey, S., Grasso, F., Payne, T.R., Tamma, V.: Text-based question difficulty prediction: a systematic review of automatic approaches. Int. J. Artif. Intell. Educ., 1–53 (2023)
2. Benedetto, L.: A quantitative study of NLP approaches to question difficulty estimation. In: Wang, N., Rebolledo-Mendez, G., Dimitrova, V., Matsuda, N., Santos, O.C. (eds.) AIED 2023. CCIS, vol. 1831, pp. 428–434. Springer, Cham (2023). https://doi.org/10.1007/978-3-031-36336-8_67
3. Benedetto, L., Aradelli, G., Cremonesi, P., Cappelli, A., Giussani, A., Turrin, R.: On the application of transformers for estimating the difficulty of multiple-choice questions from text. In: Proceedings of the 16th Workshop on Innovative Use of NLP for Building Educational Applications, pp. 147–157 (2021)
4. Benedetto, L., Cappelli, A., Turrin, R., Cremonesi, P.: Introducing a framework to assess newly created questions with natural language processing. In: Bittencourt, I.I., Cukurova, M., Muldner, K., Luckin, R., Millán, E. (eds.) AIED 2020. LNCS (LNAI), vol. 12163, pp. 43–54. Springer, Cham (2020). https://doi.org/10.1007/978-3-030-52237-7_4
5. Benedetto, L., Cappelli, A., Turrin, R., Cremonesi, P.: R2DE: a NLP approach to estimating IRT parameters of newly generated questions. In: Proceedings of the 10th International Conference on Learning Analytics & Knowledge, pp. 412–421 (2020)
6. Benedetto, L., et al.: A survey on recent approaches to question difficulty estimation from text. ACM Comput. Surv. (CSUR) (2022)
7. Culligan, B.: A comparison of three test formats to assess word difficulty. Lang. Test. **32**(4), 503–520 (2015)
8. Devlin, J., Chang, M.W., Lee, K., Toutanova, K.: BERT: pre-training of deep bidirectional transformers for language understanding. In: Proceedings of the Conference of the North American Chapter of the Association for Computational Linguistics: Human Language Technologies, Volume 1 (Long and Short Papers), pp. 4171–4186 (2019)
9. Ehara, Y.: Building an English vocabulary knowledge dataset of Japanese English-as-a-second-language learners using crowdsourcing. In: Proceedings of the Eleventh International Conference on Language Resources and Evaluation (2018)
10. Hambleton, R.K., Swaminathan, H.: Item Response Theory: Principles and Applications. Springer, Dordrecht (2013). https://doi.org/10.1007/978-94-017-1988-9
11. Hou, J., Maximilian, K., Quecedo, J.M.H., Stoyanova, N., Yangarber, R.: Modeling language learning using specialized Elo rating. In: Proceedings of the 14th Workshop on Innovative Use of NLP for Building Educational Applications, pp. 494–506 (2019)
12. Manning, C.D.: Introduction to Information Retrieval. Syngress Publishing (2008)
13. Mikolov, T., Sutskever, I., Chen, K., Corrado, G.S., Dean, J.: Distributed representations of words and phrases and their compositionality. In: Advances in Neural Information Processing Systems, vol. 26 (2013)
14. Sanh, V., Debut, L., Chaumond, J., Wolf, T.: DistilBERT, a distilled version of BERT: smaller, faster, cheaper and lighter. arXiv preprint arXiv:1910.01108 (2019)
15. Settles, B.T., LaFlair, G., Hagiwara, M.: Machine learning–driven language assessment. Trans. Assoc. Comput. Linguist. **8**, 247–263 (2020)
16. Trace, J., Brown, J.D., Janssen, G., Kozhevnikova, L.: Determining cloze item difficulty from item and passage characteristics across different learner backgrounds. Lang. Test. **34**(2), 151–174 (2017)

17. Vaswani, A., et al.: Attention is all you need. In: NIPS (2017)
18. Wang, Z., et al.: Instructions and guide for diagnostic questions: the NeurIPS 2020 education challenge. arXiv preprint arXiv:2007.12061 (2020)
19. Wolf, T., et al.: Transformers: state-of-the-art natural language processing. In: Proceedings of the 2020 Conference on Empirical Methods in Natural Language Processing: System Demonstrations, pp. 38–45 (2020)

The Role of First Language in Automated Essay Grading for Second Language Writing

Haerim Hwang(✉) (iD)

The Chinese University of Hong Kong, Hong Kong, China
haerimhwang@cuhk.edu.hk

Abstract. The advancement of AI has paved the way for human-like performance in numerous tasks. One field that has seen significant benefits from its application is the automated essay grading (AEG). Recent AEG systems have demonstrated impressive accuracy in evaluating essay quality. However, the majority of AEG systems have been developed with a primary focus on essays produced in a first language (L1) context although there are countless second language learners around the world who need their writing skills assessed for education or evaluation reasons. More importantly, most AEG systems for L2 essays do not consider the impact of L1 which is crucial in shaping linguistic features within them, possibly limiting their applicability to some essays written by learners whose L1 is not given adequate consideration. This, in turn, may bring up concerns about the equal representation and inclusion of essays from learners with various L1 backgrounds. To investigate this possibility, we developed 11 AEG systems using the XGBoost algorithm, each based on the essay data of a specific L1 group (Arabic, Chinese, French, German, Hindi, Italian, Japanese, Korean, Spanish, Telugu, Turkish), and compared these systems' performance. Furthermore, we ascertained the relative importance of linguistic features in each system. Our study showed that the AEG systems provided with essay data from different L1 groups varied in accuracy, with slightly different linguistic features being the most crucial in the system. This suggests the necessity to include more diverse and inclusive data from learners with different L1 backgrounds in the development of AEG systems.

Keywords: Automated Essay Grading · XGBoost · First Language · Diversity

1 Introduction

1.1 Automated Essay Grading

Automated Essay Grading (AEG) is computational technology that automatically grades written products. AEG holds significance in writing education due to its effectiveness in providing prompt and objective feedback, with the potential to aid in the development of students' or language learners' writing abilities [1]. Its application can also help teachers and raters save time and minimize subjectivity in essay assessments [2]. For these reasons, it has received great attention in applied linguistics, computer science, and education (for a review, see [3]). Research on AEG began in 1966 with the introduction

© The Author(s), under exclusive license to Springer Nature Switzerland AG 2024
A. M. Olney et al. (Eds.): AIED 2024, LNAI 14830, pp. 302–310, 2024.
https://doi.org/10.1007/978-3-031-64299-9_25

of Project Essay Grader created by Ajay and his team [4], which grades essays based on diction, fluency, punctuation, grammar, text length, etc. Most AEG systems developed later in the 1990s adopted pattern matching methods and statistical-based approaches. More recently, AEG systems have started to use natural language processing (NLP) and machine/deep learning techniques supplemented with a variety of content-based and/or linguistically-motivated features, achieving remarkable accuracy in evaluating essays. However, most AEG systems have largely centered on essays produced in a first language (L1), despite the fact that there is a large population of L2 learners who need assessment of their writing skills for educational or evaluation purposes. This has resulted in limited consideration for the application of AEG systems in an L2 context. Furthermore, the existing research in this area has predominantly explored the architecture or input features for the AEG system, without taking individual factors into account, such as the L2 learners' L1 that can significantly impact the writing product (see Sect. 1.2). This study, therefore, aimed to examine the possible influence of L1 on the accuracy of AEG systems by developing 11 systems based on essay data from different L1 groups and comparing these systems' accuracy.

1.2 Effects of First Language on Second Language Writing

One critical aspect to consider when assessing L2 essays is the role of the learner's L1. In fact, numerous studies on L2 production suggest that L1 significantly influences how learners engage with their L2. This was demonstrated by Jarvis, Castañeda-Jiménez, and Nielsen [5], who analyzed written narratives from L2 learners of English with five different L1 backgrounds (Danish, Finnish, Portuguese, Spanish, Swedish). They found that lexical characteristics could predict the learners' L1s. For example, L1-Finnish L2 learners of English showed a higher frequency of lexical nouns but a lower frequency of 3rd person pronouns (e.g., *he, she*), which was attributed to the nonexistence of gendered 3rd person pronouns in Finnish. Kyle, Crossley, and Kim [6] further explored this topic while including L2 proficiency as an additional factor in their analysis of TOEFL essays written by five groups of L2 learners with different L1s (Chinese, German, Hindi, Korean, Spanish). Their findings highlighted that each L1 group made somewhat systematic choices in terms of lexical and phrasal elements although this pattern was more pronounced in lower-proficiency learners than in higher-proficiency learners. For example, L1-Korean L2 learners of English did not use as many indefinite articles as other L1 groups, a pattern that could potentially be ascribed to the absence of such phenomena in their L1. In a similar vein, Paquot [7] found that learners' L1 can impact their use of lexical bundles in their L2 English. Specifically, she identified various properties of French words that contribute to this impact, such as their collocational patterns, lexico-grammatical patterns, and frequency of use. One example provided is the English lexical bundle *to be found*, which was used more frequently by L1-French L2 learners of English than learners from other L1 backgrounds, which may be due to the fact that French has a similar frame (i.e., *être + à trouver*).

In sum, previous findings highlight the significance of L1 in shaping linguistic features in L2 writing, yet existing AEG systems for L2 writing have failed to consider this factor. Consequently, these systems may not fully recognize the influence of crucial linguistic features that may vary by the learner's L1. If an AEG system relies on an exclusive dataset or does not use a balanced dataset that reflects the wide array of learners' L1s, it could lead to underrepresentation of the global learner population. This further raises concerns about the system's ability to accommodate diverse linguistic backgrounds and promote inclusivity. This study aimed to explore this potential by developing numerous AEG systems, each tailored to a specific L1 group, and comparing their performance. Furthermore, we examined the relative importance of linguistically-motivated features in each system. In the long run, considering the impact of the learner's L1 will enable AEG systems to provide more accurate grades and more meaningful feedback for L2 learners from diverse L1 backgrounds.

2 Method

2.1 Corpus

The Educational Testing Service (ETS) Corpus of Non-Native Written English [8], also referred to as the TOEFL11 corpus, was the dataset used in this research. This corpus includes 12,100 English essays, written by L2 learners of English who took the TOEFL test in 2006 and 2007 on eight different prompts, all asking to write if they either agreed or disagreed with a given statement (e.g., *It is better to have broad knowledge of many academic subjects than to specialize in one specific subject*). Every essay in the corpus was rated by at least two human raters on a 5-point scale and then given an average score, which was finally collapsed to a level of either Low ($n = 1330$), Medium ($n = 6568$), or High ($n = 4202$). We opted for this corpus for the current study due to its relatively large collection of essays composed by L2 learners of English with 11 different L1s, balanced in their numbers (1,000 essays per each L1 group). These L1s encompassed Arabic (Low: $n = 296$; Medium: $n = 605$; High: $n = 199$), Chinese (Low: $n = 98$; Medium: $n = 727$; High: $n = 275$), French (Low: $n = 63$; Medium: $n = 577$; High: $n = 460$), German (Low: $n = 15$; Medium: $n = 412$; High: $n = 673$), Hindi (Low: $n = 29$; Medium: $n = 429$; High: $n = 642$), Italian (Low: $n = 164$; Medium: $n = 623$; High: $n = 313$), Japanese (Low: $n = 233$; Medium: $n = 679$; High: $n = 188$), Korean (Low: $n = 169$; Medium: $n = 678$; High: $n = 253$), Spanish (Low: $n = 79$; Medium: $n = 563$; High: $n = 458$), Turkish (Low: $n = 90$; Medium: $n = 616$; High: $n = 394$), as well as Telugu (Low: $n = 94$; Medium: $n = 659$; High: $n = 347$), which is less commonly studied as a learner's L1 in the context of L2 production. The use of this corpus is expected to enhance the diversity, inclusivity, and representativeness of the sample.

2.2 Features

We incorporated the domain knowledge and expertise of applied linguistics in selecting 48 theoretically-motivated features that have been found to be reliable predictors of essay quality, as summarized in Table 1. Within a theoretical framework that considers complexity, accuracy, and fluency as separate constructs for evaluating learner production [9], there has been extensive L2 production research which has revealed features that can effectively predict the holistic quality of learner essays. Drawing from influential studies in the field, we opted to utilize 43 features for complexity, including 14 lexical features (three for diversity [10] and 11 for sophistication [11]), 15 syntactic features (10 for global complexity [12] and 5 for construction-based complexity [13, 14]), and 14 semantic cohesion features [15]. Regarding accuracy, we chose the spell-error ratio feature [16] for a practical reason, although we admit that this sole measure cannot fully capture linguistic accuracy (for a further discussion, see Sect. 4). Additionally, we included four fluency features (e.g., [17]) that are widely employed in AEG research.

Lexical diversity, accuracy, and three fluency features (character count, word count, sentence count) were measured in Python using lexical richness [18], language-tool-python [19], and NLTK [20], respectively. All other features were assessed using publicly-accessible tools. Lexical sophistication features were determined using the Tool for the Automatic Analysis of Lexical Sophistication [11]. While constructional diversity as well as clause count was measured using the Constructional Diversity Analyzer [13], the other syntactic features were analyzed using the Tool for the Automatic Analysis of Syntactic Sophistication and Complexity [14]. Lastly, semantic cohesion features were evaluated using the Tool for the Automatic Analysis of Text Cohesion [15].

Table 1. Theoretically-motivated features selected for the Automated Essay Grading system

Component			Complexity			Accuracy	Fluency
Subcomponent	Lexical diversity	Lexical sophistication	Global syntactic complexity	Construction-based syntactic complexity	Semantic cohesion		
	Word type frequency	Phonological neighbors (including homonyms)	Mean length of clause	Average delta p score verb (cue)-	Adjacent overlap two paragraphs all lemmas average	Spell-error ratio	Character count
	Moving average	Lexical-decision time (z-score)	Mean length of sentence	construction (outcome) (types only) – academic	Adjacent overlap two paragraphs noun lemma average		Word count
	type-token ratio	MRC (Medical Research Council Psycholinguistics Database) familiarity (all words)	Mean length of T-unit	Average lemma construction	Adjacent overlap two paragraphs argument lemma average		Clause count
	Measure of textual lexical diversity		Clauses per sentence	frequency (types only) – academic	Adjacent overlap binary two paragraphs verb lemma average		Sentence count
		Brown frequency (all words)	Dependent clauses per clause	Average faith score construction (cue)-	Adjacent overlap binary two paragraphs adverb lemma average		
		COCA fiction bigram association strength (mutual information)	Dependent clauses per T-unit	verb (outcome) (types only)	Adjacent overlap binary two paragraphs adjective lemma average		
		Orthographic neighborhood frequency	Coordinate phrases per clause	Collostruction ratio (types only) – academic	Noun synonym paragraph lemma overlap		
		COCA academic bigram association strength (approximate collexeme strength score)	Coordinate phrases per T-unit	Constructional diversity	Adjacent overlap two paragraphs pronoun lemma overlap		
		COCA fiction trigram unigram to bigram association strength (mutual information squared)	Complex nominals per clause		Verb synonym paragraph lemma overlap		
		COCA academic trigram unigram to bigram association strength (t score)	Complex nominals per T-unit		Adjacent overlap sentence verb lemma		
		COCA spoken trigram bigram to unigram association strength (approximate collexeme strength score)			Ratio of pronouns to nouns		
		SUBTLEXus frequency (content words) logarithm			Repeated content word lemmas		
					Adjacent overlap sentence content words lemma		
					Verb synonym sentence lemma overlap		

2.3   Architecture of Automated Essay Grading System

XGBoost (eXtreme Gradient Boost; [21]) is a powerful machine learning algorithm, incorporating both sequential and parallel architectures all in one. It uses an additive strategy to learn data in sequence by adding one new tree at a time. At the same time, XGBoost employs parallel computation to enhance the efficiency of computational resource usage. We have selected XGBoost due to its capacity to attain high performance in numerous classification tasks. The AEG systems based on XGBoost were constructed in Python using the xgboost package [22]. To train and validate our systems, each dataset grouped by L1, as well as the entire dataset, was divided into training and testing datasets at a ratio

of 80% and 20%, respectively. Note that we used the system based on the entire dataset
as our baseline and compared the 11 L1-specific systems against this baseline system.
This approach allows us to directly compare the performance of different L1-specific
AEG systems, as all systems are evaluated against the same reference system. Given
the imbalanced distribution of grades in the TOEFL11 corpus, the performance of our
AEG systems was evaluated through the 10-fold cross-validation method and by using
a weighted average of precision, recall, and F1.

3 Results

The AEG system developed using the entire dataset (henceforth, the baseline system)
achieved a precision of 0.76, recall of 0.75, and F1 score of 0.75. This result indicates
that our XGBoost-based AEG system exhibits a satisfactory level of accuracy. Importantly, our main analysis of the AEG systems based on the different L1-based datasets
showed noticeable differences in their performance, as outlined in Table 2. Whereas the
AEG systems developed using the L1-Chinese, L1-Japanese, L1-Korean, and L1-Turkish
datasets exhibited equivalent or superior performance in comparison to the baseline
system, those built upon the L1-Arabic, L1-French, L1-German, L1-Hindi, L1-Italian,
L1-Spanish, and L1-Telugu datasets demonstrated relatively lower performance. These
results suggest that AEG systems developed using an exclusive or unbalanced dataset
might grade learners' essays either more accurately or less accurately, depending on
their L1.

Table 2. Classification accuracy of AEG systems by writers' first language

First language of essay writers	Precision	Recall	F1 score
Arabic	0.76	0.74	0.74
Chinese	0.79	0.76	0.77
French	0.73	0.72	0.72
German	0.75	0.74	0.74
Hindi	0.75	0.74	0.74
Italian	0.76	0.74	0.74
Japanese	0.80	0.77	0.78
Korean	0.80	0.78	0.78
Spanish	0.74	0.72	0.73
Telugu	0.77	0.73	0.74
Turkish	0.77	0.77	0.75

4 Discussion

4.1 Effects of First Language on Automated Essay Grading System

Our results showed different performances of 11 AEG systems depending on the L1s of the essay writers. To investigate how these systems differ in more detail, we analyzed the feature importance scores of each AEG system. This analysis revealed that the 11 systems share some similarities but also have some key differences in the relative importance they attribute to each feature (see Fig. 1). In all of the AEG systems, character count and spell error ratio consistently ranked among the top three most significant features, along with either word type frequency or word count. However, there was a noticeable difference between the L1-specific systems that outperformed the baseline system (Chinese, Japanese, Korean, Turkish) and the L1-systems that underperformed the baseline system (Arabic, French, German, Hindi, Italian, Spanish, Telugu): It was only the former systems that displayed two or more syntactic complexity features among their top 10 important features. For example, the AEG system for the L1-Turkish dataset revealed three syntactic complexity features (i.e., complex nominals per clause, mean length of sentence, mean length of clause) as its top 10 important features.

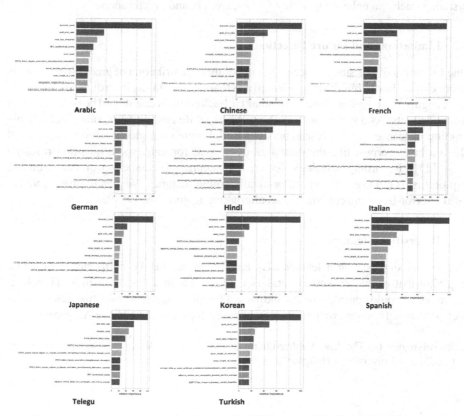

Fig. 1. Importance of the top 10 features in AEG systems by writers' first language

One possible explanation for such a difference could be the characteristics of the L1s of our L2 writers. Japanese, Korean, and Turkish are sometimes classified together as Altaic languages, though this is a subject of debate. These languages are agglutinative languages that follow a subject-object-verb word order and lack gender and number features. In addition, Chinese, as a Sino-Tibetan language, also lacks gender and number features. These four languages differ substantially from other L1s considered in this study, including Arabic in the Afro-Asiatic family, Telugu in the Dravidian family, and French, German, Hindi, Italian, and Spanish in the Indo-European family. It is therefore possible that different grammatical features in the two groups of L1s being compared here may have led to the overperformance or underperformance of the AEG system when compared to the baseline. However, this categorical interpretation should be approached with caution as our data is still far from being comprehensive, and this possibility should be tested further with other languages as L2 learners' L1s.

All in all, this study highlights the importance of involving a wide array of learner groups with different L1 backgrounds in the construction of AEG systems. Relying solely on a limited or biased dataset as the foundation for the AEG system may lead to an unfair prediction of essay grades for some L1 groups. In this regard, developers should recognize the significance of an inclusive and balanced dataset for their AEG system, which can enhance its utility for researchers and practitioners.

4.2 Limitations and Future Directions

One limitation of this study concerns the imbalanced distribution of grades in the corpus of our choice. Furthermore, despite our efforts to include datasets from diverse L1 groups, we recognize that there are numerous other L1 backgrounds that should be represented in future datasets. As a way to address these issues while ensuring fair, valid, and reliable grading of learner essays, researchers can collaborate on a global scale to create and share a huge corpus, like the databases developed for language acquisition research (e.g., [23]). Regarding our single feature for accuracy (i.e., spell-error ratio), further exploration is required on more theoretically sound features. We plan to investigate the ratio of error-free syntactic units as an alternative feature (see [9]).

4.3 Diversity Statement

Our results suggest that a learner's L1 has an impact on the performance of AEG, implying that exclusive/skewed datasets could result in unjust outcomes. Therefore, researchers should consider such an individual factor during data collection/selection for their systems. This way, we can promote the development of inclusive AI technology.

Acknowledgments. The dataset utilized in this research was made available through the library of the Chinese University of Hong Kong.

Disclosure of Interests. The authors have no competing interests to declare that are relevant to the content of this article.

References

1. Yamaura, M., Fukuda, I., Uto, M.: Neural automated essay scoring considering logical structure. In: Wang, N., Rebolledo-Mendez, G., Matsuda, N., Santos, O.C., Dimitrova, V. (eds.) ARTIFICIAL INTELLIGENCE IN EDUCATION 2023, vol. 13916, pp. 267–278. Springer, Cham (2023)
2. Li, Z., et al.: Learning when to defer to humans for short answer grading. In: Wang, N., Rebolledo-Mendez, G., Matsuda, N., Santos, O.C., Dimitrova, V. (eds.) Artificial intelligence in education 2023, vol. 13916, pp. 414–425. Springer, Cham (2023)
3. Ramesh, D., Sanampudi, S.K.: An automated essay scoring systems: a systematic literature review. Artif. Intell. Rev. **55**(3), 2495–2527 (2022)
4. Ajay, H.B., Tillett, P.I., Page, E.B.: The analysis of essays by computer (AEC-II). Final report. University of Connecticut, Storrs (1973)
5. Jarvis, S., Castañeda-Jiménez, G., Nielsen, R.: Detecting L2 writers' L1s on the basis of their lexical styles. In: Jarvis, S., Crossley, S.A. (eds.) Approaching language transfer through text classification: Explorations in the detection-based approach, pp. 34–71. Multilingual Matters, Bristol (2012)
6. Kyle, K., Crossley, S.A., Kim, Y.J.: Native language identification and writing proficiency. Int. J. Learner Corpus Res. **1**, 187–209 (2015)
7. Paquot, M.: Lexical bundles and L1 transfer effects. Int. J. Corpus Linguist. **18**, 391–417 (2013)
8. Blanchard, D., Tetreault, J., Higgins, D., Cahill, A., Chodorow, M.: TOEFL11: A corpus of non-native English. ETS Research Report Series (2014)
9. Housen, A., Kuiken, F., Vedder, I.: Dimensions of L2 performance and proficiency: Complexity, accuracy and fluency in SLA. John Benjamins, Philadelphia (2012)
10. Zenker, F., Kyle, K.: Investigating minimum text lengths for lexical diversity indices. Assess. Writ. **47**, 100505 (2021)
11. Kyle, K., Crossley, S., Berger, C.: The tool for the automatic analysis of lexical sophistication (TAALES): version 2.0. Behavior research methods **50**, 1030–1046 (2018)
12. Lu, X.: A corpus-based evaluation of syntactic complexity measures as indices of college-level ESL writers' language development. TESOL Q. **45**(1), 36–62 (2011)
13. Hwang, H., Kim, H.: Automatic analysis of constructional diversity as a predictor of EFL students' writing proficiency. Appl. Linguis. **44**(1), 127–147 (2023)
14. Kyle, K., Crossley, S.: Assessing syntactic sophistication in L2 writing: A usage-based approach. Lang. Test. **34**(4), 513–535 (2017)
15. Crossley, S.A., Kyle, K., McNamara, D.S.: The tool for the automatic analysis of text cohesion (TAACO): Automatic assessment of local, global, and text cohesion. Behav. Res. Methods **48**, 1227–1237 (2016)
16. Chen, H., Xu, J., He, B.: Automated essay scoring by capturing relative writing quality. Comput. J. **57**(9), 1318–1330 (2014)
17. Bhatt, R., Patel, M., Srivastava, G., Mago, V.: A graph based approach to automate essay evaluation. In: Gabbar, H.A., Trajkovic, L. (eds.) 2020 IEEE INTERNATIONAL CONFERENCE ON SYSTEMS, MAN, AND CYBERNETICS, pp. 4379–4385. IEEE, Toronto (2020)
18. Shen, L.: LexicalRichness: a small module to compute textual lexical richness. MIT license (2022). https://github.com/LSYS/lexicalrichness
19. language-tool-python Homepage, https://pypi.org/project/language-tool-python/. Accessed 20 January 2024
20. Bird, S., Klein, E., Loper, E: Natural language processing with Python: Analyzing text with the Natural Language Toolkit. O'Reilly, Sebastopol (2009)

21. Chen, T., Guestrin, C.: Xgboost: a scalable tree boosting system. In: Krishnapuram, B., Shah, M., Smola, A., Aggarwal, C., Shen, D., Rastogi, R. (eds.) Proceedings of the 22nd ACM Sigkdd International Conference on Knowledge Discovery and Data Mining, pp. 785–794. Association for Computing Machinery, New York (2016)
22. xgboost Homepage, https://pypi.org/project/xgboost/. Accessed 20 January 2024
23. MacWhinney, B.: The CHILDES project: Tools for analyzing talk, 3rd edn. Lawrence Erlbaum Associates, Mahwah (2000)

An Educational Psychology Inspired Approach to Student Interest Detection in Valence-Arousal Space

R. Yamamoto Ravenor[1,2]([✉])([iD]) and Diana Borza[3]([iD])

[1] Tokyo Women's Medical University, Tokyo, Japan
`yamamoto.ravenor@twmu.ac.jp`
[2] Ochanomizu University, Tokyo, Japan
[3] Babeş Bolyai University, Cluj-Napoca, Romania

Abstract. Studies on AI-based facial emotion recognition (FER) of students explore a multiplicity of algorithmic solutions, taking a categorical or a dimensional approach, but overlooking the more education-relevant and objectively measurable parameters of the latter. The "arousal" dimension aligns the degrees of emotional intensity with educational distinctions of passive-active student affect. We build on a theoretical tradition of learning where the gradual transition between these two emotional states is explained with direct reference to student interest detection. The framework proposed models student interest by passive-active emotional intensity (arousal), and only incidentally by negative-positive emotional tone (valence), attributes estimated using a resource-efficient multi-task convolutional neural network. An emotional descriptor based on a 2D histogram in the valence-arousal space is used to establish a student "baseline emotional profile" from which to run analysis on any subsequent session. This representation is explainable and graphic, allowing for simple deterministic rule-based algorithms to spot changes in arousal.

Keywords: Student interest · Facial emotion recognition · Valence and arousal · Change detection · Deep learning

1 Introduction

While FER tools can plausibly assist a teacher's efforts in the affective domain, a fundamental question remains unaddressed. What objectively measurable emotion-related variables can be associated with the affective educational achievement/s being tested? Existing FER tools for education typically compute *subjective* insights into student emotions and are founded on *general* rather than *educational* principles. Between the two approaches, the *categorical*, dealing with basic [3] and/or discrete [7] emotions, and the *dimensional* approach representing emotions along continuous dimensions such as *valence* and *arousal* [13], the latter allows for more meaningful analyses for education.

A. M. Olney et al. (Eds.): AIED 2024, LNAI 14830, pp. 311–318, 2024.
https://doi.org/10.1007/978-3-031-64299-9_26

This study draws on [6]'s classical and seminal theory of learning, which locates student *interest* in the affective continuum of educational objectives. Dramatic and gradual changes in student emotional profile, interpreted as transitions from the passive to active stance, potentially indicate interest. In our model, the transition is determined by *valence-arousal* and *time* variables, allowing for objective measurement of a student's affective state. The system we propose uses: a) a descriptor for storing and comparing student facial expressions over time to create a "default emotional profile"; b) a multi-task convolutional neural network to determine student emotional response for continuous monitoring, and c) a simple and deterministic algorithm to detect significant changes. Our method can be useful for teachers testing affective objectives, parents or students concerned with "interest", and researchers involved in related work.

2   Literature Survey

2.1   Student Interest and Educational Psychology

In educational psychology literature on affect, student "interest" is a primary affective educational objective, term commonly categorized as: active/passive, intrinsic/extrinsic [5], overt/covert, emotional/behavioural [12] etc. [13]'s model solves the semantics problem by representing emotional intensity on an axis, called *arousal*, ranging from inactive to active states, orthogonal with respect to *valence*, hedonic quality from pleasure to displeasure. In [6]'s theory, *interest* is detectable *only when active*, provided the progress, like all learning achievements, is *gradual*. While *arousal* is considered a major variable in appraising student interest, emotional *valence* is not, because the educational objective *may be* to elicit negative emotions, as when teaching about war. [19] discusses this theory and its relation with FER at greater length. To the best of our knowledge, no other learning theory locates student interest on a continuum of objectives and guides detection in ways that make sense from a FER standpoint.

2.2   Automatic Student Interest Detection

Many works studied the problem of student interest detection [10,11,15,18] and proposed diverse methodologies, such as self-reports, observational checklists/rating scales, and automated measurements, to estimate student interest. A large majority of automated measurements rely on deep learning methods for their ability to efficiently extract facial features from images or sequences.

In [15], engagement is expressed as a simple weighted average between one of the following expressions: understanding, doubt, neutral, and disgust. A convolutional neural network (CNN) based on domain adaptation is trained to distinguish between these expressions based on facial images. Following a similar approach, [10] defined "academic affective states" of interest (boredom, confusion, focus, frustration, yawning, and sleepy), and then gathered and annotated a dataset of the aforementioned expressions. Finally, a CNN is used to recognize these expressions from classroom lecture videos. In [2], the authors propose a multi-modal convolutional neural network trained on facial images, hand gestures, and body postures to predict student emotional and behavioral engagement. [1] proposed a hybrid CNN architecture operating on facial images, hand gestures, and body postures to analyze students' affective states in a classroom environment. The architecture uses two CNNs, one to analyze the affective states of a single student, and another to analyze multiple students and predict the overall affective state of the entire class. The effective state is expressed in terms of engagement, boredom, or neutral attitude. [14] proposed an optimized architecture to estimate students' behavior. The model network is pre-trained for face recognition and fine-tuned for facial expression recognition; the extracted features are suitable for predicting students' engagement levels (from disengaged to highly engaged), standard emotions, and group-level affect (positive, neutral, or negative).

Some works focus on emotion recognition in online classes. [20] proposed a face recognition technology that tracks students' status in real-time by monitoring their facial expressions during online classes and providing feedback to the teacher. [17] also proposed a similar technology to measure anger, disgust, fear, happiness, sadness, surprise, and neutral emotions, providing real-time feedback.

Existing approaches do not track the evolution of students over time, focusing on static assessment. Moreover, the absence of a standardized taxonomy leads to inconsistencies, with each approach defining its unique emotional states, hindering the comparison and synthesis of findings. Most studies define discrete values to distinguish between emotions subjectively and fail to capture all the nuances of emotions. To overcome these limitations, we use the two continuous variables, *valence* and *arousal*, to encode emotional engagement and propose a simple and effective descriptor for tracking emotional states across time. No papers were found to perform student interest detection as prescribed by any established educational psychology theory. The proposed change detection algorithm relies on the arousal level as an indicator of student interest.

3 Proposed Solution

The proposed pipeline offers a novel approach to detecting student emotions and accompanying intensities over long time periods, as depicted in Fig. 1. The system operates on recordings of onsite or online classes. An off-the-shelf face detection algorithm [8] is used to detect faces within the video stream. Subsequently, each detected face is identified by comparing it to a gallery of known faces to determine the identity of each student. Next, a multi-task neural network predicts the emotional valence and arousal of each detected participant. These two variables are stored and used to construct a dynamic emotional profile (described in Sect. 3.2) for each student. Finally, the *Change detection* module, relies on statistical measures to spot participants' deviations in arousal level from their baseline emotional profile.

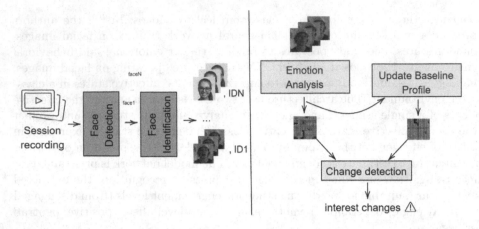

Fig. 1. Overview of the proposed solution. For each student, the *Emotion Analysis* module computes the emotional descriptor for the current session. This descriptor is used to update a "baseline" emotional profile descriptor. The *Change detection* module tracks changes in student interest.

3.1 Valence and Arousal Estimation

Inspired by [14], we employ a multi-task convolutional neural network to estimate students' valence and arousal on video frames. The backbone feature extractor is based on EfficientNet [16] to ensure a high accuracy while maintaining a low computation complexity. To enhance performance and flexibly aggregate spatial information, the backbone output undergoes adaptive average pooling and adaptive maximum pooling layers, followed by concatenation and flattening of the results. Two regression output branches use this activation map to predict arousal and valence. Each branch consists of *Batch Normalization*, *Linear* layers with 256 neurons and *Rectified Linear Unit (ReLU)*, followed by *Batch Normalization*. The final classification linear layer in each branch has a single neuron with a hyperbolic tangent (*tanh*) activation function to restrict the output to $[-1, 1]$, as illustrated in the left side of Fig. 2. Before the *Linear* layers we also apply *dropout* with probability 0.25 for the intermediate ones and 0.5 for the final classification layers. These branches are fine-tuned using the root mean square error loss (RMSE) function. The model was implemented in *pytorch*, and trained through transfer learning from the *ImageNet* dataset on AffectNet dataset samples [9] with the Adam optimizer.

3.2 Emotional Profile Descriptor

The proposed approach involves extracting the emotional state of a student (as valence and arousal values) from video frames with a frequency ν. Valence $v \in [-1, 1]$ refers to the pleasantness (values closer to 1) or unpleasantness (values closer to -1) dimension of emotions, while arousal $a \in [-1, 1]$ denotes the level of activation/intensity. These values are then stored in a 2D histogram ED

discretized into b bins, where the oX axis represents valence and the oY axis represents arousal, as depicted in Fig. 2. The resulting histogram provides a comprehensive representation of the student's emotional profile over a time interval Δ: each time a student exhibits an emotion (v_i, a_i) the corresponding bin in this descriptor is incremented. The ED can be used to visualize the distribution of emotions across the valence-arousal space. A high value of $ED(v, a)$ indicates that there is a significant number of video frames with valence v and arousal a, indicating that the student consistently expressed the corresponding emotion over an extended duration.

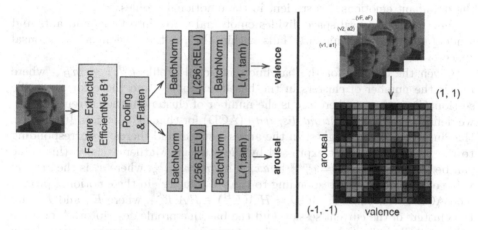

Fig. 2. Emotional profile descriptor. **Left**: A multi-task CNN is used to estimate the valence and arousal. *L(n, act)* - refers to a linear layer with n neurons and activation *act*. **Right**: The model is applied independently for each participant on all video frames and the results are stored in the emotional profile descriptor for that session.

The proposed descriptor offers several advantages over other methods. First, using a continuous space to express emotions and not a discrete taxonomy, it allows for a more nuanced understanding of emotional patterns. Second, it provides a visual representation of emotional behaviour, enabling intuitive interpretation and analysis. Third, it is computationally efficient and easily applicable to videos regardless of their duration. In addition, for each participant, a baseline or "default" emotional descriptor BD is computed allowing the teacher to grasp each student's temperament, facial particularities and expression habits inherent in human nature, using an exponential moving average with bias correction.

3.3   Spotting Dominant Emotions and Change Detection

The proposed descriptor offers a low-dimensional representation and therefore facilitates visual inspection for spotting dominant emotions and detecting pattern changes within emotional profiles. We also propose a deterministic algorithm to spot differences in arousal between two emotional profiles - either between the emotional profile for a session and the baseline description, or the profiles of two independent sessions.

The first step is to identify the "clusters" with elevated values in both emotional profiles. To eliminate noise, we apply a thresholding operation, and then, similar to an image labeling process, we select a point in the arousal-valence space that has not been considered yet and we apply breadth-first search to identify all the points belonging to the same "cluster" (i.e. connected component). The process is repeated until all the positions from the descriptor have been processed. The output of this algorithm is a set of clusters that describe the dominant emotions of a student in the emotional profiles.

The valence-arousal space divides emotional states into four quadrants and student interest predominantly falls within the positive region of the arousal space.

Given the clusters for the baseline emotional profile $E_i^B, i \leq m_B$ - where m_B is the number of clusters in the baseline profile - and those for a specific session $E_i^S, i \leq m_S$ - where m_S is the number of clusters in the current session, we define the *arousal centroid disparity* (ACD) for change detection. It relies on the cluster positioned highest in the arousal space, and therefore corresponding to the maximum interest expressed by the student. Mathematically, this value can be computed as $HA = c_a^{max_i} | max_i = argmax_{i \in m} c_a^i$, where c_a^i is the arousal value of the centroid corresponding to the ith "cluster" in the emotional profile. The ACD is computed as $ACD = HA(E^B) - HA(E^S)$, where E_S and E^B are the clusters of the current session and the baseline profile descriptor. A positive value of ACD indicates an increase in interest, while a negative value signifies a decrease.

4   Experimental Results

To train and test the proposed multi-task convolutional neural network we used AffectNet [9] dataset which comprises around 0.4 million facial images annotated with both discrete emotion categories, as well with continuous emotion descriptors (valence and arousal). The model was evaluated using the squared error (RMSE) and Concordance Correlation Coefficient (CCC) metrics (Table 1).

The proposed approach significantly outperforms both the conventional *Support Vector Regression*(SVR) model and the AlexNet CNN presented in [9]. Substantial advancements have been incorporated in the *EfficientNet* architecture since the pioneering work of AlexNet contributing to its superior performance. In [4] the authors use a YOLO-based architecture for predicting the emotional states; [4] VA refers to a model trained for predicting valence and arousal, while [4] VA+E denotes a model optimized to predict valence, arousal and discrete emotions. The proposed method surpasses [4] VA, but [4] VA+E is better by a small amount (less than 0.1). As [4] VA+E also predicts discrete emotions it has a higher ability to leverage shared representations across tasks, facilitating improved generalization and prediction accuracy.

Table 1. Valence-arousal evaluation.

Method	Valence		Arousal	
	CCC	RMSE	CCC	RMSE
[9] (SVR)	0.340	0.494	0.199	0.400
[9] (AlexNet)	0.541	0.394	0.450	0.402
[4] VA	0.826	0.282	0.556	0.237
[4] VA+E	0.845	0.269	0.606	0.228
Proposed	0.839	0.279	0.600	0.231

5 Conclusions and Future Work

In this paper, we first theorized that facial emotion recognition makes more sense for education when the analysis is continuous and focuses on the point of transition from passive to active emotional response, where student *interest* can be detected. Then we proposed a method that operates on video recordings to detect significant changes indicative of interest. The proposed method, unlike similar others, is not oblivious of the time dimension, making affective measurements more objective. The proposed system does not rely on discrete emotional classes, instead it operates on a continuous space of *valence* and *arousal*. Additionally, we proposed an emotional descriptor based on 2D histograms which can describe emotional behaviours over long periods of time. Our system calibrates itself towards the behaviour of each student as it analyses more videos, computing and updating default emotional profiles for each student. The user can, aside from recording attendance if needed, visualize each student emotional profile, and get notified of significant changes, detected through time and interpreted in terms of interest.

Acknowledgements. This work is funded by the JSPS KAKENHI Grant-in-Aid for Early-Career Scientists, under grant #JP20K13900. Grateful thanks are due to Dr. David Alan Grier for his guidance. The work done by Diana Borza was partly supported by grant SRG-UBB 32886/21.06.2023.

References

1. Ashwin, T.S., Guddeti, R.M.R.: Automatic detection of students' affective states in classroom environment using hybrid convolutional neural networks. Educ. Inf. Technol. **25**(2), 1387–1415 (2020)
2. Ashwin, T., Guddeti, R.M.R.: Unobtrusive behavioral analysis of students in classroom environment using non-verbal cues. IEEE Access **7**, 150693–150709 (2019)
3. Ekman, P.: An argument for basic emotions. Cogn. Emot. **6**(3–4), 169–200 (1992). https://doi.org/10.1080/02699939208411068
4. Handrich, S., Dinges, L., Al-Hamadi, A., Werner, P., Saxen, F., Al Aghbari, Z.: Simultaneous prediction of valence/arousal and emotion categories and its application in an HRC scenario. J. Ambient Intell. Hum. Comput. **12**(1), 57–73 (2021). https://doi.org/10.1007/s12652-020-02851-w

5. Hidi, S.: Interest, reading, and learning: theoretical and practical considerations. Educ. Psychol. Rev. **13**, 191–209 (2001)
6. Krathwohl, D., Bloom, B., Masia, B.: Taxonomy of Educational Objectives: The Classification of Educational Goals. Affective domain. Handbook II. No. v. 1 in Taxonomy of Educational Objectives: The Classification of Educational Goals, Longman Group (1964)
7. Lazarus, R.: Emotion and Adaptation. Oxford University Press, Oxford (1994)
8. Lugaresi, C., et al.: MediaPipe: a framework for building perception pipelines. arXiv preprint arXiv:1906.08172 (2019)
9. Mollahosseini, A., Hasani, B., Mahoor, M.H.: AffectNet: a database for facial expression, valence, and arousal computing in the wild. IEEE Trans. Affect. Comput. **10**(1), 18–31 (2017)
10. Pabba, C., Kumar, P.: An intelligent system for monitoring students' engagement in large classroom teaching through facial expression recognition. Expert. Syst. **39**(1), e12839 (2022)
11. Petrescu., M., Bentasup., K.: Student teacher interaction while learning computer science: early results from an experiment on undergraduates. In: Proceedings of the 15th International CSEDU, pp. 209–216. INSTICC, SciTePress (2023)
12. Renninger, A., Nieswandt, M., Hidi, S.: Interest in mathematics and science learning. Am. Educ. Res. Assoc. (2015)
13. Russell, J.A., Carroll, J.M.: On the bipolarity of positive and negative affect. Psychol. Bull. **125**(1), 3 (1999)
14. Savchenko, A.V., Savchenko, L.V., Makarov, I.: Classifying emotions and engagement in online learning based on a single facial expression recognition neural network. IEEE Trans. Affect. Comput. (2022)
15. Shen, J., Yang, H., Li, J., Cheng, Z.: Assessing learning engagement based on facial expression recognition in MOOC's scenario. Multimedia Syst. **28**(2), 469–478 (2022)
16. Tan, M., Le, Q.: EfficientNet: rethinking model scaling for convolutional neural networks. In: International Conference on Machine Learning, pp. 6105–6114. PMLR (2019)
17. Wang, W., Xu, K., Niu, H., Miao, X.: Emotion recognition of students based on facial expressions in online education based on the perspective of computer simulation. Complexity **2020** (2020)
18. Whitehill, J., Serpell, Z., Lin, Y.C., Foster, A., Movellan, J.R.: The faces of engagement: automatic recognition of student engagement from facial expressions. IEEE Trans. Affect. Comput. **5**(1), 86–98 (2014)
19. Yamamoto Ravenor, R.: AI-based facial emotion recognition solutions for education: a study of teacher-user and other categories. Adv. Mach. Learn. Artif. Intell. **4**(2), 2128–2151 (2024)
20. Zhang, M., Zhang, L.: Cross-cultural O2O English teaching based on AI emotion recognition and neural network algorithm. J. Intell. Fuzzy Syst. **40**(4), 7183–7194 (2021)

Agency in AI and Education Policy: European Resolution Three on Harnessing the Potential for AI in and Through Education

Camila Hidalgo(✉) ⓘ

Technical University of Munich, Richard-Wagner-Str. 1, 80333 München, Germany
camila.hidalgo@tum.de

Abstract. Student agency is a key educational goal increasingly being impacted by Artificial Intelligence (AI) systems integrated into the human learning process. In September 2023, the Ministers responsible for education decided to develop two policy instruments to regulate AI in education, setting teacher and learner agency as a policy goal. To examine how agency for AI regulation in the Europe is understood in policy discourse, I conducted an inductive content analysis of the documents that serve as a basis for these policy instruments. Using documents referred to in Resolution Three on Harnessing the Potential for AI in and through Education as the basis for analysis, five main areas of agency implementation were identified: ethics, AI literacy, regulations, pedagogy, and the capacity to exercise rights and attributes within each field. Key findings, three strains for agency, can guide policymakers and educational practitioners to define actors' roles and responsibility when using AI in the classroom. Different areas of implementation and attributes of agency exist, which will likely guide decision-making and design of policy instruments to assess and ensure agency when AI is employed in an educational context.

Keywords: Human Agency · AI and Education Policy · Ethics of AI in Education

1 Introduction

Different regulation paradigms to govern Artificial Intelligence (AI) are emerging globally, rooted in distinct values and incentives. Europe is striving to steer emergent technology toward human rights, democracy and the rule of law through regulation [1]. Sectorial policy instruments for AI like codes of conduct have focused on autonomous driving, health care, and public administration [2]. With the advent of more advanced and complex AI systems and data analytics in education, Resolution Three on Harnessing the Potential for AI in and through Education was established to develop policy instruments for AI in Education (AIED). At the Standing Conference of Ministers of Education in Strasbourg in September 2023, ministers agreed to develop both a legal instrument to regulate the use of AI in education, promoting and respecting European values, and recommendations to ensure teaching and learning about AI [3].

Ministers responsible for education in Europe (EEA) have set as a policy goal the active participation and agency of educators and students to teach and learn about AI in the soft-law instrument. They emphasize that "everyone [should] [understand] to an appropriate level how AI works and what potential impact it has on our lives" [3]. This policy goal echoes the research devoted to the concept of agency among scholars interested in the impact of AI on human behavior [4]. Nonetheless, the definition of agency varies across and within disciplines. Scholarly discussions of agency have suggested that there is no single definition that captures the complexity of this term and its operationalization [4, 5]. Furthermore, key policy terms, such as student and teacher agency, inform "what is measured and what is not, and where these 'standards' have been derived from" [6]. Through inductive content analysis to examine how Europe policy perceives and operationalizes the concept of agency in the context of AI in education, I have identified three strains for *agency* that refer to form and substance. First, some definitions refer to how AI systems have been developed throughout their lifecycle to serve human agency. Second, I have identified agency as human beings' capacity to exert their rights by controlling the decision-making process in the interaction with AI. Third, people should be able to understand AI's impact on their lives and how to benefit from the best of what AI offers. These three understandings entail different forms of responsibility for the actors involved in the design, development, and use of AI in education. Understanding the differences can guide lawmakers, research communities, and educational practitioners to identify the actors' roles and responsibility to ensure student and teacher agency.

2 Methods

Setting and Scope. The ministers responsible for education agreed to develop a sectorial legal instrument to govern AI in education and recommendations to ensure that teaching and learning about AI includes its impact on European values and prioritizes teachers' and learners' active participation and agency. The EU has increased its global power through digital governance based on consumer market power and technocratic capacity, thus expanding EU rules and standards regarding "which products are built and how business is conducted, in Europe and beyond" [1]—known as *the Brussels effect*. Although Europe hegemony may be contested in the current global distribution of power [7], I argue the Europe still holds regulatory influence that can provide useful insights to analyze other jurisdictions' approaches to governing AI in education.

Research Questions. As technologies gain access to more data and intelligence and perform tasks previously limited to human beings, AI is assuming a new role in human practices and society, affording new meaning to the question of agency [4, 5, 8]. However, agency and its distribution among actors in the learning loop is not new. Distribution of agency is a core aspect of formal education, with the teacher guiding the student [5]. Nonetheless, with the advent of generative AI and data analytics, agency and its implications should be reassessed so that design and development of AIED tools ensure learners' safety, security, privacy and data ownership [5]. Because student and teacher agency has emerged in the Europe policy discourse, this study examines how the concept of agency is perceived and enacted in this discourse to understand its practical implications to redefine roles and responsibilities when integrating AI into education.

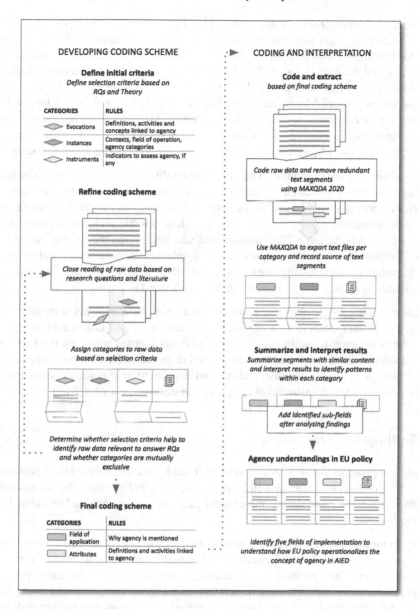

Fig. 1. Inductive qualitative analysis drawn on Puppis [9] and Hall and Steiner [10].

Data Source. This paper analyzed agency as a policy goal addressed in the policy documents and regulations identified in Resolution Three, that will serve as a basis of both a legal instrument and the recommendations of the Committee of Ministers to govern AI in education. The concept of agency was referred to explicitly in five of the twenty documents in no. 1 letter a) to f) and no. 2 letter a) to i) in Resolution Three. The five documents are the following: the Council of Europe's Committee on Artificial

Intelligence's draft regarding the development of a Framework Convention on AI, Human Rights, Democracy and the Rule of Law (CAI) [11]; the Council of Europe's report Artificial Intelligence and Education – A critical view through the lens of Human rights, Democracy and the Rule of Law (CEAI&ED) [8]; the UNESCO's Recommendation on the Ethics of Artificial Intelligence (UNESCO Re) [12], the European Commission's proposal for an Artificial Intelligence Act (AI ACT) [13, 14], and the Children's Data Protection in an Education Setting Guidelines (CDG) [15]. Concepts related to agency, such as autonomy, control, decision-making, dignity, choice, self-determination, and learner-centric approach, were variously employed throughout the five documents.

Analytical Techniques. I conducted an inductive qualitative analysis [9, 10]. The initial criteria draw on my research questions and the categories included in Klemenčič's work, namely, evocations, instances and instruments of key policy terms [6]. After an iterative process of systematic reading to develop the coding scheme and assign categories to raw data using MAXQDA2020, I summarized and interpreted the results (see Fig. 1). Hence, I classified the findings into areas of implementation: ethics, regulations, AI literacy, pedagogy, and capacity to exercise rights, then subfields for each main area and the attributes of agency within each sub-field (see Table 1).

Limitations. This article does not cover the analytical difference between the role of teacher and learner agency, their mutual interconnections and differences and policy implications for their roles. In addition, this analysis focuses solely on the Europe to gain a deeper understanding of how documents included in Resolution Three provide understandings of agency to policymakers and experts to help guide decision-making and the design of policy instruments to assess and ensure agency for AI in education.

3   Findings and Discussion

This section presents the attributes of agency in each area and sub-fields of implementation to understand how agency will be positioned as a policy goal. The main areas of implementation identified during the content analysis are summarized in Table 1 and described below.

Ethics. In most domains, AI's ability to perform human tasks perceived as restricted to humans only have raised ethical concerns [8], which suggests an anthropomorphizing of AI, extending human attributes to machines, giving AI a new role in human practices and societies. Using anthropomorphic terms to refer to AI systems could in the long-term challenge humanity's "special sense of experience and agency, raising additional concerns about, inter alia, human self-understanding, social, cultural and environmental interaction, autonomy, agency, worth and dignity" [12]. This process of attributing human characteristics to AI has been criticized, e.g., in the revised definition of *AI Systems* by the OECD, recently adopted in drafts of the AI Act and the CAI [16].

Regulations. In the December 2023 draft of the AI Act, agency is addressed under human agency and oversight as one of the general principles applicable to all AI systems in line with Article 4 a (a) [13, 14]. Operators should make their best effort to develop and use AI systems and foundation models in accordance with the high-level framework that promotes a human-centric European approach and trustworthy AI [14]. The High-Level Expert Group on AI's (HLEG) Ethics Guidelines for Trustworthy Artificial Intelligence established that human agency and oversight was one the seven key requirements that AI should meet to be deemed trustworthy [17].

In the AI act draft, human agency and oversight is a principle to be followed by the operator that develops and uses AI systems, and when possible should be incorporated in the design and use of AI models. This forthcoming binding rule defines this principle in article 4 a (a), as "AI systems should be developed and used as a tool that serves people, respects human dignity and personal autonomy and that is appropriately controlled and overseen by humans" [14]. Although the second part of the definition refers mainly to human oversight, my findings show that the terms control and oversight also refer to agency, for instance, having control over one's own learning process. The CAI draft of December 18, 2023, that will soon be finalized, sets out option C, Recital 17, that member states of the council of Europe and other signatories should promote human agency and oversight in the design, development, use and decommissioning of AI systems [11]. This convention will likely be the first global AI treaty with legal force for those state members that ratify it. As mentioned above, the OECD revised definition of *AI System* has been incorporated in the drafts of these two forthcoming binding rules, which raises the concern on attributing anthropomorphic traits to AI.

AI Literacy. Resolution Three focuses on ensuring everyone understands how AI works and its likely impact on human lives [3]. This statement refers to *AI literacy,* also known as *learning about AI*, which is comprised of two dimensions. First, learning about AI encompasses the technological dimension of learning how AI works, including the techniques and technologies; second, the human dimension refers to the AI impact on people, including human cognition, privacy and agency among others [18]. In this framework, protecting and promoting agency is one of the main aims of the human dimension of AI literacy. As Holmes *et al.* argue, "the aim of addressing the human dimension of AI literacy is to enable everyone to learn what it means to live with AI and how to take best advantage of what AI offers, while being protected from any undue influences on their agency or human dignity" [8]. The Committee of Ministers recommendation aims to ensure that teaching and learning about AI incorporates the impact of AI on the key European values of human rights, democracy and the rule of law, and to prioritize the active participation and agency of teachers and learners [3].

Pedagogy. In the relationship between AI and pedagogy, agency is regarded as a key challenge to be addressed and as a goal for the learner-centric approach to teaching and learning. This approach refers to yielding to students significant control over the learning process to maximize their agency. For child learners, capacity should be assessed differently than adults, especially when this capacity depends on how much children understand bias and fairness, informed consent as well as comprehend and contest effects of AI outcomes that impact their lives [8]. Brod *et al.* emphasized that teachers and learners should adapt their teaching and learning practices to ensure learner agency due to the increasing use of data-driven algorithms to personalize content for individual learners [4]. In the reports I analyzed, introducing AI into the classroom is perceived as reducing learners' agency. Educational technologies are often designed embedding learning theories that prioritize skills that do not necessarily promote agency, for instance, memorizing facts over thinking or critical engagement [8]. Moreover, the overuse of personalized learning may lead to data speaking on behalf of or instead of the children, overriding their right to speak for themselves [8].

Table 1. Summary of conceptual considerations of agency

Area	Sub-fields of implementation	Attributes	Policy Docs.
Ethics	Anthropomorphism of AI	Ethical issues are related to the capacity of AI systems to perform tasks which previously were limited to human beings only and the language used to name these technologies. These two factors give AI a new role in human practices and society.	UN-ESCORe [12] CEAI&ED [8]
Regulations	Principles enshrined in regulations from the HLEG's 2019 Ethics Guidelines for Trustworthy AI	AI systems shall be developed and used as a tool that serves people, respects human dignity and personal autonomy, and that is appropriately controlled and overseen by humans.	AI Act [13, 14]
		Human agency as a principle to be followed by operators developing and using AI systems; when possible, should be translated in the design and use of AI models.	AI Act [13, 14]
		Human agency as a principle to be promoted in the design, development, use and decommissioning of AI systems.	CAI [11]
AI Literacy	Aim of human dimension of AI literacy	Enable everyone to learn what it means to live with AI and how to take best advantage of what AI offers, while being protected from any undue influences on their agency or human dignity.	CEAI&ED [8]
Pedagogy	Learner-centric approach	Give children significant control over the learning process, considering that children understand AI's impact on people different than adults.	CEAI&ED [8]
	AI&ED key challenge	Pedagogical approaches embedded in most AI systems available in the market prioritize tasks that undermine learner agency and robust learning, and engineer child's development in opaque systems.	CEAI&ED [8]
Rights	Right to be heard	Right to refuse any involvement with AI classroom tools without the refusal adversely affecting their education.	CEAI&ED [8]
		Right to express own views freely in matters affecting students according to age and maturity and make significant decisions about their own education.	CEAI&ED [8]
		AI de facto exercises rights, based on data analytics, on behalf of or instead of the child making decisions.	CEAI&ED [8]
	Right to privacy & Data Protection	An essential right to protect human agency; right that is challenged by collection of "objective" learners' data by AI claiming to speak on their behalf.	UN-ESCORe [12]
		Right to be informed about how own data are collected and processed throughout educational live—digital footprint.	CDG [15]

Capacity to Exercise Rights. Agency is also conceptualized as learners' capacity to exercise fundamental rights in their educational environment. Agency is directly related to the right to privacy and data protection considered essential to protecting human agency [12]. CDG recognized that "while children's agency is vital and they must be better informed about how their own personal data are collected and processed, (…) children cannot be expected to understand a very complex online environment and to take on its responsibilities alone." [15] Therefore, the CDG emphasizes the need for legislative frameworks to address how extensive and far children's digital footprint travel throughout their educational lifetime [15]. Moreover, lawmakers and scholars have developed regulations and tools to guide the design and development of educational technologies to address children's privacy and data protection [19, 20]. However, less attention has been paid to a child's right to be heard in education. In the PREMS report, the right to be heard has two main attributes. First, the learners' and parents' right to refuse any involvement with AI classroom tools without affecting learners' right to education: learner consent and the right to opt-out [8]. Second, the inviolable right of humans to decide about their own education [8]. In this vein, the accumulation of data by AI systems may reduce learners' agency when data "speak" on behalf of children, make decisions or express their views for them. This point is particularly important to define who the agent is: the teacher, the student, or the AI. However, the aim of AI in education is precisely to enhance learning and teaching without offloading critical cognitive and physical tasks from the student [21].

4 Conclusion

This study set out to unpack the agency attributes and their fields of implementation in the policy documents that serve as a basis for the forthcoming regulatory package for AIED. Through a content analysis, I identified five main fields of implementation and diverse attributes of agency within each subfield. These categories can help decision-makers and researchers to map and systematize various conceptual considerations of agency in AIED to define the public good to be protected, promoted, and operationalized. Key policy terms like agency will likely become the conceptual tool to guide educational policy and the design of policy instruments to assess institutional practices, for instance, how to evaluate the conditions that enable individuals to enact their agency when learning and teaching about AI. Understanding agency and the implications in ethics, regulations, AI literacy, education, and capacity to exercise can guide decision-makers to establish appropriate safeguards to protect and promote human rights, democracy, and the rule of law in AI in education.

Acknowledgments. The author thanks the TUM IEAI for supporting this work and Prof. Urs Gasser for providing constructive feedback and guidance on this research project.

References

1. Bradford, A.: Digital empires: the global battle to regulate technology. Oxford University Press, New York (2023)
2. OECD: The state of implementation of the OECD AI Principles four years on (2023)

3. Council of Europe Standing Conference of Ministers of Education: Council of Europe Standing Conference of Ministers of Education "The Transformative Power of Education: Universal Values and Civic Renewal" 26th Session Resolutions. https://rm.coe.int/resolutions-26th-session-council-of-europe-standing-conference-of-mini/1680abee7f

4. Brod, G., Kucirkova, N., Shepherd, J., Jolles, D., Molenaar, I.: Agency in educational technology: interdisciplinary perspectives and implications for learning design. Educ. Psychol. Rev. **35**, 25 (2023). https://doi.org/10.1007/s10648-023-09749-x

5. Tuomi, I., Cachia, R., Villar-Onrubia, D.: On the futures of technology in education: emerging trends and policy implications. Publications Office of the European Union, LU (2023)

6. Klemenčič, M.: From student engagement to student agency: conceptual considerations of european policies on student-centered learning in higher education. High. Educ. Policy. **30**, 69–85 (2017). https://doi.org/10.1057/s41307-016-0034-4

7. Drezner, D.W.: The New Empires of the Internet Age. https://foreignpolicy.com/2024/04/20/internet-power-us-china-eu-farrell-newman-bradford-book-review/

8. Holmes, W., Persson, J., Chounta, I.-A., Wasson, B., Dimitrova, V.: Artificial intelligence and education: a critical view through the lens of human rights, democracy and the rule of law. Council of Europe, Strasbourg (2022)

9. Puppis, M.: Analyzing Talk and Text I: Qualitative Content Analysis. In: Van Den Bulck, H., Puppis, M., Donders, K., Van Audenhove, L. (eds.) The Palgrave Handbook of Methods for Media Policy Research, pp. 367–384. Springer, Cham (2019)

10. Hall, D.M., Steiner, R.: Policy content analysis: qualitative method for analyzing sub-national insect pollinator legislation. MethodsX. **7**, 100787 (2020). https://doi.org/10.1016/j.mex.2020.100787

11. Committee on Artificial Intelligence: Draft Framework Convention on Artificial Intelligence, Human Rights, Democracy and the Rule of Law. https://rm.coe.int/cai-2023-28-draft-framework-convention/1680ade043

12. UNESCO: Recommendation on the Ethics of Artificial Intelligence. UNESCO, Paris, France (2022)

13. European Commission: Proposal for a Regulation of the European Parliament and of the Council Laying down Harmonised rules on Artificial Intelligence (Artificial Intelligence Act) and Amending Certain Union Legislative Acts COM/2021/206 final. https://eur-lex.europa.eu/legal-content/EN/TXT/?uri=CELEX:52021PC0206

14. European Parliament: Amendments adopted by the European Parliament on 14 June 2023 on the proposal for a regulation of the European Parliament and of the Council on laying down harmonised rules on artificial intelligence (Artificial Intelligence Act) and amending certain Union legislative acts (COM(2021)0206 — C9-0146/2021 — 2021/0106(COD)). https://eur-lex.europa.eu/eli/C/2024/506/oj

15. Council of Europe: Children's Data Protection in an Educational Setting Guidelines. https://rm.coe.int/prems-001721-gbr-2051-convention-108-txt-a5-web-web-9/1680a9c562

16. Rotenberg, M.: Last week, the revised definition of "Artificial Intelligence system" was released by the OECD. https://www.linkedin.com/posts/marc-rotenberg_the-uncontroversial-thingness-of-ai-lucy-activity-7130205892553162752-NYFt/

17. High-Level Expert Group on Artificial Intelligence (AI HLEG): Ethics Guidelines for Trustworthy AI. European Commission, Brussels (2019)

18. Holmes, W., Bialik, M., Fadel, C.: Artificial intelligence in education promises and implications for teaching and learning. Center for Curriculum Redesign, MA, USA (2019)

19. Day, E., Pothong, K., Atabey, A., Livingstone, S.: Who controls children's education data? a socio-legal analysis of the UK governance regimes for schools and EdTech. Learn. Media Technol. (2022). https://doi.org/10.1080/17439884.2022.2152838

20. Henriques, I., Hartung, P.: Children's rights by design in AI development for education. Int. Rev. Inf. Ethics. **29**, (2020). https://doi.org/10.29173/irie424

21. Molenaar, I.: Towards hybrid human-AI learning technologies. Eur. J. Educ. **57**, 632–645 (2022). https://doi.org/10.1111/ejed.12527

Nudging Adolescents Towards Recommended Maths Exercises with Gameful Rewards

Jeroen Ooge[1]([⊠])(iD), Joran De Braekeleer[2](iD), and Katrien Verbert[2](iD)

[1] Department of Information and Computing Sciences, Utrecht University,
Utrecht, The Netherlands
`j.ooge@uu.nl`
[2] Department of Computer Science, KU Leuven, Leuven, Belgium
`joran.debraekeleer@gmail.com`, `katrien.verbert@kuleuven.be`

Abstract. E-learning systems that force learners to solve personalised exercises lower their control and possibly their motivation. To better balance learner control and automation, we created an app for practising high school equation-solving in which learners can select exercises from tailored sets and are nudged towards recommended ones with gameful rewards. Furthermore, labels indicate exercises' difficulty. A randomised controlled experiment with 154 adolescents revealed that our nudges made learners select harder exercises without negatively impacting short-term learning performance and self-reported competence. However, difficulty labels did not have such effects. In sum, our study suggests that reward-based nudging is promising to let learners voluntarily engage in more challenging learning materials while preserving selection freedom.

Keywords: Learner control · Nudging · Gamification

1 Introduction

Adaptive e-learning systems try to improve learning processes by automatically tailoring learning materials to learners' mastery levels. To challenge learners without overwhelming or boring them, exercises of suitable difficulty should keep learners within the so-called *zone of proximal development* [11]. As a result, e-learning platforms often enforce exercises that are deemed most effective by an algorithm. However, this approach can lower learners' intrinsic motivation due to reduced freedom of choice as autonomy is a core pillar in self-determination theory [5]. Alternatively, learners can be presented with multiple recommended exercises and *nudged* [2,8] towards exercises in the zone of proximal development without taking away their freedom of choice. Yet, a recent review showed that nudging has rarely been studied in the context of recommender systems [7].

J. Ooge and J. De Braekeleer—Contributed equally to this research.

A. M. Olney et al. (Eds.): AIED 2024, LNAI 14830, pp. 328–335, 2024.
https://doi.org/10.1007/978-3-031-64299-9_28

Moreover, there is untapped potential in *smart nudges*, which tailor nudges to individuals and their context [7,8]. This raises a first research question:

RQ1. How can smart nudging be operationalised in an educational recommender system to nudge learners towards recommended exercises?

Many types of nudges can leverage cognitive biases or social norms to steer people's behaviour in a desirable direction [6]. For example, incentive nudges take advantage of people's loss aversion and have been operationalised with gameful rewards to increase persuasiveness [14], which hints at a link between nudging and *gamification* [1,15]. Furthermore, salience nudges focus people's attention on what seems relevant to them. Difficulty labels for learning materials, for example, can discourage learners from processing materials indicated as difficult unless they are given task-related choice beforehand [16]. Overall, previous studies on nudging in education have shown mixed effects: while nudging often successfully changes learners' behaviour, not every nudge is equally effective in all contexts [2]. This leads to our second research question:

RQ2. How do nudges in the form of gameful rewards and difficulty labels affect chosen exercise difficulty, learning performance, and self-perceived competence?

Our work contributes to answering the above research questions. Specifically, we designed and developed a smartphone app for high school students to practice equation-solving, incorporating automated recommendations based on skill-level Elo ratings and smart gameful rewards that nudge learners towards exercises whose difficulty lies in the zone of proximal development. A randomised controlled experiment with 154 adolescents shows that gameful rewards can nudge learners towards harder exercises without decreasing their short-term learning performance or self-reported competence. Overall, we hope our work sparks more interest in personalised learning systems that motivate adolescents.

2   Materials and Methods

This section briefly introduces our app and overall study procedure. Our study was approved by the ethical committee of KU Leuven (reference G-2023-6197).

2.1   Smartphone Application

Figure 1 shows our app's general workflow. Upon choosing a topic, the app generates four exercises with varying difficulties using the algorithm described below and presents them in random order on a selection screen. There, learners first select one of four reward types, which they collect upon solving an exercise: stars to climb up in a leaderboard, coins to unlock badges, chests to collect objects, and fish to feed a virtual cat. These reward types correspond to game-design elements that are in theory most preferred by the four most common Hexad gamification user types [1]. Next, learners choose an exercise and get three chances to solve it:

they type the answer or an intermediate step and get direct feedback on whether their input is correct. Learners who find the answer are rewarded and can visit the corresponding reward interface or continue practising. This workflow results from an iterative design process with 12 young adults, detailed in [3].

Generating Exercises. Solving linear, quadratic, and cubic equations requires computational skills such as addition, multiplication, changing signs, distribution, and applying the discriminant or Horner's method. Our app captures these skills in 20 templates (e.g., $ax^2+b=c$) and generates exercises by randomly picking parameters and solutions. Exercises with difficulties slightly above learners' mastery level yet within the zone of proximal development are recommended. To estimate how difficult exercises are for specific learners, the app uses an Elo rating system [13] inspired by the variant typical for chess. Concretely, all *skills* have a global Elo rating, and learners have personal ratings for each skill, which gives a fine view of how well they master equation-solving *skills*. The rating of exercises is defined as the highest rating of the skills in their template.

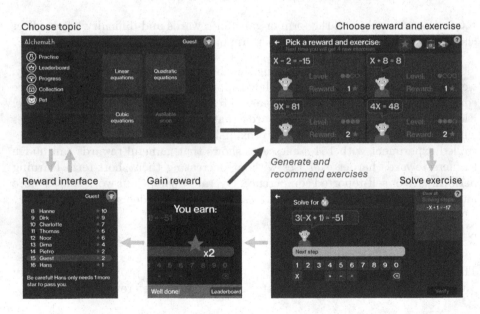

Fig. 1. Workflow in our app: learners choose a topic, reward, and exercise; after solving the latter, they are rewarded and visit the reward interface or continue practising.

2.2 Participants and Study Procedure

We asked high school maths teachers in Belgium (Flanders) to let students participate in our study during class without pressuring them and while providing exercises on paper for students who did not participate. Interested students provided informed consent, and those below 16 needed parental consent. Our study was a randomised controlled experiment with the four groups in Fig. 2, using a 2 × 2-design with the following variables:

- **Difficulty labels**: groups D1I0 and D1I1 saw exercises accompanied by coloured dots indicating their difficulty; the other groups did not.
- **Increased rewards**: groups D0I1 and D1I1 gained two rewards for recommended exercises; the other groups received one reward for all exercises.

Fig. 2. The differences between selection screens based on increased rewards and difficulty labels that differentiated the four test groups.

Participants could freely practise topics. After 15 exercises, they were referred to the post-study questionnaire in Table 1: questions Q1–Q5 measured *perceived competence* with a subscale of the Intrinsic Motivation Inventory [10] and question Q6 asked what participants looked at when choosing exercises. Afterwards, participants could continue using the app. In the background, we logged Elo rating changes and details of the exercises participants selected.

3 Results

In total, 154 adolescents participated in the study. To obtain a more focused analysis, we limited our sample to 127 adolescents between 13 and 19 years old who answered at least 10 exercises: 28 ended up in D0I0, 34 in D0I1, 29 in

Table 1. Post-study questionnaire where Q1–Q5 were scored on a 7-point scale.

ID	Question
Q1	I think I am pretty good at this task
Q2	I think I did pretty well at this activity, compared to other students
Q3	I am satisfied with my performance at this task
Q4	I felt pretty skilled at this task
Q5	After working at this task for a while, I felt pretty competent
Q6	When I chose an exercise, I particularly looked at: the reward/the exercise/the level/other; explain [in an open text field]

D1I0, and 36 in D1I1. Participants identified as female (61%), male (32%), or different (6%). During the experiment, the average participant solved 67 exercises (SD = 93) in roughly one hour, primarily practising linear equations. Only 84 participants filled out the post-study questionnaire.

Chosen Exercises. Figure 3 shows that participants with increased rewards chose harder exercises on average. A two-way ANOVA confirmed that having increased rewards significantly affected chosen difficulty ($p < 0.01$) in contrast to seeing difficulty labels ($p = 0.91$). Moreover, although borderline, there was no interaction effect between seeing increased rewards and difficulty labels ($p = 0.05$). To further test one-sided differences, we conducted pairwise t-tests with Benjamini-Hochberg correction, which confirmed that D0I1 and D1I1 chose harder exercises than D0I0 (both $p \leq 0.01$); other comparisons were insignificant (all $p > 0.12$). Finally, the answers to Q6 revealed that participants often considered rewards while selecting exercises, especially in groups with increased rewards: 75% and 70% of the participants in D0I1 and D1I1 indicated to pay attention to the rewards, respectively, compared to 36% and 47% in D0I0 and D1I0, respectively. Participants appreciated rewards because they *"motivated [them]"*, *"were fun"*, or allowed them to progress on the gameful interfaces linked to the reward types.

Short-Term Learning Performance. We investigated learning performance in two ways. First, we measured the overall change in Elo rating during the experiment. While groups with increased rewards had slightly higher Elo gains (see Fig. 3), a two-way ANOVA showed these differences were insignificant (all $p > 0.30$). Second, we defined learning performance in terms of the correctness of the given answers. For each finished exercise, we assigned a performance score of 0 if participants gave three wrong answers or gave up and $1/\#$ attempts otherwise. The average performance score in all groups was about 0.88, meaning exercises were often solved on the first attempt. A two-way ANOVA did not find significant differences across groups (all $p > 0.13$).

Impact on Self-Reported Competence. Participants in groups with and without increased rewards scored their competence for solving equations in our app with a 4.75 and 5 out of 7 on average, respectively. This difference was, however, insignificant according to a two-way ANOVA (all $p > 0.24$).

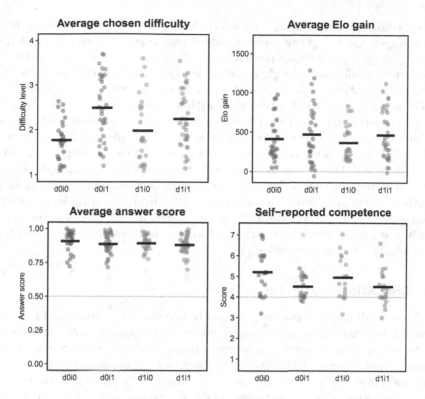

Fig. 3. Scatter plots of the three studied metrics: average chosen difficulty, learning performance measured with average Elo gain and average answer score, and self-reported competence. Group outliers are faded, and horizontal bars indicate group means.

4   Discussion and Conclusions

We explored the space between fully automated e-learning systems and systems wherein learners have full control since neither seems optimal: full automation might reduce intrinsic motivation for learning due to reduced autonomy [5], whereas full control over which exercises to solve is problematic if learners systematically under- or overestimate themselves. Furthermore, shared control can yield better learning outcomes [9] and increase learners' trust in recommender systems [12]. Our intermediate solution uses gameful rewards as nudges towards recommended exercises and displays difficulty labels to support decision-making.

4.1   Smart Nudging with Gameful Rewards Is Feasible

Few nudging interventions positively affect everyone [2]. In our case of gameful reward-based nudging in an educational recommender, the power of automatically recommending learning materials can diminish if learners are not persuaded by the rewards. As an initial step towards avoiding this pitfall, we operationalised *smart* nudging [8] by letting learners select their preferred reward type (see RQ1). Our results suggest that this nudging indeed persuades adolescents to pick more challenging exercises without negatively affecting their short-term learning performance and self-assessed competence (see RQ2). Furthermore, we found that labels indicating exercises' estimated difficulty did not yield such effects.

Future work could combine our ideas with research on personalising gamification [15] to automatically deduce learners' preferred rewards and study whether this enhances nudging effects and desirable learning goals such as performance and motivation. In addition, the trade-offs of reward-based nudging should be further explored. On the one hand, gameful rewards can be an example of transparent nudging, which is relevant to the ethical debate around nudging in the sensitive context of education for adolescents [8]. On the other hand, the motivational aspects of gameful rewards should be studied in more detail as they might mainly tap into extrinsic motivation, which has been criticised for potentially undermining intrinsic motivation [4].

4.2 Limitations and Future Work

While our findings are promising, our study had several limitations. First, most participants practised linear equations, which were relatively easy, as evidenced by the overall high performance scores. Longer-term experiments with harder exercises should verify whether our findings hold. Additionally, although the effect was not statistically significant, we are mindful that groups with increased rewards reported lower competence while all groups performed equally well. Future studies could investigate whether this undesirable phenomenon occurs in larger samples. Finally, the large importance that participants dedicated to rewards when selecting exercises might have been reinforced by the classroom context. For example, participants often whipped each other up for the leading position on the leaderboard. Future studies could study how such a heated atmosphere impacts learning outcomes and learners who feel less motivated by rewards popular among their peers.

In sum, we hope follow-up studies further explore how e-learning systems best balance automation and learner control to foster motivation and learning.

References

1. Altmeyer, M., Tondello, G.F., Krüger, A., Nacke, L.E.: HexArcade: predicting hexad user types by using gameful applications. In: Proceedings of the Annual Symposium on Computer-Human Interaction in Play, CHI PLAY 2020, pp. 219–230. Association for Computing Machinery, New York (Nov 2020). https://doi.org/10.1145/3410404.3414232
2. Damgaard, M.T., Nielsen, H.S.: Nudging in education. Econ. Educ. Rev. **64**, 313–342 (2018). https://doi.org/10.1016/j.econedurev.2018.03.008
3. De Braekeleer, J.: Gebruikers nudgen naar effectieve oefeningen. Master's thesis, KU Leuven, Faculty of Engineering Science, Leuven, Belgium (2023)

4. Deci, E.L., Koestner, R., Ryan, R.M.: Extrinsic rewards and intrinsic motivation in education: reconsidered once again. Rev. Educ. Res. **71**(1), 1–27 (2001)
5. Deci, E.L., Ryan, R.M.: Motivation, personality, and development within embedded social contexts: an overview of self-determination theory. Oxford Handbook Human Motivat. **18**(6), 85–107 (2012)
6. Dolan, P., Hallsworth, M., Halpern, D., King, D., Metcalfe, R., Vlaev, I.: Influencing behaviour: the mindspace way. J. Econ. Psychol. **33**(1), 264–277 (2012). https://doi.org/10.1016/j.joep.2011.10.009
7. Jesse, M., Jannach, D.: Digital nudging with recommender systems: survey and future directions. Comput. Hum. Behav. Rep. **3**, 100052 (2021). https://doi.org/10.1016/j.chbr.2020.100052
8. Karlsen, R., Andersen, A.: Recommendations with a Nudge. Technologies **7**(2), 45 (2019). https://doi.org/10.3390/technologies7020045
9. Long, Y., Aleven, V.: Mastery-oriented shared student/system control over problem selection in a linear equation tutor. In: Micarelli, A., Stamper, J., Panourgia, K. (eds.) Intelligent Tutoring Systems, pp. 90–100. LNCS. Springer International Publishing, Cham (2016). https://doi.org/10.1007/978-3-319-39583-89
10. McAuley, E., Duncan, T., Tammen, V.V.: Psychometric properties of the intrinsic motivation inventory in a competitive sport setting: a confirmatory factor analysis. Res. Q. Exerc. Sport **60**(1), 48–58 (1989). https://doi.org/10.1080/02701367.1989.10607413
11. Murray, T., Arroyo, I.: Toward measuring and maintaining the zone of proximal development in adaptive instructional systems. In: Cerri, S.A., Gouardères, G., Paraguaçu, F. (eds.) Intelligent Tutoring Systems, pp. 749–758. LNCS. Springer, Berlin, Heidelberg (2002). https://doi.org/10.1007/3-540-47987-2-75
12. Ooge, J., Dereu, L., Verbert, K.: Steering recommendations and visualising its impact: effects on adolescents' trust in E-learning platforms. In: Proceedings of the 28th International Conference on Intelligent User Interfaces, IUI 2023, pp. 156–170. Association for Computing Machinery, New York (Mar 2023). https://doi.org/10.1145/3581641.3584046
13. Pelánek, R.: Applications of the Elo rating system in adaptive educational systems. Comput. Educ. **98**, 169–179 (2016). https://doi.org/10.1016/j.compedu.2016.03.017
14. Petrykina, Y., Schwartz-Chassidim, H., Toch, E.: Nudging users towards online safety using gamified environments. Comput. Sec. **108**, 102270 (2021). https://doi.org/10.1016/j.cose.2021.102270
15. Rodrigues, L., Toda, A.M., Oliveira, W., Palomino, P.T., Vassileva, J., Isotani, S.: Automating gamification personalization to the user and beyond. IEEE Trans. Learn. Technol. **15**(2), 199–212 (2022). https://doi.org/10.1109/TLT.2022.3162409
16. Schneider, S., Nebel, S., Meyer, S., Rey, G.D.: The interdependency of perceived task difficulty and the choice effect when learning with multimedia materials. J. Educ. Psychol. **114**(3), 443–461 (2022). https://doi.org/10.1037/edu0000686

A Learning Approach for Increasing AI Literacy via XAI in Informal Settings

Mira Sneirson, Josephine Chai, and Iris Howley(✉)

Williams College, Williamstown, USA
{mls4,jhc4,ikh1}@williams.edu

Abstract. To achieve AI literacy, the AI community employs explainable AI (XAI), to increase AI literacy for those outside of formal educational settings. Designing and evaluating XAI remains an open question that can be guided by existing learning science research. When designers view their XAI through a learning lens, they may better define, assess, and compare explanation implementations. We surveyed and interviewed designers of interactive explanations for AI to identify how practitioners build their XAI and to better understand how a learning lens can be applied for explanations of complex AI concepts.

Keywords: AI literacy · Explainable artificial intelligence (XAI)

1 Introduction

As AI and ML algorithms significantly impact our daily lives, understanding the abilities of these algorithms, as well as their biases and flaws, is important. One solution to this need for AI literacy in the AI research community is explainable AI (XAI) [5,6] as it potentially increases AI literacy of current AI users through informal settings, rather than relying on formal educational contexts. In this article, we show that non-learning experts are already building AI explanations, often called *explainables*, and we ask how learning sciences can be applied to XAI, revealing ample opportunity for AI in Education researchers.

To do this, we created surveys customized to XAI design practitioners' products and analyzed responses and interviews for themes along goals designers hope their explainables enable users to achieve. Our findings show that XAI design practitioners struggle in evaluating the success of their artifacts and that using a learning objective framework is helpful scaffolding. Our investigation of cognitive learning objectives can be more widely applied and yield insight for designers creating complex interactive explanations.

1.1 Related Work

For the purposes of this article, we limit our scope to post-hoc XAI systems known as AI explainables [13]. These explainables often explain model predictions without specifying the underlying mechanisms by which they work [13].

A. M. Olney et al. (Eds.): AIED 2024, LNAI 14830, pp. 336–343, 2024.
https://doi.org/10.1007/978-3-031-64299-9_29

Most explainables present as animated or interactive tutorials teaching users about a specific algorithm or application of an algorithm, such as those in Fig. 1. To an AI in Education researcher, this interactive explanation may seemingly share much with intelligent tutoring systems. We expand upon this by examining how XAI practitioners design, compare, and evaluate their AI explanations, and providing a concrete framework for practitioner use.

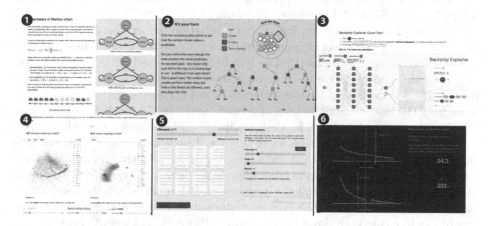

Fig. 1. Example screenshots from various XAI explainables: Exploring Hidden Markov Models (https://nipunbatra.github.io/hmm/), MLU-Explain The Random Forest Algorithm (https://mlu-explain.github.io/random-forest/), Backprop Explainer (https://xnought.github.io/backprop-explainer/), Demystifying the Embedding Space of Language Models (https://bert-vs-gpt2.dbvis.de/), Predicting What Students Know (https://www.irishowley.com/res/bkt-esperanto/index.html), (Un)Fair Machine (https://unfair-machine.netlify.app/).

The authors of [7] developed a model of the entire XAI process which includes relationships between the user, system, mental model, and task performance, with assessments for each, leading to appropriate trust and use. Considerations for measuring explanation effectiveness includes: user satisfaction, user mental model, user task performance, trust assessment, and (optionally) correctability. This XAI model includes several "tests" for evaluating XAI, which is expanded upon in the work in [12]. The authors created a literature-based taxonomy of XAI evaluation criteria which includes: faithfulness, completeness, stability, compactness, uncertainy communication, interactivity, translucence, comprehensibility, actionability, coherence, novelty, and personalization [12]. This more recent work fills in some specifics for the original XAI process model, illustrating how the evaluation criteria changes for specific XAI contexts, for specific user groups.

Much XAI research evaluating the goodness of explanations uses self-report or prediction tasks with questions like "What aspects of the news article contributed the most to this prediction?" [2]. However, examining the differences in human accuracy when assisted by decision-making systems with varying interpretability allows researchers to draw conclusions about *how* interpretability impacts accuracy, but not *why* it does, which is where a learning science lens comes in. To understand XAI users' learning process, we must go beyond self-report questions to more accurate measures of understanding [15]. In order to do this, we must have specific, measurable goals against which to evaluate an XAI. Specificity in XAI goals or objectives enables finding the best strategy for a problem [3] and to better evaluate new approaches and designs. Within the learning sciences research, one common approach to creating measurable objectives is to adopt Bloom's Revised Taxonomy of Learning Objectives [3].

Similar questions of design intent and evaluation are explored in [1] where the authors propose a learning-based approach to better assess, and compare communicative visualizations [1]. The researchers analyzed participant-selected cognitive and non-cognitive learning objectives through surveys and interviews of visualization designers. Results suggested most designers' learning objectives were from the lowest cognitive level of "recall" and "fact" knowledge dimensions [1]. We adapted this approach for designer intent and evaluation of XAI systems.

Learning objectives describe what users will be able to do after the learning experience arranged into 3 domains: cognitive, affective, and psychomotor. [4]. The *cognitive* domain is knowledge-based learning, the *affective* domain is emotion-based learning, and the *psychomotor* domain is action-based learning [4]. Objectives within the cognitive domain combine a cognitive process (verb) with a piece of knowledge (noun). Bloom's Taxonomy provides an increasing-in-cognitive complexity set of categories, and sample verbs to define a cognitive objective for each category. Refer to the literature for the full listing [4]: **Remember**: recognize, recall. **Understand**: paraphrase, exemplify, categorize, generalize, extrapolate, compare, contrast, explain. **Apply**: execute, implement. **Analyze**: differentiate, organize, attribute. **Evaluate**: detect, critique. **Create**: hypothesize, plan, produce. Bloom's Taxonomy's knowledge dimension is a range increasing in abstractness [4]: **Factual** knowledge of specific details, such as "Recall the form and parameters of a Markov chain" from Example 1 in Fig. 1. **Conceptual** knowledge integrates multiple facts. **Procedural** knowledge is skills or heuristics. **Metacognitive** knowledge is thinking about thinking.

2   Survey and Interview Studies

We still must determine if actual designers' intents also fit into these dimensions of the cognitive objective framework. To do so, we compiled a list of explainable designers from the 2018–2021 presenters from the IEEE VISxAI Workshop on Visualization for AI Explainability to participate in our IRB-approved study. According to the 5th Workshop on Visualization for AI Explainability (VISxAI)[1]: "*The goal of this workshop is to initiate a call for "explainables" / "explorables" that explain how AI techniques work using visualization.*"

Survey Study. There were 35 still accessible explainable artifacts with an average 2.4 authors per explainable. We identified 81 unique authors, 73 of which had

---

[1] https://visxai.io/.

accessible contact information. 33 unique authors completed the survey with participant backgrounds split between industry (61.8%) and academia (38.2%).

We modeled our survey items on the process in [1]. The survey asked respondents for the main goals of their explainable, to check which of a sample of cognitive objectives they had for users of their explainable, and any other learning objectives they planned. For each artifact, we provided *suggested learning objectives* that reflected our best inference of the intents of the designers based upon their artifact. Participants could also create their own learning objectives, which we refer to as *participant-suggested learning objectives*, and all did so.

32/33 respondents chose at least one of our suggested objectives as something they hoped their users would be able to do after their explainable. Labeling objectives with cognitive processes was mostly clear as the verbs in Bloom's Taxonomy could be consulted. For the knowledge dimension we developed a coding manual defining factual, conceptual, procedural, and metacognitive knowledge and had two coders label random samples until achieving a Cohen's κ coefficient of 0.71, which is considered substantial for labeling [9].

Suggested vs. Selected Learning Objectives

Verb Type	Suggested Learning Objectives (Knowledge Dimension)				Participant Selected Learning Objectives				Sel/Sug + Added
	F (Factual)	C (Conceptual)	P (Procedural)	M (Metacognitive)	F (Factual)	C (Conceptual)	P (Procedural)	M (Metacognitive)	
Remember	22	4		1	11+1	3	1		15/27 + 1
Understand	14	19	3		8+1	14+5	2+1	0+1	24/36 + 8
Apply			6				0+1	3+5	3/6 + 6
Analyze		9	3		7	3		0+3	10/12 + 3
Evaluate			3		0+1	1		0+4	1/3 + 5
Create			9			2+1		0+1	2/9 + 2

Fig. 2. Distribution of cognitive learning objectives in the surveys.

Figure 2 reflects data distribution of suggested, selected, and participant-created cognitive learning objectives. In surveys that received more than one response, we randomly selected one response to record in our data, resulting in 27 participant responses. In Fig. 2, comparing the density map of our suggested learning objectives (left), to the density map of the participant selected and suggested objectives (right), we see that participants generally agreed with our distribution of suggested objectives, except in the case of the Metacognitive knowledge dimension. These goals are generally more difficult for an outside researcher to detect, as there may not be dedicated learning activities within the explainable for Metacognitive goals such as "Build upon our experiments and formulate their own hypotheses what this might be used for..."

The most frequent objectives are classified under the Remember process and the Factual knowledge dimension (22 items). This makes sense as both Remember and Facts are the lowest cognitively complex, and is also similar to results from related work [1]. Most participants agreed with our suggested cognitive objectives classified under the cognitive process, Analyze which was chosen 83.3%. Analyze was followed by Understand at 66.7%, Remember 55.6%, Apply 50%, Evaluate 33.3%, and Create 22.2%. Most of the suggested cognitive objectives were classified under the Understand verb type, though participants more often selected Analyze objectives. The most selected knowledge dimension objective was Conceptual at 75%, and then Factual at 52.8% and Procedural at 48%.

When suggesting new objectives, the distribution of targeted objectives differed. Out of the 25 participant-suggested cognitive objectives, most participants suggested Understand (8), followed by Apply (6), Evaluate (5), Analyze (3), Create (2) and then Remember (1). Out of the 4 knowledge dimensions, most participants suggested learning objectives classified as Procedural and Conceptual (7), followed by Metacognitive (9), and Factual (2). These results show that designers of explainables prioritize supporting their users in *doing*, not just *knowing*. The remaining objectives were categorized as 10 non-learning objectives, including 3 business goals ("Reach out to the team and ask for consulting services" [P52]), 1 affective learning objectives ("This article may ignite their interest in the topic and then they may explore..." [P41]), and 1 action-based psychomotor learning objective ("Do research on <algorithm>themselves" [P92]).

Interview Study. To gain further insight into the survey responses, we performed interviews with participants who agreed to do so. 9 explainable authors agreed to a 30–45 minute semi-structured interview. The structure of our interviews was heavily influenced by prior work focusing the interview on design process, design considerations, and audience [1]. To analyze our interview data, we followed a phenomological process applying guidelines from [8]. After performing phenomenological reduction, we coded them using an iterative process with structural coding, in vivo coding, and open coding methods [14]. Our initial pass through the data was done using in vivo coding, through which we created 132 unique codes. A subsequent pass was done using open coding to further elucidate themes present throughout the data and narrow down these 119 unique codes into 5 broader thematic categories encompassing the most common codes: structure planning, play/exploration, audience, accessibility, and future goals. Finally, we used structural coding to target our research questions and to analyze the data on demographics and past experience of the interviewee. We used 6 structural codes: prior experience in XAI, teaching experience, user goals, use of learning science, intended audience, and core concepts.

All 9 participants stated that learning goals played a role in their design process, commonly working backwards from these goals. Participants used words like "*building*" to envision goals in the early stages of their process, where they create "*a stack of concepts*" (P70) that "*progressively gets more and more complicated step by step until we can get to where we actually want it to be*" (P47).

Several participants noted that they had not initially considered their goals as learning objectives. While one stated that they "*didn't write them down as learning goals*" at the time, but identifying the "*broad points I wanted people to walk away with"* (P79) was not only a key part of their planning process, but also one that they would consider to be making use of learning objectives despite not using that exact term previously. One participant proposed a solution: hiring a collaborator with a background in education, expressing *"if we now also had a set of only a publicity person, we would also have an educator person."* (P37).

A prominent theme through the interviews was learning objectives' compatability with the explorative nature of explainables. When asked whether or not they thought that further use of learning objectives would be helpful, they clarified that they *"stand by the idea of intentionally leaving it a little bit open, in terms of allowing the user to make their own conclusion"* (P9). However, these open-ended goals could be specified with higher cognitively complex learning objectives. The importance of exploration in explainables was mentioned by 7 of 9 of our participants, desiring that the user *"develop or extract a non-trivial insight playing with the material that is presented"* (P70). One participant stated having the user *"really to get a deeper sense of understanding for these individual parts by playing around with them"* (P37) was their main goal. To reckon with this, several participants balanced objectives and open-ended discovery. One designer described this balance as a *"back and forth between allowing the user to sort of interact and learn things themselves. And me sort of explaining what there is to do with it, rather than hand-holding them the entire way"* (P47).

As in the survey analysis, we also categorized each objective mentioned into the cognitive, affective, and psychomotor domains in Bloom's Revised Taxonomy [4], identifying 27 cognitive learning objectives, and 2 affective objectives. Out of these 27 objectives, the most suggested verb type was Understand (12 times by 10 unique participants). A common subtype of the Understand objective was a Summarize objective (3 times). Participants described this particular objective as being able to relay the gist of a topic *"if he's asked about [the topic] at a party"* (P103). Understand was followed by Remember (7 times), Create (3 times), Analyze (3 times), and Apply & Evaluate (1 time).

Beyond the learning objectives identified from the interviews, we observed additional non-learning objectives designers named as goals during their processes. These objectives included 1 business objective (to *"get a lot of attention"* (P80) with the result of their work), as well as desires to achieve replication of the designer's own past learning experience, and a design goal of accessibility both in terms of level of prior knowledge and of mobile/low-speed internet compatibility.

XAI designers focused on both learning and non-learning measures when evaluating their explainables. A common metric among explainables (mentioned by 6 participants) was the praise and public response generated upon release or publication. Many participants took this to be a reliable indicator of success, stating that *"other people have cited it and other people have looked at it and other people have messaged me about it...So obviously it must provide some value"* (P47). Multiple participants clarified that this metric was used due to the absence of other more formal measures: *"We're not sure about how to measure the success of these, but people like them on Twitter and that seems good"* (P79). Additional evaluation strengths authors identified included meeting their defined user goals and serving as a learning experience for the author themselves. Participants also identified 6 areas in which they felt their work fell short, including: using overly complex material, needing more time, not meeting all goals, not enough interactive components, and lacking technical robustness.

3   Conclusion

Learning objectives should lead to better XAI by providing a means to compare XAI design decisions and more rigorously evaluate the success of the XAI systems [1]. The contributions of this article to the AIED community include: investigation of the current state of informal AI literacy support, providing a broader perspective of designers' explainable design process, identifying that XAI designers struggle assessing their artifacts, exemplifying the use of learning objectives for designing XAI as helpful scaffolding, and discovering support of metacognitive goals as an important focus of explainable designer intent.

One of the most striking results is how a large number of metacognitive goals were suggested. Goals like "I hope that users will more effectively question what different algorithms they interact with are doing" (P44) are often more difficult for an outsider to infer, but these metacognitive goals are a critical piece of the rhetoric around the need for interpretable AI [13] and AI literacy. Therefore, focusing on the development and assessment of these metacognitive goals is a top priority and a fruitful avenue for AIED researchers to contribute to XAI.

Beyond cognitive objectives, many of the participant-suggested learning objectives from the survey included phrases like: "...ignite their interest...", "They might want to refer to our article whenever they are dealing with <concept>", or "Reach out to the team and ask for consulting services". These affective, reference, and business objectives (respectively) cannot be encapsulated by cognitive learning objectives, and remain important goals of XAI. The revised Bloom's Taxonomy does not include an affective dimension, and there does not appear to yet be a widely accepted holistic model that does so [4]. These affective goals are also missing from question banks for XAI, such as [11]. Communicative visualization researchers are developing affective objectives for their domain [10], and it is possible some of the affective verbs and nouns from this work could be adapted to XAI contexts. Reference objectives reflect longer term goals and also cannot be well encapsulated by XAI question banks, similar to affective objectives. Business goals made a small appearance in our results through mentions of measuring success through public use, and are also an important consideration for designers building AI explainables to increase AI literacy in informal settings.

Our studies show XAI designers are unsure how to evaluate their artifacts, most relying on their explainables' popularity on social media as a proxy for success. There is a very clear gap in how practitioners build and evaluate their post-hoc XAI, and how the research community views evaluation of post-hoc XAI. Our work opens many research opportunities for the AIED community, including: supporting XAI practitioners in developing metacognitive goals, creating an affective & reference objectives taxonomy, and investigating the relationship between exploration, spontaneity, and prescriptive XAI learning goals.

This article provides insights into how designers create XAI, focusing on their intention to change the viewer through increased AI literacy. Presenting the information in an explainable, counter-factual, or even transparent algorithm is not enough to ensure that the user correctly understands the AI model. Without that understanding, it is difficult to achieve the goals of trust and fairness for which XAI designers aim. By framing XAI goals as learning objectives, designers can evaluate whether their design was successful. In surveying and interviewing XAI designers, we demonstrated that their goals and intentions can be mapped to learning objectives. In doing so, we also discovered additional dimensions that could be added to the framework to more holistically design and evaluate XAI.

References

1. Adar, E., Lee, E.: Communicative visualizations as a learning problem. IEEE Trans. Visual Comput. Graph. **27**(2), 946–956 (2020)
2. Alvarez-Melis, D., Kaur, H., Daumé III, H., Wallach, H., Vaughan, J.W.: From human explanation to model interpretability: a framework based on weight of evidence. In: Proceedings of the AAAI Conference on HCOMP, vol. 8, p. 3. AAAI, USA (2021)
3. Ambrose, S., Bridges, M., DiPietro, M., Lovett, M., Norman, M.K.: How learning works: 7 research-based principles for smart teaching. John Wiley & Sons (2010)
4. Anderson, L.W., Krathwohl, D.R. (eds.): A Taxonomy for Learning, Teaching, and Assessing. A Revision of Bloom's Taxonomy of Educational Objectives. Allyn & Bacon, New York, NY USA (December (2001)
5. Anik, A.I., Bunt, A.: Data-centric explanations: Explaining training data of ml systems to promote transparency. In: Proceedings of the CHI Conference, pp. 1–13. ACM (2021)
6. Doshi-Velez, F., Kim, B.: Towards a rigorous science of interpretable machine learning arXiv: 1702.08608 (2017)
7. Gunning, D., Aha, D.: Darpa's explainable artificial intelligence (xai) program. AI Mag. **40**(2), 44–58 (2019)
8. Hycner, R.H.: Some guidelines for the phenomenological analysis of interview data. Hum. Stud. **8**(3), 279–303 (1985)
9. Landis, J.R., Koch, G.G.: The measurement of observer agreement for categorical data. Biometrics **33**(1), 159–174 (1977)
10. Lee-Robbins, E., Adar, E.: Affective learning objectives for communicative visualizations. IEEE Trans. on Visualiz. Comput. Grap. **29**(1), 1–11 (2023)
11. Liao, Q.V., Gruen, D., Miller, S.: Questioning the ai: informing design practices for explainable ai user experiences. In: Proceedings of the CHI Conference, pp. 1–15. ACM (2020)
12. Liao, Q.V., Zhang, Y., Luss, R., Doshi-Velez, F., Dhurandhar, A.: Connecting algorithmic research and usage contexts. In: Proceedings of the AAAI Conference on Human Computation and Crowdsourcing, vol. 10, pp. 147–159. AAAI, USA (2022)
13. Lipton, Z.C.: The mythos of model interpretability. Queue **16**(3), 31–57 (2018)
14. Saldaña, J.: The coding manual for qualitative researchers. Sage, USA (2021)
15. Sitzmann, T., Ely, K., Brown, K.G., Bauer, K.N.: Self-assessment of knowledge: a cognitive learning or affective measure? Acad. Manag. Learn. Educ. **9**(2), 169–191 (2010)

Aspect-Based Semantic Textual Similarity for Educational Test Items

Heejin Do[1]([✉]) [iD] and Gary Geunbae Lee[1,2] [iD]

[1] Graduate School of AI, POSTECH, Pohang, Republic of Korea
{heejindo,gblee}@postech.ac.kr
[2] Department of CSE, POSTECH, Pohang, Republic of Korea

Abstract. In the educational domain, identifying the similarity among test items provides various advantages for exam quality management and personalized student learning. Existing studies mostly relied on student performance data, such as the number of correct or incorrect answers, to measure item similarity. However, nuanced semantic information within the test items has been overlooked, possibly due to the lack of similarity-labeled data. Human-annotated educational data demands high-cost expertise, and items comprising multiple aspects, such as questions and choices, require detailed criteria. In this paper, we introduce a task of aspect-based semantic textual similarity for educational test items (aSTS-EI), where we assess the similarity by specific aspects within test items and present an LLM-guided benchmark dataset. We report the baseline performance by extending the STS methods, setting the groundwork for future aSTS-EI tasks. In addition, to assist data-scarce settings, we propose a progressive augmentation (ProAug) method, which generates step-by-step item aspects via recursive prompting. Experimental results imply the efficacy of existing STS methods for a shorter aspect while underlining the necessity for specialized approaches in relatively longer aspects. Nonetheless, markedly improved results with ProAug highlight the assistance of our augmentation strategy to overcome data scarcity.

Keywords: Educational Item Similarity · Semantic Textual Similarity · Dataset · Aspect-based Similarity · Natural Language Processing

1 Introduction

Analyzing the similarity between items in educational systems offers large benefits for both educators and learners. Educators can identify duplicated items or outliers when constructing exams by specifying item similarities [3]. Meanwhile, learners can receive similar question recommendations within item pools [10,12], acquiring personalized learning experiences. Highlighting the significance of gauging similarity for interactive educational systems, existing studies primarily relied on the student performance data, such as the count of correct or incorrect answers, to measure item similarity [11,16,18].

A. M. Olney et al. (Eds.): AIED 2024, LNAI 14830, pp. 344–352, 2024.
https://doi.org/10.1007/978-3-031-64299-9_30

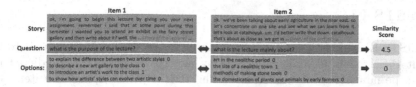

Fig. 1. An example illustrating aspect-wise similarity between two test items, where they are similar in the *Question* aspect but differ in the *Story* and *Options* aspects.

However, the semantic textual information constituting test items, which can be a direct guide to examining item similarity, has been ignored. This gap might be attributed to the lack of labeled data for directly measuring semantic similarities among educational test items, as obtaining human-annotated educational data requires expertise with high costs [12]. Generally, an educational item comprises multiple aspects, such as instructions that serve as a background to the problem, a question that triggers an answer, and options that include correct and incorrect answers. Therefore, the results can vary significantly depending on the aspect to measure similarity (Fig. 1), underscoring the need to distinguish specific aspects rather than comparing items holistically.

In this paper, we propose a task of aspect-based semantic textual similarity for educational test items (aSTS-EI), which assesses the degree of similarity between two items by specific aspects. We introduce an aSTS-EI benchmark dataset to facilitate the model evaluation, mainly targeting the under-explored English as a Foreign Language (EFL) test items. For enhanced scalability, we employed a large language model (LLM), GPT-4 [1], for similarity labeling and validated its correlation with annotations from human English experts on the test set. By extending representative pre-trained models of existing semantic textual similarity (STS) tasks, we report baseline performance on the generated dataset. Furthermore, to aid training in data-scarce settings, we introduce a progressive augmentation method (ProAug), which generates step-by-step aspects through recursive prompting, leveraging an open-sourced LLM, LLaMA2 [20].

The experimental results indicate that current STS methods can be effectively extended for relatively short *Question* aspects but require specialized methodologies for longer and complex *Options* aspects. Furthermore, the proposed ProAug method noticeably assists training, particularly demonstrating pronounced benefits in inferior aspects. Implementations across both encoder-decoder-based and contrastive-learning-based models affirm the versatile application of the generated and augmented dataset. We hope that our preliminary work could provoke the emergence of detailed aSTS-EI methods and create synergy with existing auxiliary elements such as students' response data for educational item similarity. Our datasets are available on GitHub[1].

---

[1] https://github.com/doheejin/aSTS-EI.

2   aSTS-EI

The aSTS-EI task assigns aspect-wise similarity scores for an item pair to represent their semantic similarity levels by aspects. The score is predicted in a float-type numeric value from 0 to 5. The task requires aspect-wise item pairs and the corresponding similarity labels. In our work, the aspect includes the following: *Story*, which is the background of the problem, *Question* that asks for an answer, and *Options*, which includes one correct and three incorrect choices.

Our task is motivated by a widely studied STS task, which assigns a numeric similarity score for two short sentences. Despite notable STS methods [9,17], the benchmark dataset [4] of simple daily sentences limits their direct application to the education domain. Further, the similarity between item pairs can vary greatly by the aspect (e.g., two items can be similar in terms of *Question* but totally different regarding *Options*). As aSTS-EI defines the aspect-wise similarity between two test items, more direct and refined item comparisons are achievable.

Table 1. The original TOEFL-QA and the generated aSTS-EI dataset compositions.

	Original				aSTS-EI		
	Total	Story	Question	Options	Story	Question	Options
Train	717	150	717	717	900	1,434	1,434
Valid	124	24	124	124	276	610	610
Test	122	24	122	122	276	610	610

2.1   Dataset Description

We adopt the open-sourced TOEFL-QA[2] [6,21] collection dataset for the educational test items, where each sample comprises three aspects: *Story*, *Question*, and *Options* (Table 1). *Options* has one correct and three incorrect choices, and we regard concatenated four choices as a single aspect unit. However, we exclude the *Story* in experiments as 1) it is shared among four or five *Question-Options* pairs, 2) given its lengthy context averaging 3736, treating it as an entity with specific semantic meaning for similarity is impractical, and 3) its explanatory *Lecture* or interactive *Conversation* type requires further discussions for assessment. Yet, we include it in our data construction to leave room for future studies.

Item Pair Selection. Pairs of items are configured individually for each aspect. For the *Story* aspect, where data is highly lacking, we arrange all distinct items to be paired with each other to construct the validation and test set. For the training set, we randomly select six samples for each *Story* example without

---

[2] https://github.com/iamyuanchung/TOEFL-QA

duplication to ensure an even selection of items. In *Question* and *Options*, we randomly select five samples for each example for validation and test set and select two samples for the training set without duplication (Table 1).

Similarity Annotation. Recently, many studies have discussed the potential of large language models such as GPT-3 [2] and ChatGPT [15] as an alternative to human evaluations [5,8,13]. In particular, [5] explores using LLMs to assess text quality with a 1–5 Likert scale, showing comparable results to expert human annotations. Inspired by those studies and considering future scalability, we employed GPT-4 to label the similarity between aspect-wise item pairs. To achieve multiple grading effects, we sample twice and use the average for the test set. The following is a detailed prompt used for the *Question* aspects pair similarity annotation: `"Please assess the similarity between question1 and question2 of TOEFL test item:\n Question1:{Item1 Question}\n Question2:{Item2 Question}\n Response only with the similarity score between 0-5:"`. The same prompt is used for the *Story* and *Options* aspects with different aspect names and values.

Human Evaluation. Despite the potential of LLMs, their reliability remains a point of debate. Further, more rigorous scrutiny is required for the test sets to evaluate models. Therefore, we conduct human experts annotation[3] to rate the similarity for 610 *Question* samples on a 0–5 scale and examine the correlations with GPT-4 generated labels. Following the prior work [5], we measure Kendall's τ, Spearman's ρ, and Pearson's r correlation coefficient to observe the relevance of rating trends. Table 2 exhibits positive correlations between teacher assessments and GPT-4 ratings with less than 0.001 p-values, implying that when experts assign a higher rating, GPT-4 also tends to assign a higher score.

Table 2. Correlation coefficient between the GPT-4 and human annotations.

	Kendall's τ	Spearman's ρ	Pearson's r
Question	0.250	0.319	0.522
aspect test set	(p-value<0.001)	(p-value<0.001)	(p-value<0.001)

3   Progressive Augmentation

While we present the aSTS-EI dataset in this work, real-world educational items pair data are still lacking. To further assist data-scarce settings, we propose a ProAug method (Fig. 2), leveraging the publicly available LLM, LLaMA2-13B. As aforementioned, we also exclude the *Story* aspect when applying ProAug.

---

[3] Two proficient English teachers with 100% job success are employed on the Upwork, https://www.upwork.com.

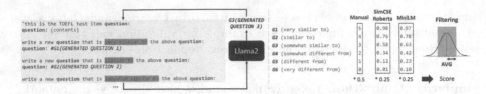

Fig. 2. Overview of the progressive augmentation method. The left describes the step-by-step generation, and the right shows the score-matching and filtering process.

Item Pair Generation. ProAug first takes a specific aspect of the original item as input and sequentially generates samples from highly similar to highly dissimilar. Assuming manual scores ranging from 0 to 5 as labels of the generated dataset, we divide the augmentation into six levels. From `very similar to` (5) to `very different from` (0), augmentation is processed step-by-step, with the output text of the current step directly combined with the next-step prompt.

Score Matching and Filtering. After obtaining item pairs per aspect, we match the pseudo similarity score for item pairs. In addition to the manual label, which is the initial score assumed for each level, two different scores are acquired. Particularly, the well-known pre-trained models SimCSE-Roberta [9] and MiniLM [22] are utilized for scoring via inference. The final pseudo score is obtained by the weighted sum with 0.5, 0.25, and 0.25 weights for manual, SimCSE, and MiniLM scores, respectively. To provide more refined scores, we conduct filtering on the pseudo-labeled scores. We only choose the aspect pairs where all three scores fall within a 1.5-threshold range from their mean as $|s_i - \frac{1}{N} * \Sigma_{i=1}^{N} s_i| < 1.5$ for $s_i \in \{s_1, s_2, s_3\}$, where s_i denotes each of three scores.

Fig. 3. The overview of T5 (a) and supervised SimCSE (b). T5 receives a two-sentence concatenated input, while SimCSE obtains separate sentence inputs with a bi-encoder.

Adaptation for Contrastive Learning. We introduce the potential utility of our dataset to acquire sentence embeddings via contrastive learning with triplet pairs by data restructuring. Unlike the conventional use of NLI data by considering *Contradictions* as negatives and *Entailment* relationships as positives, we construct triplet pairs based on the score degree. For each aSTS-EI data sample, we designated the only pair with the highest score as positive, regarding the remaining samples as negatives. As richer negative sampling generally contributes to enhanced learning [23,24], we intend to construct more negatives. For each ProAug sample, we consider the generated samples with 3, 4, and 5 manual scores as positives, while those with 0, 1, and 2 as negative candidates. Among candidates, all possible triplet pairs are matched ($ProAug_{in}$). In addition, we suggest $ProAug_{rand}$, where 4 and 5 manual-scored samples are considered positives, while negative candidates are randomly sampled from other examples.

Table 3. Experimental results with T5. *+ProAug w/o F* and *+ProAug w/ F* are separately added to the *T5(aSTS-EI)*. **Bold** denotes the best performance in the column.

Models	Question		Options	
	Pearson's r	Spearman's ρ	Pearson's r	Spearman's ρ
T5 (aSTS-EI)	0.822	0.738	0.456	0.285
+ProAug w/o F	0.803	0.732	**0.622**	**0.443**
+ProAug w/ F	**0.833**	**0.753**	0.593	0.305

Table 4. Experimental results with SimCSE. <u>Underline</u> is the second-best performance.

Models	Cosine		Manhattan		Euclidean		Dot Product	
	ρ	r	ρ	r	ρ	r	ρ	r
SimCSE (aSTS-EI)	0.547	0.550	0.645	0.556	0.637	0.550	0.548	0.556
+ProAug$_{in}$	<u>0.608</u>	<u>0.606</u>	<u>0.688</u>	<u>0.609</u>	<u>0.687</u>	<u>0.606</u>	<u>0.606</u>	<u>0.604</u>
+ProAug$_{rand}$	0.594	0.553	0.660	0.546	0.665	0.553	0.594	0.553
Total	**0.644**	**0.642**	**0.715**	**0.635**	**0.717**	**0.642**	**0.642**	**0.638**

4   Experiment

As the baseline model, we mainly employ the transformer-based pre-trained language model, T5 [17], which excels in various tasks, including the STS. It inputs two sentences with a prefix and outputs a text-type score. We fine-tune T5-large on the generated aSTS-EI dataset by setting "`similarity:`" as a prefix, followed by two aspects (Fig. 3). As the pre-trained T5 received two sentences to compare similarity as "`sentence1:`" and "`sentence2:`", we keep the same format. In addition, we explore the potential of our data in contrastive learning with the SimCSE model, which focuses on sentence representations. As SimCSE is designed for a sentence input, we applied it to the *Question* aspect, which has a single sentence, particularly using the pre-trained RoBERTa

[14] to encode input sentences. Given the triplet pair (q_i, q_i^+, q_i^-), where q_i is the i-th *Question*, q_i^+ is semantically related, and q_i^- is unrelated, their representations r_i, r_i^+, and r_i^- are obtained. The training objective is defined as $-\log(e^{\text{sim}(r_i, r_i^+)/\gamma}/\Sigma_{j=1}^{N}(e^{\text{sim}(r_i, r_j^+)/\gamma} + e^{\text{sim}(r_i, r_j^-)/\gamma}))$, where $\text{sim}(\cdot)$ is cosine similarity, N is the number of sentences in a mini-batch, and γ is the temperature hyperparameter. For T5, we used a batch size of 4, 12 epochs, 0.1 weight decay, and 1e-5 learning rate with a `A100-PCIE-40GB GPU`. For SimCSE, we set a batch size of 8, 2 epochs, 5e-5 learning rate, 0.001 temperature, and 0.2 warmup ratio.

5   Results and Discussions

Table 3 presents the results of training the T5 model on our dataset. For the *Question* aspect mostly composed of a single sentence, the generated dataset alone demonstrates competitive performance. However, for *Options*, which comprises concatenated four sentences, it shows relatively lower performance. This result might be due to current STS methods mainly focusing on similarity between short, single sentences; hence, the need for specialized models to handle longer-context aspects is emphasized. Applying our ProAug with filtering improves performance in both aspects, with a notable 13.7% increase in Pearson's r specifically for *Options*. Results of ProAug without filtering reveal that filtering strategy promotes more accurate prediction for *Question* aspect. Notably, for *Options*, not implementing any refinement yields better results, suggesting that given the inherent complexity of *Options* exhibiting inferiority with the initial data alone, the quantity might outweigh quality in augmentation.

Table 4 exhibits the results of applying our reconstructed aSTS-EI and ProAug data to the SimCSE model. Notably, the in-sample positive and negative sampling method applied ProAug data (ProAug$_{in}$) outperforms the random negative sampling used one (ProAug$_{rand}$). This result aligns with the existing research highlighting the importance of hard negative samples [7,19] over random ones. Overall, the combined use of them performed the best, showing an average 8.6% improvement over using only the initial aSTS-EI. The results suggest the synergistic use of both in-sampling and random-sampling approaches.

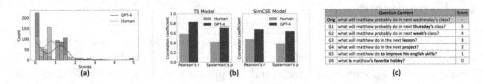

Fig. 4. (a) GPT-4 and Human score-label distributions (b) Performance comparison on Human and GPT-4 annotated test sets (c) Step-wise Generated *Question* examples.

We conducted detailed quantitative analyses of the aSTS-EI and ProAug datasets. To further examine the aSTS-EI test set, which has proven positive correlations with human annotations, we investigate the score distributions of the two values (Fig. 4a). The two score-label distributions show a similar trend across the scores. However, the performance of T5 and SimCSE models, fine-tuned with both aSTS-EI and ProAug data, notably drops when evaluated on the human-annotated test set (Fig. 4b). This suggests that our training data, also generated by LLM, might lead to a favorable assessment of the model. The result underscores future demands to explore the aSTS-EI-focused training methods or architecture rather than relying on LLM-generated training sets. In addition, we investigate whether the ProAug-generated samples align well with the intended manual similarity score degrees (Fig. 4c). As steps progress (G1G6), the generation coincides with our intentions. However, there are instances like G3 and G4 where 1-point differences are ambiguously distinguished, highlighting the need for our weighted score matching strategies over the sole use of manual scores.

Conclusion. We propose the aSTS-EI task, which considers aspect-wise nuanced semantic information between educational items to measure the similarity. Aiming to provide more practical item similarity analysis in education, we created a benchmark dataset by leveraging a language model, setting the foundation for further study. The ProAug method, via aspect-based step-by-step prompting, effectively mitigates data scarcity. We anticipate that aSTS-EI and ProAug will drive further explorations in developing more effective and practical systems.

Acknowledgments. This work was partly supported by Institute of Information & communications Technology Planning & Evaluation (IITP) grant funded by the Korea government (MSIT) (No. 2022-0-00223, Development of digital therapeutics to improve communication ability of autism spectrum disorder patients), (No. 2019-0-01906, Artificial Intelligence Graduate School Program (POSTECH)), and Smart HealthCare Program funded by the Korean National Police Agency (KNPA) (No. 220222M01).

References

1. Achiam, J., et al.: Gpt-4 technical report. arXiv preprint arXiv:2303.08774 (2023)
2. Brown, T., et al.: Language models are few-shot learners. Adv. Neural Inform. Process. Syst. (2020)
3. Cechák, J., Pelánek, R.: Experimental evaluation of similarity measures for educational items. Intern. Educ. Data Mining Soc. (2021)
4. Cer, D., et al.: SemEval-2017 task 1: Semantic textual similarity multilingual and crosslingual focused evaluation. In: SemEval-2017. ACL (2017)
5. Chiang, C.H., Lee, H.Y.: Can large language models be an alternative to human evaluations? arXiv preprint arXiv:2305.01937 (2023)
6. Chung, Y.A., Lee, H.Y., Glass, J.: Supervised and unsupervised transfer learning for question answering. In: NAACL HLT (2018)
7. Formal, T., et al.: From distillation to hard negative sampling: Making sparse neural ir models more effective. In: Proceedings of the 45th International ACM SIGIR Conference on Research and Development in Information Retrieval (2022)

8. Fu, J., et al.: Gptscore: Evaluate as you desire. arXiv preprint arXiv:2302.04166 (2023)
9. Gao, T., et al.: Simcse: simple contrastive learning of sentence embeddings. arXiv preprint arXiv:2104.08821 (2021)
10. Geng, C., et al.: A recommendation method of teaching resources based on similarity and als. J. Phys. Conf. Ser. (2021)
11. Harbouche, K., et al.: Measuring similarity of educational items using data on learners' performance and behavioral parameters: Application of new models scnn-cosine and fuzzy-kappa. Ingenierie des Systemes d'Information (2023)
12. Huang, T., Li, X.: An empirical study of finding similar exercises. arXiv preprint arXiv:2111.08322 (2021)
13. Liu, Y., et al.: Gpteval: Nlg evaluation using gpt-4 with better human alignment. arXiv preprint arXiv:2303.16634 (2023)
14. Liu, Y., et al.: Roberta: a robustly optimized bert pretraining approach. arXiv preprint arXiv:1907.11692 (2019)
15. OpenAI, T.: Chatgpt: Optimizing language models for dialogue. openai (2022)
16. Pelánek, R.: Measuring similarity of educational items: an overview. IEEE Trans. Learn. Technol. (2019)
17. Raffel, C., et al.: Exploring the limits of transfer learning with a unified text-to-text transformer. J. Mach. Learn. Res. (2020)
18. Rihák, J., Pelánek, R.: Measuring similarity of educational items using data on learners' performance. Inter. Educ. Data Mining Soc. (2017)
19. Santhanam, K., et al.: Colbertv2: effective and efficient retrieval via lightweight late interaction. arXiv preprint arXiv:2112.01488 (2021)
20. Touvron, H., et al.: Llama 2: Open foundation and fine-tuned chat models. arXiv preprint arXiv:2307.09288 (2023)
21. Tseng, B.H., et al.: Towards machine comprehension of spoken content: Initial toefl listening comprehension test by machine. In: INTERSPEECH (2016)
22. Wang, W., et al.: Minilm: deep self-attention distillation for task-agnostic compression of pre-trained transformers. Adv. Neural Inform. Process. Syst. (2020)
23. Xiong, L., et al.: Approximate nearest neighbor negative contrastive learning for dense text retrieval. arXiv preprint arXiv:2007.00808 (2020)
24. Zhan, J., et al.: Optimizing dense retrieval model training with hard negatives. In: Proceedings of the 44th International ACM SIGIR Conference on Research and Development in Information Retrieval (2021)

Adapting Emotional Support in Teams: Quality of Contribution, Emotional Stability and Conscientiousness

Isabella Saccardi[✉] and Judith Masthoff

Utrecht University, Utrecht, The Netherlands
{i.saccardi,j.f.m.masthoff}@uu.nl

Abstract. Teamwork is widely used in higher education for its learning benefits, but students often face issues while working in teams. A peer assessment tool can detect such issues and simultaneously provide support: this could be done by adding a virtual agent throughout the survey, but what should this agent say? This research describes two studies to inform the design of such an agent. We investigate the adaptation of support statements to a student filling in a peer assessment tool. This adaptation is based on the rater's Emotional Stability and Conscientiousness, and the score assigned to a teammate's quality of contribution. This adaptation is summarized in an algorithm for a peer assessment tool.

Keywords: Emotional Support · Personality · Collaborative Learning

1 Introduction and Related Works

College years are often portrayed as a time of personal growth, but this life stage can also come with new challenges, such as increased academic demands - a common source of stress among students [3]. Finding new ways to support students in their academic careers is necessary. In the present work, we focus on supporting teamwork: it is widely used due to its learning benefits [4,12], but disliked by students due to teamwork problems [2,5]. To intervene and prevent such issues, peer assessment surveys have been proposed as a way to improve grade fairness [1], but also to monitor and support students [16]. This work is based on the peer assessment survey by [16], where students rate their teammates on five dimensions of group work and a virtual agent gives adaptive feedback based on the ratings. The results are then communicated to the teachers, who intervene when necessary. This paper focuses on designing the feedback for the virtual agent, which in related work simply reacts positively to high scores and negatively to low scores. However, supportive feedback is a more nuanced process. Supportive communication means expressing encouragement and appreciation to help another cope with negative emotions, showing to the recipient that someone cares [6]. Previous research on personalization shows that people adapt support messages to the recipient's personality and stressors [9,11,18], suggesting that adaptation is a necessary characteristic of supportive feedback. In the present work, we focus on two personality traits associated with academic achievements,

A. M. Olney et al. (Eds.): AIED 2024, LNAI 14830, pp. 353–362, 2024.
https://doi.org/10.1007/978-3-031-64299-9_31

Conscientiousness and Emotional Stability [7,10]. This work builds on related work [16] by investigating what the virtual agent should say during the peer assessment survey. We aim to create a supportive feedback algorithm based on the rater's Conscientiousness and Emotional Stability levels and the rating assigned to a teammate. Similar work and methodology have investigated adaptation to Emotional Stability and ratings on Productivity [15]. We extend this by focusing on two traits and another teamwork dimension: Quality of Contribution.

2   Methods

Two experiments with the same design were conducted, one on Emotional Stability (ES) and one on Conscientiousness (Con). The design is explained below.

2.1   Materials

Validated personality stories were used to depict a student's personality. The stories reliably describe one trait at a high or low level while being neutral on the other traits (see [8,17]). They were adapted to be gender-neutral[1].

Emotional support statements. Both studies used 24 emotional support statements from a validated corpus; full details about the corpus creation and validation in [15]. It was created from a pool of 143 statements provided by 23 teachers; these statements were processed and underwent two validation studies, resulting in 69 statements reliably[2] categorized into emotional support categories. From this validated corpus, [15] selected 24 focusing on Productivity for their study. These were modified to apply to Quality of Contribution and used in our studies. Four support categories (Cat) are used: **Advice (A)**, suggesting actions to take, further divided into the subcategories of clarifying expectations between teammates (**A-Exp**), giving feedback to the other (**A-Feed**), or discussing how to improve (**A-Impr**); **Celebration (C)**, expressing happiness and celebration for the situation; **Empathy (E)**, expressing regret and sorrow for Alex's experiences, and **Supported (S)**, recommending involving a teacher. 6 statements were used for each category (2 for each A subcategory); see Table 1.

2.2   Procedure

For each study, a survey was distributed via Prolific. A story about a fictional student, Alex, was shown, depicting Alex's level on one of the two personality traits (ES for the first study, Con for the second). Participants were told that Alex was working on a ten-week project with Robin, and that after two weeks

---

[1] The stories are in supplementary materials: https://doi.org/10.5281/zenodo.10877474.

[2] A statement was reliably categorized with *Free-Marginal Kappa* $\geq 0.4(\kappa)$[14].

Table 1. Categories and statements used in the study.

Cat	Statements
A	(A-Exp1) If you have not yet done so, agree clear expectations with Robin on the quality of their contribution. (A-Exp2) I think you need to discuss with Robin the types of expectations you both have on the quality of the work, and come to some agreement. (A-Feed1) Tell Robin how you feel about the lack of quality of their contribution. (A-Feed2) I recommend you tell Robin you are quite happy with the quality of their contribution. They will be happy to hear so. (A-Impr1) Perhaps you can talk with Robin on how to improve the quality of their work. (A-Impr2) Perhaps you can talk with Robin on how to improve the quality of their work even more.
C	(C1) Delighted that you are so happy with the quality of Robin's work. (C2) Good to see the collaboration is going well! (C3) Delighted to hear this. (C4) Congratulations. (C5) Well done, keep up the good work. (C6) Congratulations on having a good teammate.
E	(E1) I'm sorry you are having some difficulties. (E2) I'm sorry you are having a tough time. (E3) Really sorry to hear that the quality of Robin's work did not meet your standards. (E4) Really sorry to hear the quality of Robin's work has not been very good. (E5) Really sorry to hear. (E6) Sorry that Robin did not do so well.
S	(S1) The teacher will talk to Robin. (S2) Please let me know if you would like me to raise this with the teacher. (S3) I will tell the teacher. (S4) I will raise this with the teacher. (S5) I will let the teacher know so that they can help you. (S6) Please let me know if you would like the teacher to talk with Robin.

Alex rated Robin on Quality of Contribution, "the quality of the work provided by an individual for the group project", with a score from 1 to 5 ($1 =$ awful, $5 =$ great). Participants had the role of a teaching assistant and gave feedback to Alex by selecting support statements from those in Table 1, presented randomly (User-As-Wizard methodology [13]). They could also add comments. Both studies had a 2x5 between-subject design, with 2 trait levels, Low(L) and High(H), and 5 Quality of Contribution Scores (Sc), 1 to 5, resulting in 10 conditions.

3   Results

In each study, we conducted a 2-way MANOVA to test the effect of trait level (L or H) and Score (1 to 5) on the chosen emotional support category. Then, we conducted a post-hoc Tukey HSD pairwise comparison to produce homogeneous subsets of scores on the number of statements used for each category. Tables 2,3 and 4 show both studies' descriptive statistics and homogeneous subsets.

Table 2. ES and Con: descriptives[a] of category occurrences per condition.

Sc	Lvl	ES									Con								
		MT	A		C		E		S		MT	A		C		E		S	
			Avg	M	Avg	M	Avg	M	Avg	M		Avg	M	Avg	M	Avg	M	Avg	M
1	L	2	1.30	1	.07	0	.60	.5	.37	0	2.5	1.40	1	.03	0	.50	0	.60	.5
	H	3	1.17	1	.00	0	.63	.5	.87	1	2	1.40	1	.03	0	.50	0	.57	0
2	L	2	1.53	1.5	.00	0	.77	0	.60	.5	2.5	1.63	1	.13	0	.37	0	.37	0
	H	2	1.53	1	.03	0	.60	0	.37	0	2	1.53	1	.03	0	.30	0	.43	0
3	L	2	1.40	1	.20	0	.43	0	.17	0	2	1.60	1	.13	0	.10	0	.27	0
	H	3	1.83	1.5	.07	0	.17	0	.57	0	1	1.30	1	.23	0	.17	0	.10	0
4	L	1.5	.67	1	1.40	1	.13	0	.03	0	1.5	.83	1	.83	1	.10	0	.10	0
	H	2.5	1.07	1	1.50	1	.03	0	.07	0	2	.67	.5	1.53	1	.00	0	.03	0
5	L	2	.57	1	1.67	2	.03	0	.07	0	1	.63	1	1.13	1	.10	0	.07	0
	H	2	.50	.5	1.73	2	.00	0	.10	0	3	.40	0	2.27	2	.00	0	.00	0

[a]Lvl = level, MT = median number of statements, Avg = Average, M = Median (cat)

Table 3. ES and Con: descriptives of subcategory occurrences per condition

SC	Lvl	ES						Con					
		A-Exp		A-Feed		A-Impr		A-Exp		A-Feed		A-Impr	
		Avg	M	Avg	M	Avg	M	Avg	M	Avg	M	Avg	M
1	L	.70	1	.20	0	.40	0	.60	.5	.20	0	.60	1
	H	.60	.5	.13	0	.43	0	.60	1	.27	0	.53	0
2	L	.60	0	.23	0	.70	1	.70	1	.30	0	.63	1
	H	.80	1	.17	0	.57	1	.73	1	.17	0	.63	1
3	L	.53	0	.27	0	.60	1	.80	1	.20	0	.60	1
	H	.83	1	.17	0	.83	1	.70	1	.13	0	.47	0
4	L	.13	0	.40	0	.13	0	.13	0	.50	.5	.20	0
	H	.20	0	.50	.5	.37	0	.07	0	.33	0	.27	0
5	L	.00	0	.53	1	.03	0	.17	0	.40	0	.07	0
	H	.00	0	.50	.5	.00	0	.00	0	.37	0	.03	0

ES Study. 300 participated aged 18-70 ($Avg = 30.1$; $SD = 9.6$), 30 per condition (15 male, 15 female). We found a main effect of Score for all categories ($F(16,877.4) = 17.25$; $p<.001$)[3]. For S, we found an effect of ES level ($F(4,290) = 5.34$; $p<.05$) and an interaction ES level x Score ($F(4,290) = 4.46$; $p<.05$).

Con Study. 300 participated aged 19-70 ($Avg = 30.4$; $SD = 8.7$), 30 per condition (15 male, 15 female). We found a main effect of Score for all categories ($F(16,877.4) = 16.71$; $p<.001$)[4]. For C, we found an effect of Con

---

[3] A: $F(4,290) = 16.43$; C: $F(4,290) = 51.69$; E: $F(4,290) = 15.99$; S: $F(4,290) = 12.21$.
[4] A: $F(4,290) = 15.46$, C: $F(4,290) = 53.29$, E: $F(4,290) = 11.61$, S: $F(4,290) = 13.39$.

level $(F(4,290) = 15.54; p<.001)$ and an interaction Con level x Score $(F(4,290) = 6.49; p<.001)$.

Table 4. ES and Con: Homogeneous subsets (SAvg = subset average).

Lvl	ES								Con							
	A		C		E		S		A		C		E		S	
	Sc	SAvg	Sc	SAvg	Sc	SAvg	Sc	SAvg	Sc	SAvg	Sc	SAvg	Sc	SAvg	Sc	SAvg
L	1,2,3	1.41	1,2,3	0.09	1,2,3	0.60	1,2	0.48	1,2,3	1.54	1,2,3	0.10	1,2	0.43	1,2,3	0.41
	4,5	0.62	4,5	1.53	1,3,4	0.39	1,3,5	0.20	1,4	1.12	4,5	0.98	2,3,4,5	0.17	2,3,4,5	0.20
					3,4,5	0.20	3,4,5	0.09	4,5	0.73						
H	2,3	1.68	1,2,3	0.03	1,2	0.62	1,3	0.72	1,2,3	1.41	1,2,3	0.10	1,2	0.40	1,2	0.50
	1,2,4	1.26	4,5	1.62	3,4,5	0.07	2,3	0.47	3,4	0.98	4	1.53	2,3,4,5	0.12	3,4,5	0.04
	4,5	0.78					2,4,5	0.18	4,5	0.53	5	2.27				

Comments reflection. Participants added 68 comments in the Con study, 57 in the ES one: we report the relevant ones for the most used categories.

Scores 1–2. 50 comments were provided for these scores. 23 stress the need for communication, as shown by the common choice of **A** for these scores. People show empathy (**E**) because Alex is struggling or because scores are low: three comments expressed concern for Low ES Alex's (e.g. *"Alex must not stress too much, such tends to happen."*) and one regret for High ES Alex. Contrasting opinions were provided about teacher's involvement (**S** category): two suggested involving teachers due to Low ES Alex's discomfort, while two others recommended speaking privately first - also recommended by three in the Con study. A few suggested that Alex's personality could influence the impartiality of the ratings. Three doubted Low ES Alex (e.g. *"Hopefully there isn't any projecting of feelings of unhappiness with self to Robin."*). One trusted High ES Alex instead (*"Alex's description makes me think that they would be objective"*). Five doubted Low Con Alex (e.g. *"I don't know if the rating was biased by the fact that Alex doesn't put the effort and hopes that someone does everything."*).

Score 3. **A** is central again, as highlighted by six (from 24) comments mentioning the need to communicate, and **S** suggested for High ES. Seven considered Alex's Con level: Low Con Alex may benefit from Robin (e.g. *"Speaking to Robin [..] would help and also highlight your weak points"*). High Con Alex may expect too much (*"Alex's description sounds like they are a good little hall monitor, it's pretty obvious that everyone else would seem average to them."*) or may inspire Robin (*"Robin doesn't seem a bad colleague at all, but could use [..] guidance"*).

Scores 4–5. High scores mean **C** and **A-Feed** for giving positive feedback to Robin. Low ES Alex does not have another reason to stress, as mentioned by three comments (e.g. *"I'm happy to see Alex was assigned with a partner that can potentially help them reduce the stress"*). 11 mention Alex Con level: Low Con Alex could be inspired (*"I hope Alex can also pick up a few points from Robin to have less procrastination"*), while High Con Alex has a suitable teammate (*"It's great to be partnered with someone who works just as hard as you do!"*).

3.1 Algorithm Creation

The following steps are taken to decide what emotional support to use when.

Step 1: Category decisions for high and low levels. For each condition, we decide how many statements to use per category based on two methods: (1) M-Dec using the Median of Table 2 and (2) Sub-Dec using the subset average of Table 4[5]. In both methods, 0 statements are used if <0.5, 1 if between 0.5 and 1.5, and 2 if ≥1.5. The results are combined into an overall decision. When results differ, decisions are guided by (a) the median number of statements overall in that condition (MT in Table 2), and (b) similarity with the other level case. E.g., for Sc 1 - Low ES, A E is chosen over A, as (a) MT is 2 and (b) A E is also more similar to A E S, which is used for Sc 1 - High ES. Table 5 shows the results[6].

Step 2: Category decisions for medium levels. Our studies investigated high and low levels as the stories express these. Decisions for medium levels rely on the validation of the stories, which were proven to depict high and low traits while depicting *a neutral level on the other traits* [8,17]. So, the High/ Low Con results inform the decision for Medium ES, and the High/ Low ES ones for Medium Con. We decide as follows. For Medium Con, a category is always used if used for (1) both High and Low ES, and either High or Low Con, or (2) both High and Low Con, and either High or Low ES. Exceptionally, it is used for Medium Con if used for High and Low ES but neither Con condition *if its exclusion for High and Low Con is plausible*. It is then also used for Medium ES[7]. Similar criteria are used for Medium ES. Table 5 (grey) shows the decisions.

Step 3: Decisions on subcategories. When A was selected, a subcategory is chosen based on the medians in Table 3. When one A is needed but all medians are 0, the subcategory with the highest average is used. When two have the same median and only one is needed, the choice is left open (indicated by "OR"). For medium levels, decisions are made using the same method as Step 2's category selection, with two changes. Firstly, for Sc 2 the "OR"s for High ES and High Con are treated as if both A-Exp and A-Impr were used. Secondly, the number of subcategories used has to match the number of As used[8], so "OR" is used in the Sc 2 and 3 medium decisions. Table 5 (A-Dec) shows the decisions.

---

[5] When a score is in multiple subsets, we use the subsets combination's average.

[6] There are two complex cases. For Sc 4 - ES High MT is 2.5 which can be 2 or 3 statements. For similarity with ES Low, A C is chosen over A 2C. For Sc 3 - ES High, the medium is 3, but 2A is chosen over 2A S as it is more similar to A, the choice for low ES. S came from combining two subsets of which one had no S.

[7] For Sc 1,2 excluding E for High, Low Con is plausible given comments that High Con may mean too high expectations and Low Con may mean unreliable ratings and bad work by Alex themselves.

[8] Which could include selecting multiple times the same subcategory.

Step 4: Decisions for combinations of personality traits. Next, we decide what to use per Score for *combinations* of ES and Con, integrating the results for the individual traits. These decisions are made per Score as follows:

R1 When one trait is Medium, the result from the other trait is used. For example for Sc 1 - Medium Con, we use A E for Low ES, and A E S for Medium and High ES, following the ES results from Table 5.
R2 When a category is used for a level of one trait and a level of the second, then it is used for the combination. For example, for Sc 3, High ES and Low Con use a second A, so this is used for the High ES-Low Con combination.
R3 When a category is absent for only one level of only one trait (say TL), and this absence is plausible, then the category is used for all combinations except those with TL (e.g., for Sc 5, the absence of a second C for Low Con is plausible, so 2C is used for all combinations except those with Low Con).

The rules are applied similarly for the subcategories. Table 6 shows the decisions.

Step 5: Decisions on the order of (sub)categories. When multiple (sub) categories are chosen, their order is based on the order in which they were most often combined. For example, for Sc 5 - Low Con and any ES, we consider all cases in those conditions where at least one A-Feed and one C co-occurred. C was used first in 19 out of 31 cases, so we opt for C; A-Feed. When orders are equivalent or highly similar, we opt for the most similar to the other cases[9].

Table 5. Steps 1–3 of algorithm decisions. *Step 1*: high and low trait level categories (Dec). *Step 2*: medium levels (grey rows). *Step 3*: A subcategories (A-Dec)

Score	Lvl	ES Sub-Dec	ES M-Dec	ES Dec	ES A-Dec	Con Sub-Dec	Con M-Dec	Con Dec	Con A-Dec
1	Low	A	A E	A E	A-Exp	A	A S	A S	A-Impr
	High	A E S		A E S	A-Exp	A S	A	A S	A-Exp
	Medium			A E S	A-Exp			A E S	A-Exp
2	Low	A E	2A S	A E	A-Impr	2A	A	2A	A-Exp, A-Impr
	High	A E	A	A E	A-Exp OR A-Impr	A S	A	A S	A-Exp OR A-Impr
	Medium			A E	A-Exp OR A-Impr			A E	A-Exp OR A-Impr
3	Low	A		A	A-Impr	2A	A	2A	A-Exp, A-Impr
	High	2A S	2A	2A	A-Exp, A-Impr	A		A	A-Exp
	Medium			A	A-Exp OR A-Impr			A	A-Exp OR A-Impr
4	Low	A 2C	A C	A C	A-Feed	A C		A C	A-Feed
	High	A 2C	A C	A C	A-Feed	A 2C	A C	A C	A-Feed
	Medium			A C	A-Feed			A C	A-Feed
5	Low	A 2C		A 2C	A-Feed	A C		A C	A-Feed
	High	A 2C		A 2C	A-Feed	A 2C	2C	A 2C	A-Feed
	Medium			A 2C	A-Feed			A 2C	A-Feed

---

[9] For Sc 5, Low Con and High and Medium ES, the preferred order is A-C-C, but C-A-A was used only once less and is preferred for consistency.

Table 6. Steps 4–6 of algorithm decisions. *Step 4:* categories (Cats) and advice (A-Dec) for the combinations of traits. *Step 5,6:* order and specific statements

Sc	Con	ES	Decisions (and Rules)				
			Cats		A-Dec		Order & Statements
1	High	Low	A	R2	A-Exp	R3	A-Exp2
		High, Medium	A S	R2 R3			A-Exp2; S6
	Low	Low	A	R2	A-Impr	Special case^a	A-Impr2
		High, Medium	A S	R2 R3			A-Impr1; S6
	Medium	Low	A E	R2 R1	A-Exp	R3	E4; A-Exp2
		High, Medium	A E S	R2 R1 R3			E4; A-Exp2; S6
2	High	High, Low	A	R2	A-Impr	R2	A-Impr1
		Medium	A S	R2 R1			A-Impr1; S6
	Medium	Any	A E	R2 R1	A-Impr	R2	A-Impr1; E3
	Low	Low	A	R2	A-Impr	R2	A-Impr2
		High, Medium	2A	R2 (R2, R1)	A-Exp, A-Impr	(R2, R1), R2	A-Impr2; A-Exp2
3	High, Medium	Medium	A	R2	A-Exp	R2	A-Exp2
		Low			A-Impr	Special case^b	A-Impr2
	High	High	A	R2	A-Exp	R2	A-Exp2
	Medium		2A	R2 R1	A-Exp, A-Impr	R2, R2	A-Exp2; A-Impr2
	Low	High, Medium	2A	R2 (R2, R1)	A-Exp, A-Impr	R2, R1	A-Exp2; A-Impr2
		Low	A	R2	A-Impr	R2	A-Impr2
4	Any	Any	A C	R2 R2	A-Feed	R2	C2; A-Feed2
5	High, Medium	Any	A 2C	R2 R2 R3	A-Feed	R2	C2; C1; A-Feed2
	Low		A C	R2 R2			C2; A-Feed2

^{a} R3 results in A-Exp for all bare Low Con which uses A-Impr as recommended for Low Con ^{b} High Con used A-Exp, Low ES A-Impr; A-Impr used in line with Sc 2.

Step 6: Decisions on specific statements. For each chosen (sub) category, we select the most used statement in the relevant conditions. For example, for Sc 2 - High Con and Low ES, A-Impr1 was used 27 times, and A-Impr2 13; we chose A-Impr1. Table 6 shows the final decisions and our algorithm.

4   Conclusion and Future Work

This paper investigates how individuals adapt supportive feedback to a student's rating on Quality of Contribution, an important aspect of teamwork, based on the rater's personality and rating provided. Two studies were conducted, resulting in an algorithm that can be used in a peer-assessment measuring teamwork, based on the survey by [16]. A peer-assessment survey is an opportunity to monitor group work and simultaneously support students during teamwork. By implementing our algorithm in the survey via a virtual agent, students may receive personalized supportive feedback based on their ratings, ES and Con levels, ultimately contributing to smoother group work. Our study also provides evidence that people adapt supportive feedback to the rating provided and to the rater's personality. Several comments considered the raters' personality while choosing the feedback. This did not always lead to different decisions on a quantitative level, but it supports adapting the feedback to personality traits and level, alongside rating. The next steps are expanding it to other teamwork attributes [16] and personality traits, starting by combining the current algorithm with [15], which used a similar methodology but focused on Productivity. Next, we will evaluate the resulting algorithm: in a peer-assessment field study with students and teachers, and in a controlled lab study using a cooperative game.

References

1. Badea, G., Popescu, E.: Instructor support module in a web-based peer assessment platform. In: 2019 23rd International Conference on System Theory, Control and Computing, ICSTCC, pp. 691–696. IEEE (2019)
2. Barfield, R.L.: Students' perceptions of and satisfaction with group grades and the group experience in the college classroom. Assess. Eval. High. Educ. **28**(4), 355–370 (2003)
3. Beiter, R., et al.: The prevalence and correlates of depression, anxiety, and stress in a sample of college students. J. Affect. Disord. **173**, 90–96 (2015)
4. Bravo, R., Catalán, S., Pina, J.M.: Analysing teamwork in higher education: an empirical study on the antecedents and consequences of team cohesiveness. Stud. High. Educ. **44**(7), 1153 1165 (2019)
5. Burdett, J.: Making groups work: university students' perceptions. Int. Electron. J. **4**(3), 177–191 (2003)
6. Burleson, B.R., Goldsmith, D.J.: How the comforting process works: alleviating emotional distress through conversationally induced reappraisals. In: Handbook of Communication and Emotion, pp. 245–280. Elsevier (1996)
7. Chamorro-Premuzic, T., Furnham, A.: Personality predicts academic performance: evidence from two longitudinal university samples. J. Res. Pers. **37**(4), 319 338 (2003)
8. Dennis, M., Masthoff, J., Mellish, C.: The quest for validated personality trait stories. In: Proceedings of the 2012 ACM International Conference on Intelligent User Interfaces, pp. 273–276 (2012)
9. Dennis, M., Masthoff, J., Mellish, C.: Adapting progress feedback and emotional support to learner personality. Int. J. AI Educ. **26**(3), 877–931 (2016)
10. Hakimi, S., Hejazi, E., Lavasani, M.G.: The relationships between personality traits and students' academic achievement. Procedia. Soc. Behav. Sci. **29**, 836–845 (2011)
11. Kindness, P., Masthoff, J., Mellish, C.: Designing emotional support messages tailored to stressors. Int. J. Hum. Comput. Stud. **97**, 1–22 (2017)
12. Laal, M., Ghodsi, S.M.: Benefits of collaborative learning. Procedia. Soc. Behav. Sci. **31**, 486–490 (2012)
13. Masthoff, J.: The user as wizard: a method for early involvement in the design and evaluation of adaptive systems. In: Fifth workshop on user-centred design and evaluation of adaptive systems, vol. 1, pp. 460–469. CiteSeer (2006)
14. Randolph, J.J.: Free-marginal multirater kappa (multirater k [free]): an alternative to Fleiss' fixed-marginal multirater kappa. In: Joensuu Learning and Instruction Symposium, Joensuu, Finland, Oct 14-15 (2005)
15. Saccardi, I., Masthoff, J.: Adapting emotional support in teams: emotional stability and productivity. In: Adjunct Proceedings of the 31st ACM Conference on User Modeling, Adaptation and Personalization, pp. 253–265 (2023)

16. Saccardi, I., Veth, D., Masthoff, J.: Identifying students' group work problems: design and field studies of a supportive peer assessment. Interact. Comput., iwad044 (2023)
17. Smith, K.A., Dennis, M., Masthoff, J., Tintarev, N.: A methodology for creating and validating psychological stories for conveying and measuring psychological traits. User Model. User Adap. Inter.. **29**(3), 573–618 (2019). https://doi.org/10.1007/s11257-019-09219-6
18. Smith, K.A., Masthoff, J., Tintarev, N., Moncur, W.: Adapting emotional support to personality for carers experiencing stress. In: International Workshop on Personalisation and Adaptation in Technology for Health - PATH 2015 in conjunction with UMAP, 30 June 2015, Dublin (2015)

Predicting Successful Programming Submissions Based on Critical Logic Blocks

Ka Weng Pan[1]([✉]), Bryn Jeffries[1,2], and Irena Koprinska[1]

[1] School of Computer Science, The University of Sydney, Sydney 2006,
NSW, Australia
kpan4528@uni.sydney.edu.au, irena.koprinska@sydney.edu.au
[2] Grok Academy, PO Box 144,Broadway, Sydney, NSW 2007, Australia
bryn.jeffries@grokacademy.org

Abstract. We propose an approach to infer the critical logic blocks contained within student submissions that can influence the passing or failing of individual exercises in introductory programming courses. Given a programming exercise, we extract critical logic blocks from the abstract syntax trees (ASTs) of its submissions and use a bag-of-words approach to train a decision tree classifier to predict the pass/failure of a submission given the logic blocks present in its AST. We apply this technique to two streams of an introductory Python programming course for high-school students, constructing decision trees for 66 programming exercises based on several thousand submissions for each exercise. We obtain classifiers with F1 scores of 92.4% for the beginners and 88.2% for the intermediate stream. These trained models are highly interpretable and can provide a visualisation of the key logic blocks that are critical in the pass/failure of a problem. We explain how the models can assist educators in understanding the common valid approaches or failed attempts carried out by students when tackling a specific problem, and they may also serve as a guide for suggesting hints to struggling students.

1 Introduction

Introductory programming courses often provide code-writing exercises to solve problems using a small set of programming constructs. While educators likely intend for a particular solution, students may achieve the required result in a variety of ways. For large student cohorts, it is challenging to assist each student in developing a correct solution. Data science methods can be used to automatically analyse data and create interpretable models mined from successful and unsuccessful historical submissions, providing insights to students and teachers.

We posit that some parts of a program are especially important in the development of a successful solution, and students may benefit from ensuring these are included as building blocks from which to proceed. In this work, we propose a novel approach that extracts logic blocks from the abstract syntax tree (AST)

© The Author(s), under exclusive license to Springer Nature Switzerland AG 2024
A. M. Olney et al. (Eds.): AIED 2024, LNAI 14830, pp. 363–371, 2024.
https://doi.org/10.1007/978-3-031-64299-9_32

of submitted code, represents the submissions in terms of these blocks and uses a decision tree (DT) classifier to predict the submission's outcome.

These interpretable models provide insights to educators about the critical logic blocks and paths in successful submissions, highlight common failures and loopholes in existing test cases, and inform educators on significant variations in students' approaches. In future work, they may also provide a means for hints to be dynamically provided to students based upon the progress they have made.

2   Related Work

There is a large body of work on data-driven methods for analysing student data on online learning platforms. This includes clustering to understand student misconceptions [3,4] or to find the characteristics of high and low performing students in terms of their behaviour on online platforms [5,6]; methods for predicting the final course mark and dropout [2,10,14]. The research on interaction and progress networks [8] is also relevant. Such networks represent the student transitions between learning tasks and can be used to identify the places where students experience difficulties. Another related stream of work is on automatic next-step hint generation [7,9,11,12], e.g., generating hints that are likely to guide students to passing more tests.

In our approach, we analyse student submissions to automatically extract the critical logic blocks that impact successful and unsuccessful submissions and the common paths of logic blocks followed by the two groups. We show the utility of our approach to provide insights to educators about successful and unsuccessful submissions and loopholes in test cases. Our work also provides the foundation for dynamic hint generation using the extracted critical logic blocks.

3   Data

We applied our method to data from the beginner and intermediate level streams of the NCSS Challenge online Python programming course, provided by Grok Academy[1], participated by primary and high school students in Australia in 2018. The data contains 41 exercises ("problems") in the beginners stream and 25 exercises in the intermediate stream. Each problem allows the participant to make multiple submissions, and automated tests provide immediate feedback of the correctness of each submission. The average percentage of passed submissions per problem is 64.5% (standard deviation: ±11.6%) for the beginners stream, and 65.0% (±13.1%) for the intermediate stream. The average number of submissions per problem is 7687 (±3444) for the beginners stream, and 4451 (±2008) for the intermediate stream.

---

[1] https://grokacademy.org/challenge.

4   Methodology

By viewing a program as a collection of logic blocks, we employ a bag-of-words (BoW) approach to vectorise submissions by the occurrences of unique logic blocks, and train a DT classifier to predict the pass/failure of a programming submission.

criterion	Block
1	var0 = 0

1,2	while var0 < 5:
	    if var0 % 2 == 0:
	        print(str(var0) + ' is even')
	    else:
	        print(str(var0) + ' is odd')
	        var0 += 1

3	while var0 < 5:
3	... var0 += 1

2,3	if var0 % 2 == 0:
	    print(str(var0) + ' is even')
	else:
	    print(str(var0) + ' is odd')

3	if var0 % 2 == 0:
2,3	... if var0 % 2 == 0: print(str(var0) + ' is even')

3	if ...:
	...
	else:
	    print(str(var0) + ' is odd')

3	print(str(var0) + ' is even')
3	print(str(var0) + ' is odd')

```
i = 0
while i < 5:
    if i % 2 == 0:
        print(str(i) + ' is even')
    else:
        print(str(i) + ' is odd')
    i += 1
```

(a) sample code

(b) logic blocks extracted from sample code

Fig. 1. Example of logic blocks extracted from a sample code

4.1   Extraction of Logic Blocks

We extract the logic blocks of a programming submission from its AST. The AST for a submission is obtained using `ast.parse` in the Python `ast` module. We define a subtree as a logic block if it satisfies one of these conditions:

1. A subtree rooted at a direct child node of the root, which corresponds to a non-indented code block of the program
2. A subtree rooted at a node with type `ast.If`, `ast.For`, `ast.While`, `ast.ClassDef`, `ast.FuncDef` or `ast.AsyncFuncDef`, which corresponds to a common composite logic encapsulating code block (it is either an if/elif/else block, a loop, a function definition or a class definition)
3. A subtree rooted at a node listed in the `body` or composes the evaluation header of a target node in criterion 2, which corresponds to a direct logic-conveying building block of a composite logic block (see criterion 2)

To minimise noise caused by variations in labelling, we normalise labels of function definitions, class definitions and user defined functions calls for the entire AST before extraction, and we normalise variable labels for each extracted subtrees after extraction. If the subtree involves subscripts, we include a copy with the subscripts masked as an additional logic block. The extracted subtrees can be converted back to code for readability using the `ast.unparse` function. An example of logic block extraction can be seen in Fig. 1.

The splitting of the AST into subtrees is inspired by [13] who construct statement subtrees for source code classification and code clone detection. However, our splitting approach and task are different.

4.2   Prediction of Submission Outcome Using Logic Blocks

We formulate the task as a machine learning classification task. Given a set of programming submissions for a problem, and their corresponding outcome status (two classes: "passed" or "failed"), we train a classifier to predict the status of a new submission for this problem. A separate classifier is created for each problem.

As an evaluation procedure we employ a 75/25% training/test split with stratification. We generate a BoW representation containing the logic blocks present in the training set for each problem, which consists of several thousand submissions. All submissions are vectorised based on the number of occurrences of each logic block. A DT classifier is trained on the training set to predict the status of the submission and the performance is evaluated on the test set. The hyperparameters of the DT classifier (maximum depth and minimum samples at leaf node) are tuned with 5-fold grid-search cross-validation. In order to enhance the readability and generalisation of the DT, we considered the following hyperparameter values for the grid-search: maximum depth: [2, 3, 4] and minimum number samples in each leaf node: [20, 40, 60].

5   Results and Discussion

5.1   Predictive Performance

Tables 1 and 2 show the results for the beginner and intermediate courses respectively, for each problem. We report accuracy, F1-score, precision, recall and also the BoW size.

The predictive performance results are very positive. For the beginners course, our method yielded a mean ($\pm$ standard deviation) accuracy of $90.4 \pm 6.71\%$, F1-score of $92.4 \pm 5.99\%$, precision of $92.4 \pm 7.10\%$ and recall of $93.1 \pm 8.74\%$.

For the intermediate course, which includes more difficult problems than the beginners course, our method achieved slightly lower but still high average accuracy of $85.2 \pm 6.92\%$, F1 score of $88.2 \pm 6.96\%$, precision of $85.9 \pm 7.98\%$, and recall of 91.2 ± 8.69.

Table 1. Classification performance for Beginners stream. N - Problem number, Acc - accuracy, F1 - F1 score, P - precision, R - recall, BoW - BoW size.

N	Acc	F1	P	R	BoW	N	Acc	F1	P	R	BoW
0	91.64	94.94	92.35	97.69	1262	21	94.83	96.49	93.62	99.55	651
1	91.23	93.12	94.69	91.60	499	22	85.89	81.57	97.93	69.89	2236
2	82.92	86.81	87.83	85.82	1047	23	97.62	97.76	98.75	96.79	2135
3	87.77	90.55	98.86	83.52	630	24	87.22	92.65	86.39	99.89	1061
4	87.17	90.70	85.64	96.40	2203	25	97.43	97.72	98.71	96.74	1182
5	94.28	96.49	93.81	99.32	376	26	94.17	96.34	94.30	98.47	702
6	95.06	95.28	98.23	92.49	2443	27	93.58	94.90	91.86	98.15	1044
7	85.64	89.53	81.98	98.60	1034	28	96.79	97.05	99.76	94.49	1034
8	93.50	95.15	99.14	91.47	964	29	99.92	99.95	99.95	99.95	380
9	89.54	92.67	89.39	96.20	1780	30	88.43	92.70	90.86	94.60	900
10	94.20	96.12	94.10	98.23	558	31	86.65	92.08	89.38	94.93	1299
11	98.13	97.71	99.63	95.87	6080	32	96.04	96.33	99.58	93.28	1116
12	81.57	85.67	77.05	96.45	6604	33	90.56	94.84	91.40	98.54	555
13	88.67	91.21	88.33	94.29	1229	34	78.75	87.13	79.16	96.89	2053
14	97.96	98.63	98.73	98.52	527	35	92.89	94.19	90.41	98.31	750
15	88.42	86.99	97.04	78.83	4133	36	89.81	93.49	89.51	97.85	2674
16	87.81	85.65	94.75	78.15	3177	37	79.67	80.38	96.47	68.89	1285
17	99.11	99.09	98.98	99.21	685	38	98.25	98.26	97.55	98.98	655
18	98.14	98.43	99.65	97.23	1817	39	71.46	78.05	68.48	90.74	3415
19	96.09	97.66	96.77	98.56	1101	40	74.92	74.22	86.76	64.84	1852
20	84.50	88.34	82.27	95.38	5488						

Table 2. Classification performance for Intermediate stream. N - Problem number, Acc - accuracy, F1 - F1 score, P - precision, R - recall, BoW - BoW size.

N	Acc	F1	P	R	BoW	N	Acc	F1	P	R	BoW
0	95.82	97.60	96.96	98.24	245	13	83.26	85.08	94.53	77.35	979
1	76.96	82.40	75.09	91.29	3073	14	79.68	79.32	77.58	81.14	6870
2	80.19	85.48	86.31	84.67	3151	15	92.09	95.78	93.17	98.55	812
3	80.18	82.74	75.03	92.21	7933	16	88.93	93.85	89.76	98.33	1417
4	82.68	87.81	80.81	96.14	7002	17	81.99	87.22	80.42	95.28	5981
5	91.00	91.71	90.00	93.49	3395	18	97.61	98.24	99.50	97.01	1097
6	80.47	87.35	81.44	94.19	1433	19	92.29	91.12	86.37	96.42	6865
7	92.49	93.92	98.85	89.47	884	20	89.48	88.35	87.44	89.28	4227
8	81.18	87.31	83.70	91.26	2547	21	80.66	81.98	71.87	95.40	13852
9	84.46	90.05	83.62	97.56	1308	22	85.71	89.87	86.62	93.36	3317
10	83.00	88.53	88.83	88.24	2609	23	96.36	97.68	99.12	96.28	896
11	83.12	89.06	83.03	96.04	1061	24	66.48	64.09	72.73	57.29	7997
12	83.81	88.06	84.00	92.52	5541						

The average BoW size is 1722 ± 1492 for the beginners course and 3780 ± 3199 for the intermediate course. The higher size for the latter reflects the greater complexity of the problems in the intermediate course.

5.2 Identifying Critical Blocks Impacting Successful and Unsuccessful Submissions

Decision trees allow educators to visualise the critical blocks that impact the correctness of student submissions. By extracting paths that lead to class `passed`, followed by many students, we can identify the diverse logic approaches in successful student submissions and the corresponding combinations of critical code blocks. Similarly, paths that lead to class `failed`, followed by many students, can reveal insights into the common struggles amongst unsuccessful submissions.

For example, for Problem 6 in the Beginners course, by inspecting a sample successful submission in Fig. 3, the problem appears to examine string to integer conversion, obtaining `input` and conditional evaluations. By observing the DT classifier for this problem in Fig. 2, we can see that the nodes of the DT captured conditional blocks, and blocks involving assignments and function calls (`int`, `input` and `print`). Tracing two of the `passed` paths in Fig. 2 reveals two sets of logic blocks that commonly appear in passed submissions and the difference between these two sets of logic blocks highlights the logical equivalence between two critical conditional blocks, where one initiates the conditional evaluation with `if var0 > 7` and the other initiates with `if var0 <= 7`.

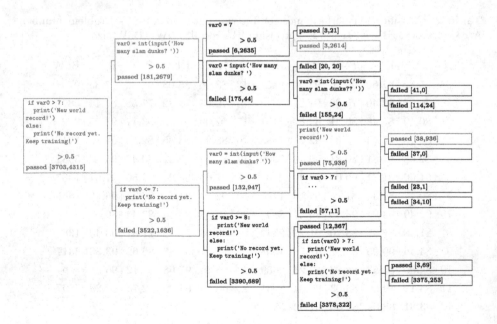

Fig. 2. DT classifier for problem 6 in Beginners course

```
dunks = int(input('How many slam dunks? '))
if dunks > 7:
  print('New world record!')
else:
  print('No record yet. Keep training!')
```

Fig. 3. Sample successful submission for Problem 6 in Beginners course

DTs can also uncover incorrect approaches that bypass the testing suite and assist educators to refine testcases. For example, for Problem 23 in the Beginners course, as shown in Fig. 4, we can see that most correct submissions included the conditional if 'Garth Nix' in var0; however, a small number of passed submissions included if 'x' in var0. This indicates an additional testcase is required to further validate the correctness of submissions. By tracing a failed path in the DT, it can be observed that students could struggle to identify the requirement of a white space in the input statement. This could indicate a crucial hint that educators can provide to students as they attempt the problem (Fig. 5).

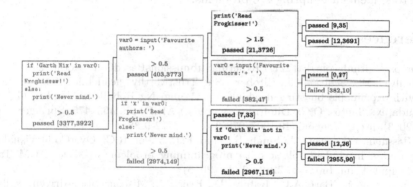

Fig. 4. DT classifier for problem 23 in Beginners course

```
authors = input('Favourite authors: ')
if 'Garth Nix' in authors:
  print('Read Frogkisser!')
else:
  print('Never mind.')
```

Fig. 5. Sample successful submission for Problem 23 in Beginners course

6 Conclusion and Future Work

We presented a novel approach that extracts meaningful logic blocks from programming submissions' ASTs, and uses a DT classifier and a BoW approach with logic block tokens to predict the pass/fail status of programming submissions. We applied the approach to a total of 66 problems across two streams of an introductory programming course and obtained an average F1 score of 92.4% for Beginners problems and 88.2% for Intermediate problems. We demonstrated the feasibility of the generated DT classifiers in assisting educators to identify sets of critical logic blocks in successful submissions, points of common failures, and loopholes in existing test suites. A limitation of our approach is that it may not be feasible for complex programming problems with intricate logic or problems with diverse possibilities of nesting or embedding logic, as the DT classifier may need to grow to big depths to capture all sets of critical logic blocks. We have made our code publicly available at [1].

In future work, we intend to evaluate the utility of the DTs in assisting in the improvement of problems. We will also incorporate the critical logic block DTs into a hint generation system, and evaluate the impact of providing these hints to students attempting the problems.

References

1. https://github.com/ka1227/PredictingSubmissionsBasedOnLogicBlocks
2. Dsilva, V., Schleiss, J., Stober, S.: Trustworthy academic risk prediction with explainable boosting machines. In: Wang, N., Rebolledo-Mendez, G., Matsuda, N., Santos, O.C., Dimitrova, V. (eds.) AIED, pp. 463–475. Springer, Cham (2023). https://doi.org/10.1007/978-3-031-36272-9_38
3. Glassman, E.L., Scott, J., Singh, R., Guo, P.J., Miller, R.C.: OverCode: visualizing variation in student solutions to programming problems at scale. ACM Trans. Comput. Hum. Interact. 22(2), 1–35 (2015)
4. Gusukuma, L., Bart, A.C., Kafura, D., Ernst, J.: Misconception-driven feedback: results from an experimental study. In: ACM Conference on International Computing Education Research, p. 160–168 (2018)
5. Koprinska, I., Stretton, J., Yacef, K.: Predicting student performance from multiple data sources. In: Conati, C., Heffernan, N., Mitrovic, A., Verdejo, M.F. (eds.) AIED 2015. LNCS (LNAI), vol. 9112, pp. 678–681. Springer, Cham (2015). https://doi.org/10.1007/978-3-319-19773-9_90
6. McBroom, J., Jeffries, B., Koprinska, I., Yacef, K.: Mining behaviors of students in autograding submission system logs. In: 9th International Educational Data Mining Society, International Conference on Educational Data Mining (EDM), Raleigh, NC, Jun 29-Jul 2, pp. 159–166 (2016)
7. McBroom, J., Koprinska, I., Yacef, K.: A survey of automated programming hint generation: the hints framework. ACM Comput. Surv. 54(8), 1–27 (2021)
8. McBroom, J., Paassen, B., Jeffries, B., Koprinska, I., Yacef, K.: Progress networks as a tool for analysing student programming difficulties. In: Australasian Computing Education Conference, p. 158–167 (2021)
9. Piech, C., et al.: Deep knowledge tracing. Adv. Neural Inform. Process. Syst. 28, 505–513 (2015)

10. Polito, S., Koprinska, I., Jeffries, B.: Exploring student engagement in an online programming course using machine learning methods. In: Rodrigo, M.M., Matsuda, N., Cristea, A.I., Dimitrova, V. (eds.) AIED 2022, pp. 546–550. Springer, Cham (2022). https://doi.org/10.1007/978-3-031-11647-6_112

11. Price, T.W., et al.: A comparison of the quality of data-driven programming hint generation algorithms. Int. J. Artif. Intell. Educ. **29**, 368–395 (2019)

12. Rivers, K., Koedinger, K.R.: Data-driven hint generation in vast solution spaces: a self-improving python programming tutor. Int. J. Artif. Intell. Educ. **27**, 37–64 (2017)

13. Zhang, J., Wang, X., Zhang, H., Sun, H., Wang, K., Liu, X.: A novel neural source code representation based on abstract syntax tree. In: International Conference on Software Engineering, pp. 783–794 (2019)

14. Zhang, V., Jeffries, B., Koprinska, I.: Predicting Progress in a Large-Scale Online Programming Course. In: Wang, N., Rebolledo-Mendez, G., Matsuda, N., Santos, O.C., Dimitrova, V. (eds.) AIED 2023, pp. 810–816. Springer, Cham (2023). https://doi.org/10.1007/978-3-031-36272-9_76

Student At-Risk Identification and Classification Through Multitask Learning: A Case Study on the Moroccan Education System

Ismail Elbouknify[1]([✉]) [iD], Ismail Berrada[1], Loubna Mekouar[1] [iD],
Youssef Iraqi[1] [iD], EL Houcine Bergou[1] [iD], Hind Belhabib[2], Younes Nail[2],
and Souhail Wardi[2]

[1] College of Computing, Mohammed VI Polytechnic University,
Ben Guerir, Morocco
{ismail.elbouknify,ismail.berrada,loubna.mekouar,
youssef.iraqi,elhoucine.bergou}@um6p.ma
[2] Ministry of National, Education, Preschool, and Sports, Rabat, Morocco

Abstract. Early identification of students at-risk of dropping out or failing is a critical challenge in education. In this paper, we address this imperative by deploying multitasking models with a dual purpose: early identification and subsequent classification into success, failure, or dropout. The multitask models used include the deep multitask model, as well as Extreme Gradient Boosting (XGBoost) and Light Gradient Boosting Machine (LightGBM) based multitask models. We provided a comparative analysis of the performance of traditional machine learning, deep learning, and multitasking models. Our proposed approach has been applied to the Moroccan education system using a proprietary dataset provided by the Moroccan Ministry of National Education, Pre-school, and Sports. The results show that the multitasking model demonstrates superior accuracy and effectiveness in both tasks, while also reducing training and prediction time.

Keywords: Students At-Risk · Multitask Learning · Deep multitask Learning · Artificial Intelligence in education

1 Introduction

The issue of students at-risk is of paramount importance as it represents a significant challenge in the education system. Any student can be at-risk of dropping out or failing at any level of their education. This problem affects students' academic progress and has implications for educational institutions and the learning environment as a whole. By identifying and addressing students at-risk, efforts can be directed toward creating a more inclusive and effective education system.

A. M. Olney et al. (Eds.): AIED 2024, LNAI 14830, pp. 372–380, 2024.
https://doi.org/10.1007/978-3-031-64299-9_33

Artificial Intelligence (AI) is being used in many aspects of education, such as generating content [12], and implementing recommender systems [10]. In addition, AI is being used to predict student outcomes, which includes identifying students at-risk [13] and predicting students performance [17]. This involves the application of both Machine Learning (ML) models such as Decision Tree (DT), Support Vector Machine (SVM) [16] and Deep Learning (DL) models such as Convolutional Neural Network (CNN), Long Short-Term Memory (LSTM) [18], using student academic, and demographic information [2]. Moreover, human-AI collaboration holds great promise for improving educational outcomes [15]. This includes a critical consideration of the ethical dimensions of pedagogical decision-making [5], while recognizing the potential for unintended consequences [4].

This paper addresses the problem of identifying at-risk students in the Moroccan education system, multitasking models were used to identify students at-risk and then classify them into success, failure, or dropout categories. The main goal is to provide valuable insights and support for at-risk students. The main contributions of the paper are (1) Implementation of multitasking models to perform two main related tasks: identification and classification of students at-risk. (2) The use of a real data set on education from the Moroccan Ministry of National Education, Pre-school Education, and Sport. (3) Extensive comparison between multiclass classification models and the multitask models. (4) Use Explainable Artificial Intelligence (XAI) to find the features that influence the prediction of a particular class, this enhances the interpretability and transparency of the model's decision-making process.

The rest of this paper is organized as follows: Sect. 2 explores related work on predicting students at-risk. Section 3 outlines the proposed approach. Section 4 provides detailed information about the dataset used, and presents the results. Finally, Sect. 5 presents the conclusions drawn from this study.

2   Related Work

AI is being used in several areas of education, including the early identification of students at-risk [14]. The primary aim of using AI is to accurately identify at-risk students and thereby propose early interventions aimed at providing them with the necessary support [6]. Some papers focus on analyzing the risks associated with academic failure [1], while others concentrate on examining the risk of dropping out [14]. For early identification of at-risk students, many papers in the field use ML models such as DT, Random Forest (RF), LightGBM, and XGBoost, for their analyses [16]. Conversely, others opt for DL models such as Artificial Neural Network (ANN), CNN, LSTM [18] to enhance the accuracy and efficiency of student risk identification processes. Indeed, the use of diverse historical student data, including academic, socio-economic, and personal dimensions [6], enables a comprehensive analysis of dropout risk factors, fostering nuanced understanding and informed interventions [6]. However, there is a notable lack of research addressing the identification of at-risk students in real-world education systems. Furthermore, there is often insufficient discussion of actionable strategies to effectively support these at-risk students. This study aims to fill these gaps by exploring not only methods for identifying at-risk students but also how we can understand the model's decision, to propose actionable strategies. In summary, Table 1 provides a concise overview of some of the important related research.

Table 1. Summary of some important work on the prediction of students at-risk

Ref	Type of Risk	Features category	Dataset size (Student)	Models	Best model performance
[14]	Dropout	Academic, demographic	3 million	DT, XGBoost, LightGBM, Categorical Boosting	Precision=0.71, recall=0.91, Geometric Mean Score=0.91, F1-score=0.59
[6]	Dropout	Academic, demographic, sosio-economic	29,972	DT, Logistic Regression (LR), Adaptive Boosting (AdaBoost), RF, XGBoost	Area Under Curve (AUC)=0.68, Precision=0.92, recall=0.72, Kolmogorov-Smirnov score=0.86
[11]	Dropout	Academic	78,722	ANN	Accuracy=0.81, F1-score=0.83
[16]	Dropout	Academic, demographic	60,010	XGBoost, LightGBM, DT, LR, RF	Accuracy=0.94, precision=0.81, recall=0.78, F1-score=0.79
[7]	Failure	Academic	2,419	SVM	AUC=0.79
[1]	Failure	Academic, demographic	32,593	Gradient Boosting Machine, Generalized Linear Model ANN	Accuracy=0.89, precision=0.91, recall=0.86, F1-score=0.91, AUC=0.93

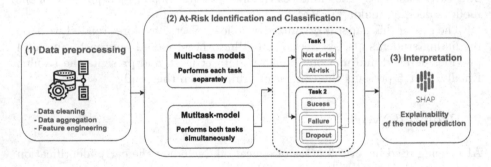

Fig. 1. Overview of the proposed approach

3 Proposed Approach

3.1 Overview

This study, shown in Fig. 1, focuses on identifying and classifying at-risk students, the proposed approach consists of 3 main components **(1) Data preprocessing:** where data is cleaned, aggregated, and structured into rows. **(2) Students' at-risk identification and classification:** where both multiclass models and multitask models are designed to perform two related tasks: identify at-risk students and classify them into success, failure, or dropout categories using one year of historical data. **(3) Interpretation:** where the SHapley Additive exPlanations (SHAP) [8] technique is used to explain the model's predictions.

3.2 Data Pre-processing

Data pre-processing is a fundamental step in preparing raw data for analysis and modeling. Several key steps were taken to ensure the quality and reliability of the dataset. Firstly, different tables from the database were merged to create organized rows. The data were then aggregated and organized by year for each student. Addressing missing values was also a priority, with ministry expertise based on enhancing dataset completeness. Finally, rigorous error correction and careful feature engineering were performed to improve model performance.

3.3 Student At-Risk Identification and Classification

In this study, we aim to perform two related tasks simultaneously: identifying at-risk students and classifying them into success, failure, or dropout categories. We used XGBoost and LightGBM as base models for multitask learning models called LightGBM Multitask Model (MT-LightGBM) and XGBoost Multitask Model (MT-XGBoost), and the deep multitask model Multitask Artificial Neural Network (MT-ANN). We compare multitask models with multiclass models that have been used to perform each task separately and evaluate their effectiveness in different classification scenarios. The multiclass models used include MT-ANN architecture, In addition, the multiclass machine learning models used include DT, RF, XGBoost, and LightGBM[1].

3.4 Interpretation of the Model Predictions

XAI has been applied in various fields such as healthcare [3], and education [6], to increase transparency and interpretability of AI models. This study explores the interpretation of the model predictions to gain insight into student prediction measures. We used the SHAP technique [8], which provides a method for understanding the contribution of each feature to the prediction by assigning each feature an important value for a particular prediction. This analysis provides valuable insights that can help suggest interventions to support students.

4 Experiments and Results

In this section, we present the dataset used in this study, and the obtained results for multitask models. Additionally, we provide a comparison between multiclass and multitask classification on the test dataset.

---

[1] The architectures and hyperparameters of the models are available at https://shorturl.at/hrPVY.

4.1 Data

The dataset used in this study was provided by the Moroccan Ministry of National Education, Pre-school, and Sports[2]. The study focused on middle schools in the Fez-Meknes region, for five years from 2016 to 2020. It takes the form of a database containing academic and demographic information. Table 2 shows some features of the dataset used[3]. The dataset contains three labels that categorize students' situations: **(1) Success**: denotes progression to the next academic level, **(2) Failure**: denotes remaining at the same level, **(3) Dropout**: denotes students who discontinue their studies. Table 3 shows the statistics for student labels by level in the dataset used in this study, for each task. Table 4 shows the dataset gender distribution across labels. We can observe that the male gender has a high risk rate compared to the female gender. To ensure rigorous evaluation of the models, the dataset was split by randomly selecting 20% of the data from each academic year for testing.

Table 2. Features of the Dataset

Type	Category	Example of features
Academic	General information	Scholarship
		Number of days missed (authorized)
	Grade	Average grades for each subject
		Coefficient of each subject
	Class	Number of students in the class
		Average grade of the class
	Schools	Establishment name
		Province
Demographic	Personal	Student gender
	Information	Student birthplace

Table 3. Student Labels Statistics

Level	# of Students	Task 1		Task 2		
		# At-risk	# Not At-risk	# Success	# Failure	# Dropout
Middle School 1	349800	101309	248491	248491	51896	49413
Middle School 2	286548	114507	172041	172041	76468	38039
Middle School 3	299191	79549	219642	219642	43288	36261

---

[2] https://www.men.gov.ma/.

[3] The full dataset features are available at https://shorturl.at/fsuIR.

Table 4. Dataset Gender distribution across labels

	Total data	At-risk	Dropout	Failure
Male	53.2%	64.4%	66.1%	62.9%
Female	46.8%	35.6%	33.9%	37.1%

4.2 Results

Table 5 summarises the performance of multitasking models using accuracy, as well as macro-averaged precision, recall, and F1 score. We can observe that the multitask model achieves commendable results in both tasks. In Middle School 1, the accuracy of the MT-LightGBM is 0.82, the recall is 0.76, the precision is 0.78, and the F1 score is 0.77 for Task 1. For Task 2, the accuracy is 0.80, the recall is 0.63, the precision is 0.70, and the F1 score is 0.65. This indicates that the multitasking models are good at handling both tasks simultaneously. Moreover, the MT-LightGBM model outperforms the other multitask models.

Table 6 shows the comparison between multitask and multiclass models, We can observe that LightGBM outperforms other multiclass models, and MT-LightGBM outperforms multitask models. In particular, multitask models show superior performance and less training and prediction time compared to their single-task counterparts. This suggests that multitasking models speed up training and prediction, and effectively facilitate the sharing of information between tasks.

Table 5. Multitask models evaluation on both tasks

Level	Model	Task 1				Task 2			
		Accuracy	Recall	Precision	F1-score	Accuracy	Recall	Precision	F1-score
Middle School 1	MT-ANN	0.80	0.74	0.76	0.75	0.78	0.60	0.67	0.62
	MT-XGBoost	0.82	0.76	0.78	0.77	0.79	0.62	0.70	0.65
	MT-LightGBM	**0.82**	**0.76**	**0.78**	**0.77**	**0.80**	**0.63**	**0.70**	**0.65**
Middle School 2	MT-ANN	0.78	0.76	0.76	0.76	0.74	0.64	0.67	0.66
	MT-XGBoost	0.80	0.78	0.79	0.79	0.76	0.67	0.71	0.69
	MT-LightGBM	**0.80**	**0.79**	**0.79**	**0.79**	**0.77**	**0.68**	**0.71**	**0.69**
Middle School 3	MT-ANN	0.83	0.70	0.79	0.78	0.80	0.51	0.65	0.51
	MT-XGBoost	0.84	0.73	0.80	0.75	0.82	0.56	0.69	0.60
	MT-LightGBM	**0.84**	**0.73**	**0.80**	**0.75**	**0.82**	**0.57**	**0.69**	**0.60**

Table 6. Comparison of both task models using multiclass and multitask classification for Middle School 2 on test dataset

Type of classification	Model	Classification performance								Training time (second)	Prediction latency (second)
		Task 1				Task 2					
		Accuracy	Recall	Precision	F1-score	Accuracy	Recall	Precision	F1-score		
Multiclass	DT	0.76	0.75	0.75	0.75	0.73	0.62	0.66	0.64	**5.09**	**0.02**
	RF	0.77	0.76	0.76	0.76	0.73	0.61	0.68	0.63	315.96	4.31
	XGBoost	0.80	0.78	0.79	0.79	0.76	0.67	0.70	0.68	457.43	0.56
	LightGBM	**0.80**	**0.79**	**0.79**	**0.79**	0.76	0.67	0.71	0.68	47.94	5.58
	ANN	0.78	0.76	0.76	0.76	0.74	0.62	0.67	0.64	230.10	15.04
Multitask	MT-ANN	0.78	0.76	0.76	0.76	0.74	0.64	0.67	0.66	145.01	14.55
	MT-XGBoost	0.80	0.78	0.79	0.79	0.76	0.67	0.71	0.69	385.61	**0.54**
	MT-LightGBM	**0.80**	**0.79**	**0.79**	**0.79**	**0.77**	**0.68**	**0.71**	**0.69**	**40.33**	5.46

4.3   Interpretation of the Model Predictions

Figure 2 shows the top 7 features that influence each class prediction according to the SHAP technique[4]. In the visualization of the SHAP values, the length of each vertical bar corresponds to the magnitude of the SHAP values and indicates the impact of the feature on the model predictions. Features highlighted in blue represent lower values or negative impacts on the prediction, while those highlighted in red represent higher values or positive impacts on the prediction. We can observe that some common features influence the model's predictions across classes including overall grades average, age at the current academic level, and ranking in class. Education professionals can use these interpretations to analyse the factors that influence student performance, enabling the development of interventions to retain students and improve educational outcomes.

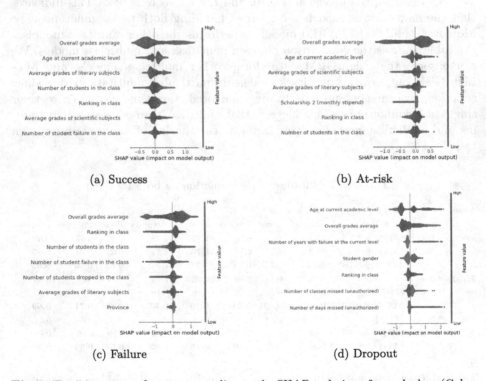

(a) Success   (b) At-risk

(c) Failure   (d) Dropout

Fig. 2. Top 7 important features according to the SHAP technique for each class (Color figure online)

---

[4] https://shap.readthedocs.io/.

5   Conclusion

In conclusion, this study highlights the critical role of AI in addressing the challenge of identifying and categorizing students at-risk in education. The study emphasizes the importance of early detection for timely intervention and highlights the benefit of multitasking models to improve model performance and speed up training and prediction, and how XAI can help us understand and identify the factors influencing the model's predictions for effective intervention. Future work will focus on developing recommendation systems [9] that use AI to help students persist in their studies and improve their academic performance. By harnessing the power of AI, these recommendation systems can provide personalized guidance and support to students, contributing significantly to their overall educational success.

References

1. Al-Shabandar, R., Hussain, A.J., Liatsis, P., Keight, R.: Detecting at-risk students with early interventions using machine learning techniques. IEEE Access **7**, 149464–149478 (2019)
2. Baltà-Salvador, R., Olmedo-Torre, N., Peña, M.: Perceived discrimination and dropout intentions of underrepresented minority students in engineering degrees. IEEE Trans. Educ. **65**(3), 267–276 (2022)
3. Elbouknify, I., Bouhoute, A., Fardousse, K., Berrada, I., Badri, A.: CT-xCOV: a CT-scan based Explainable Framework for COVid-19 diagnosis. In: 2023 10th International Conference on Wireless Networks and Mobile Communications (WINCOM), pp. 1–8 (2023)
4. Holmes, W., et al.: Ethics of AI in education: towards a community-wide framework. Int. J. Artif. Intell. Educ. **32**, 504–526 (2021)
5. Isotani, S., Bittencourt, I.I., Walker, E.: Artificial intelligence and educational policy: bridging research and practice. In: Wang, N., Rebolledo-Mendez, G., Dimitrova, V., Matsuda, N., Santos, O.C. (eds.) AIED 2023. CCIS, vol. 1831, pp. 63–68. Springer, Cham (2023). https://doi.org/10.1007/978-3-031-36336-8_9
6. Krüger, J.G.C., de Souza Britto Jr, A., Barddal, J.P.: An explainable machine learning approach for student dropout prediction. Expert Syst. Appl. **233**, 120933 (2023)
7. Liao, S.N., Zingaro, D., Thai, K., Alvarado, C., Griswold, W.G., Porter, L.: A robust machine learning technique to predict low-performing students. ACM Trans. Comput. Educ. (TOCE) **19**(3), 1–19 (2019)
8. Lundberg, S.M., Lee, S.I.: A unified approach to interpreting model predictions. In: Advances in Neural Information Processing Systems, vol. 30, pp. 4765–4774. Curran Associates, Inc. (2017)
9. Mekouar, L., Iraqi, Y., Damaj, I.: A global user profile framework for effective recommender systems. Multimedia Tools Appl. **83**, 50711–50731 (2023)
10. Mekouar, L., Iraqi, Y., Damaj, I., Naous, T.: A survey on blockchain-based recommender systems: integration architecture and taxonomy. Comput. Commun. **187**, 1–19 (2022)
11. Olive, D.M., Huynh, D.Q., Reynolds, M., Dougiamas, M., Wiese, D.: A quest for a one-size-fits-all neural network: early prediction of students at risk in online courses. IEEE Trans. Learn. Technol. **12**(2), 171–183 (2019)

12. Olney, A.M.: Generating multiple choice questions from a textbook: LLMs match human performance on most metrics. In: AIED Workshops (2023)
13. Pereira, F.D., et al.: Early dropout prediction for programming courses supported by online judges. In: Isotani, S., Millán, E., Ogan, A., Hastings, P., McLaren, B., Luckin, R. (eds.) AIED 2019. LNCS (LNAI), vol. 11626, pp. 67–72. Springer, Cham (2019). https://doi.org/10.1007/978-3-030-23207-8_13
14. Rodríguez, P., Villanueva, A., Dombrovskaia, L., Valenzuela, J.P.: A methodology to design, develop, and evaluate machine learning models for predicting dropout in school systems: the case of Chile. Educ. Inf. Technol. **28**, 10103–10149 (2023)
15. Shimmei, M., Bier, N., Matsuda, N.: Machine-generated questions attract instructors when acquainted with learning objectives. In: Wang, N., Rebolledo-Mendez, G., Matsuda, N., Santos, O.C., Dimitrova, V. (eds.) AIED 2023. LNCS, vol. 13916, pp. 3–15. Springer, Cham (2023). https://doi.org/10.1007/978-3-031-36272-9_1
16. Song, Z., Sung, S.H., Park, D.M., Park, B.K.: All-year dropout prediction modeling and analysis for university students. Appl. Sci. **13**(2), 1143 (2023)
17. Sosnovsky, S., Hamzah, A.: Improving prediction of student performance in a blended course. In: Rodrigo, M.M., Matsuda, N., Cristea, A.I., Dimitrova, V. (eds.) AIED 2022. LNCS, vol. 13355, pp. 594–599. Springer, Cham (2022). https://doi.org/10.1007/978-3-031-11644-5_54
18. Yin, S., Lei, L., Wang, H., Chen, W.: Power of attention in MOOC dropout prediction. IEEE Access **8**, 202993–203002 (2020)

Investigating the Predictive Potential of Large Language Models in Student Dropout Prediction

Abdelghafour Aboukacem(✉) , Ismail Berrada, El Houcine Bergou ,
Youssef Iraqi , and Loubna Mekouar

College of Computing, Mohammed VI Polytechnic University, Ben Guerir, Morocco
{abdelghafour.aboukacem,ismail.berrada,elhoucine.bergou,
youssef.iraqi,loubna.mekouar}@um6p.ma

Abstract. In the landscape of educational analytics, the usage of Machine Learning (ML), Deep Learning (DL), and Survival Analysis (SA), for student dropout prediction often encounters challenges in effectively detecting dropout cases and explaining dropout reasons. This is due to many challenges such as data imbalance, data processing issues, cold start problem, and the limited explainability of the predictive models. This paper explores the usage of Large Language Models (LLMs) for dropout prediction to tackle the previous challenges. We introduce an approach that leverages the adaptability and contextual understanding of LLMs to discern subtle indicators of potential dropout risks. In particular, we employ a Retrieval Augmented Generation (RAG)-assisted few-shot learning paradigm and prompt engineering to transfer the knowledge of LLMs. An intensive experimentation of our approach has been conducted using real-life Moroccan student data containing academic, demographic, and socio-economic information, to predict yearly school dropouts. Our findings highlight that LLMs outperform baseline ML models while showing the ability to produce textual analysis of students' data. Thus, LLMs have a promising potential to be employed as student dropout prediction assistants in educational institutions hoping to mitigate this phenomenon.

Keywords: Student dropout prediction · Large Language Models · Few-shot learning · Retrieval Augmented Generation

1 Introduction

In recent years, the educational field has witnessed a major paradigm shift with the integration of advanced technologies and data-driven approaches. For instance, data-driven approaches are incorporated into decision-making to include suitable, reliable, and effective technologies and strategies for better learning quality [14]. Moreover, the recent advancements of Natural Language Processing (NLP) have vastly improved automated grading systems for all types

© The Author(s), under exclusive license to Springer Nature Switzerland AG 2024
A. M. Olney et al. (Eds.): AIED 2024, LNAI 14830, pp. 381–388, 2024.
https://doi.org/10.1007/978-3-031-64299-9_34

of student evaluations [10]. Finally, data-driven methods also cover predictive modeling of student performance [18]. This predictive modeling is the building block of Dropout Early Warning Systems (DEWS) which aim to predict, explain, and prevent students' dropout. It is undeniable that students' dropout is a pressing concern that continues to set back the effectiveness of educational institutions worldwide. Dropout rates significantly impact both individual students and the broader social structure, affecting not only academic performance but also long-term career prospects and economic stability [7]. As an effort to mitigate this issue, several works have been conducted to incorporate predictive techniques for students' dropout learning environments such as online [6], traditional universities [4], as well as the K-12 learning system [9]. Furthermore, due to the differences we find in these learning environments, i.e. the learning modality and period, a variety of dropout formulations can be adopted. For example, in online learning where the learning period is shorter, time-frame dropout prediction, e.g. "A student is considered a dropout if they are not active for the next 10 consecutive days", is widely adopted [15]. Other dropout formulations that have been adopted include but are not limited to semester-wise dropout [13] and gradual dropout prediction throughout different stages of a course [1].

In what concerns the employed predictive techniques, Machine Learning (ML) models such as eXtreme Gradient Boosting (XGBoost), Decision Trees (DT), Support Vector Machine (SVM), are commonly utilized [5,8].

Nevertheless, several challenges can impede the utilization of previously mentioned techniques, including:

1. **Data Quality**:
   - **Missing Values**: Handling missing data involves either deleting records or imputing values, which may lead to biased insights.
   - **Data Balance**: Unequal representation of dropout versus non-dropout instances results in imbalanced learning.
2. **Data Scarcity**: Limited data accessibility renders predictive modeling nearly infeasible.
3. **Feature Heterogeneity**: Variations in recorded information across educational grades necessitate tailored predictive models for each grade, posing challenges in maintenance and adaptability.
4. **Dropout Risk Factor Identification**: Understanding the reasons for dropout is crucial. However, explainability tools like SHAP may provide misleading insights, and feature importance metrics only shed light on model behavior, not the underlying causes for student dropout.

On the other hand, the recent emergence of Large Language Models (LLMs) gave birth to new applications in education. From text generation, and grading, to data analysis [2], the use of LLMs is rapidly expanding. Models such as Open AI's ChatGPT have shown capabilities of reasoning in many fields [19]. By learning from huge text corpora in a self-supervised manner, these models learn to create a representation of the world as we know it. The LLMs can be limited when the task is specific, thus, some guidance is required by the user to help steer them in the right direction. Thus, to further expand the knowledge of these models with up-to-date data, context retrieval techniques englobed in Retrieval Augmented Generation (RAG) [12] can be of use to provide the LLMs with the most relevant context to help answer a query.

In this paper, we investigate the use of LLMs to solve the previously mentioned Student Dropout Prediction (SDP) problems. We present an LLM based system that leverages the reasoning power of small, open source LLMs. In addition, to adapt these models to predict students' dropout, we use them in a Few-Shot Learning (FSL) paradigm by providing them with historic dropout and non-dropout cases. Furthermore, to enhance the context provided to the LLMs, we incorporate a similarity-based example retrieval which is the RAG part of our solution. Finally, we extensively evaluate our framework on real-life Moroccan data and compare it to the most commonly used ML models we find in the literature.

We summarize the contributions of our work in the following:

- The first exploration of small, yet powerful, open-source LLMs for SDP: to the best of our knowledge, this is the first work that investigates the use of small instruction LLMs to predict students' dropout.
- Enhanced dropout predictions through FSL and prompting: we demonstrate the capabilities of LLMs in assessing the risk of students' dropout through RAG-assisted FSL. We also evaluate the effect of the number of examples provided as context to the LLMs.
- Superior performance and unexplored territory: we demonstrate the predictive power of LLMs that requires but a few examples to learn from, revealing their superior performance compared to common baseline ML models in the literature.
- Better data flexibility: we showcase the flexibility of LLMs in handling missing values and emphasize the novelty of exploring instructional LLMs in the context of students' dropout, an area previously uninvestigated in current research.

The rest of this paper is structured as follows, Sect. 2 delves into the adopted method, Sect. 3 showcases our experimental setting, Sect. 4 discusses the obtained results. Finally, we conclude and suggest future work in Sect. 5.

2 Proposed Approach

Our method relies on FSL, a subset of ML utilized for tasks like classification and entity recognition in scenarios of data scarcity and transfer learning. FSL trains learners on a small, yet representative dataset, aiming for robust performance. Our approach operates as follows:

- Student data, aimed at predicting dropout likelihood, undergoes vectorization and mapping via the Student Record Processor, treating the student as a "query".

- The Similarity Retriever selects examples, or "shots," from pre-vectorized learning data based on the vectorized query data.
- The retrieved shots and query data are then processed by the Student Record Processor and sent to the Prompt Generator.
- The Prompt Generator generates prompts using a pre-specified template, which are then fed to the LLMs to generate responses.

Figure 1 showcases the general architecture of the proposed method. A detailed description of what each component accomplishes can be found on GitHub.[1]

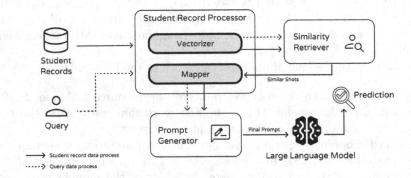

Fig. 1. RAG-assiseted FSL process for dropout prediction of a new student (query) using historic student records

3 Experimental Design

3.1 Dataset

For our experimental work, an anonymized data set was provided by the Kingdom of Morocco Ministry of National Education Preschool and Sports.[2] The dataset contains academic, demographic, and socio-economic information about students from the first year of primary school to the last year of high school. Moreover, the data concerns the public sector of the Fes-Meknes region. Finally, the data spans from 2015/2016 to 2020/2021 school years. More details about the dataset can be found on GitHub.[3]

3.2 Prompt Engineering

Using the right prompt can drastically change the responses of LLMs [16]. Thus, we mapped the meaning of the values we have in the dataset into a template that represents shot and query data in a JSON-like format. We refer to this template as the *structured* prompt template shown in Fig. 2.

---

[1] https://shorturl.at/qEHY0.
[2] https://www.men.gov.ma/.
[3] https://shorturl.at/cfgAT.

```
I will give you the academic and demographic data of a primary school student and I want you to classify the student as a Dropout or
a Non Dropout for next year. I will provide you with historical examples of students who dropped out and others who did not drop out
to use as a basis for your analysis.
These are examples of students who dropped out:
Student 1:
        Sex: Female, Age: 11,
        Handicap: no handicap,
        Failed last year: No, Final grade: 8.57, Class ranking: 1,
        Current level fails: 0,
        Has financial aid: Yes,
        Last year's status: passed,
        Absent days: 3,
        Absent classes: 0,
        Lives in a boarding school: No,
        Living area: rural,
These are examples of students who did not drop out:
...
Classify the following student into two classes "Dropout" or "Non Dropout"
Target student:
        Sex: Female, Age: 11,
        Handicap: no handicap,
        Failed last year: No, Final grade: 8.86,
        Class ranking: 1,
        Current level fails: 0,
        Has financial aid: Yes, Last year's status: passed,
        Absent days: 3,
        Absent classes: 0,
        Lives in a boarding school: No,
        Living area: rural
```

Fig. 2. A snippet of a 2-shot **structured** prompt.

Moreover, the following instruction is used as a system instruction: *"You are an assistant in student dropout detection who classifies a student as a dropout or non-dropout based on historical students. If the situation is unclear, classify the student as a dropout. Ensure that your analysis concludes with a definitive decision presented in the precise format: Decision: Dropout/Non-Dropout."*. The goal in defining this system instruction is to customize the personality of the LLMs and to ensure that we can easily extract the LLMs' decisions.

3.3 Experimental Setting

For our experiment, we used two open-source smaller instruction LLMs: Zephyr 7b-beta [17] and Mistral 7b [11]. These models were at the time of conducting this work the best 7b models on the Hugging Face Open LLM Leaderboard [3]. Using our prompt template, we construct prompts with the following numbers of shots: 2, 4, 6, 8, and 10. We then evaluate the LLMs on all the generated prompts. To compare the performance, we trained ML models: XGBoost, DT, and Random Forest (RF) on the data of the *whole* shot pool and evaluated on the same queries. Finally, to evaluate the quality of the predictions we opt for class-specific precision and recall as well Area Under the Receiver Operating Characteristic (ROC) Curve (AUC) which gives an idea about the performance of the models regardless of class imbalance.

4   Results and Discussion

Tables 1 presents the results of the LLMs on the **structured** prompts across the selected shot range. Additionally, the table showcases the performance of

Table 1. Results of Zephyr 7b-beta and Mistral 7b in all considered shot ranges using **structured prompts** compared to ML models

Model	N_shots	Accuracy	ND Precision	ND Recall	D Precision	D Recall	AUC
Zephyr	2	0.44	**0.95**	0.41	0.10	**0.77**	0.59
Zephyr	4	0.58	**0.95**	0.57	0.13	0.68	0.63
Zephyr	6	0.57	**0.95**	0.56	0.13	0.72	0.64
Zephyr	8	0.58	**0.95**	0.57	0.13	0.73	0.65
Zephyr	10	**0.67**	**0.95**	**0.67**	**0.16**	0.68	**0.67**
Mistral	2	**0.82**	0.93	**0.86**	**0.21**	0.40	0.63
Mistral	4	0.71	0.94	0.72	0.16	0.56	0.64
Mistral	6	0.68	**0.95**	0.68	0.15	0.61	**0.65**
Mistral	8	0.74	0.94	0.76	0.17	0.54	**0.65**
Mistral	10	0.48	**0.95**	0.45	0.11	**0.76**	0.61
XGB	N/A	**0.91**	0.92	**0.99**	**0.62**	0.10	0.55
DT	N/A	0.88	**0.93**	0.94	0.27	**0.24**	**0.59**
RF	N/A	0.90	0.92	0.97	0.37	0.16	0.57

ML models for comparison. From the table, we notice that all of the LLMs outperform the ML models in dropout recall which ranges between 40% (Mistral on 2-shot prompts) and 77% (Zephyr 7b-beta on 2-shot structured prompts). On the other hand, ML models scored a recall range between 10% (XGBoost) and 24% (DT). Furthermore, the performance of the LLMs on the non-dropout students is mostly comparable to that of the ML models in terms of precision. Finally, we generally see that as the number of shots grows, the performance of the LLMs gets better in terms of AUC which reaches 67% (Zephyr 7b-beta on 10-shot prompts).

However, some performance drawbacks can be highlighted. First, dropout precision of the LLMs reached a low of 10% in the case of Zephyr-beta on 2 shot prompts and reached a high of 21% in the case of Zephyr-beta on 10 shot prompts which is hardly comparable to ML models with values between 27% (DT) and 62% (XGB). This means the LLMs tend to produce more false positives. This is due to the system prompt we used, we emphasized the importance of not missing a dropout case. One other smaller shortcoming of the LLMs is non-dropout recall. Although the values are respectable, between 41% and 86%, they are not as effective as ML models. This is once again ties back to our system prompt which makes the LLMs produce more false positives.

To exhibit the reasoning power of the LLMs, Fig. 3 showcases the reasoning power of Zephyr 7b-beta to analyze and compare the query student's data to the provided examples. For an educational institution seeking to help students stay on track, an analysis like that might be useful to quickly detect where the possibility of dropout comes from. It is crucial to remember that the LLMs were not fine-tuned, let alone trained, on this classification task. The displayed performance was a result of nothing but a few examples to learn from. ML models do not have the ability that the LLMs have to generalize, during inference time, using nothing but a few examples. Furthermore, for missing values, added or deleted features, or even different grade levels, a single LLM can be adapted through easy-to-adapt prompts to cover all different scenarios.

After analyzing the historical data and comparing it to the target student's information, I conclude that the target student is more likely to drop out based on the following factors:

1. Age: The target student is 12 years old, which is older than the majority of students who did not drop out.

2. Absent days: The target student has a higher number of absent days compared to the students who did not drop out.

3. Financial aid: The target student receives financial aid, which is a factor that has been associated with dropouts in previous studies.

4. Uncertainty: While the target student's academic performance and class ranking are similar to those who did not drop out, there is still some uncertainty as to whether they will continue their education.

Based on these factors, I classify the target student as a Dropout with the following decision: Decision: Dropout

Fig. 3. Example response from Zephyr 7b-beta

5   Conclusion and Perspectives

For educational institutions, detecting possible dropout cases is crucial to the success of the general educational system. In this paper, we argue that the LLMs have their place, as a useful tool, in detecting students' dropout. The models showcased better predictive performance than baseline ML models. Furthermore, these models provided student data analysis based on a few examples to learn from.

That being said, the method we proposed has some limitations that are worth noting. First of all, the most noticeable limitation is inference time which grows exponentially with the size of the prompt and the number of desired tokens to be generated. The computation time is indeed tolerable in the case of one or a few response generations, but the same cannot be said for evaluation tasks such as the one conducted in this work. Moreover, formulating the right prompts to get clear answers from the LLMs is a difficult task since there is no fail-proof method that can be utilized.

As a proposition for future work, fine-tuning these models on training data could help improve the predictive performance. Moreover, evaluating and fine-tuning these models on data from different grade levels could confirm the possibility of having a single model as a dropout detection agent. Finally, another interesting topic to investigate is the quality of dropout reason identification provided by the LLMs.

References

1. Adnan, M., et al.: Predicting at-risk students at different percentages of course length for early intervention using machine learning models. IEEE Access **9**, 7519–7539 (2021)
2. Alqahtani, T., et al.: The emergent role of artificial intelligence, natural learning processing, and large language models in higher education and research. Res. Soc. Adm. Pharm. **19**(8), 1236–1242 (2023)
3. Beeching, E., et al.: Open LLM Leaderboard (2023)
4. Berka, P., Marek, L.: Bachelor's degree student dropouts: who tend to stay and who tend to leave? Stud. Educ. Eval. **70**, 100999 (2021)
5. Coleman, C., Baker, R., Stephenson, S.: Brightbytes: a better cold-start for early prediction of student at-risk status in new school districts (2019)
6. Coussement, K., Phan, M., De Caigny, A., Benoit, D.F., Raes, A.: Predicting student dropout in subscription-based online learning environments: the beneficial impact of the logit leaf model. Decis. Support Syst. **135**, 113325 (2020)
7. De Witte, K., Cabus, S., Thyssen, G., Groot, W., van den Brink, H.M.: A critical review of the literature on school dropout. Educ. Res. Rev. **10**, 13–28 (2013)
8. Del Bonifro, F., Gabbrielli, M., Lisanti, G., Zingaro, S.P.: Student dropout prediction. In: Bittencourt, I.I., Cukurova, M., Muldner, K., Luckin, R., Millán, E. (eds.) AIED 2020. LNCS (LNAI), vol. 12163, pp. 129–140. Springer, Cham (2020). https://doi.org/10.1007/978-3-030-52237-7_11
9. Du, X., Yang, J., Hung, J.L.: An integrated framework based on latent variational autoencoder for providing early warning of at-risk students. IEEE Access **8**, 10110–10122 (2020)
10. Erickson, J.A., Botelho, A.F., McAteer, S., Varatharaj, A., Heffernan, N.T.: The automated grading of student open responses in mathematics. In: Proceedings of the Tenth International Conference on Learning Analytics & Knowledge. LAK '20, New York, NY, USA, pp. 615–624. Association for Computing Machinery (2020)
11. Jiang, A.Q., et al.: Mistral 7B (2023)
12. Lewis, P., et al.: Retrieval-Augmented Generation for Knowledge-Intensive NLP Tasks (2021)
13. Meca, I., Rabasa, A., Sobrino, E., Juan, L.E.J.: Early warning methodology for dropping out of university degrees. In: Eighth International Conference on Technological Ecosystems for Enhancing Multiculturality. TEEM'20, New York, NY, pp. 245–249. USA Association for Computing Machinery (2021)
14. Mekouar, L., Bader, M., Belqasmi, F.: The super-node topology in collaborative learning. In: Proceedings of the 22nd Annual Conference on Information Technology Education. SIGITE '21, New York, NY, USA, pp. 67–68. Association for Computing Machinery (2021)
15. Mubarak, A.A., Cao, H., Hezam, I.M.: Deep analytic model for student dropout prediction in massive open online courses. Comput. Electr. Eng. **93**, 107271 (2021)
16. Nori, H., et al.: Can Generalist Foundation Models Outcompete Special-Purpose Tuning? Case Study in Medicine (2023)
17. Tunstall, L., et al.: Zephyr: Direct Distillation of LM Alignment (2023)
18. Yagci, M.: Educational data mining: prediction of students' academic performance using machine learning algorithms. Smart Learn. Environ. **9**(1), 11 (2022)
19. Yu, F., Zhang, H., Tiwari, P., Wang, B.: Natural language reasoning, a survey (2023)

Towards Automated Multiple Choice Question Generation and Evaluation: Aligning with Bloom's Taxonomy

Kevin Hwang[1](✉) ⓘ, Kenneth Wang[1](✉) ⓘ, Maryam Alomair[2] ⓘ, Fow-Sen Choa[2] ⓘ, and Lujie Karen Chen[2](✉) ⓘ

[1] Glenelg High School, Glenelg, MD 21737, USA
{khwang8265,kwang6166}@inst.hcpss.org
[2] University of Maryland, Baltimore County, Baltimore, MD 21250, USA
{maryama4,choa,lujiec}@umbc.edu

Abstract. Multiple Choice Questions (MCQs) are frequently used for educational assessments for their efficiency in grading and providing feedback. However, manually generating MCQs has some limitations and challenges. This study explores an AI-driven approach to creating and evaluating Bloom's Taxonomy-aligned college-level biology MCQs using a varied number of shots in few-shot prompting with GPT-4. Shots, or examples of correct prompt-response pairs, were sourced from previously published datasets containing educator-approved MCQs labeled with their Bloom's taxonomy and were matched to prompts via a maximal marginal relevance search. To obtain ground truths to compare GPT-4 against, three expert human evaluators with a minimum of 4 years of educational experience annotated a random sample of the generated questions with regards to relevance to the input prompt, classroom usability, and perceived Bloom's Taxonomy level. Furthermore, we explored the feasibility of an AI-driven evaluation approach that can rate question usability using the Item Writing Flaws (IWFs) framework. We conclude that GPT-4 generally shows promise in generating relevant and usable questions. However, more work needs to be done to improve Bloom-level alignment accuracy (accuracy of alignment between GPT-4's target level and the actual level of the generated question). Moreover, we note that a general inverse relationship exists between alignment accuracy and number of shots. On the other hand, no clear trend between shot number and relevance/usability was observed. These findings shed light on automated question generation and assessment, presenting the potential for advancements in AI-driven educational evaluation methods.

Keywords: automated question generation · GPT-4 · Bloom's taxonomy · large language models · multiple choice question generation

---

K. Hwang and K. Wang—Contributed equally.

A. M. Olney et al. (Eds.): AIED 2024, LNAI 14830, pp. 389–396, 2024.
https://doi.org/10.1007/978-3-031-64299-9_35

1 Introduction

Multiple-choice questions (MCQs) are indisputably a useful assessment tool in education [12, 13, 21]. Their effectiveness lies in their capacity for easy and efficient grading, allowing educators to swiftly evaluate many responses. This immediate provision of feedback is invaluable for enhancing learning outcomes as it enables students to make timely improvements in their areas of weakness. Well-designed MCQs possess the ability to assess knowledge across different levels of Bloom's Taxonomy [3], a framework that classifies different cognitive skills and abilities that students use to learn into levels, where each "Bloom level" progressively assesses higher levels of thinking skills [3]. By aligning questions with different levels of Bloom's Taxonomy, instructors can control the cognitive depth of their questions, catering to the diverse learning needs of their students and encouraging critical thinking among learners [12].

The conventional manual method of generating and assessing questions has been demanding in terms of labor, often necessitating substantial human input and expertise. Furthermore, depending on a restricted pool of questions may result in the repetition of questions across different assessments, potentially undermining the security of the assessment. In recent years, automated question generation, particularly using Large Language Models (LLMs), has gained prominence as a method to streamline the question creation process. Yet, the potential to consistently generate high-quality MCQs aligned with Bloom's Taxonomy is an area that remains unexplored.

In this experiment, we investigated the potential of an AI-driven process for the creation and evaluation of college-level biology MCQs. This process consisted of three main parts. Firstly, we generated 250 questions with GPT-4, using 5 different shot variations, and aligning to 5 levels of Bloom's taxonomy. Secondly, to attain a ground truth, we had three human evaluators with a minimum of 4 years of experience in biology evaluate a random sample of 150 of these 250 questions on their alignment to Bloom's taxonomy as well as their usability and relevance in the classroom and to the subject. Lastly, we compared GPT-4's results with the ground-truth results as well as our automatic evaluation pipeline's results to the ground-truth results.

This study seeks to address three research questions: RQ1 investigates the extent to which GPT-4 can generate usable and relevant questions that are aligned with the prompted Bloom's Taxonomy levels, RQ2 explores how the performance of GPT-4 differs between the number of shots provided to it, and RQ3 evaluates how well the automated evaluation system models actual human feedback.

1.1 Related Work

There has been a long history of automatic MCQ generation, starting with [6] who proposed the usage of NLP word-tagging to automatically create fill-in-the-blank (cloze) MCQs. [6]'s findings were improved upon by [15] and [16] who created methods of MCQ generation that select a relevant sentence to the tested topic and use what [15] coined as "'wh' phrases" to replace specific words in a sentence and turn them into questions. Moreover, previous studies have demonstrated the diverse applications of Pretrained Language Models (PLMs). As an example, [24] employed GPT-3, prompted with question-answer pairs sourced from a biology textbook to generate MCQs and free-response questions (FRQs), and concluded that generated questions were ready to be used in classrooms. As another example, [4] fine-tuned T5 using a combination of educational question datasets, evaluating the generated questions for linguistic quality and similarity to human-generated questions. They indicated that AI-generated questions could easily be repurposed by teachers. These approaches highlight the versatility of PLMs in generating educational questions.

In light of the existing research landscape, our study aims to address a notable gap in the literature concerning educational question generation, as none of the aforementioned methods have specifically targeted the alignment of questions to Bloom's taxonomy. In our literature review, we have identified only two existing methods that have attempted to generate questions aligned with Bloom's taxonomy. [14] employs a template-based question generation system, using keyword identification and pattern matching to generate questions based on templates aligned with Bloom's taxonomy, and [9] utilizes Bloom's taxonomy as a contextual template in prompt engineering. However, the former does not evaluate questions after generation, providing no insight into the alignment between the targeted Bloom level and the actual Bloom level of the generated questions. Moreover, none of these methods explore the generation of multiple-choice questions or use GPT-4 to generate questions. Our paper fills this research gap by addressing the capability of GPT-4 to generate multiple choice questions that align with Bloom's taxonomy using a varied few-shot prompting approach.

2 Methods

2.1 Question Generation Strategy

We utilized few-shot prompting, a technique employing shots, or examples of a task being carried out, to help GPT-4 better understand its task [7]. Our shots were composed of a context, Bloom level, and a resultant multiple-choice question with one answer and three distractors. Most shots were sourced from a subset of questions from the EduQG dataset [11], which had their Bloom's taxonomy level labeled. To compensate for the lack of Evaluate-type questions in the few-shot dataset, we zero-shot prompted GPT-3.5 (gpt-3.5-turbo) using the full textual content of chapters from OpenStax Biology 2e [19] and Chemistry 2e [19], and asked an expert human evaluator to classify the Bloom levels of questions generated, and also to find an excerpt from the large context given to GPT-3.5 that was specifically relevant to the generated question – this was to mirror the format present in EduQG and to ensure that all Bloom levels would have an ample amount of shots. Moreover, we asked the human evaluator to evaluate quality by asking them whether they would use the question in a classroom setting, which is in line with how all shots used in our prompting have been either created or noted as usable in the classroom by trained educators.

Shots were vectorized using OpenAI's Ada embeddings model [10] and were stored in a FAISS vector store [8] for quick lookups. To select shots for any given input prompt, we used a maximal marginal relevance search to optimize for a vectorized shot's similarity to the vectorized input context as well as introduce diversity among selected shots which has been proven to improve LLM performance in [25].

We prompted GPT-4 (gpt-4-1106-preview), OpenAI's state-of-the-art generative model [1], and we used a relatively high temperature of 0.9 (on OpenAI's scale of 0–2) to promote model creativity.[1] The choice of GPT-4 was primarily due to its state-of-the-art performance on traditional benchmarks [1]. The choice of the temperature 0.9 over other relatively high temperatures was arbitrary and more work needs to be done on how temperature affects the performance of GPT-4 on this task. Prompting and formatting of generated questions were handled using LangChain [5] and OpenAI's Python library [18]. In our research, we varied the number of shots and compared the results of using 0, 1, 3, 5, and 7 shots in our prompting. For each number of shots, we prompted GPT-4 with two randomly selected chapter summaries from OpenStax Concepts in Biology [19], and we asked GPT-4 to generate five MCQs for each of the levels of Bloom's taxonomy. The Create level was excluded because the MCQ format makes it challenging, if not impossible, for students to demonstrate their ability to synthesize new ideas by solely selecting from the provided responses. In total, we generated 250 questions, with 50 questions for each number of shots and 50 questions for each level of Bloom's taxonomy.

2.2 Question Evaluation Strategy

We evaluated generated MCQs using both automated and manual methods and then assessed how well the former could replicate the results of the latter.

For the automated evaluation, we used a rules-based model developed by [17] to detect 19 item-writing flaws (IWFs), which are common mistakes in MCQ generation [2]. The model would reject any question if it had more than one item writing flaw, in accordance with [23].

For the manual evaluation, we utilized an adjudication system involving three expert human evaluators. Firstly, we took a random subset of 150 of the 250 questions, stratified on the number of shots and Bloom's Taxonomy level (30 for each shot number and 30 for each Bloom level; not exclusive). Then, we had two of the human evaluators evaluate the same 150 questions on 3 aspects: Bloom Level, usability on a scale of 1–4 (1 being not usable at all, 2 being usable with major edits, 3 being usable with minor edits, and 4 being perfectly usable); and relevancy, on the same scale as usability. This scale mirrors the one found in [9] (for usefulness) for its insights into question generation performance. In this case, usability was defined as how functional a question is in a classroom setting, and relevancy was defined as how applicable a question was to its input context. We had the two human evaluators evaluate independently, unaware of the other's evaluations and GPT-4's target Bloom level, to avoid confirmation bias. After the two human evaluators had completed evaluation of the 150 questions, we had a third human evaluator adjudicate any discrepancies between the evaluations of the two human evaluators on the 150 questions, resulting in a final, singular set, which we could use to compare GPT-4 and our automated evaluation system against.

---

[1] The prompts used for generation can be found here (https://tinyurl.com/kh-aied).

2.3 Demographics and Diversity of Evaluators

Including the three human evaluators from the evaluation of GPT-4 generated questions, and the single human evaluator for the selection of shots, we had 4 total human evaluators, with experience spanning from elementary school to graduate school. Two of them are K-12 STEM educators with 15 + years and 16 + years of domain experience each, one of them is a PhD student with 4 + years of domain experience, and the last one is a professor with 26 + years of domain experience. We chose evaluators of varying educational experiences to introduce different perspectives in the question evaluation process. The distribution of our evaluators' different teaching experiences is noted for its insight into their diversity. We did not factor in traditional demographics, like race, sex or political affiliation, due to our evaluations not being on any dividing racial, sexual, or political lines.

3   Results and Discussion

RQ1: To What Extent Can GPT-4 Generate Usable and Relevant Multiple-Choice Questions that Align with Bloom's Taxonomy?

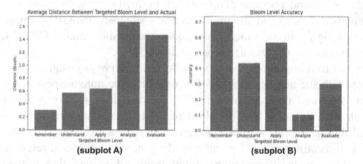

Fig. 1. Subplot A shows the average distance (in levels) between the actual Bloom level of questions and their targeted Bloom levels; Subplot B shows the alignment accuracy (accuracy of alignment between GPT-4's target level and the actual level of the generated question by the targeted Bloom level given to GPT-4[2].

We found that GPT-4 struggles the most with generating Analyze and Evaluate questions, and we observe no clear relationship between the targeted Bloom level and the alignment accuracy. However, we note that as the targeted Bloom level progresses higher up Bloom's taxonomy the average distance between the targeted level and the human-evaluated level generally increases, as shown in Fig. 1. This relationship has a Spearman correlation coefficient of *0.90* and a p-value of *0.037*, indicating a statistically significant strong positive monotonic relationship [22]. The result demonstrates that,

---

[2] Due to space limits, the confusion matrix is made available here (https://tinyurl.com/kh-aied-conf-mat).

as GPT-4's targeted Bloom level gets higher, it gets progressively further away from assessing the correct level of thinking skills.

We observe that the questions generated by GPT-4 tend to be highly usable and relevant with minimal to no editing required, regardless of the targeted Bloom level. This is evident from human ratings, whose mean usability and relevance scores for each Bloom level consistently exceed three. Indeed, 82.67% of questions generated by GPT-4 have been marked as usable or usable with minor edits by their human evaluator and 91.33% of questions were marked as relevant or relevant with minor edits, indicating GPT-4 is proficient in generating relevant, real-world applicable questions, even if the Bloom level of the question itself is not consistent with what was asked for. The high usability and relevance of questions mirror results from previous literature generating FRQs aligned with Bloom's Taxonomy [9].

RQ2: How does the Number of Shots Given to GPT-4 Affect its Question Generation Performance?

In terms of Bloom-level alignment accuracy, we note that an increase in the number of shots results in a decrease in performance. Though we originally hypothesized that few-shot prompting would better GPT-4's question generation performance, as noted in [7], we observe a statistically significant strong negative monotonic relationship between number of shots and Bloom-level alignment accuracy, denoted by a Spearman correlation coefficient of -0.95 and a p-value of 0.014 [22]. Following [20], the inverse relationship between number of shots and alignment accuracy suggests that GPT-4 misunderstands what it is looking for in the shots. Thus, adding more shots makes GPT-4 learn more irrelevant information, decreasing its performance.

In terms of usability and relevance, we do not observe any statistically significant relationship with the number of shots. For usability, we note a Spearman correlation coefficient of 0.2 and a p-value of 0.75 [22]. For relevance, the correlation coefficient is 0.36 with a p-value of 0.55. Though there exists no strong correlation, we observe that, for all shot numbers, the questions generated have a mean usability and relevance above three, indicating that GPT-4 can reliably generate usable and relevant questions (or questions that are usable and relevant with minor edits) regardless of the number of shots provided to it.

RQ3: How Well does the Automated Evaluation System Model Actual Human Feedback?

To assess the accuracy of the automated evaluation system, we dichotomized the usability score of the 150 human-rated questions into a binary: a rating of 1-2 (zero usability and usability major edits respectively) was dichotomized into a rating of "not acceptable", and a rating of 3-4 (usability with minor edits and strong usability, respectively) was dichotomized into a rating of "acceptable."

Comparing the automated evaluation system with the human-evaluated, ground-truth usability evaluations, we find that it has an accuracy of 43.33% and an F-Score of 0.52, which is consistent with a random guess proportional to the class distribution.

We also attempted to find the effect of IWFs on question usability. To do this, we compared the automated IWF detection model created by [17] against the human-evaluated, ground-truth usability using Spearman's rank correlation coefficient [22]. We found that the p-value associated with the Spearman coefficient was *0.94*, meaning that there is no significant correlation between the two. We hypothesize that this is because, while IWFs critique the format of the question, it seems as though educators place more significance on the content of the questions themselves, and they may overlook some IWFs in determining usability.

4 Conclusion

The results of RQ1 indicate that GPT-4 is proficient at generating questions that are usable and relevant in classroom settings regardless of the target Bloom level. However, it struggles with generating questions that align with the targeted Bloom levels, especially for Analyze and Evaluate type questions. Moreover, as the Bloom levels increase, so does the average distance from the target Bloom level. These errors with alignment accuracy underscore the need to explore different approaches, like fine-tuning or template-based generation. The results of RQ2 show that GPT-4 is yet again proficient at creating usable and relevant questions, regardless of the number of shots given to it. However, the observations strongly indicate that an increase in shots negatively impacts the Bloom level alignment accuracy. This may be because GPT-4 misunderstands what it is supposed to be looking for in its shots, and further work needs to be done helping GPT-4 better understand how to interpret the shots given to it. The results of RQ3 indicate that previous literature creating quality evaluation systems based on detecting IWFs may not necessarily align with human evaluation of classroom usability. More work needs to be done bridging the gap between machine and human-evaluation, either by modeling human evaluation with IWF features, or developing a new method of quality evaluation altogether. It is of utmost importance to develop a reliable machine-evaluation system to serve as an automated guardrail for future generation approaches.

In the future, to avoid high API and computational costs, we want to explore the viability of using local, open-source LLMs as well as smaller LLMs, to generate MCQs. Furthermore, we'd like to explore fine-tuning GPT-4 for multiple-choice question generation, comparing differences between fine-tuning and few-shot prompting. Another plausible direction is measuring the impact of prompt engineering approaches besides few-shot prompting, such as having GPT-4 utilize chain-of-thought reasoning.

Disclosure of Interests. The authors declare no competing interests.

References

1. Achiam, J., et al.: Gpt-4 technical report (2023). arXiv preprint arXiv:2303.08774
2. Breakall, J., Randles, C., Tasker, R.: Development and use of a multiple-choice item writing flaws evaluation instrument in the context of general chemistry. Chem. Edu. Res. Pract. **20**(2), 369–382 (2019)
3. Bloom, B.: A taxonomy of cognitive objectives. McKay, New York (1956)
4. Bulathwela, S., Muse, H., Yilmaz, E.: Scalable educational question generation with pre-trained language models. International Conference on Artificial Intelligence in Education. Springer Nature Switzerland, Cham (2023)

5. Chase, H.: LangChain [Computer software] (2022). https://github.com/langchain-ai/langchain

6. Coniam, D.: A preliminary inquiry into using corpus word frequency data in the automatic generation of English language cloze tests. Calico Journal, 15–33 (1997)

7. Dong, Q., et al.: A survey for in-context learning. arXiv preprint arXiv:2301.00234 (2022)

8. Douze, M., et al.: The Faiss library. arXiv preprint arXiv:2401.08281 (2024)

9. Elkins, S., et al.: How useful are educational questions generated by large language models? International Conference on Artificial Intelligence in Education. Springer Nature Switzerland, Cham (2023)

10. Greene, R., Sanders, T., Wang, L., Neelakantan, A.: New and improved embedding model (2022). https://openai.com/blog/new-and-improved-embedding-model

11. Hadifar, A., Bitew, S.K., Deleu, J., Develder, C., Demeester, T.: EduQG: a multi-format multiple-choice dataset for the educational domain. IEEE Access **11**, 20885–20896 (2023)

12. Javaeed, A.: Assessment of higher ordered thinking in medical education: multiple choice questions and modified essay questions. MedEdPublish **7**, 128 (2018)

13. Klůfa, J.: Multiple choice question tests–advantages and disadvantages. Recent Advances in Educational Technologies (2018)

14. Kusuma, S.F., Alhamri, R.Z.: Generating Indonesian question automatically based on Bloom's taxonomy using template based method. Kinetik: Game Technology, Information System, Computer Network, Computing, Electronics, and Control, 145–152 (2018)

15. Majumder, M., Saha, S.K.: A system for generating multiple choice questions: With a novel approach for sentence selection. In: Proceedings of the 2nd workshop on natural language processing techniques for educational applications, pp. 64–72 (2015)

16. Mitkov, R., Le An, H., Karamanis, N.: A computer-aided environment for generating multiple-choice test items. Nat. Lang. Eng. **12**(2), 177–194 (2006)

17. Moore, S., et al.: Assessing the quality of multiple-choice questions using GPT-4 and rule-based methods. European Conference on Technology Enhanced Learning. Springer Nature Switzerland, Cham (2023)

18. OpenAI Python Library. OpenAI (2020). GitHub, https://github.com/openai/openai-python. 24 Sept. 2023

19. OpenStax | Free Textbooks Online with No Catch. https://openstax.org/. Accessed 24 Sept. 2023

20. Reynolds, L., McDonell, K.: Prompt programming for large language models: Beyond the few-shot paradigm. In: Extended Abstracts of the 2021 CHI Conference on Human Factors in Computing Systems, pp. 1–7 (2021)

21. Riggs, C.D., Kang, S., Rennie, O.: Positive impact of multiple-choice question authoring and regular quiz participation on student learning. CBE—Life Sciences Education **19**(2), ar16 (2020)

22. Spearman, C.: The proof and measurement of association between two things (1961)

23. Tarrant, M., et al.: The frequency of item writing flaws in multiple-choice questions used in high stakes nursing assessments. Nurse Education Today **26**(8), 662–671 (2006)

24. Wang, Z., et al.: Towards human-like educational question generation with large language models. In: International conference on artificial intelligence in education. Springer International Publishing, Cham (2022)

25. Ye, X., et al.: Complementary explanations for effective in-context learning (2022). arXiv preprint arXiv:2211.13892

Enhancing Student Dialogue Productivity with Learning Analytics and Fuzzy Rules

Adelson de Araujo[✉], Jara Martens[✉], and Pantelis M. Papadopoulos[✉]

Department of Learning, Data, and Technology, University of Twente, Enschede, The Netherlands
a.dearaujo@utwente.com, j.martens@student.utwente.nl, p.m.papadopoulos@utwente.nl

Abstract. This study explores the use of the Collaborative Learning Agent for Interactive Reasoning (Clair) in a digital collaborative learning activity where interaction takes place via chat. Clair is designed to adaptively facilitate productive student dialogue using "talk moves" based on the Academically Productive Talk (APT) framework, a popular approach in related conversational agent studies. In this paper, we detail how Clair, powered by learning analytics, machine learning, and a fuzzy rule-based system, can adaptively trigger talk moves in student dialogue. In an experimental study conducted with $n = 9$ university student dyads, we assess the impact of Clair's presence on student dialogue productivity. We analyzed the within-subjects differences (with/without Clair) in four key goals of student dialogue productivity: the frequency of (a) students sharing thoughts, (b) orienting and listening, (c) deepening reasoning, and (d) engaging with others' reasoning. Our findings indicate a notable improvement in deepening reasoning ($p = .047$), highlighting Clair's capability to prompt students to engage in more critical thinking and elaborate on their ideas. Yet, the impact on other goals was less pronounced, suggesting the complexity of facilitating all goals of productivity. This paper demonstrates the potential of integrating learning analytics and fuzzy rules into triggering approaches for collaborative conversational agents, offering a novel approach to adaptively trigger talk moves in student dialogue. The results also underline the need for further refinement in the design and application of such systems to comprehensively support productive student dialogues in collaboration settings.

Keywords: conversational agents · collaborative learning · student dialogue · academically productive talk · learning analytics

1 Introduction

In the 21st century, *Critical Thinking, Creativity, Collaboration*, and *Communication* skills—collectively known as the 4Cs—have become essential for navigating the fast-changing global landscape. However, productive dialogue, where students build on each other's ideas constructively (Chi and Wylie, 2014), often requires explicit guidance; otherwise, participation can be uneven and superficial (Gillies, 2019). For example, an eloquent student contributes more while the partner just agrees, an idea presented is too vague, or students may seldom build on each other's contributions.

Teachers play a key role in fostering the 4Cs and encouraging productive student dialogue. The Academically Productive Talk (APT), or 'Accountable Talk' (Michaels et al., 2008), is a method that enhances student dialogue through structured 'talk moves', reflective prompts that teachers can make to guide students, such as *"Could someone summarize what we have discussed so far?"* (i.e., Recapping). In student dialogue, these content-independent strategies encourage peer interaction and reasoning, fostering a deeper understanding of the subject being discussed. Yet, effective APT application requires teachers to skillfully listen to conversations and pose timely questions. Thus it can be challenging, time-consuming, and not feasible for the teachers to handle multiple groups at once.

To address this, researchers are exploring APT strategies delivered by collaborative conversational agents (CCAs) as scalable solutions to facilitate student dialogue. For instance, Tegos et al. (2016) found that the presence of a CCA in student dialogue was linked to a higher frequency of explicit reasoning behaviors and more balanced participation. Nguyen (2023) found a positive effect, compared to a control condition, on students' transactive exchanges, i.e., explanations that are directed towards building on a partner's contribution. Adamson et al. (2014) reported on several studies, demonstrating that CCAs' effects can largely depend on the audience and learning material. These studies employed learning analytics, e.g., through natural language processing, and a rule-based system to trigger interventions in student groups. Yet, studies demonstrating whether and how CCAs promote productive dialogue are limited.

Building on the foundations laid by previous studies, we developed a CCA that combines various learning analytics with a fuzzy rule-based approach to improve the flexibility required to trigger a variety of talk moves in student dialogue. This paper uniquely contributes to existing knowledge by (i) describing in detail our novel CCA triggering technique and (ii) reporting on a new case study evaluating our CCA across various dimensions of productive dialogue among university students.

1.1 Research Question

By timely prompting students with a variety of talk moves, we anticipate that the presence of CCAs can encourage productive student dialogue. Michaels and O'Connor (2015) elaborated on essential goals of teacher guidance that align with observable student behaviors, which are particularly relevant for assessing the impact of CCAs. These goals are outlined as the Four Goals for Productive Discussions (FGPD):

1. Helping students share their own thoughts.
2. Helping students orient to and listen carefully to one another.
3. Helping students deepen their reasoning.
4. Helping students engage with each other's reasoning.

Accordingly, this paper presents a case study formulated around the following research question:

RQ: To what extent does the collaborative conversational agent make written student dialogue more productive compared to when the agent is not present in terms of (a) sharing their thoughts, (b) orienting and listening to one another, (c) deepening their reasoning, and (d) engaging with each other's ideas?

2   Collaborative Learning Agent for Interactive Reasoning (Clair)

In designing collaborative conversational agents for online student dialogues, finding the right type and time for a talk move can be challenging (Adamson et al., 2014; Nguyen, 2023; Tegos et al., 2016). Clair uses eight talk moves (see labels and examples in Table 1 in Sect. 2.2) crafted based on APT guidelines (Michaels et al., 2015). Clair's talk moves can either target the last student who spoke, another discussant, or both students. Clair uses three alternative phrasings of each talk move to avoid repetition. By employing various talk moves, we aim to comprehensively support students in the FGPDs based on what happens in the dialogue.

Figure 1 illustrates the process for using learning analytics and fuzzy rules to trigger talk moves in student dialogue. The process starts with Clair's configuration, which requires the fuzzy rules chosen and configuration details of the task at hand including the topic keywords.

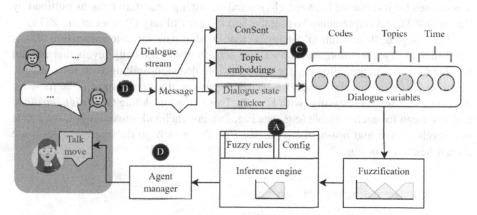

Fig. 1. Flowchart of Clair's internal components. Clair's triggering mechanism must be prepared with configuration adjusted to the task at hand and the fuzzy rules (A). Subsequently, Clair is ready to receive messages (B), calculate its dialogue variables (C), and send talk moves to the dialogue (D).

2.1   Dialogue Variables

To make decisions on the timing of interventions, Clair first employs learning analytics instruments hereby called 'dialogue variables'. In the current version of Clair, there are a total of twelve dialogue variables, calculated for each message. Eight of them are probabilities of collaborative behavior categories, outputs of a ConSent model, powered by the pre-trained multilingual Universal Sentence Encoder (mUSE, Yang et al., 2019) which we evaluated in previous work and found satisfactory to moderate levels of reliability (de Araujo et al., 2023b). Three of these are related to the message' *'focus'* (at κ = 0.60), which are Domain (L1_DOM), Coordination (L1_COO), and Off-task (L1_OFF); whereas the five others are related to its *'intent'* (κ = 0.61), which are Informative (L2C_IN), Argumentative (L2C_AR), Asking for information (L2C_AI), Active motivating (L2C_AM), and None of the specified (L2C_NOS).

There are four other dialogue variables. The first two are about how students addressed the topic: Topic similarity (TSIM) measures semantic similarity to a list of topic keywords (transformed into mUSE embeddings; Yang et al., 2019) and Topic accumulation (TACC) measures the ratio of the speaker's accumulated TSIM value compared to dialogue partners. The other two are metrics from the dialogue state: Time spent (TIME) measures the time since the dialogue started and Messaging speed (PACE) measures the proportion of messages per unit of time. For more information and examples on these dialogue variables, we refer to our previous work (de Araujo et al., 2024).

2.2 Triggering Mechanism

The triggering mechanism, based on a fuzzy expert system, decides whether to select a talk move in response to a chat message and its dialogue variables. This approach was chosen for its balance between clarity and managing uncertain data, as outlined by Zadeh (1983), and implemented using the Scikit-Fuzzy library (Warner et al., 2019).

The first step, 'fuzzification', changes clear-cut dialogue variable values into "fuzzy sets" such as *high*, *medium*, and *low*. These sets represent intensity levels that are not fixed but cover a range, allowing for more flexible decision-making with the dialogue variables. To tailor these fuzzy sets for each dialogue variable, we analyzed chat message data using K-Means clustering (with K = 3). This helped us define what *high*, *medium*, and *low* mean for each variable (e.g., see Fig. 2a). For each talk move output, we define two levels: *active* and *not-active* (e.g., see Fig. 2b) which guide whether a response should be chosen or not.

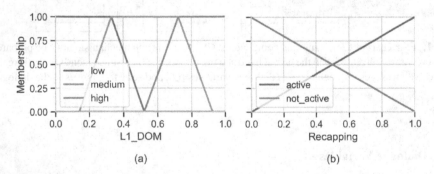

Fig. 2. Membership functions for the dialogue variable L1_DOM (a) and the talk move Recapping (b).

After the fuzzification step, the 'inference engine' step is responsible for determining outputs for each talk move by relying on fuzzy rules applied to the inputs that indicate the likelihood that each talk move should be triggered. The fuzzy rules are in the form of "IF-THEN" statements that relate to states of inputs with a desired level of output for each talk move. More specifically, fuzzy rules are defined in the form of "*IF $(x_1$ is $A_1)$ and ...$(x_n$ is $A_n)$ THEN $(y_1$ is $B_1)$ and...$(y_m$ is $B_m)$*", where A_i and B_j are fuzzy sets to describe the x_i dialogue variable and y_j talk moves, respectively. To ensure that the rule base is simpler to interpret and adjust, we currently employ one rule to determine when each talk move's *active* level should be *high*, i.e., moments when Clair could intervene, and another general rule to determine when all talk moves' *not-active* level should be *high*, i.e., moments when Clair should not intervene at all.

The triggering mechanism's final step is translating the fuzzy output into a binary decision. This step, usually known as 'defuzzification', is implemented by Clair's 'agent manager'. The agent manager initially chooses the primary talk move candidates based on the active values (higher than 0.75). Candidates used in the last three interventions are excluded to avoid repetition, and talk move frequency is monitored, favoring less used ones when active values and frequencies match. In a tie, the selection is random and the final utterance variation is randomly chosen and sent to the dialogue.

The fuzzy rules for all talk moves utilized in the current version of Clair are detailed in Table 1. The formulation of rules and refinement were conducted by expert evaluation and interviewing secondary school teachers (de Araujo et al., 2023a).

Table 1. Clair's talk moves with examples and associated fuzzy inference rules used in the triggering mechanism.

Talk move	Example	Fuzzy inference rule
Recapping	*"Can someone give a brief summary of what we've covered so far?"*	L1_DOM is *medium/low* and L2C_AR is not *high* and TSIM is *low* and TIME is *high*
Add-on	*" < discussant >, could you add some new perspective to what < speaker > just said?"*	(L1_DOM is *high* or L1_COO is *high*) and L2C_AR is *high* and TSIM is not *low* and TACC is *high*
Rephrasing	*" < discussant >, can you rephrase what < speaker > said so that everyone is on the same page?"*	L1_DOM is *high* and L2C_AR is *high* and TSIM is *high* and TACC is *high*
Agree/Disagree	*" < discussant >, can you explain to < speaker > if there is something you disagree with?"*	(L1_DOM is *high* or L1_COO is *high*) and L2C_AR is *high/medium* and TSIM is *medium* and TACC is not *low*
Linking contributions	*" < discussant >, can you link your ideas to what < speaker > said?"*	L1_DOM is *high* and (L2C_IN is *high* or L2C_AR is *medium*) and TSIM is *medium* and TACC is *high*

(continued)

Table 1. (*continued*)

Talk move	Example	Fuzzy inference rule
Build on prior knowledge	" < speaker >, can you explain to < discussant > how this fits into the bigger picture?"	L1_DOM is *high* and (L2C_IN is *high* or L2C_AR is *high*) and TSIM is *high* and TACC is *low* and TIME is *high/medium*
Example	" < speaker >, can you give an example or a real-life scenario for < discussant > that can help illustrate the concept better?"	L1_DOM is *high* and L2C_AR is *high* and TSIM is *medium* and TACC is *medium/low*
Expand reasoning	" < speaker >, can you explain to < discussant > how you got this idea?"	L1_DOM is *high* and (L2C_IN is *high* or L2C_AR is *high/medium*) and TSIM is *high/medium* and TACC is *medium/low*

3   Case Study

To evaluate Clair's impact and address RQ1, we conducted a within-subjects repeated-measures study with university students on a previously covered topic. In total, 18 Psychology Bachelor's students volunteered to participate (13 females, 5 males; age range: 19 to 23, M = 21.6, SD = 1.10). Participants received an anonymized username and were randomly assigned into pairs ($n = 9$).

After login, the participants provided their consent and demographical information in a form. The activity started with a familiarization phase that lasted 8 min, engaged the pairs in chat discussion of a topic on climate change, and helped the participants to get a better understanding of the overall task goal for the upcoming phases.

Next, participants discussed the two main topics (i.e., classical and operant conditioning). Each topic was discussed for 15 min and finished with the pair's final answer in the chat. In the second topic, all participants interacted with Clair. Finally, the participants filled out a questionnaire indicating their satisfaction with the task and with Clair which was reported on a preliminary report by Martens (2023).

3.1   Data Analysis

To analyze our RQ, we employ learning analytics with the dialogue variables from Con-Sent models, i.e., message's focus and intent, to measure FGPDs. Using these variables, we applied sequential pattern mining to track target behaviors, determining outcomes by the frequency of behavior-indicating patterns in dialogues. We analyzed dialogues in n-message windows, identifying patterns within these message sequences. Conditions for a pattern match within these sequences are assessed in three steps, and the behavior does not need to appear in consecutive messages. We found $n = 7$ windows more consistently capture FGPD-related patterns, including three condition-meeting messages and four peripheral ones. Smaller windows, e.g., $n = 5$, seem to not capture behaviors of Goals 3 and 4.

In particular, the pattern conditions used the following proxies to operationalize dependent variables of FGPD: Goal 1's pattern involves identifying sequences of task-related informative and argumentative statements, indicating thought sharing; Goal 2 focuses on recognizing responses to peers' task-related questions; Goal 3 on expanding one's own task contributions; and Goal 4 on collaborative argument discussion. Thresholds were adopted using each dialogue variable's limit of *high*, *medium*, and *low* from the triggering mechanism. Ultimately, each goal's outcome variable is measured as the count of unique pattern matches. This approach is further described in our previous work (de Araujo et al., 2024).

3.2 Results

Wilcoxon's signed rank test indicated varied within-subjects impacts of Clair on the FGPDs. We observed a statistically significant improvement in Goal 3 (deepening reasoning) from Phase 1 to Phase 2 ($p = .046$, $r_{rb} = 0.905$; for $\alpha = .05$). This suggests that Clair effectively facilitated students in elaborating and expanding their reasoning over the dialogue. However, the results for Goals 1 (sharing thoughts, $p = .159$), 2 (orienting and listening, $p = .891$), and 4 (engaging with others' reasoning, $p = .074$) did not show significant differences between phases.

4 Discussion and Conclusion

Results analysis suggests that tools like Clair can partly aid productive student dialogue. In particular, the presence of Clair can help an audience of university students to deepen their reasoning more often than when Clair is absent. The significant impact can be attributed to Clair's ability to prompt students to build on each other's contributions and explain their reasoning. This is aligned with findings from Tegos et al. (2016), who found an increase in students' explicit reasoning behavior in the presence of a similar agent. Also, our result could be explained by the high responsiveness level of university-level participants (85% talk moves responded, 6% acknowledged, 9% ignored).

Educators may currently consider these tools as a supplement rather than a replacement for comprehensive guidance toward online productive dialogue. The results might not generalize to other audiences, e.g., K-12, as university students are usually more skilled in discussing with each other. In addition, our case study had a small sample size and further research may be required to uncover Clair's impact.

Regarding theory-building, we demonstrated a clear need for further research and development to more effectively support all FGPDs. Assuming human teachers can help student dialogue in all FGPDs, developing more advanced, human-like interaction features in CCAs could potentially facilitate their effectiveness. Furthermore, exploring the customization of talk moves, e.g., to the needs of different student audiences, may increase the chances of impactful results. Future research could explore incorporating large language models, to comprehensively guide productive student dialogue.

References

Adamson, D., Dyke, G., Jang, H., Rosé, C.P.: Towards an agile approach to adapting dynamic collaboration support to student needs. Int. J. Artif. Intell. Educ. **24**(1), 92–124 (2014). https://doi.org/10.1007/s40593-013-0012-6

Chi, M.T.H., Wylie, R.: The ICAP framework: linking cognitive engagement to active learning outcomes. Educational Psychologist **49**(4), 219–243 (2014). https://doi.org/10.1080/00461520.2014.965823

de Araujo, A., Papadopoulos, P. M., McKenney, S., & de Jong, T. (2023a). Supporting Collaborative Online Science Education with a Transferable and Configurable Conversational Agent. *15th International Conference on Computer-Supported Collaborative Learning (CSCL)*

de Araujo, A., Papadopoulos, P.M., McKenney, S., de Jong, T.: Automated coding of student chats, a trans-topic and language approach. Computers and Education: Artificial Intelligence **4**, 100123 (2023). https://doi.org/10.1016/j.caeai.2023.100123

de Araujo, A., Papadopoulos, P.M., McKenney, S., de Jong, T.: A learning analytics-based collaborative conversational agent to foster productive dialogue in inquiry learning. J. Comp. Ass. Learn. (2024). In press

Gillies, R.M.: Promoting academically productive student dialogue during collaborative learning. Int. J. Educ. Res. **97**, 200–209 (2019). https://doi.org/10.1016/J.IJER.2017.07.014

Martens, J.: Artificial Intelligence in Education: AI Conversational Agent for Online Collaborative Learning (2023). http://essay.utwente.nl/95259/

Michaels, S., O'Connor, C.: Conceptualizing talk moves as tools: Professional development approaches for academically productive discussions. In: Resnick, L.B., Asterhan, C., Clarke, S.N. (eds.) Socializing intelligence through talk and dialogue, pp. 347–362. American Educational Research Association (2015)

Michaels, S., O'Connor, C., Resnick, L.B.: Deliberative discourse idealized and realized: Accountable talk in the classroom and in civic life. Stud. Philos. Educ. **27**(4), 283–297 (2008). https://doi.org/10.1007/s11217-007-9071-1

Nguyen, H.: Role design considerations of conversational agents to facilitate discussion and systems thinking. Computers & Education **192**(104661) (2023). https://doi.org/10.1016/J.COMPEDU.2022.104661

Tegos, S., Demetriadis, S., Papadopoulos, P.M., Weinberger, A.: Conversational agents for academically productive talk: a comparison of directed and undirected agent interventions. Int. J. Comput.-Support. Collab. Learn. **11**(4), 417–440 (2016). https://doi.org/10.1007/s11412-016-9246-2

Warner, J., Sexauer, J., Scikit-fuzzy, Hörteborn, A.: JDWarner/scikit-fuzzy: Scikit-Fuzzy version 0.4.2 (2019). https://doi.org/10.5281/ZENODO.3541386

Yang, Y., et al.: Multilingual Universal Sentence Encoder for Semantic Retrieval, 87–94 (2019). ArXiv:1907.04307

Zadeh, L.A.: The role of fuzzy logic in the management of uncertainty in expert systems. Fuzzy Sets Syst. **11**(1–3), 199–227 (1983). https://doi.org/10.1016/S0165-0114(83)80081-5

HiTA: A RAG-Based Educational Platform that Centers Educators in the Instructional Loop

Chang Liu[1]([⊠])(iD), Loc Hoang[2], Andrew Stolman[2], and Bo Wu[1,2]([⊠])(iD)

[1] Colorado School of Mines, Golden 80401, USA
{liuchang,bwu}@mines.edu
[2] HiTA AI Inc., Santa Clara, USA
{loc,astolman}@hita.ai

Abstract. Large Language Models (LLMs) have the potential to influence the current educational framework by providing answers to students without educators' involvement. Herein, we have developed a new system, called HiTA, that places educators at the center of AI-assisted learning loops. The key concept involves using LLMs as teaching assistants to amplify educators' expertise with their pedagogical supervision. Built on retrieval-augmented generation, HiTA incorporates advanced features such as a user-friendly interface, a multi-mode prompt processing subsystem, and a customization subsystem that provides course-specific support. Compared to general LLMs, HiTA provides pedagogically oriented and course-adherent responses. Our system has been deployed to 6 educators and 400 students spanning entry-level to advanced courses at the Colorado School of Mines for two semesters. In the Fall 2023 semester, 270 students posted over 14,000 questions within approximately two months. From the 47 feedback forms collected post-semester, over 97% of survey respondents found HiTA helpful, and more than 80% considered it more beneficial than ChatGPT.

Keywords: Education · LLM · ChatGPT

1 Introduction

Traditional educational methods often fail to provide instant help to all students instantly due to limited resources and diverse needs. Large Language Models (LLMs) have addressed some of these issues thanks to their capabilities in areas like conversational generation and information retrieval [1,3]. LLMs are already used by students for homework assistance and knowledge exploration [2,6]. However, challenges persist in effectively using LLMs for educational purposes.

LLMs can sometimes generate hallucinated or even harmful information [8]. Their effectiveness is also constrained by the training data which may be outdated or contain problematic content [1,13]. Although previous research has used LLMs in education, the LLMs often failed to fully leverage educators' expertise

© The Author(s), under exclusive license to Springer Nature Switzerland AG 2024
A. M. Olney et al. (Eds.): AIED 2024, LNAI 14830, pp. 405–412, 2024.
https://doi.org/10.1007/978-3-031-64299-9_37

[7,15]. Ethical concerns also arise since students might directly obtain answers for their assignments [5]. These challenges highlight the gap between LLMs and their applications in education. Herein, we develop a new system, HiTA, based on these observations. Our primary contributions include:

1) We developed HiTA, an LLM-powered system that reorients the AI-assisted instructional loop to better leverage educators' expertise. This system functions as a Virtual Teaching Assistant (VTA) that enables educators to supervise LLM-generated responses, control the teaching styles of VTAs, manage course materials, and monitor interactions between students and VTAs.
2) We built HiTA based on Retrieval-Augmented Generation (RAG), and it incorporates features such as a customization subsystem, a multi-mode prompt processing subsystem, and a user-friendly interface.
3) We conduct a pilot study with 6 instructors and 400 students from 4 courses in Computer Science at the Colorado School of Mines for two semesters. In Fall 2023, 270 students submitted over 14,000 queries. In the post-semester survey, over 97% of 47 respondents found HiTA helpful, and over 80% considered it more beneficial than ChatGPT. In Computer Organization, the section initially using HiTA outperformed the other until the midterm. Performance between the sections aligned after HiTA was made available to both post-midterm. These preliminary findings provide insight into the effectiveness of HiTA.

2   Related Works

LLMs have shown strong performance in various domains [3]. Recent developments significantly broaden the potential applications of LLMs in education [1,13]. Indeed, LLMs are increasingly used by students in higher education [2] with many concerns about their impact on the current educational paradigm [9,16]. Several studies have been conducted to explore the application of LLMs in education [5,7]. Pardos et al. introduced an open-source tutoring system that offers hints and scaffolds to students in algebra classes using LLMs [14]. This adaptive system tutors students according to their abilities based on educator-defined questions and required skills. Kazemitabaar et al. presented CodeAid, a system designed to provide timely feedback for students learning programming [10]. The authors conducted educator interviews to gain further insights of this system.

Moreover, RAG is used to address hallucination issues and improve performance [11]. This technique retrieves content based on a similarity search between the embedding of questions and a knowledge database. Liu et al. developed a set of LLM-powered tools that serve as tutors for students in the CS50 class [12]. This RAG-based system provides subject-matter guidance and leads students to solutions rather than providing them directly. EduChat is an LLM-powered chatbot system designed for educational purposes [4]. It is pre-trained and fine-tuned on an education-specific corpus to acquire adequate knowledge. Built on RAG, EduChat can provide up-to-date responses by accessing online resources.

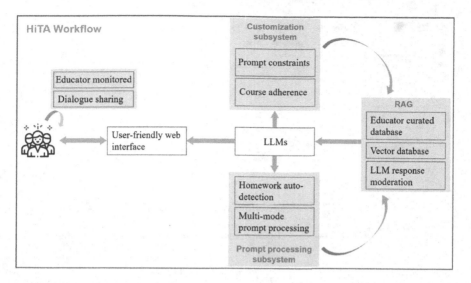

Fig. 1. HiTA System Workflow: Students interact with the user interface. LLMs generate responses based on the questions, chat history, retrieved context from RAG, and specific rules in multiple modes. The generated answers are presented via the interface.

These studies show that LLMs can effectively assist educators and complement some aspects of their work. However, they mainly focus on interactions with students in specific subjects and have not fully integrated educators into the tutoring process to leverage their expertise comprehensively.

3 Methodology

3.1 Overview

Figure 1 shows the workflow of HiTA, which is divided into three phases. The initial phase includes pre-teaching preparations, involving the creation of response generation rules, management of course materials, and construction of databases for RAG. The second phase processes student inquiries, dispatching questions to appropriate modes via homework auto-detection or user selections. Responses are generated based on the questions, chat history, context retrieved from RAG, and pre-defined rules from educators. The post-teaching phase concentrates on the storage of chat histories and user interactions through dialogue sharing.

3.2 System Features

Customization Subsystem. RAG ensures that LLMs consistently access relevant data from course materials. To further incorporate educators' efforts and deliver pedagogically oriented responses, we developed a customization subsystem. Educators can upload text-based rules via the instructor tools. These rules

408   C. Liu et al.

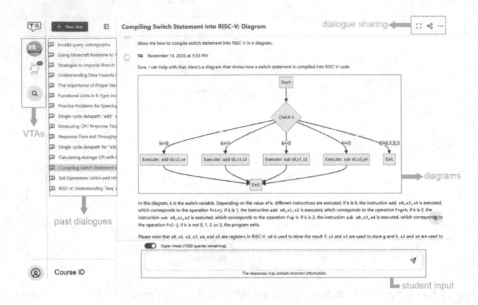

Fig. 2. Example of User Interface: HiTA responds to a student's question with texts and a diagram; students can select appropriate VTAs, view past dialogues, pose questions, and choose to share these dialogues.

define the response styles of LLMs and prevent irrelevant or harmful answers. In the responses, we have implemented a feature that displays reference links to relevant course materials to enable quick access with a single click. Additionally, HiTA can generate diagrams within the answers to help students develop a clear understanding of key concepts and identify knowledge gaps.

Multi-mode Prompting Subsystem. We developed a multi-mode prompt processing subsystem to cater to diverse educational scenarios. The general mode addresses a broad range of common course-related questions, while the homework mode is specifically designed for homework-related inquiries. Instead of providing direct answers, this mode offers step-by-step hints to lead students toward the correct solutions. The practice mode supports formative assessments and provides crucial feedback to both educators and students by generating a variety of quizzes closely aligned with current course materials. To prevent misuse, we implemented a homework auto-detection feature that automatically identifies and dispatches homework-related questions based on similarity searches over embeddings of student questions and homework materials.

User Interface. We developed the user interface using the React/Next.js framework and adopted a chat-style format to facilitate ease of interaction. Figure 2 shows an example of student conversations with a diagram included in the response. Moreover, educators can efficiently manage course materials, upload specific rules, and review chat histories using instructor tools with just a few clicks.

3.3   Educational Ethics

LLMs may generate misleading or even harmful responses, both of which are critical concerns in educational settings where inappropriate responses could negatively impact students' learning interest. HiTA implements three main strategies to mitigate this risk. First, it utilizes RAG to ensure that relevant course materials are forwarded to LLMs. Second, we employ GPT-3.5-turbo and GPT-4 APIs from OpenAI, which include a built-in moderation process to filter out harmful content. Lastly, HiTA enables students to immediately report inappropriate and incorrect responses. To maintain educational integrity and avoid data privacy issues, HiTA provides references from course materials in its responses.

4   Experiments and Results

HiTA has been used by 6 instructors and 400 students in multiple courses at Colorado School of Mines. In Fall 2023, HiTA was primarily deployed in two courses: Computer Science for STEM and Computer Organization with the latter being more advanced. Our analysis mainly focuses on data from this semester.

4.1   Student Performance Comparison

The Computer Organization course was divided into Sections A and B. There were 83 students in Section A and 50 in Section B. Initially, our system was made available to students in Section A and was later extended to both sections after the midterm exam. The average midterm grade in Section A was 86.68, exceeding that of Section B by over 6 points. Following the introduction of our system, the academic performance of the two sections became more aligned: the performance gap between the sections for the final exam reduced to 2.27 points. Although numerous other factors may have contributed to the closure of this performance gap, we view this as an encouraging indication that HiTA is correlated with positive student outcomes.

4.2   System Usage Analysis

In Fig. 3, we present the total and average question numbers for each student relative to the usage days in the Fall 2023 semester. It reveals that some students are significantly more engaged than others: the top 30 students accounting for approximately 55% of all inquiries. Furthermore, 5 students consistently submitted more than five questions per day after registering with our system.

4.3   Feedback Survey

In Spring 2023, we deployed HiTA to 130 students in the Advanced Computer Architecture and Operating Systems courses and invited them to complete a feedback survey. Of these, 36 students submitted responses. Based on the results,

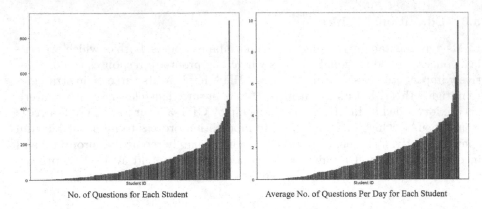

No. of Questions for Each Student   Average No. of Questions Per Day for Each Student

Fig. 3. Total and Average Numbers of Student Questions Relative to the Usage Days.

96.7% of the respondents believed HiTA generates helpful responses. When comparing our system with ChatGPT, 53.6% of the respondents considered HiTA more useful for their studies, and 48% regarded it as safer than ChatGPT.

In the Fall 2023 semester, 270 students from the Computer Science for STEM and Computer Organization courses submitted over 14,000 questions. In the post-semester feedback survey, 97.9% of 47 respondents found our system beneficial. Additionally, 91.5% considered our VTAs to be at least as helpful as actual TA office hours. Compared with ChatGPT, 80.9% of respondents rated our system as more useful, and 55.3% considered it safer to use.

4.4   Response Accuracy and Comparison

In order to assess the accuracy of responses generated by HiTA, we randomly selected 50 dialogues each from the Computer Science for STEM course and the Computer Organization course. We asked a teaching assistant to manually review these responses and categorize them as either correct or incorrect. Correct answers are defined as clear and able to provide at least partial assistance to student questions whereas incorrect ones contain evidently ambiguous or erroneous information. In the Computer Science for STEM course, 96% of dialogues were deemed correct and capable of offering some help to students. In the Computer Organization course, 94% of the dialogues were classified as correct.

We illustrate the difference between the generated responses from the HiTA platform and ChatGPT using a specific homework question, which is as follows: *What are the smallest negative and the largest positive integers in 8-bit two's complement notation? Write your solution for each in both binary and base 10.*

HiTA: *To determine the smallest negative and the largest positive integers in 8-bit two's complement notation, you need to understand how two's complement representation works. In an n-bit two's complement system: The most significant bit (MSB) is the sign bit. If the MSB is 0, the number is positive; if it is 1, the number is negative. The range of representable numbers is from -2^{n-1} to*

$2^{n-1} - 1$. *For an 8-bit two's complement system: The smallest negative number has a 1 as the MSB and 0 s for the rest of the bits. This represents -2^7. The largest positive number has a 0 as the MSB and 1 s for the rest of the bits. This represents $2^7 - 1$. Can you use this information to write down the binary and base 10 representations for the smallest negative and largest positive integers in 8-bit two's complement notation? Remember, for the binary representation, you should have a total of 8 bits. Once you've tried, I can provide further guidance if needed.* **source 1**[1]

ChatGPT (GPT-4): *In 8-bit two's complement notation, the range of representable numbers is from -2^7 to $2^7 - 1$. This is because one of the 8 bits is used for the sign, leaving 7 bits for the magnitude of the number. Two's complement notation allows for a straightforward representation of negative numbers and simplifies binary arithmetic operations. The smallest negative integer can be represented by setting the sign bit to 1 and all other bits to 0. In 8-bit two's complement, this is: Binary: 10000000 Base 10: $-2^7 = -128$ The largest positive integer can be represented by setting the sign bit to 0 and all other bits to 1. In 8-bit two's complement, this is: Binary: 01111111 Base 10: $2^7 - 1 = 127$ So, in 8-bit two's complement notation: The smallest negative integer is -128 (binary: 10000000) The largest positive integer is 127 (binary: 01111111)*

Observe that HiTA's response provides essential hints rather than direct answers to questions. This feature offers greater educational suitability compared to ChatGPT, which tends to provide correct results. Moreover, HiTA could provide links to specific course materials used to generate the response for reference.

5 Discussions and Future Work

Our future work primarily focuses on three aspects. First, we plan to conduct a large-scale test involving over 1000 students across various majors and universities in Spring 2024. Second, our analysis indicates that a minority of students posted a considerable volume of questions, and approximately 16% posed fewer than ten questions which highlights the challenge of increasing student engagement. Third, while HiTA currently enables educators to set specific rules and review chat histories, there is potential to further support educators by integrating features such as automated homework grading.

6 Conclusion

We introduce HiTA as an LLM-powered conversational platform designed for educational purposes. Built on RAG, HiTA functions as a VTA to integrate educators' expertise into the instructional loop. HiTA has been deployed to 400 students across various courses and received positive feedback that underscores

---

[1] **source 1** denotes a reference link to relevant content in course materials.

its educational advantages compared to general LLMs. Additionally, analysis of student performance in midterm and final exams in the Computer Organization course provides preliminary evidence of HiTA's effectiveness.

References

1. Bubeck, S., et al.: Sparks of artificial general intelligence: early experiments with GPT-4. arXiv preprint arXiv:2303.12712 (2023)
2. Chan, C.K.Y., Hu, W.: Students' voices on generative AI: perceptions, benefits, and challenges in higher education. Int. J. Educ. Technol. High. Educ. 20(1), 43 (2023)
3. Chang, Y., et al.: A survey on evaluation of large language models. ACM Trans. Intell. Syst. Technol. 15, 1–45 (2023)
4. Dan, Y., et al.: Educhat: a large-scale language model-based chatbot system for intelligent education. arXiv preprint arXiv:2308.02773 (2023)
5. Extance, A.: ChatGPT has entered the classroom: how LLMS could transform education. Nature 623(7987), 474–477 (2023)
6. Intelligent.com (2023). https://www.intelligent.com/one-third-of-college-students-used-chatgpt-for-schoolwork-during-the-2022-23-academic-year/. Accessed 05 Feb 2024
7. Jeon, J., Lee, S.: Large language models in education: a focus on the complementary relationship between human teachers and ChatGPT. Educ. Inf. Technol. 28(12), 15873–15892 (2023)
8. Ji, Z., et al.: Survey of hallucination in natural language generation. ACM Comput. Surv. 55(12), 1–38 (2023)
9. Kasneci, E., et al.: ChatGPT for good? On opportunities and challenges of large language models for education. Learn. Individ. Differ. 103, 102274 (2023)
10. Kazemitabaar, M., et al.: CodeAid: evaluating a classroom deployment of an LLM-based programming assistant that balances student and educator needs. arXiv preprint arXiv:2401.11314 (2024)
11. Li, H., Su, Y., Cai, D., Wang, Y., Liu, L.: A survey on retrieval-augmented text generation. arXiv preprint arXiv:2202.01110 (2022)
12. Liu, R., Zenke, C., Liu, C., Holmes, A., Thornton, P., Malan, D.J.: Teaching cs50 with AI (2024)
13. OpenAI: ChatGPT (2023). https://openai.com/chatgpt/. Accessed 30 Nov 2023
14. Pardos, Z.A., Tang, M., Anastasopoulos, I., Sheel, S.K., Zhang, E.: OATutor: an open-source adaptive tutoring system and curated content library for learning sciences research. In: Proceedings of the 2023 Chi Conference on Human Factors in Computing Systems, pp. 1–17 (2023)
15. Zawacki-Richter, O., Marín, V.I., Bond, M., Gouverneur, F.: Systematic review of research on artificial intelligence applications in higher education-where are the educators? Int. J. Educ. Technol. High. Educ. 16(1), 1–27 (2019)
16. Zhai, X.: ChatGPT user experience: implications for education. Available at SSRN 4312418 (2022)

Towards Convergence: Characterizing Students' Design Moves in Computational Modeling Through Log Data with Video and Cluster Analysis

Adelmo Eloy[1,2(✉)] ⓘ, Aditi Wagh[3] ⓘ, Tamar Fuhrmann[2] ⓘ,
Roseli de Deus Lopes[1] ⓘ, and Paulo Blikstein[2] ⓘ

[1] Universidade de Sao Paulo, Sao Paulo, SP 05508-010, Brazil
adelmo.eloy@usp.br
[2] Teachers College, Columbia University, New York, NY 10027, USA
[3] Massachusetts Institute of Technology, Cambridge, MA 02319, USA

Abstract. Programming computational models is foundational for students to better understand scientific phenomena, mirroring the methods employed by real scientists. Previous research has used qualitative and quantitative methods to investigate engagement in computational practices through modeling. This paper compares two methods of using log data – alongside video analysis or through cluster analysis - to analyze sixth-grade students' design moves in a block-based environment for science learning. We compare patterns identified through each approach and elaborate on their commonalities and differences. Our findings highlight how the methods add to each other, providing insights for future research at the intersection of computer science and science education.

Keywords: Computational Modeling · Log Data Analysis · Mixed-Methods Approaches · Cluster Analysis

1 Introduction

Programming computational models is foundational for students to better understand scientific phenomena, mirroring the methods employed by real scientists. Particularly, when engaging with agent-based modeling (ABM) with the support of computing technologies, students translate their scientific hypotheses into entities, properties and rules in a way a computer can interpret (e.g., [1]). Within the ABM paradigm, researchers have adopted block-based programming approaches to lower the threshold for learners to construct computational models (e.g., [2]), enhancing the learning of key computer science practices such as modularity [3] and debugging [4]. In this paper, we compare two methods using log data to characterize sixth-grade students' design moves as they built agent-based models in a block-based environment in science class. For brevity, we refer to log data analysis with video analysis as "Method 1" and log data analysis through cluster analysis as "Method 2". In this paper, we answer the following research

A. M. Olney et al. (Eds.): AIED 2024, LNAI 14830, pp. 413–421, 2024.
https://doi.org/10.1007/978-3-031-64299-9_38

question: *what are the relationships between outcomes from Method 1 and Method 2 when investigating design moves made by students while programming?*

2 Related Work

Previous research has investigated students' behaviors and practices in programming computational models. For instance, [5] characterized middle-school students' learning behaviors during science model-building activities. Similarly, [6] defined four conceptual stages to describe agent-based programming practices from over 1,500 projects in an online community. In this paper, we investigate students' computational practices through a unit of analysis we call "design move", consisting of every action or sequence of actions students perform in their code, while programming a computational model.

To do this, we build on prior work that has leveraged log data from students' interactions with computational modeling environments, with varying goals and approaches. For example, [7] employed quantitative techniques to investigate learning outcomes of engineering students programming in NetLogo to model scientific phenomena. Through extensive log data analysis, they identified patterns in students' coding style and behavior based on metrics such as compilation frequency and code size. Similarly, [4] characterized middle school students' modeling behaviors based on a set of quantitative features extracted from the log data, such as the ratio of total simulation runs to total actions performed and the average number of actions between simulation runs. Five features were defined and used to inform the clustering of 29 students through hierarchical clustering. We built on this by performing cluster analysis of students' design moves using a similar set of parameters.

Finally, our work is informed by prior studies that explore multiple techniques to investigate multiple techniques for data analysis in computational modeling. For instance, [8] combined process mining and natural language processing to study students' regulation of collaborative design of computational models in middle school science activities. Likewise, [3] explored conceptual learning and computational practices in a middle school classroom through coding models for evolution. They used video and log data analysis, specifically using Levenshtein Distance to describe the nature of code changes over time. These approaches informed our work integrating log and video analysis to interpret students' design moves, and creating data visualizations to support pattern recognition and communicate our findings.

3 Methods

3.1 Setting and Data Sources

Our data came from 67 6th-graders from three schools in California. They engaged in a unit on ink diffusion over 6–8 class periods, following the same activities with slight variation. Students experimented with ink diffusion in hot and cold water, drew paper models to illustrate differences in diffusion rates, and then used a modeling environment to create and share computational models. They assessed their models against real-world data and watched a video about diffusion. To analyze design moves in Method 1, we chose one pair per class with the longest screen recording. In Method 2, we selected data from 31 active student accounts used either individually or in pairs.

In this study, we use MoDa ("Modeling + Data", https://moda.education), a block-and agent-based computational modeling environment designed for middle school science classrooms [9]. MoDa includes three main areas (Fig. 1) that influenced our analysis. In the Design area, students can code models by using a drag-and-drop interface with domain-specific blocks. In the Simulation area, students can set up and run their models, adjusting phenomenon-relevant parameters (e.g., temperature). In the Data area, students can play videos which, in this unit, showed ink spread in hot and cold water.

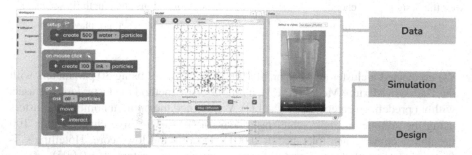

Fig. 1. MoDa's Design, Simulation, and Data areas used to categorize and analyze log data.

3.2 Data Analysis

We analyzed students' interaction logs with MoDa using Methods 1 and 2, with design moves as the unit of analysis. A design move is an individual or sequence of actions performed in the Design area, which both influences and is influenced by actions in the Simulation and Data areas."

Method 1. Method 1 consisted of combining log data visuals to inform qualitative analysis. By creating visualizations from log data via a Python script, we identified visual patterns and examined video segments corresponding to each visual pattern. The patterns resulting from this analysis were termed as design moves, which are defined based on students' actions in the videos and the corresponding the log data. We used the definition of design moves to categorize 167 instances from the three students under analysis. A detailed description of this method can be found in [10].

Method 2. Method 2 consisted of defining parameters based on the log data available to characterize our unit of analysis (design moves) and employing a clustering method to identify similar groups in our sample and interpret their main features.

1. We first defined 30 quantitative parameters to describe each design move, based on the data logs in MoDa. These parameters are summarized in Table 1 and were applied to 1906 design moves from the 31 students; a detailed description of parameters is available upon request.

Table 1. Description of parameters for characterizing design moves

Dimension	Level	# of parameters	Description
During the design	General	2	# of actions, length (in seconds),
	Action	4	% of actions (e.g. create, delete)
Before the design	General	3	type, # of actions and length (in seconds)
	Action	6	% of actions (e.g. set, go, play video)
After the design	General	3	type, # of actions and length (in seconds)
	Action	9	% of actions (e.g. set, go, play video)

2. We conducted a cluster analysis on of 1906 design moves using 30 parameters with the Gaussian Mixture Model (GMM), a probabilistic method to cluster large datasets without predefined categories (e.g., [11]). We defined the optimal number of clusters using the silhouette score, achieving a peak of 0.143 with five clusters (max 100 iterations; convergence tolerance = 0.001). Despite the low score, Hotelling's T-Squared test revealed significant differences among the clusters ($p < 0.005$).

3. We analyzed cluster characteristics by assessing parameter values, sizes, and distributions, and highlighted key differences using select parameters.

Comparing Methods 1 and 2. After obtaining results from each method, we compared them using a convergence matrix and validated their relationship with a Chi-Square test, revealing associations and interdependencies between design moves and clusters. We then refined our understanding of design moves by examining sample video cases.

4   Findings

4.1   Characterizing Design Moves Based on Method 1

Table 2 describes six design moves identified through Method 1, along with their frequency among a sample of three students. For a more detailed description, see [10].

4.2   Characterizing Design Moves Based on Method 2

Table 3 presents the clusters identified through Method 2, the frequency and two key features for each cluster, with a full table available upon request. The "# of actions" column shows the average number of actions in each cluster, while the "Main actions" column details the most common actions performed.

Table 2. Design moves' description based on Method 1.

Design move	Frequency	Description
Exploratory	46 (28%)	2–4 changes in the code and iteratively testing them
Debugging	59 (35%)	1–2 changes in the code and testing them after facing an error or a malfunction in the simulation
Fine-tuning	25 (15%)	1–2 changes in the code and testing them out after successfully running their simulation
Data-informed	3 (2%)	1–4 changes in the code after interacting with video data
From-scratch	5 (3%)	5+ changes in the code, usually starting over a blank screen, with no simulation or data actions in between,
Remixing	2 (1%)	2–4 changes in the code by copying, pasting, and adapting a chunk of existing code in their model
Other	27 (16%)	Cases that did not fit with any of the design moves above

Table 3. Clusters for design moves based on Method 2.

Cluster	Frequency	# of actions	Main actions
C1	933 (49%)	4.3	adding (44%), deleting (21%)
C2	285 (15%)	4.3	adding (32%), rearranging (30%)
C3	285 (15%)	1.0	rearranging (100%)
C4	267 (14%)	1.5	parameters (100%)
C5	136 (7%)	2.7	parameters (40%), adding (25%)

4.3 Comparing Results from the Two Methods

We compared results from Methods 1 and 2 using a contingency table to explore the relationship between variables (Table 4). The first column lists design moves from Method 1, and the subsequent columns show their distribution across clusters from Method 2, based on the 167 instances. Statistical analysis revealed a significant relationship between design moves and the clusters ($\chi2\ (30) = 143.52$, $p < 0.001$). Cramér's V coefficient indicated a large effect size of 0.46, suggesting a substantial relationship.

Finally, Table 5 presents a summary of our findings on design moves in computational modeling with MoDa, comparing results from Methods 1 and 2.

Table 4. Contingency table associating results from Methods 1 and 2.

Design move	C1	C2	C3	C4	C5	Total
Exploratory	39	2	0	2	3	46
Debugging	38	4	14	3	0	59
Fine-tuning	6	3	0	14	2	25
Data-informed	0	0	0	0	3	3
From-scratch	2	2	0	0	1	5
Remixing	2	0	0	0	0	2
Other	6	5	11	2	3	27
Total	93	16	25	21	12	167

Table 5. Summary of the comparison between results from Methods 1 and 2.

Design moves	Clusters	Findings from comparing results from Methods 1 and 2
Fine-tuning	C4	Fine-tuning design is predominantly associated with C4, though few design move cases are observed in other clusters
Data-informed	C5	Data-informed design is exclusively associated with C5, while 75% of the instances in C1 are attributed to other design moves
Exploratory	C1	Exploratory design is mainly linked to C1, with over half of the instances in this cluster related to other design moves
Debugging	C1, C3	Debugging design is primarily associated with C1 and C3
Remixing	C1	Remixing design is exclusively linked to C1, while C1 has instances linked to nearly all design moves
Other	C3	Nearly half of the instances in C3 are associated with "Other", which is distributed across various clusters
From-scratch	-	From-scratch design is not linked to any specific cluster
-	C2	C2 is not associated with any specific design move

4.4 Towards Convergence: How Methods 1 and 2 Can Inform Each Other

We examined discrepancies between results from Methods 1 and 2 using samples from students' video data, to illustrate how each method can inform each other. We offer two examples from a student pair: the first showing how video data detail specific goals of each design move, and the second showing how cluster analysis uncovers patterns missed by qualitative analysis alone. Given names are pseudonyms.

Comparing Different Instances of C5. Though all data-informed moves fell into C5, this cluster also included instances from other design moves. For example, in Harry and Oakley's work, they designed a model in which ink particles changed water's color upon interacting. Harry ran the model simulation and video data side-by-side for their teacher, who observed that "your ink is staying in, like, one spot." Harry responded, "I should make it move; it should move"; they modified the code, adding the "ask water particles" and "move" blocks (Move #1). After additional adjustments, they called the teacher and ran the video and simulation again. the teacher reminded them they already knew particles could not infect each other (based on a prior experiment) and asked them to "work on creating the scientific model, when the particles are bouncing off each other." Harry and Oakley started a new model and added a "create water particles" block (Move #2). Moves #1 and #2 are examples of design moves preceded by interactions with video and simulation, which might have led them to be clustered into C5. However, Move #1 directly addressed discrepancies between the simulation and the video data (data-informed design), while Move #2 was prompted by the teacher's suggestion to start a new model, and not by data (an instance we labeled as "Other").

Comparing Debugging Moves in Clusters C1 and C3. Debugging moves mostly fell into two clusters: C1 and C3; Harry and Oakley's work provides examples that illustrate such variation. After making water particles move and bounce, Harry added a "create ink particles" under the "Go" bracket (Move #3), which is executed continuously. They then ran the simulation, and Oakley noticed "You're hitting all the area now." Realizing a mistake, Harry said, "Oops I put it in the", and fixed the error by dragging "create ink particles" from "Go" to "on mouse click", a bracket executed by a mouse click. After noticing the ink particles remained static, they added "ask ink particles" and "move" blocks under "on mouse click" and reran the model (Move #4). Moves #3 and #4, both categorized as debugging moves, were placed in clusters C3 and C1, respectively. Move #3 involved nuanced code modifications, such as dragging blocks, typical of C3. In contrast, Move #4 entailed more structural code changes, like adding an "create particles" block, typical of C1. In addition, analyzing the behaviors leading to debugging moves in C3 revealed a pattern in unlabeled instances ("other"), typically involving block adjustments in the Design area without clear intent.

5 Discussion and Future Research

Our findings illustrate how different methodological approaches to using log data can complement each other when investigating students' design moves in block- and agent-based computational modeling. Joint analysis of log data visuals and video data uncovered contextual insights beyond cluster analysis. Specifically, it revealed varied student motivations for interacting with video data, from passively watching them to actively using it to inform code modifications (Data-informed move). Likewise, students could rearrange their code done to visually organize the blocks or make an actual change to the code to solve an issue (Debugging move). These insights inform us of ways to refine the definition for each move (e.g., "exploratory" design exclusively involving new blocks or combinations), and new parameters for better cluster analysis. In contrast, results from cluster analysis refined our understanding of students' design moves defined through qualitative analysis. For instance, distinct design moves into the same cluster (e.g. Exploratory and Remixing design) or distributed across clusters (e.g. From-scratch design) led us to reconsider their relevance within a larger data sample. In addition, classifying unmatched cases (i.e. "other") into clusters highlighted their similarities with known design moves, aiding in preventing future misclassifications.

Our findings are subject to limitations. Both methods are subject to interpretive bias, such as relying on qualitative video analysis from a short sample to frame design moves and relying on a subset of parameters to interpret clusters. Additionally, the unbalanced classroom representation in Methods 1 and 2 could bias the analysis toward practices from specific classrooms. Further analysis focusing on class comparison may help address this limitation. Nevertheless, our preliminary results reinforce the multiple benefits of integrating quantitative learning analytics techniques with qualitative methods, as highlighted by previous research [12]. Going forward, we aim to extend this analysis to new rounds of mixed-methods and students' data to be able to provide more robust definitions of student's design moves in computational modeling.

Acknowledgments. This study is supported by funding from the National Science Foundation under Grant No. 2010413. We thank the science teachers for using MoDa in their classrooms, Jacob Wolf for supporting data collection, and João Eloy for supporting log data processing.

Disclosure of Interests. The authors have no competing interests to declare that are relevant to the content of this article.

References

1. Wilensky, U., Reisman, K.: Thinking like a wolf, a sheep, or a firefly: Learning biology through constructing and testing computational theories—an embodied modeling approach. Cogn. Instr. **24**(2), 171–209 (2006)
2. Wilkerson, M., Wagh, A., Wilensky, U.: Balancing curricular and pedagogical needs in computational construction kits: Lessons from the DeltaTick project. Sci. Educ. **99**(3), 465–499 (2015)
3. Wagh, A., et al.: Anchor code: modularity as evidence for conceptual learning and computational practices of students using a code-first environment. In: Proceedings of the 12th International Conference on Computer Supported Collaborative Learning. International Society of the Learning Sciences (2017)
4. Hutchins, N., Biswas, G., Grover, S., Basu, S., Snyder, C.: A systematic approach for analyzing students' computational modeling processes in C2STEM. In: Proceedings of the International Conference on Artificial Intelligence in Education. AIED 2019, pp. 116–121. Springer, Cham (2019)
5. Zhang, N., Biswas, G., Hutchins, N.: Measuring and analyzing students' strategic learning behaviors in open-ended learning environments. Int. J. Artif. Intell. Educ. **32**, 931–970 (2022)
6. Chen, J., Wilensky, U.J.: Measuring young learners' open-ended agent-based programming practices with learning analytics. In: Proceedings of AERA Annual Meeting (2023)
7. Blikstein, P.: Using learning analytics to assess students' behavior in open-ended programming tasks. In: Proceedings of the 1st International Conference on Learning Analytics and Knowledge, pp. 110–116 (2011)
8. Emara, M., Hutchins, N.M., Grover, S., Snyder, C., Biswas, G.: Examining student regulation of collaborative, computational, problem-solving processes in open-ended learning environments. J. Learn. Analy. **8**(1), 49–74 (2021)

9. Wagh, A., et al.: MoDa: designing a tool to interweave computational modeling with real-world data analysis for science learning in middle school. In: Proceedings of Interaction Design and Children, pp. 206–211 (2022)

10. Eloy, A., et al.: Decomposing students' design moves when programming agent-based models. To appear in: Proceedings of the 18th International Conference of the Learning Sciences. International Society of the Learning Sciences (2024)

11. Scheidt, M., et al.: Engineering students' noncognitive and affective factors: group differences from cluster analysis. J. Eng. Educ. **110**(2), 343–370 (2021)

12. Sherin, B., Kersting, N., Berland, M.: Learning analytics in support of qualitative analysis. In: Proceedings of the 13th International Conference of the Learning Sciences. International Society of the Learning Sciences (2018)

Mars, Minecraft, and AI: A Deep Learning Approach to Improve Learning by Building

Samuel Hum[(✉)], Evan Shipley[(✉)], Matt Gadbury[(✉)], H Chad Lane[(✉)], and Jeffrey Ginger[(✉)]

University of Illinois at Urbana-Champaign, Urbana, IL 61820, USA
{hum3,evanjs3,gadbury2,hclane,ginger}@illinois.edu

Abstract. Middle school students learned about astronomy and STEM concepts while exploring *Minecraft* simulations of hypothetical Earths and exoplanets. Small groups ($n = 24$) were tasked with building feasible habitats on Mars. In this paper, we present a scoring scheme for habitat assessment that was used to build novel multi/mixed-input AI models. Using Spearman's rank correlations, we found that our scoring scheme was reliable with regards to team size and face-to-face instruction time and validated with self-explanation scores. We took an exploratory approach to analyzing image and block data to compare seven different input conditions. Using one-way ANOVAs, we found that the means of the conditions were not equal for accuracy, precision, recall, and F1 metrics. A post hoc Tukey HSD test found that models built using images only were statistically significantly worse than conditions that used block data on the metrics. We also report the results of optimized models using block only data on additional Mars bases ($n = 57$).

Keywords: Scoring Scheme · Artificial Intelligence · Habitat Building · Informal Learning · Minecraft

1 Introduction

Minecraft has become one of the most popular games in the world, with over 140 million monthly users, and 21.21% of daily user traffic originating from the U.S. [7]. Researchers have shown that even with little experience learners master controls quickly and can effectively engage content when *Minecraft* is used in STEM learning environments [3]. Given the ubiquity of the game, as well as ample opportunities to conduct research on learning and motivation [3], *Minecraft* deserves attention regarding effectiveness for promoting positive student outcomes in formal and informal learning environments.

The data used in this study comes from the What-if Hypothetical Implementations in Minecraft (WHIMC) project that uses *Minecraft* as a vehicle for understanding student interest and motivation in exploring STEM content, primarily Astronomy and Earth Science. Learners are presented with a variety of

A. M. Olney et al. (Eds.): AIED 2024, LNAI 14830, pp. 422–430, 2024.
https://doi.org/10.1007/978-3-031-64299-9_39

"what-if" scenarios, such as "What if Earth had a colder sun?". Working through these counterfactual examples of phenomena in science has shown promise of enhancing learning above and beyond studying strictly factual information [4]. A major highlight of the camp experience is the collaborative build phase, where students work with peers to construct habitats on Mars that scientists or explorers might inhabit. A *Minecraft* habitat or base is an assigned region, sized under a few hundred blocks in any dimension, where participants work in teams to build. This paper proposes a novel scoring scheme for work products such as the Mars habitats in *Minecraft* and a novel method for work product assessment in educational video games. We propose the use of multi/mixed-input models in *Minecraft* that takes data from in-game images and materials groups used for their bases to assess learning. Although multimodal data has been used in educational research and video game environments, AI models using different forms of data have not been applied to student work products [6]. Thus, our paper is guided by the following questions and hypotheses:

RQ1. What criteria constitute a reliable scoring scheme for Mars habitat builds in *Minecraft*, and to what extent does the scoring scheme hold across differences in group sizes and amount of face-to-face time with instructors? H1: A comprehensive scoring scheme based on inclusion of essential aspects of habitability will contribute to a reliable scoring scheme. Group size and face-to-face time with instructors will not contribute to significant differences between scores.

RQ2. What, if any, relationship do Mars habitat scores have with learning outcomes? H2: Higher scores on Mars habitats will positively correlate with knowledge assessment scores.

RQ3. What type or combination of data input accessible in Minecraft should be used for artificially intelligent detectors of habitat quality? H3: Incorporating models built using a combination of multiple forms of data concatenated together will outperform models solely built using one form of data. These models can extrapolate information from different aspects of the bases for more accurate predictions that would be impossible for models only built using a single type of data.

2 Background

In this study, we are interested in analyzing what are called, "student work products", referring to specific, task-driven designs and creations, also called "artifacts". In the end of a learning activity, a student has created a product manifesting their conceptual understanding of the content they interacted with throughout a learning experience. Work products as means of assessing learner knowledge and creativity emerged from Constructionism and the idea that knowledge is produced through students' creative and collaborative work [2]. This notion has been fully embraced by the Maker Movement and the desired approach to better understand what tools and activities are contributing to learning and other desirable outcomes (*i.e.* creativity) [5]. In *Minecraft*, an observational study examining learner understanding of urban planning, found that learners from a small town in Brazil incorporated their own interpretations of what matters in a habitat work product and included additional spaces they deemed important, such as playgrounds [1]. Assessing student *Minecraft* builds to provide in-activity learner support, however, has unique challenges. It is an ill-defined domain and there is no one right way to build a habitat. Students may incorporate structures that have personal meaning that may arise from student prior experiences [1], which can be difficult for humans, and as a result, computers to interpret. We seek to understand how AI models can be designed around digital making assessment using an exploratory approach and expand the literature regarding student support in open-ended learning activities.

3   Methods

3.1   Participants

A total of $n = 48$ middle school age students are included in this study (31% female) with an average age of 11.96 years old from camp data collected in 2022. All students participated in 1-week summer camps held in three distinct locations in the West, Midwest, and East United States. Demographic breakdown is as follows: 30% Caucasian, 23.75% African-American, 21.25% preferred not to answer, 12.5% Hispanic, 2.5% Asian, 1.25% American Indian, and 7% Other. A total of $n = 24$ bases were analyzed for the scoring scheme. To optimize our AI models using the scoring scheme, additional bases were collected in 2023. We used data from a total of $n = 131$ students that made 57 bases across 16 camps (25.69% female, 2.75% non-binary). Participants had an average age of 11.49 years old. Of the students that entered demographic information on our survey, the breakdown is as follows: 54.4% Caucasian, 20% African-American, 8% preferred not to answer, 8% Hispanic, 4% Asian, and 5.6% Other. Consent to participate was obtained from at least one parent/guardian and verbal assent was assessed at the beginning of each camp. Participant familiarity with building in Minecraft varied considerably.

3.2   Materials

Participants were all provided with a laptop, mouse, and an individual loaner (anonymous) account to play *Minecraft: Java Edition*. Participants used the same account for each session of their respective after school program or summer camp. All maps explored by participants were created by our lab and represent simulations of "What if" questions, such as "What if Earth was a moon to a larger planet?", as well as known exoplanets, planets outside of our solar system (*e.g.* Kepler 186-f). Design of worlds was done in consultation with an astrophysicist and each features extreme conditions, such as high winds, widespread volcanic activity, freezing temperatures, or low gravity, which can be seen or measured using in-game science tools. As part of the camp curriculum, participants complete self-explanation questions following their exploration of each in-game world. Each world has three total questions, scored on a scale of 0 to 3, each showing a level of astronomy explanation and comprehension of the material presented.

3.3 Procedure

During the final sessions of camps participant groups were formed based on seating arrangements, existing friendships or by researcher assignment. They were prompted with an introductory video and presentation and then challenged to design a habitat for humans to work and survive on Mars. Participants were invited to employ knowledge they gained from exploring previous hypothetical Earths and exoplanets to inform how to respond to extreme conditions on Mars. Groups had around 3 hours to collaborate and build their habitats and presented them to peers and parents on the last day, explaining the problems they addressed and how they solved them, as well as what made their habitat special.

3.4 Data Analysis

Habitat Scoring Scheme. The scoring scheme for the Mars habitats was drafted in consultation with a professor of astronomy. It was designed to account for the scientific considerations that could be represented in student bases during the Mars habitat challenge. In total, 11 categories were outlined: area where the base is built, atmosphere regulation within the base, combating different levels of gravity, communications facilities, food and water considerations, health and wellness facilities, transportation (*e.g.* rovers, rocket launchpad, etc.), power generation (*e.g.* nuclear reactors), protection from radiation, rounded structure shape, and storage for supplies. Each category was scored using a three-tier system that was later used as labels for the AI classifiers described in later sections. These tiers are classified from least score to highest score as "Basic", "Intermediate", and "Mastered", each representing a level of application and mastery that the participants show during the Mars habitat activity. The "Basic" tier awards 0 points for the category and is represented in habitats with the concept not being present or present but highly unrealistic. The "Intermediate" tier awards a number of points halfway between 0 and the maximum for each category and is represented in habitats with the concept being present within the habitat, but unfinished. The final "Mastered" tier awards the maximum number of points per category, and represents that the team integrated the concept clearly and accurately reflects what scientists would realistically use when making a real habitat on Mars.

To complete the scoring for all 24 habitats, two researchers reviewed each habitat and scored them. One researcher scored all of the habitats. The other was trained by scoring five bases individually. The remaining 19 habitats were double scored independently. Comparing scores, an average agreement of 93% emerged, with a calculated Cohen's kappa of $\kappa = 0.87$, indicating excellent agreement.

Artificial Intelligence Architecture. The habitat scores were used as labels to train AI models. The architectures for the models were chosen around the capabilities of the learning environment and the habitat scoring categories outlined above. Plug-ins on the *Minecraft* server automatically collect instances where students place or remove blocks (referred to here as block data, including the type of material, its coordinates and state) and can take screenshots of the world in-game. It is impossible, however, for a model to predict all of the categories solely from one input type (food sources cannot be interpreted from aerial images, area of the base cannot be interpreted from underground images, shape cannot be interpreted by block data, etc.). Thus, we took an exploratory approach to determine which frameworks and data sources work best for *Minecraft*.

We designed three baseline models for the three input types: aerial images, underground images, and block data. For both aerial images and underground images, we used a convolutional neural network (CNN) that consisted of 2 2-dimensional convolutional layers, batch normalization layers, dropout layers, and 2 dense layers. Images in our dataset were resized to 128x128 and each pixel value was normalized. For block data, the columns for the dataset were the types of blocks used for all of the groups and for each cell were the number of the block type used by the group normalized. We used a multi-layered perceptron (MLP) for the block data, which consisted of 3 dense layers and dropout layers.

There were a total of four multi/mixed-input classifiers that consisted of all permutations of the input types. To concatenate the models into a single classifier, we used late fusion to concatenate the output layers together to use as input to a dense connected layer to get the final classification for the category. Before concatenation, we used the same model architectures in the multi/mixed-input classifiers as the ones used in the baseline models. Each model was used to predict the score on the habitat scoring scheme for a single category.

AI Model Comparison. A total of $n = 21$ bases were used to compare different input types for the AI models, 3 were omitted from the dataset due to missing block data. To compare the seven AI frameworks described above, we used 5-fold cross-validation. To handle dataset imbalances we used class weighting and to prevent overfitting we used early stopping. We then ran an Analysis of Variance (ANOVA) to determine whether the means of the seven models for all of the categories were identical and a post hoc Tukey's Honest Significant Difference (HSD) test to determine which pairwise mean comparisons between conditions yielded significant differences.

4   Results

4.1   Habitats and Learning

To demonstrate that the scoring process was reliable for all teams, two Spearman's rank correlations were performed comparing habitat scores to team sizes and face-to-face time. There were non-significant Spearmans rank correlations between team size and habitat scores ($r[22] = -0.03$, $p = 0.89$) and face-to-face camp instruction time and habitat scores ($r[22] = 0.13$, $p = 0.54$). To assess validity, there was a significant positive correlation between group mean self-explanation score and habitat scores, $r(22) = 0.51$, $p = 0.01$.

4.2 AI Model Comparison

Model Metrics. One-way ANOVAs were conducted to compare the conditions on accuracy, precision, recall, and F1 scores. The ANOVAs for all four metrics were significant: accuracy ($F[6, 378] = 9.82$, $p < 0.01$), precision ($F[6, 378] = 6.99$, $p < 0.01$), recall ($F[6, 378] = 3.42$, $p < 0.01$), and F1 score ($F[6, 378] = 6.93$, $p < 0.01$). Table 1 shows the Tukey HSD test results of the comparisons between conditions on the metrics.

Table 1. Post hoc Tukey HSD mean differences for the conditions on the metrics (A = Aerial, U = Underground, B = Blocks). *** $p < 0.001$, ** $p < 0.01$, * $p < 0.05$

Comparison	Accuracy	Precision	Recall	F1
U vs. A	.04	.01	.01	.01
B vs. A	.22***	.18**	.09	.16**
A+U vs. A	.05	.02	−.01	.01
A+B vs. A	.20***	.13	.08	.13*
U+B vs. A	.19***	.15*	.09	.14*
A+U+B vs. A	.25***	.19***	.12*	.18***
B vs. U	.18**	.17**	.08	.15**
A+U vs. U	.01	.01	−.02	−.001
A+B vs. U	.16**	.12	.07	.11
U+B vs. U	.16*	.14*	.08	.12
A+U+B vs. U	.21***	.19***	.11	.17**
A+U vs. B	−.17**	−.16**	−.10	−.15**
A+B vs. B	−.02	−.05	−.01	−.03
U+B vs. B	−.03	−.03	−.004	−.02
A+U+B vs. B	.03	.02	.03	.02
A+B vs. A+U	.15*	.11	.09	.12
U+B vs. A+U	.14*	.13	.10	.12
A+U+B vs. A+U	.20***	.18**	.13*	.17**
U+B vs. A+B	−.01	.02	.01	.01
A+U+B vs. A+B	.05	.06	.04	.05
A+U+B vs. U+B	.06	.05	.03	.05

4.3 Model Optimization

A total of $n = 57$ bases were used for feature selection and finding the optimal amounts of layers and nodes for block-only models. We optimized the block-only models because they did not perform significantly worse than our best models and they could reduce deployment issues such as server lag and collecting non-representative image data using automated image capturing. For feature selection, we removed highly correlated building materials, using $r \geq .95$ as a cutoff, and kept a single column to represent the removed columns. Then, using a randomized search with 50 iterations with varying numbers of hidden layers (minimum of 1 and maximum of 4 layers) and nodes in each layer (minimum of 25 and maximum of 500 nodes), we ran a 5-fold cross validation for each habitat scoring category (results shown in Table 2).

Table 2. Results of the randomized search 5-fold cross validation for the best models regarding accuracy for each category.

Feature	Accuracy	Precision	Recall	F1
Oxygen Production/Atmosphere Regulation	.72	.57	.45	.81
Radiation Protection	.62	.57	.41	.64
Power Generation	.77	.79	.58	.77
Communication	.74	.78	.57	.67
Shape of Structure	.62	.54	.45	.72
Area Built	.76	.87	.58	.72
Transportation	.61	.57	.47	.64
Combating Lack of Gravity	.56	.55	.55	.54
Food and Water	.49	.48	.35	.60
Supplies	.67	.63	.47	.65
Health and Wellness	.62	.57	.47	.64
Average	.65	.63	.47	.68

5   Discussion

In this paper we discuss our research to develop a scoring scheme to assess student Mars habitats and a habitat classifier based on that scoring scheme. Consistent with our first two hypotheses, we conclude that the scoring method for the Mars habitats provides a fair and reliable means of assigning a numerical score to any given base built by camp participants. Because team size and face-to-face time had a non-significant impact on habitat scores, we can infer that varied contexts do not substantially impact learners' abilities to construct meaningful habitats. This was further proven by correlating the mean self-explanation score with the habitat scores. The higher a group was able to score on the questions, the better their habitat scored overall. This indicates that when participants integrate more astronomy knowledge from the exploration phase into their builds, habitats tend to be more comprehensive and accurate.

Contrary to our third hypothesis, a model built with a combination of data sources (aerial and underground images) was significantly outperformed by the model using only block data. This result is surprising, since this approach provides the same visual information that human scorers are given when assessing habitats. As noted previously, students may build representations of structures based on their prior experiences. Our findings indicate that although student builds appear different based on student background, structures that serve specific purposes use similar materials. Thus, utilizing block data models provides the necessary context to understand such builds. The dataset on which we base our study, however, is comprised mostly from Caucasian male participants, and it is unclear how generalizable our models are. To ensure our models are able to accurately assess build from diverse populations it is important that we continue to conduct member checks (such as the student habitat presentations), have scorers with backgrounds that align with our participants, and gather more data from underrepresented students in our dataset.

6 Future Work

Our habitat analysis and assessment models have been recently integrated into pedagogical agents on our *Minecraft* server. They will be used in upcoming studies to provide real-time habitat feedback with students teams during camps, which we believe will help to alleviate teacher burden and support student learning. Other future directions include providing students with a progress checklist during the activity (to identify and work on what they may have missed), designing collaborative building agents to expedite work, and integrating the system into curriculum as a stealth assessment or tool to inform student support needs.

Acknowledgments. The materials used in this study are based upon work supported by the National Science Foundation and Directorate for Education and Human Resources under Grants 1713609 and 1906873.

References

1. de Andrade, B., Poplin, A., Sousa de Sena, Í.: Minecraft as a tool for engaging children in urban planning: a case study in Tirol town, Brazil. ISPRS Int. J. Geo-Inf. **9**(3), 170 (2020)
2. Harel, I.E., Papert, S.E.: Constructionism. Ablex Publishing (1991)
3. Lane, H.C., et al.: Triggering stem interest with minecraft in a hybrid summer camp (2022)
4. Nyhout, A., Ganea, P.A.: Scientific reasoning and counterfactual reasoning in development. Adv. Child Dev. Behav. **61**, 223–253 (2021)
5. Papavlasopoulou, S., Giannakos, M.N., Jaccheri, L.: Empirical studies on the maker movement, a promising approach to learning: A literature review. Entertain. Comput. **18**, 57–78 (2017)

6. Sharma, K., Giannakos, M.: Multimodal data capabilities for learning: what can multimodal data tell us about learning? Br. J. Edu. Technol. **51**(5), 1450–1484 (2020)
7. Woodward, M.: Minecraft user statistics: How many people play minecraft in 2023? (2023). https://www.searchlogistics.com/learn/statistics/minecraft-user-statistics/

Characterising Learning in Informal Settings Using Deep Learning with Network Data

Simon Krukowski[1]($\boxtimes$) (iD), H. Ulrich Hoppe[2] (iD), and Daniel Bodemer[1] (iD)

[1] University of Duisburg-Essen, Duisburg, Germany
{simon.krukowski,daniel.bodemer}@uni-due.de
[2] RIAS Institute, Duisburg, Germany
uh@rias-institute.de

Abstract. Online Citizen Science (CS) projects represent informal settings in which volunteers can learn and discuss about different areas of research while participating in scientific activities. In such settings, however, volunteer involvement is geared by project needs and individual learning occurs more as a side-effect. Data-driven, longitudinal studies examining such learning impacts are scarce. We study the user activity in the Chimp&See discussion forum on Zooniverse through the lens of social network analysis (SNA) to detect emerging user roles and evolutionary changes in behaviour indicative of learning. We explore the potential of structural network embeddings to identify similarities in relational patterns in comparison to externally assigned roles. Our analyses show that explicit roles such as "moderator" exhibit a high proximity in the embeddings, and that external promotions in the form of assigned role changes are preceded by a convergence of the corresponding behavioural patterns towards the ones of already established moderators, which is indicative of a profile change based on engagement and ensuing skill acquisition. Implications and potential applications are discussed.

Keywords: Informal learning · Behavioral analysis · Social network analysis · Graph embeddings · Citizen science

1 Introduction and Background

Informal settings such as online citizen science (CS) projects allow non-professional "citizen scientists" or "volunteers" to participate in scientific activities, supposedly leading to improving their scientific literacy and domain knowledge. Accordingly, informal learning can be attributed to participation in CS as a secondary effect, and recent research shows positive effects on scientific skills and domain knowledge, which in turn result in better contribution quality and more intrinsic motivation to participate [1]. Similar to MOOCs, many online CS projects offer a structured discussion forum which acts as the primary space for information exchange and discussion. Building on the notion of communal

© The Author(s), under exclusive license to Springer Nature Switzerland AG 2024
A. M. Olney et al. (Eds.): AIED 2024, LNAI 14830, pp. 431–438, 2024.
https://doi.org/10.1007/978-3-031-64299-9_40

presence, we argue that volunteers in CS projects learn collaboratively by asking questions, discussing their findings and building relationships with other users [1,9]. In these contexts, different roles of users can either be explicit (e.g., publicly visible in the forum) or implicit, manifesting through behavioural patterns that are observable. Changes in role patterns can be indicative of skill acquisition and learning. To this end, the role of moderators in shaping the dynamics of the forum is of particular importance, as they manage the community e.g., by introducing new users or facilitating knowledge production [11].

A typical feature of online communities are freely accessible community archives that are open for research purposes. Based on these data sources, it is evident that volunteers in CS projects change their explicit role over time and get, e.g., promoted to moderators [8]. Such external promotions can be seen as strong indicators for the recognition of achievement and skill acquisition. In this sense, promotions or role changes can be used as "ground truth" for learning achievements. While not all potentially suitable community members can be recruited for higher roles due to personal choice or position limits, we can use the promoted ones with their trajectories as references for successful learning. This gives us a data-driven, behavioural and potentially longitudinal approach to studying learning effects, complementing existing studies that infer learning in CS settings based on self-reported subjective data [1].

User behaviour associated with learning effects has often been studied in the context of MOOC forums, where user-user relations are extracted from threaded discussions and then examined using social network analysis (SNA) [3,7,10]. Here, the identification of implicit roles and interaction profiles is of particular interest, as the progression through such roles can serve as an indicator of evolving competencies. Recent network embedding techniques [4] combine SNA with deep learning to generate low-dimensional latent representations of network structures that allow for the detection of structural and role-oriented similarities as one of the potential outputs. Notably, actors with similar interaction patterns do not necessarily need to be densely connected among themselves, and accordingly, role detection [2] should not be confounded with subcommunity detection.

In a novel approach, this work exploits and explores the usage and applicability of structural network embeddings in the analysis of role progressions using the moderator profiles as reference points and anchors for validation. To this end, it addresses two challenges. Firstly, the methodological challenge of applying and studying structural network embedding techniques. Secondly, it also contributes to widening the scope of AI-based analytics approaches towards informal learning scenarios in online communities and social media (cf. [6]), where the specific challenge is to characterise learning progressions in the absence of explicit learning goals and mastery criteria. For this purpose, we rely on the identification of implicit role profiles as behavioural patterns in networks. Thus, we formulate the following two research questions:

RQ1: Can network embeddings accurately reflect structurally similar users based on their communication behaviour in online CS forums?

RQ2: Can we use these embeddings as an additional layer of analysis to characterise behavioural changes as indicative of learning?

2   Method

For our analyses, we collected 136,378 comments from 37,605 discussions from the "Talk page" of the Zooniverse *Chimp&See* project spanning a period of 54 months (01/2019 - 06/2023), where volunteers identify animals in wildlife footage. Each comment contains additional information such as the explicit role of the user (e.g., volunteer, moderator), which is assigned by the project owners and is publicly visible in the forum. In total, 6 highly active users changed their role over the course of the project and got promoted to volunteer-moderators, on average after 22.5 months ($SD = 10.02$) of activity in the forum. To study the interactions, we extracted multi-relational directed networks with users as nodes and interactions as edges, corresponding to monthly time frames based on three types of interactions (commenting, replying and mentioning). The pipeline can be seen in Fig. 1. In total, we have 54 distinct networks, amounting to 545 nodes and 95,048 edges ($M = 1827.85$, $SD = 1026.73$ on average per time frame), including only users who had been active in the time frame (degree > 1).

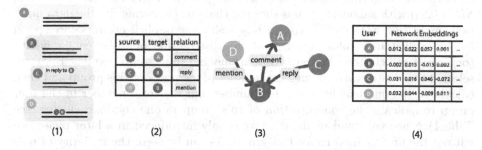

Fig. 1. Our pipeline showing the discussion structure (1), the resulting edges (2) and multirelational networks (3) which are used to create network embeddings (4).

The network embeddings were created using Node2Vec (see [5]), and we configured the algorithm to ensure the embeddings represent structural equivalence by associating nodes based on similar structural roles. Thus, apart from basic and SNA-based attributes (e.g., username, role, degree etc.), we also have a 128-digit long vector for every user, representing their topological features (i.e., interactions) within the respective time frame. The primary advantage of using such structural network embeddings as opposed to standard SNA techniques is that we can measure structural similarities and differences among users by calculating distances in the embedding space, allowing us to track changes and convergences in behaviour. To address our RQs, we first examine how well the obtained embeddings represent communication behaviour by using k-Means clustering to create clusters of users with similar topological features (i.e., implicit roles) as reflected by their embeddings and study their properties. We use an Euclidean distance measure to quantify changes in the similarity between particular users or user groups (e.g., moderators) over time as indicators of learning progression. For groups of users, we average the embeddings per group before calculation. Since the created networks are directed, we can distinguish between active and passive interactions, reflected by the out-degrees (active) and in-degrees (passive) of the users. Building on this distinction, we calculated the degree ratio by dividing out/in-degree (after incrementing each by 1 to avoid zero division). For this measure, high values indicate that the users interact with others more actively (e.g., give replies) than passively (e.g., receive replies).

3   Results

In a first step, we calculated the out/in degree ratio for the different explicit user roles (see Table 1). It shows that it is highest for moderators and lowest for volunteers, confirming related findings regarding the role of moderators as community managers in CS projects [11]. To test whether the embeddings differentiate users based on their behaviour, we examined the above-mentioned k-Means clusters, since these should lead to distinct clusters characterising different implicit types of users. Across all time frames, we observed between 2 and 6 clusters ($M = 4.18$, $SD = 0.94$) with silhouette scores showing that the clustering fits the data quite well ($M = 0.72$, $SD = 0.25$). To test how well these implicit clusters correspond to the explicit user roles, we checked the distribution of roles (e.g., moderator) across the clusters: For each time frame and user group (explicit role), we selected the cluster with the highest number of members of this group and then calculated the quotient between this number and the total number in the user group to indicate the concentration of this group in one common cluster (see Table 1). Since volunteer-moderators were only introduced in a later phase, we left out the first 27 months for this group. As can be seen, the majority of user roles is located in one common cluster ($M = 73.76\%$, $SD = 16.62\%$), particularly volunteers and scientists. Thus, the implicit (behavioural) clusters largely correspond to the explicit user roles. We then analysed each of the respective time frames and displayed the network and corresponding embeddings in an interactive dashboard using t-SNE (t-distributed Stochastic Neighbor Embedding) for visualisation in a lower dimensional space (see Fig. 2). To understand the differences among the clusters, we consider their metrics and distribution of user types and visualise the corresponding network and embeddings. This can be seen in exemplary form for project month 34 in Table 2 and the interactive dashboard (Fig. 2). This time frame was chosen because the community development was sufficiently advanced, yet most role changes (5 out of 6) were still to occur.

It shows that cluster 2 is characterised by a high degree-ratio, indicating more active than passive interactions, while cluster 0, 1 and 3 show the opposite. The dashboard also exemplifies this, additionally showing the structural similarity of the users, as the orange nodes (cluster 2) appear to be similar to each other in

Table 1. Mean out/in degree ratio and standard deviation for the different explicit user types with number of distinct users (N) and concentration in one common cluster.

Explicit role	Degree ratio	N	Common cluster
moderator	4.47 (8.59)	5	56.17%
scientist	1.42 (3.48)	11	80.13%
team	0.63 (1.03)	23	73.51%
volunteer	0.60 (1.30)	526	90.43%
volunteer-moderator	2.01 (1.67)	6	68.54%

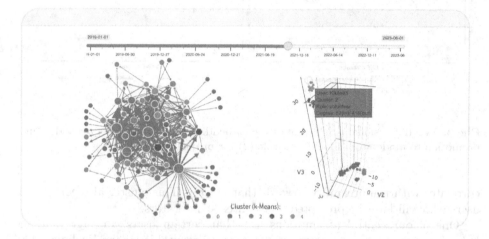

Fig. 2. Interactive dashboard showing the network and three-dimensional t-SNE plot for month 34 with nodes coloured by k-Means cluster

the embedding space, yet are not necessarily closely connected in the network. A similar picture emerges across all time frames, with 139.47 ($SD = 206.97$) users per cluster on average and a degree ratio ranging from 0.19 to 6.86. The high standard deviation indicates significant differences between the clusters, and if we rank these clusters by degree ratio (higher ranks = higher degree ratio), we see that the share of moderators increases in the higher ranks, while it decreases for volunteers (1% moderators and 93% volunteers in rank 1 with $N = 6,599$ and 15% moderators and 69% volunteers in rank 5 with $N = 160$. Note the higher N values due to aggregation across time frames). Thus, our approach identifies structurally similar users based on their behaviour quite well. The use of embeddings becomes particularly interesting when considering the distance between users and relating it to other attributes, such as the degree or a future promotion. As can be seen in the two-dimensional t-SNE (Fig. 3), highly active users (as measured by degree) get placed closer to each other and resemble a cluster, which mirrors the findings described above. However, the degree also

Table 2. Clusters (k-Means) with metrics and user roles for project month 34

Cluster id	In- degree	Out- degree	Degree- ratio	N	Volunt.	Mod.	Scientist	Team	Volunt.- mod.
0	15.31	5.57	0.40	20	13	1	2	4	0
1	1.85	0.00	0.35	501	482	0	8	11	0
2	109.09	151.32	1.38	11	5	3	0	2	1
3	98.50	16	0.17	2	1	1	0	0	0
4	26.29	13.83	0.54	11	8	0	1	2	0

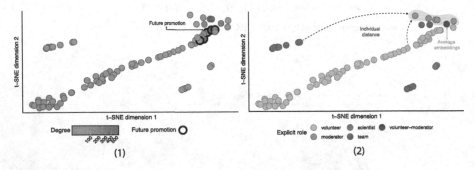

Fig. 3. Two-dimensional t-SNE plot of the embeddings coloured by degree and future promotion to moderator (1) and user role (2) for project month 34

correlates with the proximity towards that cluster. This cluster also contains 5 users who will later be promoted to volunteer-moderators.

One of our central assumptions was that certain users converge in their behaviour towards expert users (moderators), measurable by the Euclidian distance in the embedding space. Thus, we averaged the embeddings for the moderator user group and computed the individual distances towards it (cf. Figure 3). In general, this distance correlates negatively with project time ($r = -.23$, $p < .001$), indicating that it decreases over time, as users become more similar in their behaviour to moderators. This effect is more pronounced for promoted users ($r = -.29$, $p < .001$) than for unpromoted users ($r = -.22$, $p < .001$), and descriptively, the distance is smaller for promoted ($M = 1.08$, $SD = 1.13$) than unpromoted users ($M = 1.59$, $SD = 1.39$), yet slightly fails to reach significance ($p = .058$) as determined by a Wilcoxon test due to the small $N = 6$ for promoted moderators ($W = 889$, $r = .08$). Of particular interest is the distance towards the moderators prior to the role change, which is higher before ($M = 1.08$, $SD = 1.13$) than afterwards ($M = 0.44$, $SD = 0.52$). If we examine this over time along with the degree ratio (Fig. 4), it can be seen that while the distance decreases, the degree ratio increases. This indicates a convergence in behaviour, manifesting through more structural similarity to the moderators as well as more active than passive interactions (degree ratio). Although we can also see an expected slight decrease in distance to moderators for active but unpromoted users, it is less pronounced and does not go along with an increase in degree ratio.

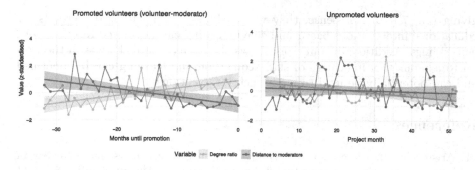

Fig. 4. Standardised distance to moderators and degree ratio for promoted vs. unpromoted volunteers

4   Discussion and Outlook

In a novel approach, we explored the potential of extending SNA with network embeddings for the analysis of behavioural patterns as indicators of informal learning. By measuring structural similarity through proximity in embedded space, we can distinguish users who behave similarly as opposed to being densely connected to each other. We found that the embeddings do indeed reflect different clusters of users with similar behavioural patterns (e.g., high degree ratio), and that these clusters often comprise a relatively high percentage of users with specific externally assigned roles like moderators (RQ1). This corroborates findings indicating that moderators play a central role in community management as they actively interact with other users, e.g., by linking external resources [11].

For RQ2, we used the behavioural profile of moderators as a reference point and calculated the distance between individual users and this reference. We expected users to progress in their behaviour over time as they become a part of the community, aligning their profiles to those of more experienced users such as moderators. Our results show that this is indeed the case, and particularly so for users who are later promoted to moderators. We can interpret this convergence in behaviour towards moderators in the time prior to their promotion as a quantitative indicator that goes along with increased competence and cumulative skill acquisition. The increase in degree ratio indicates more active than passive interactions, which supports this claim.

The results of the exploratory study reported here serve as an initial demonstration of feasibility with a heuristic validation. Based on the results presented, we plan to expand the sample across multiple CS projects and their corresponding discussion forums and possibly include different reference profiles beyond the one of moderators. Additional variables like message content can also be used for prediction tasks that enrich the analysis of user profiles as learning indicators.

Our study shows that topological information captured in interaction networks of forum discussions can be used to create structural network embeddings reflecting user behaviour. In the future, these embeddings can potentially be used alongside content-based measures like topic-relatedness or help-seeking/help-giving (cf. [7]). In this sense, they widen the scope of AI-based methods for educational interventions based on behavioural indicators derived from network embeddings. Additionally, our work shows that behavioural changes in the run-up to a promotion indicative of skill acquisition and learning can be identified and can potentially be used to guide such interventions.

References

1. Aristeidou, M., Herodotou, C.: Online Citizen Science: a systematic review of effects on learning and scientific literacy. Citizen Sci. Theory Practice **5**(1), 11 (2020). https://doi.org/10.5334/cstp.224
2. Borgatti, S.P., Everett, M.G.: Notions of position in social network analysis. Sociol. Methodol. **22**, 1–35 (1992)
3. Chen, B., Poquet, O.: Networks in learning analytics: where theory, methodology, and practice intersect. J. Learn. Anal. **9**(1), 1–12 (2022)
4. Chen, H., Perozzi, B., Al-Rfou, R., Skiena, S.: A tutorial on network embeddings. arXiv preprint arXiv:1808.02590 (2018)
5. Grover, A., Leskovec, J.: node2vec: scalable feature learning for networks. In: Proceedings of the 22nd ACM International Conference on Knowledge Discovery and Data Mining - SIGKDD, pp. 855–864. ACM (2016). https://doi.org/10.1145/2939672.2939754
6. Haythornthwaite, C.: Analytics for informal learning in social media. In: Lang, C., George, S., Wise, A., Gašević, D., Merceron, A. (eds.) The Handbook of Learning Analytics, 2nd edn., pp. 163–172. SoLAR (2022). doi: https://doi.org/10.18608/hla22
7. Hecking, T., Chounta, I.A., Hoppe, H.U.: Investigating social and semantic user roles in MOOC discussion forums. Proceedings of the Sixth International Conference on Learning Analytics & Knowledge, pp. 198–207 (2016)
8. Krukowski, S., Amarasinghe, I., Gutiérrez-Páez, N.F., Hoppe, H.U.: Does volunteer engagement pay off? an analysis of user participation in online citizen science projects. In: Wong, L.H., Hayashi, Y., Collazos, C.A., Alvarez, C., Zurita, G., Baloian, N. (eds.) Collaboration Technologies and Social Computing, pp. 67–82. Springer International Publishing, Cham (2022)
9. Mugar, G., Østerlund, C., Jackson, C.B., Crowston, K.: Being present in online communities: learning in citizen science. In: Proceedings of the 7th International Conference on Communities and Technologies, pp. 129–138 (2015). https://doi.org/10.1145/2768545.2768555
10. Rabbany, R., Elatia, S., Takaffoli, M., Zaïane, O.R.: Collaborative learning of students in online discussion forums: a social network analysis perspective. In: Peña-Ayala, A. (ed.) Educational Data Mining. SCI, vol. 524, pp. 441–466. Springer, Cham (2014). https://doi.org/10.1007/978-3-319-02738-8_16
11. Rohden, F., Kullenberg, C., Hagen, N., Kasperowski, D.: Tagging, pinging and linking - user roles in virtual citizen science forums. Citizen Sci. Theory Practice **4**(1), 19 (2019). https://doi.org/10.5334/cstp.181

Exploring Teachers' Perception of Artificial Intelligence: The Socio-emotional Deficiency as Opportunities and Challenges in Human-AI Complementarity in K-12 Education

Soon-young Oh[1] and Yongsu Ahn[2]([⊠])

[1] Michigan State University, East Lansing 48823, USA
ohsoon@msu.edu
[2] University of Pittsburgh, Pittsburgh 15260, USA
yongsu.ahn@pitt.edu

Abstract. In schools, teachers play a multitude of roles, serving as educators, counselors, decision-makers, and members of the school community. With recent advances in artificial intelligence (AI), there is increasing discussion about how AI can assist, complement, and collaborate with teachers. To pave the way for better teacher-AI complementary relationships in schools, our study aims to expand the discourse on teacher-AI complementarity by seeking educators' perspectives on the potential strengths and limitations of AI across a spectrum of responsibilities. Through a mixed method using a survey with 100 elementary school teachers in South Korea and in-depth interviews with 12 teachers, our findings indicate that teachers anticipate AI's potential to complement human teachers by automating administrative tasks and enhancing personalized learning through advanced intelligence. Interestingly, the deficit of AI's socio-emotional capabilities has been perceived as both challenges and opportunities. Overall, our study demonstrates the nuanced perception of teachers and different levels of expectations over their roles, challenging the need for decisions about AI adoption tailored to educators' preferences and concerns.

Keywords: Teachers' perception of AI · Teacher-AI complementarity · Teachers' role · Human-AI complementarity

1 Introduction

Teachers are composites that play an array of roles in schools. These roles encompass not only educational responsibilities such as teaching, guiding, and communicating with students and parents but also administrative duties ranging from document management, event coordination, and engagement in the decision-making process [5]. The multifaceted nature of these roles necessitates that teachers possess a variety of cognitive abilities, for instance, as observers and mentors who provide guidance to students and as workers who efficiently process and organize information about students and the school community.

As artificial intelligence (AI) continues to advance and integrate into our society, discussions have emerged regarding its potential role in schools and its impact on teachers. Despite the overall benefits of introducing AI to schools, it is crucial to understand how teachers perceive the complementarity of AI. Such understanding – whether they seek support in specific tasks or express resistance and skepticism – can not only help identify potential barriers or opportunities to adopt AI capabilities but also tailor AI integration to their needs and concerns. However, in what *capacities* do teachers anticipate AI acting as an assistant or collaborator in their duties, or automating and complementing their tasks to alleviate their burden? What *tasks* do teachers believe AI can excel at or struggle with, shaping the dynamics of teacher-AI complementarity differently? While various studies explore teachers' perceptions of AI, the majority focus on classrooms [3] or the educational roles of teachers [1,2]. As highlighted in existing work [5], teachers' roles and workload in the school scene span a broader spectrum of educational activities as well as administrative tasks, potentially leading to a variety of AI integration in education.

As an initial step toward fostering effective teacher-AI complementarity in schools, our research investigates teachers' expectations regarding AI capabilities across a range of eleven key teacher roles. Through a survey involving 100 teachers and in-depth interviews with 12 teachers, our findings demonstrate a diverse range of AI roles envisioned by teachers, spanning from document processing automation to roles as curriculum planners, decision-makers, and even leaders. The thematic analysis unveils a nuanced perspective: while teachers recognize AI's advanced intelligence, they highlight its deficiency in socio-emotional capabilities. Specifically, the AI's deficit of socio-emotional capabilities is perceived as opening up both opportunities and challenges. On one hand, teachers express concerns that AI's inability to interpret nuanced student communication-both verbal and non-verbal-could hinder guidance and impede interpersonal growth by blurring the lines between human and AI interactions. On the other hand, the absence of emotion in AI is regarded as advantageous, positioning it as a fair and impartial entity capable of undertaking various tasks, ranging from task allocation to final decision-making within educational settings.

Contributions. Overall, our study broadens the discussion on AI capabilities in education across a multitude of teachers' roles. The contributions include: (1) **Teachers' task classification and cognitive mapping**: We develop a two-level classification of teachers' tasks in K-12 schools. These tasks are then mapped to cognitive abilities, allowing us to characterize them as a combination of different types of cognitive involvement. (2) **Techers' perception and imaginaries of future education**: We highlight the diverse array of roles spanning from automated document processing to tutoring and decision-making, underscoring AI's roles in future education beyond the current boundaries of AI advancements, including generative AI and large language models.

2   Study Method

We employ a mixed method 1) to quantitatively derive the perceived AI complementarity using the survey method, and 2) to qualitatively investigate teachers' thoughts and sense-making of AI's potential opportunities and challenges in complementing their roles.

Teacher Task Classification. Our initial step involves filling a notable gap in the existing literature by creating a comprehensive classification of teacher tasks, drawing from classifications found in previous studies [4–7]. We sought advice from six education experts with doctoral degrees in educational administration and technology. The classification from our foundational work (Fig. 1) organizes teacher tasks into two levels, with the Level 1 (L1) covering broad domains, specifically educational and administrative tasks, and the Level 2 (L2) specifying a range of eleven sub-tasks, capturing both overarching job domains and the intricate details of individual task elements.

Survey Design. To quantitatively measure teachers' perception of AI's roles, we utilized the Analytic Hierarchy Process (AHP), a method for deriving priorities through pairwise comparisons among factors within a decision-making hierarchy [8]. In this survey design, item pairs are presented for pairwise comparison, asking participants to score them on a scale ranging from −9 to 9. The lowest score indicates a strong preference for the left item over the right, while the highest score indicates the opposite preference. The AHP analysis results in per-task weights as normalized relative preferences summing up to 1. We refer to these weights as the Perceived AI Complementarity Score (PCS), the degree to which a task is perceived as more suitable for AI to complement human teachers.

Recruitment and Sampling. We recruited 100 elementary school teachers in South Korea over two months, from August to September 2020. Snowball sampling was employed to construct a nationwide sample of teachers from various schools. We ensured the quality of all responses through two rounds of consistency ratio (CR) checks, maintaining a threshold of 0.1. Subsequently, 12 participants were selected for one-on-one, semi-structured interviews to explore detailed rationales behind their responses. Interviewees were purposefully chosen to represent diverse levels of experience and backgrounds. This included a balanced mix of regular teachers (50%) and head teachers (50%), and years of experience distributed as follows: ≤ 5 years (16.7%), 6–10 years (25%), 11–20 years (25%), and > 20 years (33.3%).

Task-Ability Association. With teachers' perception of tasks based on the survey responses, we provide a perspective of viewing tasks along the axis of cognitive abilities with the following question: What cognitive abilities do teachers perceive AI as being able (or not able) to complement? To conduct the analysis, we follow three analytical steps: 1) Task-ability mapping (Fig. 1B): We mapped the relationship between eleven teacher tasks in our classification and fourteen cognitive abilities defined in [9], which were derived from AI, animal, and psychological studies. This mapping involved annotation tasks on whether each task required specific cognitive abilities, represented in Fig. 1B by green (indicating required) and gray (indicating not required). 2) Task group identification (Fig. 1C): Utilizing the binary mapping between tasks and abilities, we performed a clustering analysis using the k-means method (k = 4) to identify task groups of eleven teacher tasks based on their similarity in cognitive abilities and computed aggregate PCS scores. This allows us to investigate which cognitive abilities involved in tasks are perceived as having more AI complementarity.

Interview Procedure and Analysis. The one-on-one interview sessions were designed to allow participants to share their detailed thoughts on the complementarity between teachers and AI while answering a predetermined set of questions. The questions include: 1) Can you elaborate on your response in the survey regarding AI's complementarity in each task? 2) What opportunities and challenges do AI face in teacher-AI collaboration? 3) To what extent can AI complement human teachers in each task, and will AI assist, complement, or replace the role of humans?

To capture the essence of participants' thoughts, we conducted a thematic analysis of the transcribed audio recordings from the interviews. Two coders, each with expertise in AI and education, transcribed and analyzed the interview, focusing mainly on two points: 1) What specific tasks do teachers perceive AI can or cannot perform better? 2) For those tasks, what opportunities or values and challenges or adversities do teachers perceive AI may encounter?

3   Analysis Results

3.1   Teachers' Perception: Administrative Affairs Perceived as the Most Viable Tasks for AI

The PCS scores derived from the AHP analysis (Fig. 1A) in the L1 tasks indicate a preference for AI's involvement in school administration (sum of PCS in L2 tasks: 0.820) over educational tasks (0.180). In the L2 tasks, four subtasks in school administration emerged as the top priority, ranking from first to fourth among all tasks. Specifically, administrative affairs were identified as the highest priority (PCS: 0.437), followed by policy administration (PCS: 0.172), educational administration (PCS: 0.138), and external relations (PCS: 0.109), underscoring a belief in AI's greater efficacy in administrative roles.

3.2   Task-Ability Mapping: Socio-emotional Capabilities Perceived as Low AI Proficiency

As presented in Fig. 1C, we identified four different groups of tasks named Perceptual & Instructional, Reflective, Socio-emotional, and Analytical based on the

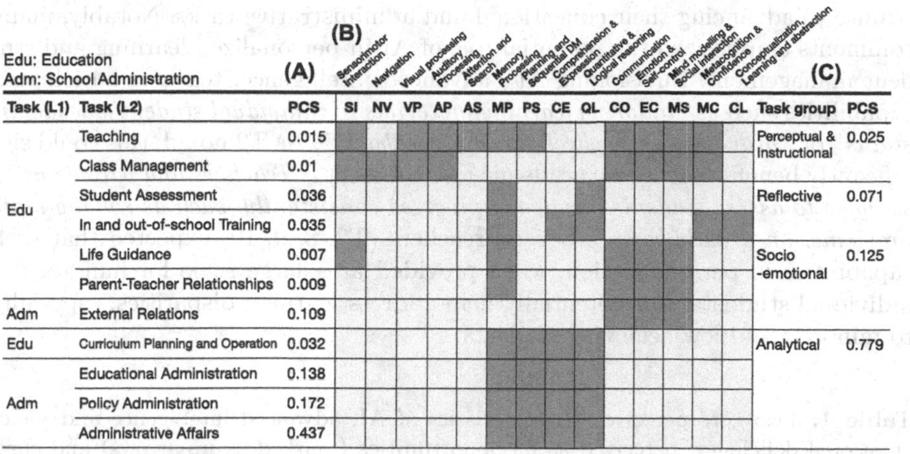

Fig. 1. The overview of quantitative analysis and results. (A) Task-wise perceived AI complementarity scores (PCS) from the survey using AHP method: Higher scores indicates AI being perceived as capable of complementing humans. (B) Task-ability mapping between tasks and abilities. (C) Task groups identified from clustering the tasks using the mapping in (B).

k-means method. By averaging the PCS in each task group, we found that AIs are least perceived as complementing human teachers in teaching and class management that involve perceptual and instructional capabilities. The socio-emotional task group was also perceived as having low complementary. Especially, two of the specific tasks, life guidance and parent-teacher relationships as individual tasks, obtained the lowest perceived complementarity score, showing the challenging nature of socio-emotional capabilities as AI-complementary tasks. On the other hand, analytical tasks were dominantly perceived as the most capable complementarity duties AI can take on.

3.3 Opportunities and Challenges of AI: Advanced Intelligence and Socio-emotional Deficiency

The thematic analysis of in-depth semi-structured interviews highlights two focal themes of teachers' perception on characteristics of AI capabilities: (1) advanced intelligence and (2) lack of socio-emotional capabilities. We show that the 68 meaningful comments mentioned by 12 teacher interviewees (referred to as T1-12) (Table 1), as opportunities (marked as up-arrows) and challenges (marked as down-arrows) of teacher-AI complementarity, largely fall into either of these two characteristics of AI, particularly highlighting AI's socio-emotional deficiency perceived as both its strengths and limitations in the educational contexts.

Advanced Intelligence in Personalized Learning and Automation. First, teachers highlighted various aspects of teacher tasks where AI could contribute to advancing their educational and administrative tasks. Notably, many comments emphasized the potential role of AI in personalized learning and student management. Envisioning AI's advanced intelligence, teachers anticipated capabilities such as *"tailoring learning materials to individual student's academic status"* or *"providing daily check-ups and feedback."*. As T2 noted, this could significantly benefit teachers in academic management, *"Teachers and parents often struggle to assess students' academic progress consistently, such as solving math problems, on a daily basis."*. Several teachers (T2, 8, 9, 10) expected that such capabilities in personalization, when provided at a large scale for hundreds of individual students, can potentially help address learning disparities, especially to minority and low-achieving students.

Table 1. Two core perceived characteristics of AI, advanced intelligence and socio-emotional deficiency, perceived as its opportunities (marked as up-arrows) and challenges (down-arrows) in K-12 education.

Task	Advanced intelligence perceived as opportunities	Socio-emotional deficiency perceived as opportunities and challenges
Teaching	(↑) Personalized learning (↑) 1-on-1 education (Equity)	(↓) Emotion/intent recognition (↓) Social interaction (↓) Non-verbal communication (↓) Persuasion, Guidance
Life guidance	(↑) Personalized management	(↑) Consistent communication management (Fairness)
Class management	-	
Parent-teacher Relationships	-	
Student Assessment	(↑) Personalized feedback	(↓) Behavioral feedback (↑) Automated scoring (Fairness, Efficiency)
In and out-of-school training	(↑) AI-powered instruction and personalized training	-
Curriculum planning and operation	-	(↑) Data-driven planning (Fairness)
Educational administration	(↑) Automation (Efficiency)	(↑) Data-driven decision making and task assignment (Fairness, Deauthorization)
Policy administration		
Administrative affairs		
External relations	-	(↓) Negotiation, Interpersonal relationship

Furthermore, the majority of administrative tasks were perceived as significant opportunities with the help of automation. Teachers found that some tasks, such as document processing or budget planning (T7), are typically *"fixed and standardized"* (T2, 10, 12) or *"handled on an annual or monthly basis"* (T7). Educational administrations such as curriculum planning were also expected for AI to not only automate tasks, but also advance the planning to be more contextualized to each school environment based on learning a variety of data from both national and school levels (T9, 11).

Lack of Social-emotional Capabilities as Challenges in Teaching and Life Guidance. Simultaneously, teachers tended to perceive AI as lacking socio-emotional capabilities, particularly when it comes to tasks that involve guiding students during teaching and class activities. Participants stated that such tasks often pose one of the greatest challenges even for human teachers, resulting from tricky interactions, as T9 commented, *"During class, many students claim that they are paying attention while actually being distracted and engaged in unrelated activities, or pretending that they understood everything."* Several teachers highlighted that such situations require the ability to recognize nuanced verbal and non-verbal communications and respond in a way that motivates and guides students to stay focused in class. During the interview, these complexities were often described as *"there are too many variations"* (T1, 10), as *"[it] needs a significant level of adaptability with experiences and careful observations"* (T11). Most of the teachers in the study were skeptical about whether AI can truly *"grasp these subtle cues, relationships among students such as jealousy, and handle issues like parental complaints"* (T9), given that these interactions often have no obvious answers and involve a lack of available and unstructured data (T1).

Teachers found it especially challenging in elementary school, where students before the age of 18 ego development in human-human communication, *"I'm worried that children, still navigating their ego development, may struggle to differentiate between human and AI interaction, potentially leading to confusion about their own identity."* (T12), or *"Because interpersonal relationships are crucial, children's feelings of isolation could intensify."* (T1)

Additionally, teachers recognized AI's potential in facilitating connections with the local community and parents, with educators emphasizing the indispensability of human-to-human rapport. As one teacher stated, *"Building relationships with the local community requires genuine human interaction"* (T9), and *"[such tasks] should involve a political aspect within subtle dynamics of relationships with the local community using negotiation or political skills"* (T10).

Lack of Socio-emotional Capabilities as Opportunities for Fair and Non-authoritative Decision-Making. Despite AI's perceived limitation on teaching and guiding due to AI's socio-emotional deficiency, this was, at the same time, an opportunity for AI to step into schools as a fair and nonauthoritative agency at various levels of teachers' tasks. Participants, as members of an educational organization, felt that *"decisions in schools often get swayed by* *those with a louder voice, or who've taken on senior roles."* (T4) or experienced conflicts between them in task assignment, *"When a human teacher handles it, there's often a lot of conflict about who's responsible for certain tasks in school management, like whether this is my job or yours."* (T9) For instance, they suggested that AI could provide feedback such as, *"[AI] could say like, "in the past, making choices like this drew a lot of criticism." It could provide a basis for judgment."* (T4)

T9 shared more radical imaginaries of AI roles as a decision maker and leader to foster equal and non-authoritative cultures.

> *"Leveraging AI is about simply assigning it a leader role. So, if we prompt AI to take on the principal's role and let it allocate suitable roles to teachers, we could prevent conflicts and establish a more horizontal structure."*

Additionally, teachers, as graders and mentors, expressed a challenge in achieving fair evaluation during grading and providing feedback. T1 articulated this concern, saying,

> *"In grading, there are situations where emotions come into play. I sometimes wish for partial credit when a student answers incorrectly, especially if they've usually been good. So, there are moments when you think, 'Hmm, maybe they deserve some points here,' or the other way around. And that's where I see artificial intelligence might be really helping out, making evaluations more objective and fair."*

4   Conclusion

This paper investigates teachers' perceptions of AI capabilities across various teacher tasks. Through survey data and in-depth interviews, we uncover a spectrum of opportunities and challenges associated with AI, characterized by its advanced intelligence yet socio-emotional limitations. Our study expands the discourse on the future of AI across a range of teacher tasks, with an in-depth discussion of teachers' perspectives on AI capabilities beyond recent developments in generative AI.

References

1. Woodruff, K., Hutson, J., Arnone, K.: Perceptions and barriers to adopting artificial intelligence in K-12 education: a survey of educators in fifty states (2023)
2. Felix, C.: The role of the teacher and AI in education. International Perspectives On The Role of Technology in Humanizing Higher Education, pp. 33-48 (2020)
3. Kim, N., Kim, M.: Teacher's perceptions of using an artificial intelligence-based educational tool for scientific writing. Front. Educ. **7**, 142 (2022)
4. Mintzberg, H.: The Structuring of Organizations Prentice-Hall. Englewood Cliffs, NJ (1979)

5. Bidwell, C.: The school as a formal organization. In: Handbook of Organizations (RLE: Organizations), pp. 972–1022 (2013)
6. Hoy, W., Miskel, C.: Educational administration: Theory, research, and practice. *(No Title)* (2008)
7. Kwon, H.: A study on how to establish job standards for elementary school teachers. J. Korean Teacher Educ. **27**, 191–214 (2010)
8. Saaty, T.: A scaling method for priorities in hierarchical structures. J. Math. Psychol. **15**, 234–281 (1977)
9. Tolan, S., Pesole, A., Martinez-Plumed, F., Fernandez-Macias, E., Hernandez-Orallo, J., Gomez, E.: Measuring the occupational impact of ai: tasks, cognitive abilities and ai benchmarks. J. Artif. Intell. Res. **71**, 191–236 (2021)

Who Pilots the Copilots?
Mapping a Generative AI's Actor-Network to Assess Its Educational Impacts

Francesco Balzan[1]([⊠])[iD], Monique Munarini[2][iD], and Lorenzo Angeli[3][iD]

[1] University of Bologna, Via Zamboni 33, 40126 Bologna, Italy
`francesco.balzan3@unibo.it`
[2] University of Pisa, Largo B. Pontecorvo, 56127 Pisa, Italy
`monique.munarini@phd.unipi.it`
[3] University of Trento, Via Sommarive 9, 38123 Povo (TN), Italy
`lorenzo.angeli@unitn.it`

Abstract. Generative AI (GenAI) is praised as a transformative force for education, with the potential to significantly alter teaching and learning. Despite its promise, debates persist regarding GenAI impacts, with critical voices highlighting the necessity for thorough ethical scrutiny. While traditional ethical evaluations of GenAI tend to focus on the opacity of AI decision-making, we argue that the true challenge for ethical evaluation extends beyond the models themselves, and to the socio-technical networks shaping GenAI development and training. To address this limitation, we present an evaluation method, called Ethical Network Evaluation for AI (ENEA), which combines Latour's Actor-Network Theory—used to map network dynamics by tracing actors' interests and values—with Brusseau's AI Human Impact framework, which identifies ethical indicators for evaluating AI systems. By applying ENEA to GenAI "copilots" in education, we show how making Actor-Networks visible lets us unveil a great variety of dilemmas, guiding ethical auditing and stakeholder discussions.

Keywords: AI for Education · Actor-Network Theory · AI human impact · socio-technical systems · GitHub Copilot · Programming classes

1 Introduction

> Here is the question I wish to raise to designers: where are the visualization tools that allow the contradictory and controversial nature of matters of concern to be represented?

Bruno Latour [12]

The field of Science and Technology Studies (STS) sees AI systems as socio-technical constructs integrating technical components with social elements [7].

In this paper, we argue that the heterogeneous nature of AI systems gives rise to two kinds of intractability for their ethical assessment: technical intractability, stemming from the AIs' opaque technical aspects, and socio-technical intractability, arising from the transparency (i.e., invisibility) of the socio-technical networks behind AI development and training. While much attention has been given to technical intractability, our focus is on illuminating the socio-technical aspect.

For this purpose, we propose ENEA (Ethical Network Evaluation for AI), a method combining Actor-Network Theory (ANT) with James Brusseau's AI Human Impact framework, to systematically map the flow of interests and values of stakeholders involved in the development and training of GenAI chatbots operating as "copilots"[1] in educational settings. We pose three directions to guide our inquiry a) focusing on mapping interests and values, b) integrating descriptive mapping with a normative framework, and c) understanding the educational impacts of GenAI copilots.

In Sect. 2 we briefly present the paper's positioning in the literature (i.e., Actor-Network Theory and AI ethical evaluation); in Sect. 3, we present our analytical model, ENEA, from its structure to a worked example on GitHub Copilot (Sect. 4). To close the contribution, Sect. 5 presents the current limitations of our approach and avenues for future work.

2   Background

Actor-Network Theory (ANT), conceptualised by Michel Callon and Bruno Latour, is a framework adept at unravelling the complexities of socio-technical systems. ANT explores the interplay between technology and social processes [6,14,18] through the definition of "actants", emphasising the interdependent nature of technological and human agency [13]. In ANT, artefacts embody and transmit values and interests in the form of "prescriptions" enacted by the agents deploying such technology. ANT's focus is on evaluating and elucidating the role of "mediators": actants producing unpredictable outputs from inputs, which Latour characterised as being so embedded in our networks to become unquestioned, accepted facets of the status quo [13]. Much like our elementary mental states and cognitive processes [15], mediators and their prescriptions are transparent—they operate below the threshold of attention. ANT aims to unravel the mediators' transparent interactions and prescriptions, making it suitable for addressing the issue of AI's socio-technical intractability. Yet, tracing the actants' values and interests is only a descriptive step. Can we empower ANT with normative capacities to guide the ethical assessment of GenAI in education?

Various authors have tried to augment ANT with normative capabilities [1,4,11]. Latour himself proposed the concept of "matters of concern" which, in contrast to "matters of facts", allows us to conceive technical systems as embedded within complex socio-technical networks, entangled in social, political, and

---

[1] Of which a famous example is GitHub Copilot.

ethical relationships. We contend that AI technologies, especially in their application in education, should be understood not as "matters of fact", but as "matters of concern". In the language of this paper, shifting to a "matters of concern" approach mirrors the attention shift from the AI systems' technical intractability to their socio-technical intractability. In other words, conceiving and assessing AI systems as "matters of concern" is the act of following the flow of mediators and their prescriptions, making their transparent networks visible.

We enable this shift of attention by integrating ANT with indicators from James Brusseau's AI human impact framework for ethical assessment. Most ethical assessments of AI emphasise challenges of technical intractability, originating from the models' design (e.g., algorithmic biases [7]) and performance [10] without tackling socio-technical intractability - i.e., treating AI systems as "matters of fact" rather than as "matters of concern". While some proposed strategies aim to extend internal ethical auditing processes to stakeholders involved in the development of the AI system [17], others suggested contextualizing the AI system within the socio-technical network in which it will be deployed [7], we claim that they do not fully address socio-technical intractability. This limitation stems from a narrow conceptualization of actors and agents, which overlooks how these components propagate prescriptions and influence the final product. In essence, the values and interests embedded within these network components remain partially unexplored due to their transparency, highlighting a gap in the current methodologies for ethical assessment. To bridge this gap, our evaluation framework, called ENEA, draws from James Brusseau's AI Human Impact framework [5], as it offers a powerful synthesis of the key ethical issues connected to AI in both research and policy contexts.

3   The ENEA Framework

The AI and Education (AIED) field amply explored the impact of Generative AI (GenAI) on learning. Studies range from investigating its potential to enhance or impede human learning [8] to its influence on meta-learning and critical thinking [2]. In this paper, we aim to discuss some educational implications of GenAI copilots: AI-powered tools designed to support users in various tasks by generating human-like responses, suggestions, or content based on user input. We argue that their mode of side-by-side interaction carries its own ethical dilemmas due to the configuration of GenAI copilots as "actors" that require human-like interaction, which can only be effectively addressed and evaluated if GenAI copilots are conceived as "matters of concern".

The fundamental premise of ENEA lies in the understanding that, in ANT, values and interests can be framed as prescriptions that propagate through the actants' interactions in the network. ENEA thus becomes a tool to reveal the complex interactions between technological advancements, ethical considerations, and educational experiences that establish or replicate power relations and social dynamics. ENEA aims to shed light on how each copilot carries its own prescriptions, and how these prescriptions may affect the educational space.

To do so, we need to address socio-technical intractability. As a first step, we propose a formalisation-visualisation of ANT that represents actants and their inter-relations as vertices and edges in a directed graph. This mapping highlights the heterogeneous nature of the GenAIs' Actor-Networks (the vertices), and visualises the flow of prescriptions (the edges). In line with the broader ANT tradition, we focus on discussing the structure of the Actor-Network rather than specifying precise inclusion/exclusion criteria. While the choice of specific vertices and edges that we present here is deductive, we could also be more inductive, adding actants, prescriptions and connections based on findings from the literature. We argue, though, that the "veracity" of the model would not change, as Actor-Networks are contingent constructions rather than static objects.

Our graphs, "ENEA maps", are structured around two boundaries. The first boundary separates where the GenAI is built (its "upstream") from where it is used (its "downstream"). This boundary reflects the distinction between designers and users, with consequences that are amply explored in interaction design [3]. The second boundary is internal to the upstream, and separates design and development from training data. Each GenAI thus becomes a double boundary object [20], mediating between the upstream and the downstream, but also between designers/developers and data gathered from people. Ultimately, ENEA aims to shed light on the prescriptions that teachers and students using copilots may receive from deep within the Actor-Network. Some of these prescriptions, we argue, may give rise to potential issues in autonomy, dignity, equity, performance and accountability, summarised in Table 1. The result of an ENEA

Table 1. Ethical indicators from AI Human Impact, their applicability to GenAI in education, and how to use ENEA to analyse a Copilot for that indicator.

Indicator	Applicability to GenAI in Education	ENEA analysis
Autonomy	Students and teachers should be able to co-define their own rules, without the AI carrying external interference	Does the copilot embed any "rules" (prescriptions) in its design? Who is the source of those rules? Follow the ENEA map to check if any of those rules reach the student-teacher relationship (passing through the Copilot)
Dignity	Students should be able to use AI tools exclusively to learn and enhance their knowledge, without the AI tools serving others' purposes	What other ends might be embedded in the Copilot's design? Who is the origin of those ends? Follow the ENEA map to check whether any of those ends reach the students (passing through the Copilot)
Performance	The output of AI tools should be consistently functional, effective, and efficient	What may alter the quality of the Copilot's outputs? Who affects the quality? Follow the ENEA map to check whether any sources of performance (other than the students' input) reach the Copilot block
Accountability	Students and teachers should recognise when their tools make decisions, how and why these decisions are taken, and know whose responsibility it is for their tools' outputs	Who or what is the origin of the Copilot's text suggestion? Can students recognise this? Follow the ENEA map to see if the students can reach the origin of the text suggestions
Equity	All students in classes using Copilot should be able to successfully leverage the tool	Are there cases in which, within the same class, some students may be able to productively use the Copilot, while others receive fewer benefits? In the ENEA map, check whether the students' block can be subdivided into sub-blocks (of students who would receive different benefits)

analysis, then, is the flagging of further investigation on the deployment of the GenAI copilot according to the relevant indicator.

4   Worked Example: Applying ENEA to GitHub Copilot

In computing education, Yilmaz et al. [22] report that copilot-like GenAI systems like ChatGPT can boost computational thinking, self-efficacy, and motivation among programming students. At the same time, other studies focusing on GitHub Copilot, a GenAI system specifically tuned for code generation, suggest that GenAI's productivity benefits may lead to superficial understanding and reduced problem-solving and creativity [9, 21]. Takerngsaksiri et al. [21], in particular, caution that relying on GenAI for coding might foster dependency on autocompletion features rather than developing authentic coding proficiency.

We suggest that, while these studies offer valuable perspectives on the immediate educational effects of GenAI copilots, they lack a comprehensive analysis of the potential impacts emerging from deeper within the copilot's Actor-Network. In this section, we propose a worked example of the ENEA analysis of GitHub Copilot to highlight how prescriptions propagate within its specific Actor-Network, while generating ethical tensions. We will first present the structure of GitHub's Actor-Network, and then conduct a brief ENEA Analysis following the principles outlined in Table 1.

We provide a first approximation of GitHub Copilot's Actor-Network in Fig. 1. In that, the downstream is a schematic representation of the social dynamics that compose the educational setting, while the upstream represents the far more complex situation of Copilot's construction. On the training side, we acknowledge that Copilot's training data is, per GitHub's own documentation[2], largely based on GitHub projects and StackExchange answers; on the design and development, we summarise the network of organisations that design the model and their parent organisations.

Even in this simplified representation, some actants and relationships are noteworthy in their crossing of boundaries: on the training side, different *programming languages* gather different communities on GitHub and StackExchange, and the popularity of programming languages creates impacts on performance and equity, as discussed below; on the design and development side, *Microsoft* holds a crucial role as a common controlling agent for the two companies that develop Copilot, GitHub and OpenAI[3]. Microsoft's position in the design and development Actor-Network means that one single actant can affect Copilot both through its base model and through its control over the company deploying it.

There also are two elements that cross the ENEA structural boundaries. The "maximisation of return on investment" (RoI maximisation) prescription

---

[2] See https://github.com/features/copilot (FAQs section, General question #4. (Accessed 2024/01/26).

[3] See https://openai.com/our-structure and the profiles of GitHub's leadership at https://github.com/about/leadership (Accessed 2024/01/26).

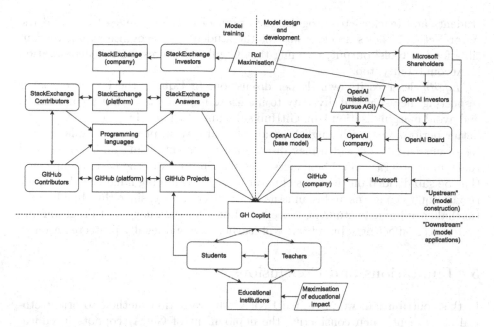

Fig. 1. The general ENEA map for Copilot. Rounded rectangles represent human actants, rectangles represent non-human actants, parallelograms represent prescriptions. Arrows show the propagation of prescriptions.

connects to actants on both sides of Copilot's upstream, and creates an alignment between the model's training set and its design and development, blurring the upstream boundary. Students also enable boundary-crossing when they publish their code on GitHub, which may eventually become part of the model's data set. In this way, students, who would normally be passive "users" of an education technology [19] acquire a direct avenue to become part of their own educational tool's upstream.

As for the analysis of GitHub Copilot, in Table 1, we defined *Autonomy* as "giving rules to oneself" [5]. We should then check whether the self-determination of students and teachers is preserved in the educational context. Since there is a path that connects the RoI maximisation and OpenAI Mission to the students and teachers that pass through GH Copilot, we can claim that Copilot requires further investigation for Autonomy. We can apply a similar logic for *Dignity*: RoI Maximisation and the pursuit of AGI are not just rules, but also ends, and they trickle down to the students passing through Copilot, meaning that Copilot should also be thoroughly assessed for Dignity. As for *Performance*, the popularity of programming language is a substantial source of performance[4], which reaches the Copilot block, other than the students' input. GitHub Copilot should also be assessed for Performance. In *Accountability*, there is no clear path for the

---

[4] See https://docs.github.com/en/copilot/using-github-copilot/getting-started-with-github-copilot (Accessed 2024/01/26).

students and teachers to recognise the origin of Copilot's suggestions, and the Actor-Network shows no clear way for the students to recognise who is responsible for the tools' output, meaning that Copilot deserves further investigation in Accountability, too.

Equity deserves its own deeper discussion. In Equity, the ultimate goal is respecting inclusion and diversity to integrate individuals within a community, and avoid marginalisation [16]. GitHub's documentation acknowledges Copilot's better performance with English prompts[5] and certain programming languages (as discussed above), which may disadvantage students using underrepresented (natural or programming) languages. The student block in the ENEA map could thus be sub-divided on the basis of natural or programming language. This gives rise to Equity concerns and is an issue of intersectionality, since this dual linguistic challenge impacts students twice based on their language and programming preferences, affecting a broad range of learners beyond legally protected groups.

5   Limitations and Conclusions

In this contribution, we presented ENEA, an evaluation method to orient ethical assessment when considering the deployment of GenAI copilots in education. ENEA aims at challenging the characterisation of AI systems as inherently intractable, a narrative supported by the major AI companies[6] to create a discourse that sees AI systems as superhumanly autonomous and sophisticated. Our work on ENEA challenges the public perception of AI-generated content as an objective, distilled essence of the collective human knowledge which, we argue, conceals the links between the AI tools' functioning and venal interests that exist in its Actor-Network.

Current limitations of our work are linked to the tradition ENEA comes from: the visualisations we propose, along with many of the considerations that it lets us draw, are contingent and easily subject to individual biases. We argue, however, that this is not necessarily a shortcoming, but rather should be seen as a feature of ENEA, and an immediately-acknowledged step towards disclosure of inevitably-present personal stances in AI assessment.

ENEA can aid in pinpointing problematic actants or prescriptions in an Actor-Network, and help plan interventions. With due adaptations, we see ENEA as potentially useful beyond educational settings. We hope that ENEA can become a tool for the scientific and educational communities to highlight points of attention, focus collective action, and ultimately build GenAI systems that respect all the involved humans and non-humans.

---

[5] See for example https://docs.github.com/en/copilot/github-copilot-in-the-cli/about-github-copilot-in-the-cli (Accessed 2024/01/26).

[6] Effectively summarised in the TESCREAL acronym: Transhumanism, Extropianism, Singularitarianism, Cosmism, Rationalism, Effective Altruism, and Longtermism.

Acknowledgments. L.A. thanks Fabio Gasparini for the many insightful conversations and comments. F.B. was supported by Future AI Research (FAIR) PE01, SPOKE 8 on PERVASIVE AI funded by the National Recovery and Resilience Plan (NRRP).

Disclosure of Interests. The authors have no competing interests to declare that are relevant to the content of this article.

References

1. Akrich, M.: The description of technical objects (1992)
2. Barana, A., Marchisio, M., Roman, F.: Fostering problem solving and critical thinking in mathematics through generative artificial intelligence. In: 20th International Conference on Cognition and Exploratory Learning in the Digital Age. PRT (2023)
3. Bardzell, J., Bardzell, S.: The user reconfigured: on subjectivities of information. In: Proceedings of the Fifth Decennial Aarhus Conference on Critical Alternatives. CA 2015, pp. 133–144. Aarhus University Press, Aarhus N, August 2015
4. Bowker, G.C., Star, S.L.: Sorting Things Out: Classification and Its Consequences. MIT Press, Cambridge (2000)
5. Brusseau, J.: AI human impact: toward a model for ethical investing in AI-intensive companies. J. Sustain. Financ. Invest. **13**(2), 1030–1057 (2023)
6. Callon, M., Latour, B.: Don't throw the baby out with the bath school! A reply to collins and yearley. In: Science as Practice and Culture, pp. 343–368. University of Chicago Press (1992). https://doi.org/10.7208/9780226668208-013
7. Dignum, V.: Responsible Artificial Intelligence: How to Develop and Use AI in a Responsible Way. Springer, Cham (2019). https://doi.org/10.1007/978-3-030-30371-6
8. Ernst, N.A., Bavota, G.: AI-Driven development is here: should you worry? IEEE Softw. **39**(2), 106–110 (2022)
9. Finnic-Ansley, J., Denny, P., Becker, B.A., Luxton-Reilly, A., Prather, J.: The robots are coming: exploring the implications of OpenAI codex on introductory programming. In: Proceedings of 24th Australasian Computing Education Conference. ACE 2022, pp. 10–19. Association for Computing Machinery, February 2022
10. Hickman, S.E., Baxter, G.C., Gilbert, F.J.: Adoption of artificial intelligence in breast imaging: evaluation, ethical constraints and limitations. Br. J. Cancer **125**(1), 15–22 (2021)
11. Introna, L.D.: Ethics and the speaking of things. Theory Culture Soc. **26**(4), 25–46 (2009)
12. Latour, B.: A Cautious Prometheus? A Few Steps Toward a Philosophy of Design, p. 2. Universal Publishers (2008). https://sciencespo.hal.science/hal-00972919
13. Latour, B.: Reassembling the Social: An Introduction to Actor-Network-Theory. OUP Oxford, September 2007
14. Law, J.: After ant: complexity, naming and topology **47**, 1–14 (1999)
15. Metzinger, T.: Phenomenal transparency and cognitive self-reference **2**(4), 353–39 (2003). https://doi.org/10.1023/B:PHEN.0000007366.42918.eb
16. Minow, M.: Equality vs equity **1**, 167–193 (2021). https://doi.org/10.1162/ajle_a_00019
17. Raji, I.D., et al.: Closing the AI accountability gap: defining an end-to-end framework for internal algorithmic auditing. In: 2020 Conference on Fairness, Accountability, and Transparency. ACM, January 2020

18. Rydin, Y.: Actor-network theory and planning theory: a response to booelens. Plan. Theory **9**(3), 265–268 (2010)
19. Selwyn, N.: On the Limits of Artificial Intelligence (AI) in Education **10**, 3 (2024). https://doi.org/10.23865/ntpk.v10.6062
20. Star, S.L., Griesemer, J.R.: Institutional ecology, 'translations' and boundary objects: amateurs and professionals in Berkeley's museum of vertebrate zoology, 1907–39. Soc. Stud. Sci. **19**(3), 387–420 (1989)
21. Takerngsaksiri, W., Warusavitarne, C., Yaacoub, C., Hou, M.H.K., Tantithamthavorn, C.: Students' perspective on AI code completion: benefits and challenges, October 2023
22. Yilmaz, R., Karaoglan Yilmaz, F.G.: The effect of generative artificial intelligence (AI)-based tool use on students' computational thinking skills, programming self-efficacy and motivation. Comput. Educ. **4** (2023)

Author Index

A. M. Olney et al. (Eds.): AIED 2024, LNAI 14830, pp. 457–461, 2024.
https://doi.org/10.1007/978-3-031-64299-9